Editorial Board Members' Collection Series: Smart Energy Systems Design

Editorial Board Members' Collection Series: Smart Energy Systems Design

Collection Editors

Surender Reddy Salkuti
Dibin Zhu

MDPI

Basel • Beijing • Wuhan • Barcelona • Belgrade • Novi Sad • Cluj • Manchester

Collection Editors
Surender Reddy Salkuti
Woosong University
Daejeon
Republic of Korea

Dibin Zhu
Shanghai Jiao Tong University
Shanghai
China

Editorial Office
MDPI AG
Grosspeteranlage 5
4052 Basel, Switzerland

This is a reprint of the Topical Collection, published open access by the journal *Designs* (ISSN 2411-9660), freely accessible at: https://www.mdpi.com/journal/designs/topical_collections/25IX5BL2G9.

For citation purposes, cite each article independently as indicated on the article page online and as indicated below:

Lastname, A.A.; Lastname, B.B. Article Title. *Journal Name* **Year**, *Volume Number*, Page Range.

ISBN 978-3-7258-2705-3 (Hbk)
ISBN 978-3-7258-2706-0 (PDF)
https://doi.org/10.3390/books978-3-7258-2706-0

Contents

designs

Editorial

Editorial Board Members' Collection Series: Smart Energy System Design

Surender Reddy Salkuti

Department of Railroad and Electrical Engineering, Woosong University, Daejeon 34606, Republic of Korea; surender@wsu.ac.kr

1. Introduction

The collection series presents various emerging approaches for designing growing renewable energy (RE), energy storage (ES), and smart transportation with electric vehicles (EVs) in the power and automobile industries. This editorial discusses various design, infrastructure, and technological aspects of energy storage and electric vehicles. In recent years, smart energy system development has gained momentum; these systems reduce greenhouse gas (GHG) production with no tailpipe emissions and improve energy efficiency and air quality, making them environmentally beneficial and crucial for climate change. EV utilization and energy storage not only pose challenges to smart energy systems but can also act as catalysts for their enhancement. Emerging technological advances create the capability to transform the transportation industry as well as grid energy demands. Power generation from renewable energy resources (RESs) plays a major role in producing sustainable green power and reducing dependency on fossil fuels; in turn, this leads to various technical, economic and environmental benefits.

The renewable energy and EV market is set to expand in accordance with governmental initiatives and ambitious plans. EV charging stations play an important role in the effective utilization of EV ecosystems, enabling easy access to charge the EVs in different places including commercial, residential and public areas. These charging stations integrate EVs into the existing power grid. Effective load management strategies, such as smart meters with the option of time-of-use pricing, smart and dynamic charging, and fast charging technologies, play an important role in handling the increased demand from charging stations, especially in peak load demand hours. Therefore, the proper location and sizing of charging stations encourages faster EV adoption, thus promoting green transportation practices. The intermittent nature of wind speed, solar irradiation and EV loads introduces uncertainties to the power grid. Therefore, energy storage devices such as fuel cells, batteries, etc., must be used for energy balancing. EV charging patterns provide additional load fluctuations to the power grid; they also vary based on charging infrastructures and user behaviours, which may further increase the challenges in maintaining grid stability. One of the biggest challenges for EV users is the driving anxiety that comes from predicting the single-charge driving range.

Smart energy systems integrating multiple energy sectors are considered a promising paradigm for providing a comprehensive and optimized solution for an achievable, affordable, and sustainable energy system in the near future. Although extensive studies on the definition, implementation, and optimization of these systems have been conducted, designing and managing smart energy systems remain critical challenges. Thus, this collection aims to bring together the most recent research on a variety of exciting topics such as modern power grid control and operation; renewable energy grid–microgrid power generation; intelligent transportation systems; energy harvesting and wireless power transfer technologies; innovations and challenges in the transition to net zero energy; electronic equipment related to energy systems; distributed energy resource (DER) power converter technologies;

Citation: Salkuti, S.R. Editorial Board Members' Collection Series: Smart Energy System Design. *Designs* **2024**, *8*, 120. https://doi.org/10.3390/designs8060120

Received: 5 November 2024
Accepted: 11 November 2024
Published: 12 November 2024

energy storage and electric vehicles; the application of intelligent control strategies in energy systems; and big data, artificial intelligence and cybersecurity in power systems.

El Harouri et al. [1] focused on the large-scale integration of more affordable and environmentally friendly energy demands for residential consumers by considering the state of charge (SoC) of EVs and the amount of power supplied by the PV system. In this work, a combination of fuzzy logic controller (FLC) and proportional integral (PI) controller is developed to control the residential system with the help of a bi-directional buck-boost converter; this controller is effective in minimizing the current and voltage ripples in the system. A rule-based energy management system (EMS) strategy is developed to control the flow of energy between various components of residential systems. All the possible power transfer scenarios between home-to-vehicle and vehicle-to-home modes of operation are examined and validated in this work. The obtained results on the proposed EMS show a reliable and stable solution which can handle or reduce the grid intervention.

Sambhi et al. [2] evaluated the proposed hybrid energy generating system for economic, environmental, and technical analysis. This analysis is performed on power generation from RERs (solar energy) and hydrogen-based energy by considering on- and off-grid operating conditions. The proposed hybrid energy system has a load of 400 kWh/day, including 74.27 kW of peak load, and a hydrogen load of 10 kg/day, including 1.86 kg/h of peak load. Sensitivity analysis is also performed, including on the size of the PV system, number of batteries, CO_2 emissions, renewable fraction, and hydrogen tank capacity. In the future, this work could be continued by including government subsidies and energy and government tariffs.

Maklakov et al. [3] developed a novel reactive power compensation (RPC) approach by using a regenerative controlled synchronous motor drive (SMD) and voltage source inverter. This work presented the mathematical formulation and block diagrams of three-level neutral point clamped (3L-NPC)-based SMD models. All the works reported in the literature do not consider reactive power generation when designing these models. This work presented the possibility of RPC with 3L-NPC-based SMD. This was tested on the main drive of a metal plate hot rolling mill. The obtained results show that, during the breakdown, an SMD has a reactive power equivalent to 16% of the total rated power and is used as a part of an industrial smart grid.

Rashid et al. [4] presented a comprehensive review of applications of various phase change materials (PCMs) such as solar stills, air heaters, collectors, and chimneys. As solar irradiation is uncertain, the phase change materials are used in various types of solar energy systems. These materials are used in solar applications for thermal energy storage. This work focused on the current practical implications of PCMs in various solar thermal systems. The maturity of PCMs and their incorporation into solar power applications for thermal energy storage are discussed. Recent technical advancements in PCMs make them suitable for use in producing solar energy, despite their limited availability and high costs.

Liu et al. [5] present a systematic review of electrical energy pricing and its impact on the energy landscape. Modern hybrid power networks are evolving by integrating various intelligent control methods and communication and information systems, which motivates us to move towards smart electrical power grids. This work provides the various methods, frameworks, and recent developments related to energy trading and pricing. The main objective of this work is to review energy trading for microgrid-to-microgrid, peer-to-peer pricing related to the centralized power grid, as well as different pricing approaches. It provides an overview of and deep insights into the current pricing frameworks and their various limitations in present market operations.

Viswa Teja et al. [6] proposed a maximum power point tracking (MPPT) control algorithm by using an artificial neural network (ANN) for a solar battery-powered EV. The proposed system's performance is examined by analysing the data under various load conditions receiving constant variables like torque and speed. A solar PV system is required to meet the energy requirements in a day. The proposed ANN-based MPPT is used to regulate a solar PV module and increase its output. Using the proposed controller can increase solar PV module efficiency where torque ripples and reduced currents are examined on the opposite end. The proposed approach provides various advantages for EV applications which require optimal torque and consistent velocity to satisfy different loading conditions in a day.

Gómez et al. [7] present a comprehensive review assessing hydrogen as an effective alternative for reducing GHG emissions, especially in the transportation sector (on-board applications). Fuel cell EVs (FCEVs) have significant potential for storing and converting chemical energy into electrical energy with zero emissions and thus for solving the climate crisis and making the roads and environment as clean as possible. The proposed work analysed the modern technology and innovations behind hydrogen storage and FCEV techniques for use in the road transportation sector. For this, FCEVs must have a storage capacity of (5-10) kgs of hydrogen to achieve a driving range of 500 kms. Chemical and physical energy storage (ES) approaches are more valuable to the hydrogen economy; however, compressed hydrogen storage is the most advanced approach for the transportation sector in on-board applications.

Park et al. [8] present and review the field experiment results of a Korea Electric Power Corporation (KEPCO) project, the aim of which was to create a new scheme called transmission system operator (TSO)—distribution system operators (DSOs)—distributed energy resource aggregators (DERAs), and namely the TSO-DSO-DERA interaction. This experiment was performed using a prequalification algorithm, which can be used to test realistic grid operating conditions. The obtained results confirm that the proposed algorithm is very effective and robust in nature, that it can be used to solve real-world grid problems and that it can be useful for DSOs in providing a reliable power supply to distribution networks. The proposed, prequalification approach can also support DERA's participation in the wholesale electricity market.

Hu et al. [9] present a detailed overview of electric grid modelling, including uncertainties, their impact on the system and their mitigation. In power systems, uncertainties arise from RERs, loads and EV loads; they play significant roles in maintaining grid stability, reliability, and energy and economic efficiency. The emerging power networks utilize an advanced metering infrastructure, which reduces the risk of cyberattacks. This work presented a comprehensive review of the uncertainties related to climate factors, asset management, and cyberattacks; it also described various techniques to mitigate these issues. This work also presented scenario-based approaches for demand response and energy storage systems, as well as reserves for regulating service to mitigate modern power system uncertainties. This work explored trade-off strategies to improve the reliability and computational speed which influence the planning and operation of future hybrid power grids.

Marpaung et al. [10] present a power generation technology based on a solid-state thermos electric strategy to generate electrical energy from a low-temperature geothermal-based heat source. The aim of this work is to design, operate and test a small-scale, portable power generation system which can produce power by using a heat source with a low-temperature steam or water phase. The various parameters utilized for developing this device are electric voltage and current, hot- and cold-side temperatures, load resistance and measurement time. The proposed device can act as an alternative to renewable energy and can be operated to produce stable power.

Gomez-Redondo et al. [11] analyse and present a comparative review of an AC microgrid. This work focused on reviewing the standard grid connection methods that are applicable to microgrids and control schemes. AC microgrids utilize hierarchical control systems to provide grid-connected and islanded modes of operation. With the recent advancements in battery technologies and the increasing popularity of DC loads with solar PV systems, the future adoption of hybrid microgrids is a promising solution. This work also presents an experimental setup with parallel inverters as a control scheme for testing fault detection techniques.

2. Conclusions

In recent years, the integration of renewable-based distributed energy resources (DERs), such as wind turbines, solar photovoltaic (PV) units, and electric vehicles (EVs), is growing rapidly, as these technologies have substantial advantages. Energy storage systems (ESSs) can be considered the optimal solution for facilitating renewable power and EV integration. Renewable energy resources (RERs), ESSs, and EVs are becoming increasingly popular, as they are environmentally friendlier and more sustainable. Future advancements and the continually advancement of EV technology may lead to the development of EVs with longer driving ranges with dynamic or single-charge options. The purpose of this collection, "Smart Energy System Design", is to provide a comprehensive overview of smart energy system design; the latest advancements, opportunities, challenges, and potential applications of energy storage; renewable energy; EVs; big data; artificial intelligence; internet of things; and machine learning technologies. All of the manuscripts reviewed in this Editorial Board Collection series are of practical importance for the development of smart energy systems.

Funding: This research work was funded by "WOOSONG UNIVERSITY's Academic Research Funding—2024".

Conflicts of Interest: The author declares no conflicts of interest.

References

1. El Harouri, K.; El Hani, S.; Naseri, N.; Elbouchikhi, E.; Benbouzid, M.; Skander-Mustapha, S. Hybrid Control and Energy Management of a Residential System Integrating Vehicle-to-Home Technology. *Designs* **2023**, *7*, 52. [CrossRef]
2. Sambhi, S.; Sharma, H.; Bhadoria, V.; Kumar, P.; Fotis, G.; Ekonomou, L. Technical and 2E Analysis of Hybrid Energy Generating System with Hydrogen Production for SRM IST Delhi-NCR Campus. *Designs* **2023**, *7*, 55. [CrossRef]
3. Maklakov, A.S.; Nikolaev, A.A.; Lisovskaya, T.A. Control over Grid Reactive Power by Using a Powerful Regenerative Controlled-Speed Synchronous Motor Drive. *Designs* **2023**, *7*, 62. [CrossRef]
4. Rashid, F.L.; Al-Obaidi, M.A.; Dulaimi, A.; Bahlol, H.Y.; Hasan, A. Recent Advances, Development, and Impact of Using Phase Change Materials as Thermal Energy Storage in Different Solar Energy Systems: A Review. *Designs* **2023**, *7*, 66. [CrossRef]
5. Liu, J.; Hu, H.; Yu, S.S.; Trinh, H. Electricity Pricing and Its Role in Modern Smart Energy System Design: A Review. *Designs* **2023**, *7*, 76. [CrossRef]
6. Viswa Teja, A.; Razia Sultana, W.; Salkuti, S.R. Performance Explorations of a PMS Motor Drive Using an ANN-Based MPPT Controller for Solar-Battery Powered Electric Vehicles. *Designs* **2023**, *7*, 79. [CrossRef]
7. Gómez, J.A.; Santos, D.M.F. The Status of On-Board Hydrogen Storage in Fuel Cell Electric Vehicles. *Designs* **2023**, *7*, 97. [CrossRef]
8. Park, J.-S.; Kim, B.-H. Field Experiment for a Prequalification Scheme for a Distribution System Operator on Distributed Energy Resource Aggregations. *Designs* **2023**, *7*, 134. [CrossRef]
9. Hu, H.; Yu, S.S.; Trinh, H. A Review of Uncertainties in Power Systems—Modeling, Impact, and Mitigation. *Designs* **2024**, *8*, 10. [CrossRef]
10. Marpaung, H.; Supriyadi; Lasmi, N.K.; Singarimbun, A.; Srigutomo, W. Design and Application of Low-Temperature Geothermal Thermoelectric Power Generation (Lotemg–TPG) in Sari Ater Hot Spring, Ciater, Subang, West Java, Indonesia. *Designs* **2024**, *8*, 60. [CrossRef]
11. Gomez-Redondo, M.; Rivera, M.; Muñoz, J.; Wheeler, P. A Systematic Literature Review on AC Microgrids. *Designs* **2024**, *8*, 77. [CrossRef]

designs

MDPI

Article

Hybrid Control and Energy Management of a Residential System Integrating Vehicle-to-Home Technology

Khadija El Harouri [1], Soumia El Hani [1,*], Nisrine Naseri [1], Elhoussin Elbouchikhi [2,*], Mohamed Benbouzid [3,4] and Sondes Skander-Mustapha [5]

1 Department of Electrical Engineering, Energy Optimization Diagnosis and Control, STIS Research Center, ENSAM, Mohammed V University in Rabat, Rabat 10100, Morocco
2 ISEN Yncréa Ouest, Nantes Campus, LABISEN, 33, Avenue du Champ de Manoeuvre, 44470 Carquefou, France
3 Institut de Recherche Dupuy de Lôme (UMR CNRS 6027 IRDL), University of Brest, 29238 Brest, France
4 Logistics Engineering College, Shanghai Maritime University, Shanghai 201306, China
5 Laboratory of Electrical Systems, Ecole Nationale d'Ingénieurs de Tunis, LR11ES15, University of Tunis El Manar, Tunis 1002, Tunisia
* Correspondence s.elhani@um5r.ac.ma (S.E.H.); elhoussin.elbouchikhi@isen-ouest.yncrea.fr (E.E.)

Abstract: Electric vehicles (EV) and photovoltaic (PV) systems are increasingly becoming environmentally friendly and more affordable solutions for consumers. This article discusses the integration of PV and EV in a residential system to meet the requirements of residential loads taking into account the PV supplied power, availability and the state of charge (SOC) of EVs. A hybrid control model has been proposed to control the residential system. The combined PI-Fuzzy logic controller is employed to control the buck-boost bi-directional converter. The DC-AC grid-side converter is controlled by the ADRC controller. The effectiveness of PI-Fuzzy logic controller in reducing voltage and current ripples and ADRC controller in rejecting disturbances is demonstrated in each case. A rule-based energy management strategy has been proposed to control the flow of energy between the components of the residential system. The suggested energy management system (EMS) covers every scenario that might occur. Whether the EV is linked to the home or not, and also takes into account the owner using the EV in an emergency situation. The EV operates in two modes, Home-to-Vehicle (H2V) mode and Vehicle-to-Home (V2H) mode, depending on the power produced by the PV and the conditions related to the EV. All possible scenarios are tested and validated. The simulation results show that the proposed EMS is a reliable solution that can reduce the power grid intervention.

Keywords: Vehicle-to-Home; Home-to-Vehicle; energy management systems; residential system; rule-based algorithm; PI-Fuzzy logic control; ADRC control

Citation: El Harouri, K.; El Hani, S.; Naseri, N.; Elbouchikhi, E.; Benbouzid, M.; Skander-Mustapha, S. Hybrid Control and Energy Management of a Residential System Integrating Vehicle-to-Home Technology. *Designs* **2023**, *7*, 52. https://doi.org/10.3390/designs7020052

Academic Editor: Joshua M. Pearce

Received: 10 February 2023
Revised: 13 March 2023
Accepted: 21 March 2023
Published: 4 April 2023

1. Introduction

Since its appearance, the automobile has become more and more a main necessity for human beings, which allows them to make life easier and minimize travel time. However, in recent years the automobile sector has posed certain problems, such as gas emissions, the greenhouse effect, dependence on oil, which continues to become scarce, etc. These circumstances impose an orientation towards new technologies to overcome these problems [1]. In the upcoming years, electric vehicles (EVs) powered by renewable energies, can play a significant role in the transition to sustainable modes of transportation, reduce CO_2 emissions, and improve local air quality. Currently, the rate of EV development is exponentially increasing, more than 16.5 million electric automobiles were in circulation in 2021, a tripling in only three years as shown in Figure 1 [2]. The main forces behind the recent global growth of EVs are the tightening of CO_2 emission limits that took place in 2020 and 2021, as well as the increase in the price of fuel. The pace of sales was also aided by the increase in tax breaks and the increase in purchase subsidies [2].

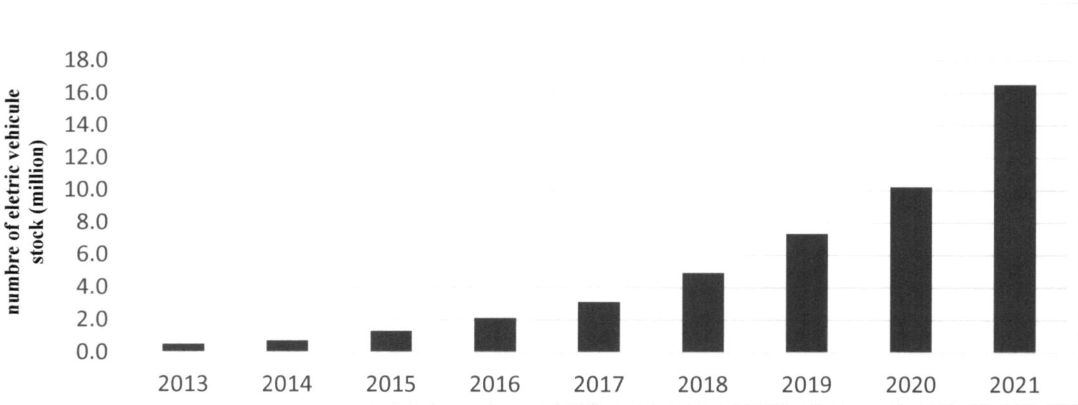

Figure 1. Evolution of global EV.

The rapid development of EVs has led to a sharp increase in the demand for electricity from these electric vehicles, which has led to thinking about a set of solutions to remedy this increase in energy demand. Several research directions are addressed on this topic, such as the integration of EV in micro-grids [3,4], the Vehicle-to-Grid (V2G) and Vehicle-to-Vehicle (V2V) concepts [5–9], charging management of EVs in parking lots [10,11], the impact of EV on the grid [12,13], the charging infrastructure [14], battery degradation [15] and the charging economics concept [16]. The integration of EVs in residential systems is an important aspect of addressing the problem of high energy demand, with the optimal integration of renewable energy systems (RES) and energy storage systems (ESS). This integration makes it possible to exploit the maximum power of renewable energy sources, as well as to use the EV as a temporary storage system in the house and as a source of energy production in order to decrease the budget of EV operation and home consumption, and to relieve the public electricity grid.

Most EVs are used for a short time around 5% of the day, and for the rest of the day around 95% of the remaining time [17], EVs are parked in workplace parking lots or at home. Thus, we can benefit from them during those times when they are parked. The EV can use its battery to store energy from renewable energy sources or to help storage systems to store that energy; in this case, the vehicle is used as a storage system. During low energy production, the EV can intervene to help meet the requirements of home loads by sending the missing energy, this is the Vehicle-to-Home (V2H) concept. During on-peak periods when the price of electricity from the public grid is particularly high, the EV can be used as a source of energy, and it has the ability to sell the energy stored in its battery to the electrical grid; this is the V2G concept. To manage these operating modes, plan the charging of the EV and manage the energy exchange in the residential system between the sources and the loads, the EMS make it possible to meet these requirements. Charging is said to be "normal" when the car charges to full capacity as soon as it is connected to the grid. The EMS makes it possible to modify this "normal" behavior, with the use of different operating modes such as V2H, and V2G to minimize the cost of the owner's electricity bill while meeting his personal constraints in terms of the EV's availability and autonomy.

This work presents a residential system connected to the electrical grid. It consists of PV as the primary source, a battery system for energy storage, an EV, and residential loads. A hybrid control has been suggested in this article to control system components. The charge and discharge of the EV are managed by the combined PI-Fuzzy logic control, which combines the advantages of both conventional PI controls and fuzzy logic. The injection and the use of the grid are controlled by the Active Disturbance Rejection Control (ADRC). The energy management in this residential system is carried out using the rule-based method. The electrical grid is used as a backup source, and the PV is the main source that must be exploited to the maximum. In the case when the energy generated through the PV system reached its maximum, it provides the power to the residential loads and charges the batteries and the electrical vehicle battery, and the excess PV power is fed to the grid. Additionally, when the PV system's power output is low, the battery and the EV are used to supply residential loads. The grid maintains the shortage when the power of the components of the system is insufficient to meet the loads. Several scenarios are studied in this work, taking into account certain conditions such as energy generated through the PV generator, the demand for residential loads, the availability of the EV, EV SOC.

The main contributions of this paper include:

- A hybrid control system that includes PI-Fuzzy logic, and ADRC.
- A combined PI-Fuzzy logic control, which combines the benefits of a classical PI controller and a fuzzy logic controller.
- An energy management system that addresses all possible scenarios, considering the emergency use of the EV by the owner.
- The use of the V2H concept to relieve the electricity grid.

The outline of this article is as follows: Section 2 displays a description of the residential system. The presentation of the proposed control laws and the energy management system are provided in Section 3. Section 4 shows the simulation results and a discussion of the main scenarios and operation modes. Finally, Section 5 provides conclusions.

Motivation and Novelty

This paper proposes a hybrid control model and energy management strategy that examines possible scenarios, for energy exchange in a residential system including PV system, EV, and energy storage system, residential loads and electricity grid. The proposed EMS makes it possible to exploit the maximum power of PV system. It is primarily used as the main source, and the grid as a backup source. Furthermore, the V2H and H2V concepts have been taken into account. Compared with the existed articles, this paper combines and provides a complete description of all possible operational scenarios with or without an EV connection, to relieve the power grid, considering the emergency use of the EV by the owner.

2. System Description

The proposed residential system consists of a PV system with battery storage system, EV, residential loads, and utility grid as shown by Figure 2. The PV system is considered as the main source of energy, the battery system acts as a method for momentary storage. Whereas the EV is used as a load in the context of the H2V concept, or energy source distributed as part of the possibility of income from V2H. Finally, the main grid is used as a backup source.

At this home, the energy control system that is in charge of managing the energy flow in the home communicates with the various elements that compose our residential system, the load and all energy sources. This is in order to know at any time the state of each element and information on the energy produced by the PV system, the energy consumption, the availability of the EV, etc.

Figure 2 presents the configuration of the residential system model proposed in this work. The PV system is connected to the DC bus using the Boost converter, that achieves the MPPT control. Additionally, the EV is coupled to the DC bus by the bi-directional Buck-Boost converter, which carries out the charging and discharging control of the EV. In H2V mode, the EV operates in charging mode (EV used as consumer). In V2H mode, the EV operates in discharging mode (EV used as power source). Likewise, the storage battery is coupled to the DC Bus by the Buck-Boost converter, which performs the charge and discharge control of the battery. On the other side, the electrical grid is coupled to the DC Bus using the bi-directional DC-AC converter. This converter can be operated as an inverter, in the case of energy fed to the power grid. Or as a rectifier to supply the load, in order to meet residential load requirements. Similarly, the residential loads are coupled to the DC bus by the DC-AC converter.

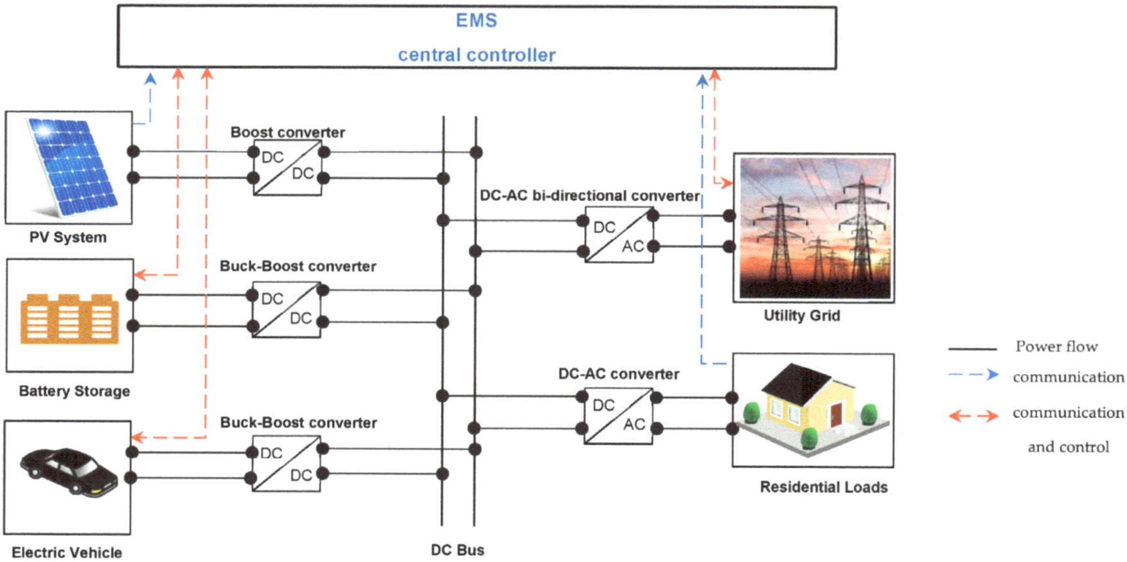

Figure 2. The configuration of the proposed system.

Residantial System Parameters

In this house, all the components composing this residential system namely the PV system, EV, battery storage, electrical grid and the residential loads are connected to the DC bus, which is set to 900 V.

The model of the PV system used can generate a maximum power of 63 kW at 25 °C and 1000 W/m^2. The EV used is of the Renault ZOE type with a battery capacity of 22 kWh [18], the storage battery is of 50 Ah capacity [19], as well as a residential load of 26 kW and a nominal voltage electrical grid of 220/380 V, 50 Hz are considered in this work. Table 1 displays the characteristics of these components.

Table 1. Residential system component parameters.

PV System	EV	Battery Storage	Residential Load
Module: SunPower SPR-305-WHT Type: 305 W @ 1000 W/m^2, 25 °C Number of series per string: 5 Number of parallels per string: 40 Maximum Power: 63 kW	Vehicle name: Renault ZOE Battery capacity: 22 kWh Rated voltage: 400 V	Capacity: 50 Ah Single module voltage: 12 V Number of series connected modules: 34 Rated voltage: 400 V	Type: AC Power: 26 kW

3. Control Strategy of the Proposed Residential System

The general structure of the overall control system of the proposed residential system is presented in Figure 3. The control system detects the power produced by the PV, the power demanded by the residential loads, the availability of EV, the state of charge and the charging power of the storage battery and the EV, the voltage from the DC bus to generate the control signals for each power converter, in order to control how much energy is exchanged in the residential system between the loads and the sources.

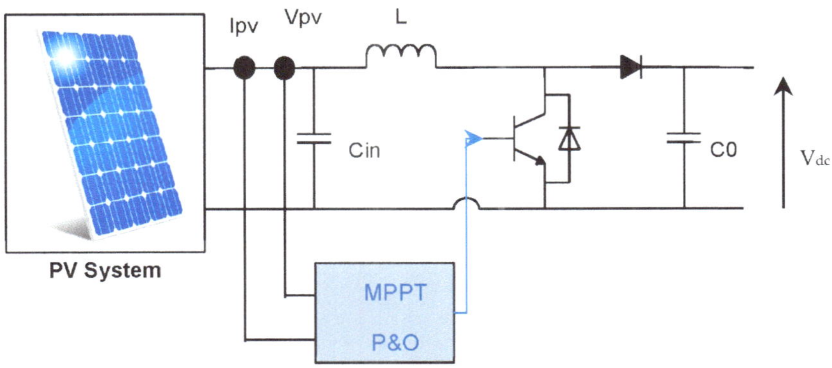

Figure 3. PV system control design.

3.1. Local Controllers Design

3.1.1. PV System Control

In order to maximize the power available in PV arrays, Maximum Power Point Tracking (MPPT) techniques based on the perturb and observe (P&O) algorithm are employed to maintain the operating point of the PV array at its maximum power point [20,21].

The goal of the technique is to maintain the PV output power close to the maximum power point (MPP). To achieve this, the resulting I_{pv} current and V_{pv} voltage values of the PV generator are recorded, then the power is determined. Subsequently, the variation of the power P_{pv} and the voltage V_{pv} are calculated to know the location of this point $(V_{pv}(t),P_{pv}(t))$ in the P–V curve and the sign of the derivatives are positive or negative, a positive power derivative means that the operating point is approaching the MPP. Then, the sign of the voltage drift is tested to know the direction of the search.

The search direction determines whether the control signals of pulse width modulation. (PWM) is in the increase or decrease stage of the duty cycle.

3.1.2. EV Control

A bi-directional buck-boost converter is used between the EV and the DC bus to determine when the EV battery is charged and discharged [22,23].

Conventional PI controllers are widely used in most of the energy systems due to their simple design, easy implementation, affordable price and their robustness. However, for non-linear systems and in case of parameters variation PI control is not effective. On the other hand, the fuzzy logic control (FLC) is an effective control in case of system parameters variations, the uncertainty of the inputs, the non-linearity of the systems and it presents a reduced response time. Therefore, to control the converter we have developed a hybrid PI-Fuzzy logic control, that combines the benefits of traditional PI controller with the advantages of the fuzzy logic controller, in order to obtain a robust and flexible controller to effectively manage EV charging and discharging.

A PI-Fuzzy logic controller controls the EV charging and discharging using a reference current that is determined by the EMS based on the operating mode of EV as shown by Figure 4. Similarly, a bidirectional buck/boost converter based a PI-Fuzzy logic controller is utilized to control charging and discharging of the battery energy storage, in the same way.

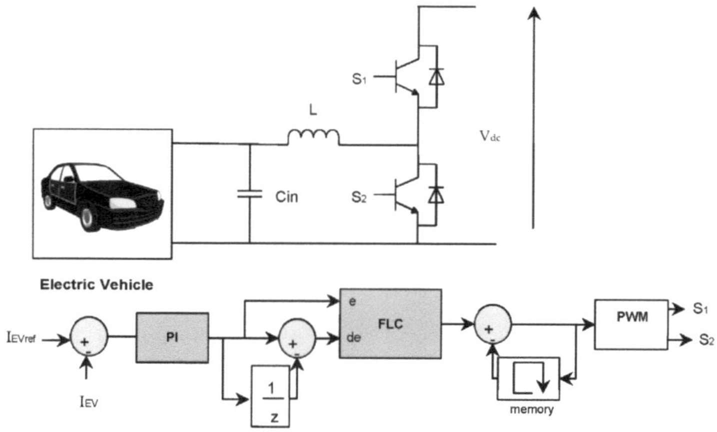

Figure 4. EV control design.

Table 2 provides the FLC rule base that was used in this research work.
The Membership functions for e and de applied in this work are:
NB: is Negative Big
NS: is Negative Small
ZZ: is Zero
PS: Positive Small
PB: is Positive Big.
The fuzzy rules used in this work are:
When E is PB and DE is NB ⇒ the output is ZZ
When E is PB and DE is NS ⇒ the output is PS
When E is PB and DE is ZZ ⇒ the output is PS
When E is PB and DE is PS ⇒ the output is PB
When E is PB and DE is PB ⇒ the output is PB
Where e and de are the inputs of the FLC, they respectively present the error and the change in error.

Table 2. Fuzzy rule table [24].

E			**DE**		
	NB	NS	ZZ	PS	PB
PB	ZZ	PS	PS	PB	PB
PS	NS	ZZ	PS	PS	PB
ZZ	NS	NS	ZZ	PS	PS
NS	NB	NS	NS	ZZ	PS
NB	NB	NB	NS	NS	ZZ

3.1.3. Grid-Side Converter Control

The grid-side converter is implemented using the bi-directional DC-AC converter located between the electrical grid and the DC bus. The control of this converter is based on ADRC control, which controls the flow of energy from or to the electrical grid as depicted by Figure 5. This converter operates in two ways, either rectifier or inverter depending on the situations. When we want to use the power of the electrical grid, the converter is therefore operated as a rectifier, the current is flowing from the converter to the DC bus. Contrarywise, when we want to supply the excess power to the grid, the converter is therefore operated like an inverter, the current is flowing from the DC bus to the main grid [25].

ADRC is a robust control method, it is employed in this work to control the grid currents and the DC bus voltage.

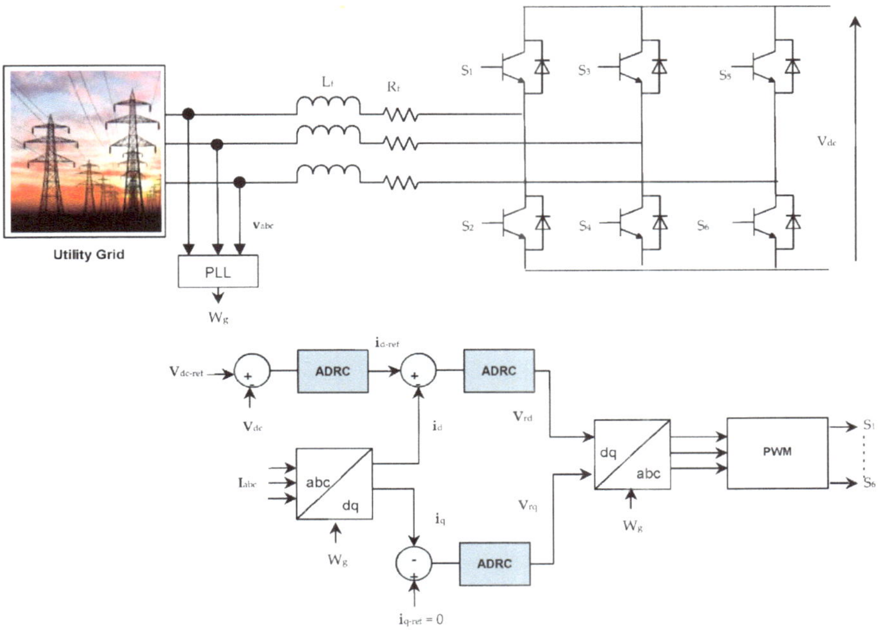

Figure 5. The grid-side converter control design.

3.2. Rule-Based Energy Management System

To manage the energy between the demand and generation in residential system, the rule-based algorithm is used to implement an energy management system. All energy management parameters are provided in Table 3.

Table 3. Energy system parameters.

P_{PV}	Photovoltaic power
P_L	Load power
P_{EV}	Electric vehicle power
P_{bat}	Storage battery power
P_{grid}	Grid power
P_{fg}	Power from the grid
P_{tg}	Power to the grid

The proposed rule-based algorithm makes it possible to plan the use of the components of the residential system according to the power produced, the power requested and the availability of EV. The electric vehicle has two operating cases. The first case when the EV is connected to the home, in this case the electric vehicle participates in the residential system and can operate in two modes. The H2V mode, in this mode the vehicle is considered as a load, it consumes energy. The V2H mode, in this mode the vehicle is considered as an energy source, it provides energy. The second case of operation is when the EV is not connected at home, in this case the electric vehicle does not participate in the residential system.

The PV system is used as a main source in this work and the electrical grid as a backup source. If the PV generator produces more energy, the energy from the PV is used to satisfy the residential loads demand, charge the home storage batteries and the EV battery (if connected) to be used when needed, and the excess PV energy is fed to the grid. Moreover, when the power generated by the PV generator is low, the storage battery and the EV are used to supply the residential loads, the power grid intervenes to ensure continuity of service if the power from the storage battery and the EV and the PV is not enough to satisfy the demand for residential loads.
Problem formulation:

$$P_L(t) - P_V(t) - P_{bat}(t) - P_{EV}(t) - P_{grid}(t) = 0 \quad (1)$$

The proposed control system consists of several possible scenarios following two cases:
Case 1: EV is connected to home

- Scenario 1: This scenario was regarded as a standard situation, PV powers residential loads and charges the EV (H2V) and storage battery and excess PV power is fed to the grid.

$$P_V(t) > P_L(t) + P_{bat}(t) + P_{EV}(t) \quad (2)$$

$$P_{tg}(t) = P_V(t) - P_L(t) - P_{bat}(t) - P_{EV}(t) \quad (3)$$

- Scenario 2: PV energy is enough to power residential loads and charges the EV (H2V) and storage battery, no excess energy.

$$P_V(t) = P_L(t) + P_{bat}(t) + P_{EV}(t) \quad (4)$$

- Scenario 3: the PV energy is not enough to supply the residential loads. The storage battery is used to satisfy the demand for residential loads and charges the EV (H2V) if it is discharged.

$$P_V(t) + P_{bat}(t) = P_L(t) + P_{EV}(t) \quad (5)$$

- Scenario 4: PV power and storage battery power are not enough to power residential loads, the EV intervenes to meet load demand (V2H).

$$P_L(t) = P_{PV}(t) + P_{bat}(t) + P_{EV}(t) \quad (6)$$

- Scenario 5: PV power, storage battery power and EV (V2H) power are not enough to power residential loads. In this case, the grid intervenes to ensure the load demand is met.

$$P_{fg}(t) = P_L(t) - P_{EV}(t) - P_{bat}(t) - P_{PV}(t) \quad (7)$$

- Scenario 6: PV power and storage battery power are not enough to power residential loads, and the EV is discharged. In this case, the grid intervenes to ensure load demand is satisfied.

$$P_{fg}(t) = P_L(t) + P_{EV}(t) - P_{bat}(t) - P_{PV}(t) \quad (8)$$

Case 2: EV is not connected to home

- Scenario 7: PV energy is enough to power residential loads and charges the storage battery, and excess PV power is supplied to the grid.

$$P_{PV}(t) > P_L(t) + P_{bat}(t) \tag{9}$$

$$P_{tg}(t) = P_V(t) - P_L(t) - P_{bat}(t) \tag{10}$$

- Scenario 8: PV energy is enough to power residential loads and charges the storage battery, no excess energy.

$$P_V(t) = P_L(t) + P_{bat}(t) \tag{11}$$

- Scenario 9: the PV energy is not enough to supply the residential loads. The storage battery is used to meet the demand of the loads.

$$P_{bat}(t) = P_L(t) - P_{PV}(t) \tag{12}$$

- Scenario 10: PV power and storage battery power are not enough to power residential loads. The grid intervenes to ensure the load demand is satisfied.

$$P_{fg}(t) = P_L(t) - P_{bat}(t) - P_{PV}(t) \tag{13}$$

Figure 6 depicts the algorithmic flowchart of the proposed EMS describing different operating modes.

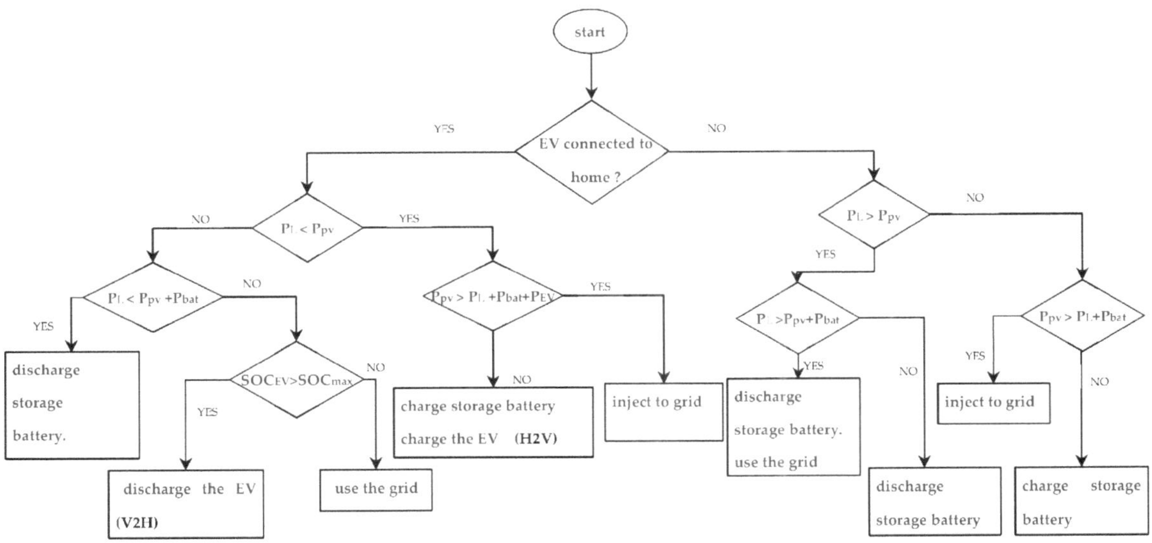

Figure 6. Proposed residential EMS flowchart.

4. Simulation Results and Discussion

To verify the effectiveness of the proposed residential EMS and associated control strategies, all the derived sceanrios have been simulated and results presented and discussed. This is in order to get a thorough investiagation that allow to demonstrate the interest of renewables and V2H technology in the residential sector.

The EV can be operated in V2H mode only when it is available (connected to home) and when its charge level is higher than the SOCmax, which is set in this study at 50%, because the EV must always be available and charged with an acceptable state of charge as soon as the user needs it, it is necessary to think of the emergency case of use of the vehicle by the owner.

4.1. EV Connected to Home

4.1.1. Case 1

Figure 7 demonstrates the outcomes of the first scenario. The power produced by the PV generator is of higher value and equal to 63 kW. It is able to supply the residential loads with 26 kW power and charge the EV with 7.5 kW charging power and charge the storage battery with a power of 7.3 kW, and there remains about 22.2 kW it is supplied to the electrical grid.

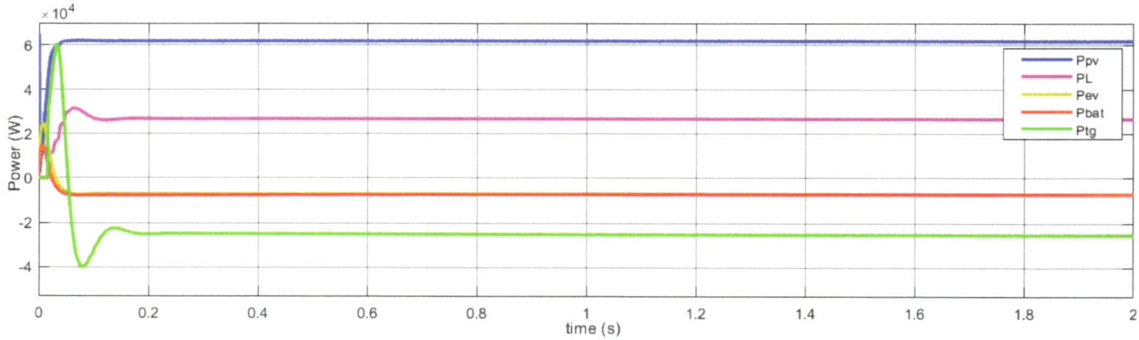

Figure 7. Residential system power flow with V2H.

A negative value of the power $\left(P_{grid}, P_{bat}, P_{EV}\right)$ means that the element receives energy (charging the battery/charging the EV/injecting energy into the grid). Contrarywise, the element is providing energy (by depleting the battery, discharging the EV, or using grid electricity) if the power value was positive.

In this case the energy produced through the PV is sufficient to meet the requirements of the household loads, to charge the EV and the storage battery, and the rest of the power is injected into the electricity grid. Figure 8a shows EV charging and Figure 8b shows battery charging, and the DC bus voltage is shown in Figure 9.

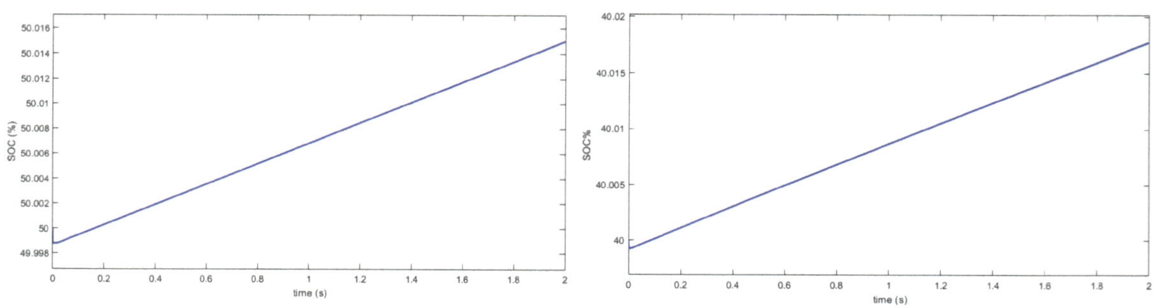

(a) (b)

Figure 8. EV and battery charging from PV. (**a**) EV SOC. (**b**) battery SOC.

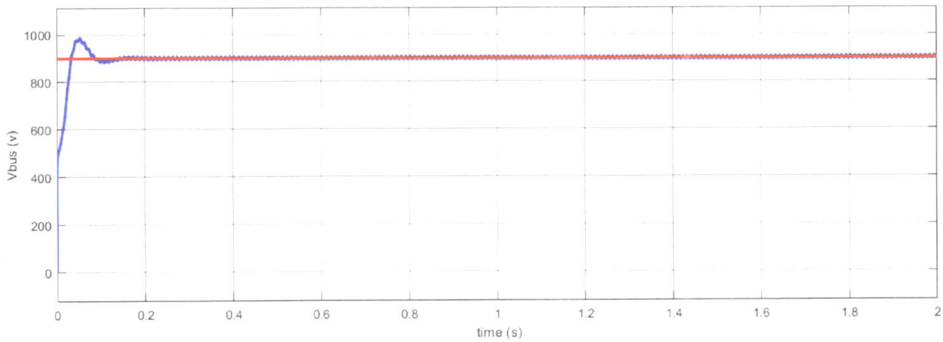

Figure 9. DC Bus voltage.

4.1.2. Case 2

In this scenario, the energy generated through the PV generator is of low value, the PV power and the storage battery power are not enough to supply the residential loads. Moreover, since the EV is connected to the house and its charge level is greater than 50%, so it intervenes to meet the demand for household loads, and the electricity grid intervenes to supply the load remaining power of around 8.5 kW as shown in Figure 10.

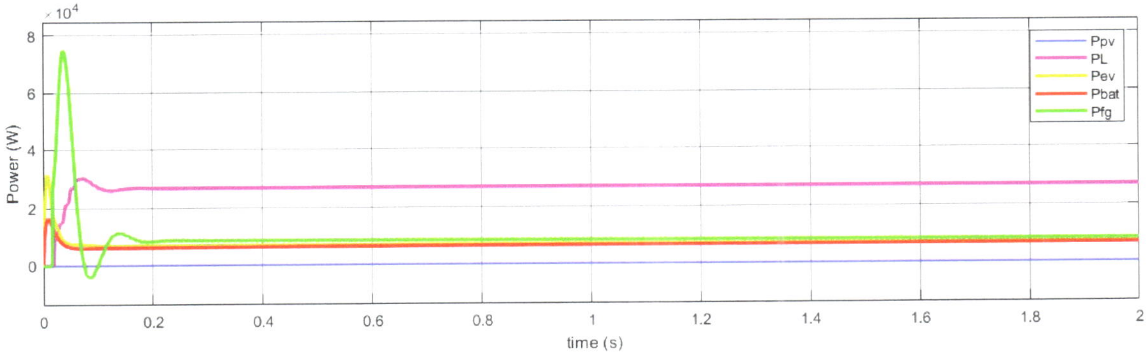

Figure 10. Power distribution in the system.

The EV operates in this case in V2H mode. It makes use of the energy contained in its battery to power household loads as depicted by Figure 11.

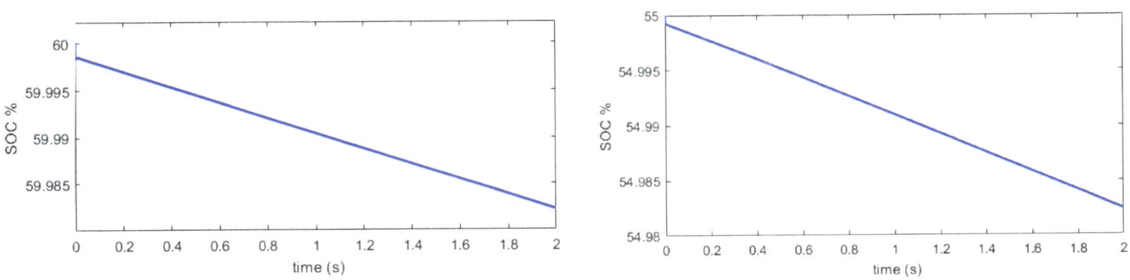

(**a**) (**b**)

Figure 11. EV and battery discharging to meet the demand of the loads. (**a**) EV SOC. (**b**) storage battery SOC.

4.1.3. Case 3

As shown by Figure 12, the PV power and the storage battery power are not enough to supply the residential loads, and the EV is discharged. In this case, the grid intervenes to ensure the load requirements are satisfied, and the EV operates in the H2V mode.

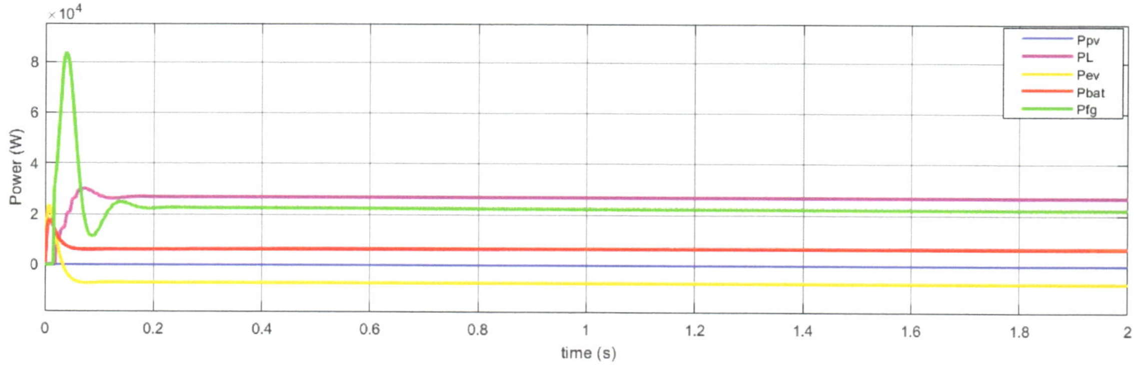

Figure 12. Power distribution in the system.

Figure 13 represents the SOC of the EV and the storage battery. The EV operates in H2V mode, it charges through power from storage battery, PV system, and utility grid.

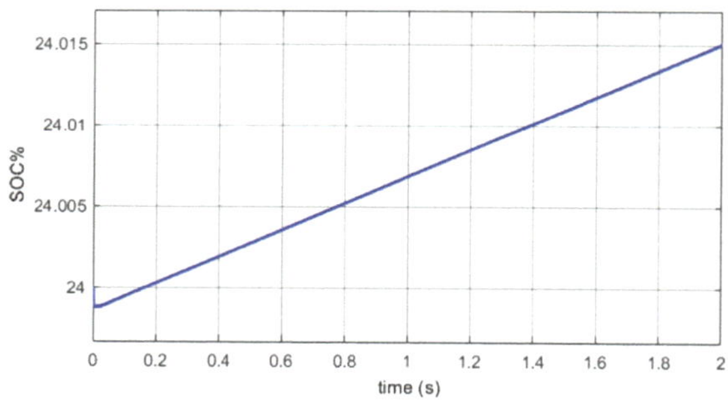

(a)

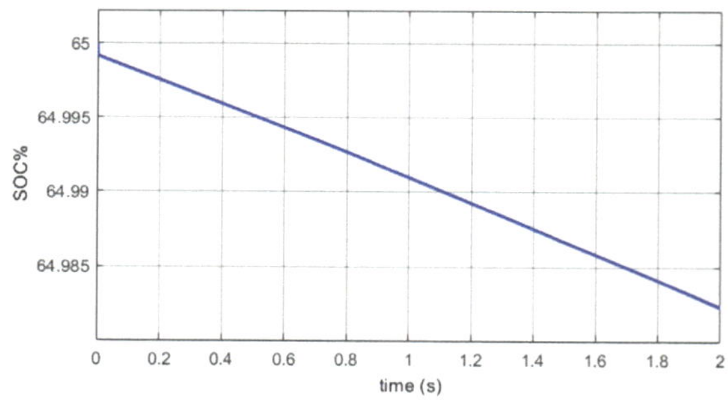

(b)

Figure 13. SOC of the EV and the storage battery. (**a**) EV SOC. (**b**) Storage battery SOC.

In this case, the EV cannot operate in V2H mode because it is discharged. The grid power used is about 22.5 kW, compared to the previous case (case 2), the grid power has been used was just 8.5 kW. Therefore, the simulation results presented in Figure 12 show that the use of EV in V2H mode makes it possible to relieve the electrical grid.

4.2. EV Is Not Connected to Home

PV power and storage battery power are not enough to supply residential loads, and the EV is not connected to the house, in this case the grid intervenes to ensure the continuity of service. In this scenario the EV is not connected to the house, it does not intervene in the residential system.

Figure 14 shows that the power of the EV is at zero because the EV is not connected to the home, and the power produced by the PV is low. Therefore, the storage battery, which will be discharged to satisfy the demand of the household loads as shown by Figure 15, and the electricity grid intervenes to satisfy the load requirements. In this scenario, the power of the grid used is approximately 16 kW. On the other hand, in the same scenario but with the intervention of the EV, the grid uses just a power of 8.5 kW (Figure 10). This means that the participation of the EV allows to reduce the use of the grid, which leads to the relief of the grid.

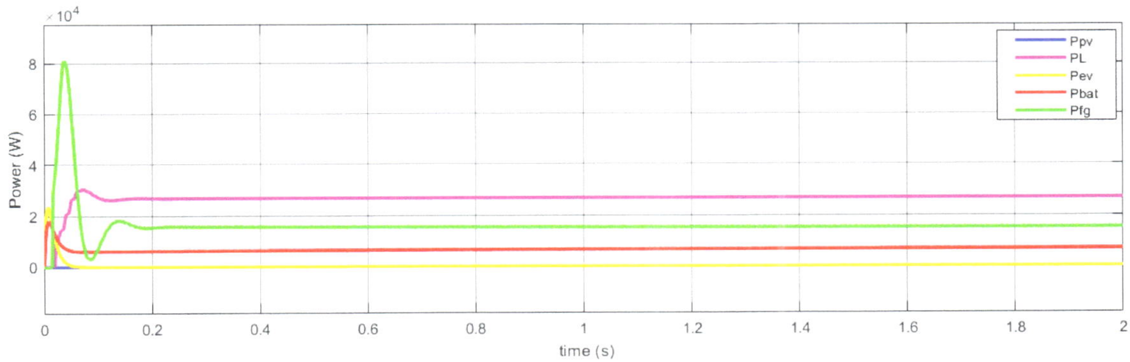

Figure 14. Power distribution in the system.

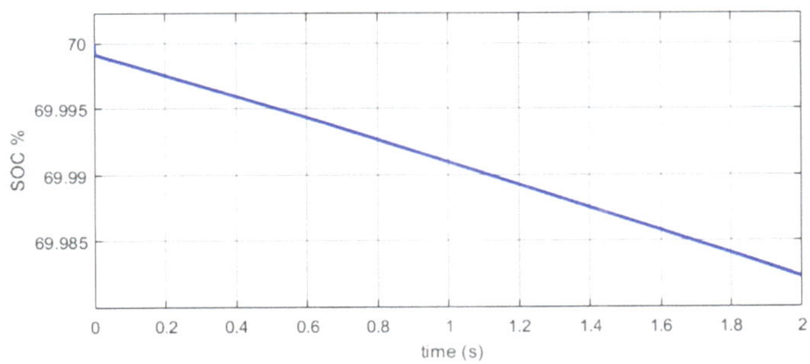

Figure 15. State of charge of the storage battery.

5. Summary and Conclusions

The V2H concept is considered an advanced and useful technology, as it can contribute to reducing residential electricity consumption bills and relieving the grid in times of high electricity demand. Most EV owners can benefit from this concept when the EVs are parked at workplace or at home. This work deals with the integration of EV in a residential system operated in a grid-connected mode, and it comprises the PV system, battery storage system, and home loads.

A hybrid control has been suggested to control the different elements of the system, the EV converter is controlled through the combined PI-Fuzzy logic control, which combines the two controls, the conventional PI and the fuzzy logic, the bidirectional AC–DC converter is controlled by ADRC control. The results show the efficiency of the both controllers, the PI-Fuzzy logic controller in reducing voltage and current ripples, and the ADRC controller in rejecting disturbances.

To manage the energy exchange between residential system elements, an energy management system based on the rule-based method is proposed. The suggested EMS addresses all possible scenarios, considering the emergency use of the EV by the owner. All these possible scenarios have been tested and validated.

The simulation results show that the proposed EMS with a residential system integrating V2H technology is a reliable solution that can better handle peak energy demand while simultaneously reducing power grid intervention. The PV is considered as main source, battery for storage, utility grid is used as a backup source. The EV in this work operates in two modes, V2H mode and H2V mode depending on operating conditions. When the energy produced through the PV generator is higher, this power is used to meet the requirements of residential loads and charge the storage batteries and the EV battery to be used when needed (EV operates in H2V mode), and the excess PV energy is injected into the grid. On the contrary, when the energy generated through the PV generator is low, the storage battery and the EV are used to supply the residential loads (EV operates in V2H mode), the electricity grid intervenes to satisfy the load if the power of the battery storage and EV and PV is not enough.

Author Contributions: Conceptualization, K.E.H., S.E.H. and E.E.; Funding acquisition, E.E.; Investigation, N.N.; Methodology, S.E.H., E.E. and M.B.; Software, K.E.H. and N.N.; Supervision, S.E.H., E.E., M.B. and S.S.-M.; Validation, S.E.H. and E.E.; Writing—original draft, K.E.H. and N.N.; Writing—review & editing, S.E.H., E.E., M.B. and S.S.-M. All authors have read and agreed to the published version of the manuscript.

Funding: This research received no external funding.

Data Availability Statement: No new data were created or analyzed in this study. Data sharing is not applicable to this article.

Conflicts of Interest: The authors declare no conflict of interest.

References

1. Wu, Y.; Zhang, L. Can the development of electric vehicles reduce the emission of air pollutants and greenhouse gases in developing countries? *Transp. Res. Part D Transp. Environ.* **2017**, *51*, 129–145. [CrossRef]
2. IEA. Global EV Outlook 2022—Securing Supplies for an Electric Future. Glob. EV Outlook 2022. p. 221. Available online: https://iea.blob.core.windows.net/assets/ad8fb04c-4f75-42fc-973a-6e54c8a4449a/GlobalElectricVehicleOutlook2022.pdf (accessed on 9 February 2023).
3. Liu, Z.; Chen, Y.; Zhuo, R.; Jia, H. Energy storage capacity optimization for autonomy microgrid considering CHP and EV scheduling. *Appl. Energy* **2018**, *210*, 1113–1125. [CrossRef]
4. Bimenyimana, S.; Wang, C.; Nduwamungu, A.; Asemota, G.N.O.; Utetiwabo, W.; Ho, C.L.; Niyonteze, J.D.D.; Hagumimana, N.; Habineza, T.; Bashir, W.; et al. Integration of Microgrids and Electric Vehicle Technologies in the National Grid as the Key Enabler to the Sustainable Development for Rwanda. *Int. J. Photoenergy* **2021**, *2021*, 1–17. [CrossRef]
5. Chaurasiya, S.; Mishra, N.; Singh, B. A 50kW Bidirectional Fast EV Charger with G2V V2G/V2V Capability and Wide Voltage Range. In Proceedings of the 2020 IEEE 5th International Conference on Computing Communication and Automation (ICCCA), Greater Noida, India, 30–31 October 2020; pp. 652–657. [CrossRef]
6. Sovacool, B.K.; Kester, J.; Noel, L.; Zarazua de Rubens, G. Actors, business models, and innovation activity systems for vehicle-to-grid (V2G) technology: A comprehensive review. *Renew. Sustain. Energy Rev.* **2020**, *131*, 109963. [CrossRef]
7. Li, X.; Tan, Y.; Liu, X.; Liao, Q.; Sun, B.; Cao, G.; Li, C.; Yang, X.; Wang, Z. A cost-benefit analysis of V2G electric vehicles supporting peak shaving in Shanghai. *Electr. Power Syst. Res.* **2020**, *179*, 106058. [CrossRef]
8. Taghizadeh, S.; Hossain, M.J.; Poursafar, N.; Lu, J.; Konstantinou, G. A Multifunctional Single-Phase EV On-Board Charger with a New V2V Charging Assistance Capability. *IEEE Access* **2020**, *8*, 116812–116823. [CrossRef]
9. Koufakis, A.M.; Rigas, E.S.; Bassiliades, N.; Ramchurn, S.D. Offline and Online Electric Vehicle Charging Scheduling with V2V Energy Transfer. *IEEE Trans. Intell. Transp. Syst.* **2020**, *21*, 2128–2138. [CrossRef]
10. Alinejad, M.; Rezaei, O.; Habibifar, R.; Azimian, M. A Charge/Discharge Plan for Electric Vehicles in an Intelligent Parking Lot Considering Destructive Random Decisions, and V2G and V2V Energy Transfer Modes. *Sustainability* **2022**, *14*, 12816. [CrossRef]
11. Mohammad, A.; Zamora, R.; Lie, T.T. Transactive Energy Management of PV-Based EV Integrated Parking Lots. *IEEE Syst. J.* **2021**, *15*, 5674–5682. [CrossRef]
12. Wang, L.; Qin, Z.; Slangen, T.; Bauer, P.; Van Wijk, T. Grid Impact of Electric Vehicle Fast Charging Stations: Trends, Standards, Issues and Mitigation Measures-An Overview. *IEEE Open J. Power Electron.* **2021**, *2*, 56–74. [CrossRef]
13. Stiasny, J.; Zufferey, T.; Pareschi, G.; Toffanin, D.; Hug, G.; Boulouchos, K. Sensitivity analysis of electric vehicle impact on low-voltage distribution grids. *Electr. Power Syst. Res.* **2021**, *191*, 106696. [CrossRef]
14. Das, H.S.; Rahman, M.M.; Li, S.; Tan, C.W. Electric vehicles standards, charging infrastructure, and impact on grid integration: A technological review. *Renew. Sustain. Energy Rev.* **2020**, *120*, 109618. [CrossRef]
15. Liu, K.; Hu, X.; Yang, Z.; Xie, Y.; Feng, S. Lithium-ion battery charging management considering economic costs of electrical energy loss and battery degradation. *Energy Convers. Manag.* **2019**, *195*, 167–179. [CrossRef]
16. Patil, H.; Kalkhambkar, V.N. Grid Integration of Electric Vehicles for Economic Benefits: A Review. *J. Mod. Power Syst. Clean Energy* **2021**, *9*, 13–26. [CrossRef]
17. Ouramdane, O.; Elbouchikhi, E.; Amirat, Y.; Le Gall, F.; Gooya, E.S. Home Energy Management Considering Renewable Resources, Energy Storage, and an Electric Vehicle as a Backup. *Energies* **2022**, *15*, 2830. [CrossRef]

18. Berjoza, D.; Jurgena, I. Effects of change in the weight of electric vehicles on their performance characteristics. *Agron. Res.* **2017**, *15*, 952–963.
19. Kamal, T.; Karabacak, M.; Hassan, S.Z.; Fernández-Ramírez, L.M.; Riaz, M.H.; Riaz, M.T.; Khan, M.A.; Khan, L. Energy management and switching control of PHEV charging stations in a hybrid smart micro-grid system. *Electronics* **2018**, *7*, 156. [CrossRef]
20. Thakran, S.; Singh, J.; Garg, R.; Mahajan, P. Implementation of P&O Algorithm for MPPT in SPV System. In Proceedings of the 2018 International Conference on Power Energy, Environment and Intelligent Control (PEEIC), Greater Noida, India, 13–14 April 2018; pp. 242–245. [CrossRef]
21. Bhattacharyya, S.; Kumar P, D.S.; Samanta, S.; Mishra, S. Steady output and fast tracking MPPT (SOFT-MPPT) for P&O and InC algorithms. *IEEE Trans. Sustain. Energy* **2021**, *12*, 293–302. [CrossRef]
22. El Harouri, K.; El Hani, S.; Elbouchikhi, E.; Benbouzid, M.; Mediouni, H. Grid-connected plug-in electric vehicles charging stations energy management and control. *Int. J. Energy Convers.* **2019**, *7*, 49–57. [CrossRef]
23. Aboudrar, I.; El Hani, S.; El Harouri, K.; Martins, J.; Goncalves, R.J. Reactive Power Compensation by ADRC in Vehicle to Grid Application during Grid Fault Conditions. In Proceedings of the 2020 International Conference on Electrical and Information Technologies (ICEIT), Rabat, Morocco, 4–7 March 2020; pp. 1–6. [CrossRef]
24. Sahin, M.E.; Okumus, H.I. Fuzzy logic controlled buck-boost DC-DC converter for solar energy-battery system. In Proceedings of the 2011 International Symposium on Innovations in Intelligent Systems and Applications, Istanbul, Turkey, 15–18 June 2011; pp. 394–397. [CrossRef]
25. Liu, Y.; Wen, J.; Xu, D.; Huang, Z.; Zhou, H. The decoupled vector-control of PMSM based on nonlinear multi-input multi-output decoupling ADRC. *Adv. Mech. Eng.* **2020**, *12*, 1–12. [CrossRef]

designs

MDPI

Article

Technical and 2E Analysis of Hybrid Energy Generating System with Hydrogen Production for SRM IST Delhi-NCR Campus

Shilpa Sambhi [1], Himanshu Sharma [1,*], Vikas Bhadoria [2], Pankaj Kumar [1], Georgios Fotis [3] and Lambros Ekonomou [4,*]

[1] Department of Electrical and Electronics Engineering, SRM Institute of Science and Technology, Delhi-NCR Campus, Ghaziabad 201204, Uttar Pradesh, India; shilpasambhi@gmail.com (S.S.); pankajkumar903106@gmail.com (P.K.)
[2] Industry Integration Cell, Shri Vishwakarma Skill University, Palwal 121102, Haryana, India; vikasbhadoria@gmail.com
[3] Department of Electrical and Electronics Engineering Educators, ASPETE—School of Pedagogical and Technological Education, 141 21 N. Heraklion, Greece; gfotis@gmail.com
[4] UBITECH Energy Sprl, 367 Avenue Louise, B-1050 Brussels, Belgium
* Correspondence: himanshuresearch58@gmail.com (H.S.); lekonomou@ubitech.eu (L.E.)

Abstract: This work intends to perform technical and 2E (economic & environmental) analysis for the proposed hybrid energy generating system for a part load at SRM IST at the Delhi-NCR campus, India. The investigation has been done for electricity generation and hydrogen production through renewable energy sources, mainly solar energy. It is in line with the Indian Government's initiatives. The proposed hybrid system has to meet the electric load demand of 400 kWh/day with a peak load of 74.27 kW and hydrogen load demand of 10 kg/day with a peak demand of 1.86 kg/h. The analysis has been performed for both on-grid and off-grid conditions. As a result, optimum results have been obtained off-grid condition, with $0.408 per kWh cost of energy, $16.6 per kg cost of hydrogen, low O&M cost ($21,955 per year), a high renewable fraction (99.8%), and low greenhouse emissions (247 kg/year). In addition, sensitivity analysis has been performed between—(1) the solar PV array size & the number of battery strings, with NPC, renewable fraction & CO_2 emissions as sensitivity variables, and (2) reformer capacity & hydrogen tank capacity, with NPC as sensitivity variable.

Keywords: hybrid power plant; cost of energy; net present cost; green hydrogen; reformer; electrolyser

Citation: Sambhi, S.; Sharma, H.; Bhadoria, V.; Kumar, P.; Fotis, G.; Ekonomou, L. Technical and 2E Analysis of Hybrid Energy Generating System with Hydrogen Production for SRM IST Delhi-NCR Campus. *Designs* **2023**, *7*, 55. https://doi.org/10.3390/designs7020055

Academic Editors: Surender Reddy Salkuti and Dibin Zhu

Received: 18 March 2023
Revised: 6 April 2023
Accepted: 7 April 2023
Published: 9 April 2023

1. Introduction

Most economists and international organizations have realized the importance of sustainable energy in people's lives. It may ensure a better quality of life regarding health and economic independence in various countries. The world's electricity consumption is estimated to grow by 2.3% per year from 2015 to 2040 [1]. Worldwide power plants still use fossil fuels to fulfil about 80% of the electricity demand, threatening the rising levels of greenhouse gas, carbon dioxide (CO_2), levels [2]. The rise in CO_2 levels increases the average temperature globally, adversely affecting the global climate. In December 2015, world leaders decided to put a limit on the increase in global level to 2 °C at the Conference of the Parties, in short, called COP21 [3]. The burning of fossil fuels not only produces CO_2 but also other harmful gases like sulphur dioxide (SO_2), nitrogen oxides (NOx), and particulate matter. To accelerate the shift towards renewable energy sources (RES) for electricity generation, many countries like Denmark, Sweden, Norway, Italy, New Zealand, Canada, and Great Britain have started imposing carbon tax [4]. Due to recent technological advancements and high availability, solar energy is the most promising means to produce electricity [5]. The countries experiencing high wind speeds also use wind energy to produce electricity [6]. Another RES is hydro-kinetic energy, which uses speed of naturally flowing in river streams or man-made waterways to produce electricity [7].

Different strategies, like using locally available renewable sources for electricity generation, combined heat & power installations, and designing buildings with energy efficiency, have been adopted by European Union to reduce greenhouse gas [8]. European Union has set targets for national emissions in a year, electricity generation using RES and carbon capture & storage [9]. A Canadian company, Carbon Engineering Ltd., has developed an air-to-fuel technique to capture CO_2 directly from the atmosphere [10]. Electricity generated from RES has been used to split hydrogen from water. In the next step, hydrogen reacts with captured CO_2 to produce hydrocarbon fuels like diesel, and the entire process emits very little pollutants into the atmosphere. RES are unpredictable and intermittent. The availability of RES depends upon the local weather conditions of the selected location. Electricity generation from RES can be used in combination with grid supply or as a stand-alone system. In the case of stand-alone systems, hybrid energy-generating plants are economically feasible compared to systems using a single energy source [11]. This is because hybrid systems are less costly in electricity production, and the risk of power shortage is reduced compared to a single energy source. It was forecasted in 2015 by the Fraunhofer Institute for Solar Energy Systems that the capacity of solar PV may exceed 30TW peak power by 2050 [12]. Wind farms are now connected to the grid, and the electricity generation from wind turbines is available in the range of megawatts [13].

An important component in hybrid energy plants is the provision for storage of excess generated electricity. This excess electricity can be stored in the form of— (1) electrochemical energy (battery bank), (2) chemical energy (hydrogen), and (3) electrical energy (supercapacitors) [14]. Battery bank maintains power quality and provides short-term energy storage because of the self-discharge problem [15]. Hydrogen provides long-term storage (in months & years) for high volumes (in terawatt-hours) [16]. The improvement in electrolysers & their related components and the decrease in the cost of solar PV & wind technology are the main factors for motivating the concept of a 'hydrogen economy' [17]. As a result, hydrogen technology has been adopted by Japan, South Korea, China, and Germany [18].

India has taken the initiative of the National Green Hydrogen Mission to become a major exporter of hydrogen, with an incentive plan of $2.11 billion. This initiative is Strategic Interventions for Green Hydrogen Transition Programme (SIGHT) [19]. Under SIGHT, two distinct mechanisms exist: (1) to manufacture electrolysers with a 60 to 100 GW capacity in India and (2) to produce green hydrogen. The regions capable of producing/utilizing large-scale hydrogen will be developed as Green Hydrogen Hubs. The SIGHT program has three phases—(1) to establish policies and guidelines and initiate pilot projects (2022–2024), (2) to bring down manufacturing costs & promote incentives for large-scale production, to scale up production of green hydrogen (2025–2026), and (3) to focus on other distinctive hydrogen applications [19].

Hydrogen energy is the upcoming renewable energy system used in hybrid power plants to store excess electrical energy from the grid or RES. It has gained popularity because of the advancement in fuel cell technology [20]. Generally, a hydrogen energy system consists of three main apparatus for three process steps required for hydrogen production—(1) an electrolyser is required for producing hydrogen from the excess electricity generated from RES, (2) the hydrogen produced from step 1 is stored in a storage tank, and (3) the electricity is reproduced from stored hydrogen using a fuel cell [21]. During the process of production and reproduction of hydrogen, no pollutants are released into the atmosphere. Furthermore, hydrogen energy systems are efficient as the stored hydrogen does not degrade with time, and it attracts low maintenance costs for storing hydrogen for a considerably long time [22].

It is imperative to design the optimum size of the hybrid plant so that system components are fully utilized with minimum investment. This mainly depends on—(1) the availability of RES, (2) load demand, and (3) the weather conditions of the site [23]. To achieve the optimum system reliability and cost, one or a combination of different sizing methods like the graphical construction method, iterative method and artificial intelligence method can be applied [24]. Some work has been done in many parts of the world to either reduce the cost of energy or provide electricity to places remotely located from the primary grid, along with the production of hydrogen [25]. Iran, China, Ecuadorian Amazon, Turkey, Egypt, UAE, Iraq, Afghanistan, and the USA have been these countries. Table 1 summarizes a study in these countries for hybrid energy plants prioritizing hydrogen production at various locations.

Table 1. Summary of different hybrid energy plants prioritizing hydrogen production.

Hybrid Plant Design	Study Location	On-Grid/ Off-Grid	COE ($/kWh)	Ref.
PV/HK/FC/Battery/HT/EL	Ecuadorian Amazon	Off-grid	0.184	[26]
PV/WT/HK	Iran	Off-grid	0.1155	[27]
SOFC-WGS-TSA-PEMFC	China	Off-grid	0.0395 to 0.0405	[28]
PV	Turkey	Off-grid	0.6124	[29]
PV/WT	Egypt	Off-grid	0.137 to 0.219	[30]
PV/FC	UAE	Off-grid	0.145	[31]
PV/FC	Iraq	Off-grid	0.195	[32]
PV/WT/BGG	Iran	Off-grid	0.128 to 0.223	[33]
PV/WT/BGG/FC	Iran	Off-grid	0.164 to 0.233	[34]
		On-grid	0.096 to 0.125	
PV/BM	China	Off-grid	0.6137	[35]
WT	Afghanistan	Off-grid	0.063 to 0.079	[36]
PV/WT/HK	Iran	Off-grid	0.1155	[37]
PV	Turkey	Off-grid	0.6124	[38]
WT	Afghanistan	Off-grid	0.0529 to 0.1135	[39]
PV/WT/battery/EL	USA	Off-grid	0.5	[1]

Legends: PV—photovoltaic, HK—hydro kinetic, FC—fuel cell, HT—hydrogen tank, EL—electrolyser, WT—wind turbine, SOFC—solid oxide fuel cell, WGS—water gas shift reaction, TSA—thermal swing adsorption, PEMFC—proton exchange membrane fuel cell, BGG—biogas generator, BM—bio mass.

This study has been conducted at SRM IST, Delhi NCR educational campus. A hybrid energy generating system (HEGS) has been proposed for meeting the load of 15 classrooms in an academic building. The geographical location of this campus has been marked in Figure 1 as 28.8377° north and 77.5826° east [40]. This area experiences approximately 300 solar days. During the winter season, it experiences minimum temperature reaches around 1 °C for 3 to 4 days. The data for average monthly solar radiation (ALLSKY INSOL), clearness index, diffuse horizontal irradiance (DIFF), direct normal radiance (DNR), average monthly temperature and average daylight hours for the Delhi-NCR region has been taken from NASA (National Aeronautics and Space Administration)—PDAV (Power Data Access Viewer) tool [41]. Figure 2 presents the ALLSKY INSOL & clearness index, and Figure 3 compares the selected region's ALLSKY INSOL, DIFF & DNR. DIFF is the light scattered at the earth's surface due to dust particles and moisture content in the atmosphere [42]. DNR is location specific and measured at the earth's surface, but it does not include the scattered radiation due to atmospheric components [43]. The average monthly temperature and daylight hours for this region are shown in Figure 4. The proposed HEGS has been analyzed for both on-grid and off-grid operations. Time series analysis and sensitivity analysis have been performed to validate the energy balance and to understand the impact of variation in the size of solar PV array, number of battery strings, reformer capacity & hydrogen tank capacity on NPC, renewable fraction, and CO_2 emissions.

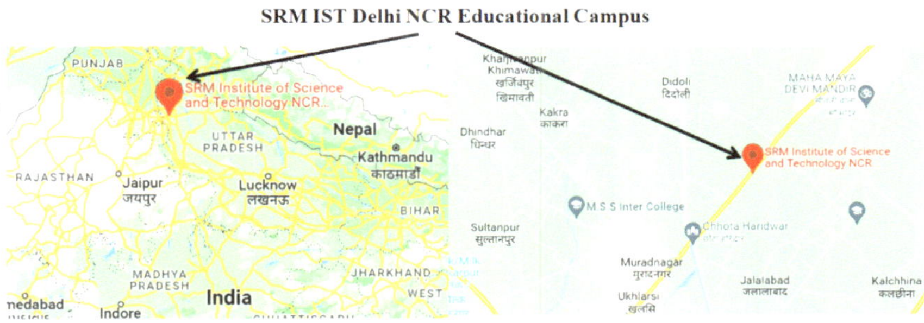

Figure 1. Map showing the location of SRM IST, Delhi-NCR Educational Campus.

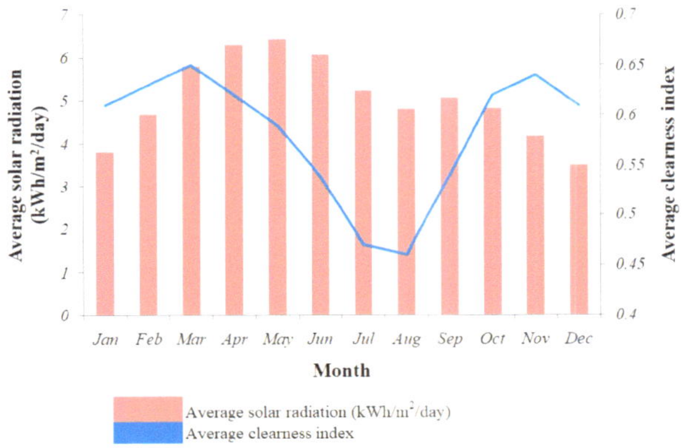

Figure 2. ALLSKY INSOL & clearness index of the selected region.

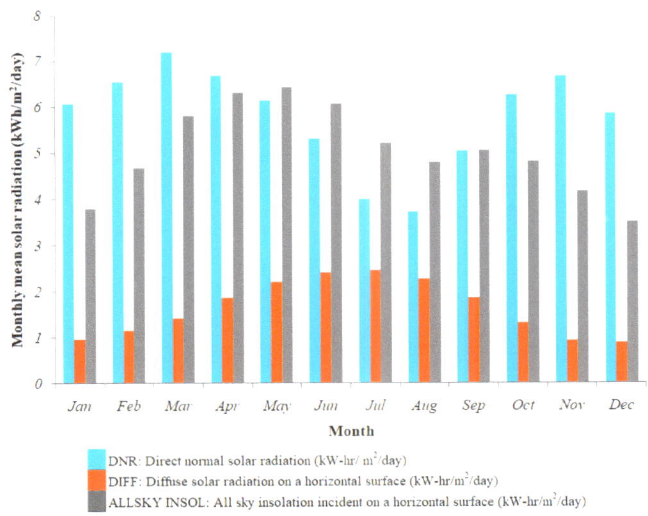

Figure 3. Comparison of ALLSKY INSOL, DIFF and DNR of the sun's radiation of the selected region.

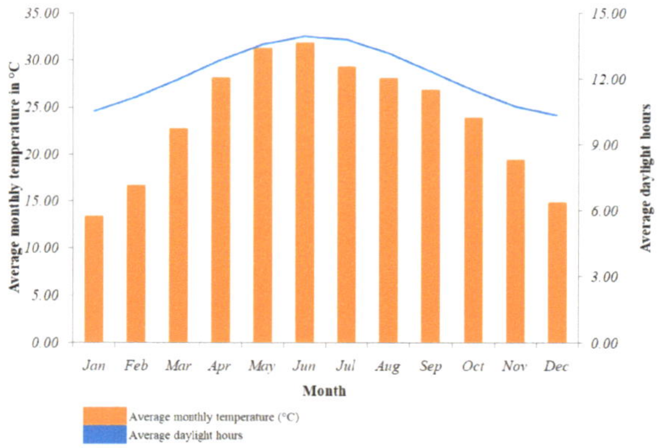

Figure 4. Average monthly temperature and daylight hours of the selected region.

The organization of the paper is as follows: in Section 1, the introduction and previous related work are discussed; the components of HEGS are discussed in Section 2; the sizing and economics of HEGS are discussed in Section 3; Section 4 provides the results and discussions; sensitivity analysis is provided in Section 5, and in Section 6, the conclusion is drawn.

2. The Hybrid Energy Generating System (HEGS)

A general model of renewable HEGS has been shown in Figure 5. In this model, solar energy is captured to generate electricity and hydrogen, which can be stored and utilized to serve the load as the dispatch strategy decides. The electrical energy can be stored in a battery bank, and hydrogen can be stored in a hydrogen tank. Hydrogen provides long-term energy storage; it has a high capacity but a low delivery rate. Therefore, to meet the short-span surges in load demand, it is required to have short-term storage of electricity as provided by the battery bank. It has a low capacity but a high delivery rate.

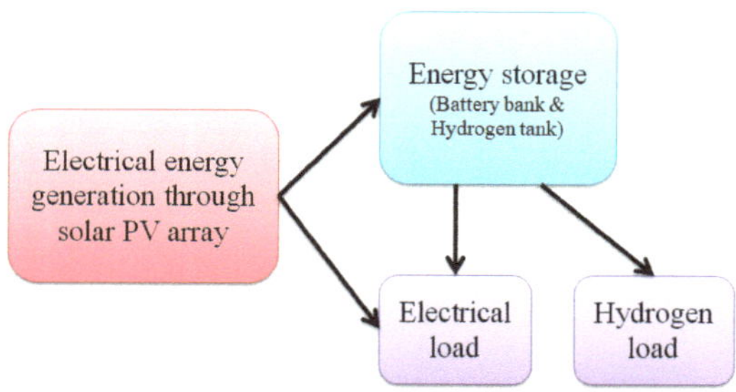

Figure 5. Renewable HEGS general model.

To serve the selected load, the proposed HEGS modelled in HOMER software has been shown in Figure 6. This HEGS consist of the following components—(1) solar PV array, (2) diesel generator (DG), (3) battery bank (BT), (4) electrolyser (ETL), (5) reformer (RFM), (6) hydrogen tank (HT), and (7) Grid. While choosing HEGS components, it is imperative to understand that oversizing the system will be expensive, while an under-sized system will fail to meet the load demand. Therefore, the simulation of the HEGS model has been done in HOMER software to minimize net present cost (NPC) for 24 h uninterrupted supply to load. The system with minimum NPC will also have a minimum cost of energy (COE). The constraints considered during the simulation were—(1) the 80% derating factor of the PV array, (2) the minimum load ratio of DG is 25%, (3) the state of charge (SOC) of the battery is from 20% to 100%, (4) annual purchase capacity of the grid is 100 kW, (5) HEGS will not purchase electricity from grid between 9 AM to 4 PM and (6) HEGS lifetime is 25 years.

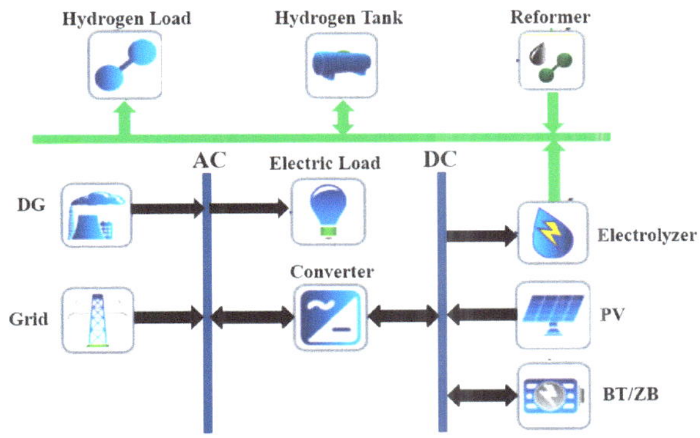

Figure 6. Modelling of proposed HEGS.

2.1. Energy Generation

The solar energy is captured by a PV array and converted to electrical energy. The size of the solar PV array is decided by the desired output voltage and current. The generic plat-plate PV module has been used to model the proposed HEGS. Suppose Y_{PV} is solar array output power from in kW. In that case, f_{PV} is the solar array derating factor in %, G_T is the solar radiation incident on the solar array in kW/m^2, $G_{T,STC}$ (1 kW/m^2) is the incident radiation from the sun, α_P is the power temperature coefficient in % per °C, T_c is the PV cell temperature in the current time step in °C, $T_{c,STC}$ is PV cell temperature under standard conditions (25 °C), then PV array power output, P_{output}, has been calculated using (1) [44]—the technical details solar PV array has been given in Table A1 in Appendix A.

$$P_{output} = Y_{pv} f_{pv} \left(\frac{G_T}{G_{T,STC}} \right) [1 + \alpha_p (T_c - T_{c,STC})] \tag{1}$$

When the electricity generated from the PV array and the energy stored in the battery bank cannot meet the load demand, a diesel-based generator operates to fulfil that deficiency. The fuel consumed by DG depends on the electricity generated. If F_0 is the coefficient of fuel curve intercept in L/h/kW, F_1 is the fuel curve slope in L/h/kW, Y_{gen} is the generator-rated capacity in kW, and P_{gen} is the generator power output in kW. Fuel consumed by DG, F, has been calculated using (2) [45].

$$F = F_0 Y_{gen} + F_1 P_{gen} \tag{2}$$

HOMER software has plotted the fuel curve for 50 kW DG with a slope of 0.333 L/h/kW and an intercept coefficient of 2.5 L/h, as shown in Figure 7. To fit the line to the data points, HOMER uses the linear least-squares method. The fuel consumption during the idle state is represented by the y-axis intercept, known as no-load fuel consumption. The slope of this curve is called marginal fuel consumption. If ρ_{DG} is the density of the diesel in kg/m^3, HLD is the lower heating response of the diesel in MJ/kg, and then the efficiency has been calculated using (3) [46]. The efficiency curve has been plotted from zero to rated output, as shown in Figure 8. The technical details DG has been given in Table A1 in Appendix A.

$$\eta_{DG} = \frac{3.6 P_{gen}}{F \rho_{DG} H_{LD}} \tag{3}$$

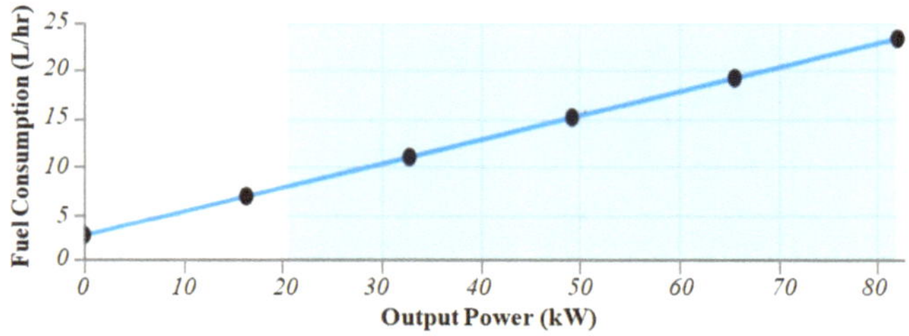

Figure 7. DG fuel curve.

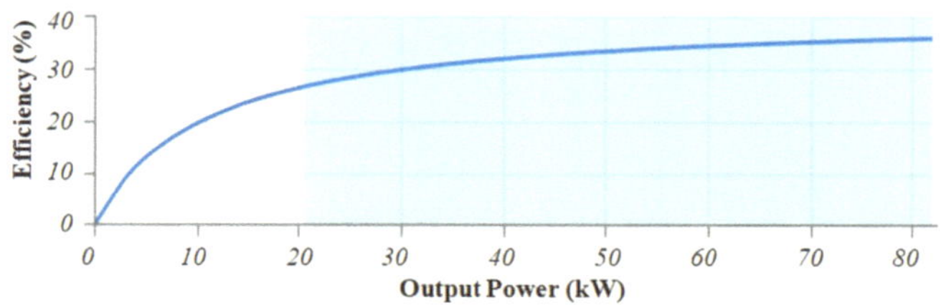

Figure 8. DG efficiency curve.

On connecting to the grid, the reliability of HEGS is enhanced. The electrical energy is generated mainly from the PV array. The excess energy is utilized in two ways—(1) it is stored in a battery bank to serve the load with the PV array during peak hours, and (2) to operate the electrolyser to produce hydrogen. DG is operated if electricity supplied by the PV array and battery bank is insufficient to serve the load. If the PV array, battery bank, and DG cannot serve the load, energy is purchased from the grid. The energy purchase and sell-back prices have been taken as $0.084 per kWh [47]. This price has been taken the same for all weekdays. Grid pollutant emissions per kWh are estimated to be 632 g of CO_2, 2.74 g of SO_2 and 1.34 g of NOx.

2.2. Energy Conversion

Hydrogen is produced by water electrolysis using electricity generated from HEGS. This hydrogen can be stored in a hydrogen tank for supplying to hydrogen load. Electrolyser efficiency is the ratio of the energy content (higher heating value) of the hydrogen produced to the amount of electricity consumed. Simply put, it is the efficiency of electrolysers converting electricity into hydrogen. The electrolyser efficiency ranges from 90% to 95%, and the hydrogen purity ranges from 99.8% to 99.9998% [48]. If N is the number of cells in the stack, Vt is the thermo-neutral voltage in V, Pi is the total input power to the stack in W, and I is the input current through the cell in A, the energy efficiency of the electrolyser, η_E, cell can be calculated as (4) [49].

$$\eta_E = \frac{NV_tI}{P_i} \tag{4}$$

A natural gas-based reformer is used to produce hydrogen. Natural gas contains methane, which produces hydrogen after undergoing thermal processes like a steam-methane reformation. In steam reforming, high-temperature steam (700–1000 °C) reacts with methane. This reaction produces hydrogen carbon monoxide and a very small amount of CO_2. The steam-methane reaction is shown in (5) [50]. In the final step, known as pressure-swing adsorption, pure hydrogen is obtained by removing CO_2 and other impurities.

$$CH_4 + H_2O(+\text{heat}) \rightarrow CO + 3H_2 \tag{5}$$

2.3. Energy Storage

After serving the electrical and hydrogen loads, the excess electricity generated can be stored in a battery bank. Whenever the PV array, DG or grid cannot meet the load demand, the electrical energy stored in the battery bank can be utilized to serve the load. Moreover, it helps to increase the renewable fraction by turning off DG during the daytime or lowering energy purchases from the grid. As the battery bank ensures an uninterrupted power supply to load, it becomes imperative to monitor battery SOC. It protects the battery from overcharging and goes beyond the minimum SOC required, adversely affecting the life of the battery. If Q(t) is the current capacity and Q(n) is the nominal capacity, then battery SOC can be calculated as provided in (6) [51].

$$\mathrm{SOC(t)} = \frac{\mathrm{Q(t)}}{\mathrm{Q(n)}} \tag{6}$$

If no loss of energy is assumed during the charging and discharging of the battery, the SOC related to the battery current is calculated as provided in (7) [1].

$$\mathrm{SOC(t)} = \mathrm{SOC(0)} + \frac{1}{\mathrm{C}}\int_0^t \mathrm{I} \cdot \left(\mathrm{V_{BT}}, \mathrm{t'}\right) \cdot \mathrm{dt'} \tag{7}$$

C is the capacity of the battery in kWh, I is the battery current in A, and V_{BT} is the battery voltage in V.

The life of the battery depends upon two independent factors—(1) lifetime productivity and (2) float life of storage. For example, if suppose N_{BT} is the number of batteries in a battery bank. In that case, Q_{LP} is single battery lifetime productivity in kWh, Q_{SP} is annual storage productivity in kWh/year, and $L_{BT,f}$ is storage float life in years. Then, the life of the battery bank (L_{BT}) can be calculated using (8) [23]. The technical details battery bank is given in Table A1 in Appendix A.

$$L_{BT} = \begin{cases} \frac{N_{BT}\ Q_{LP}}{Q_{SP}} & \text{If limited by productivity} \\ L_{BT,\ f} & \text{If limited by lifetime} \\ \mathrm{MIN}\left(\frac{N_{BT}\ Q_{LP}}{Q_{SP}},\ L_{BT,\ f}\right) & \text{If limited by productivity and lifetime} \end{cases} \tag{8}$$

The storage of hydrogen can be considered in a similar way to that of a battery. Depending upon the hydrogen tank capacity, the hydrogen is taken in ("charging phenomena") and delivered ("discharging phenomena") to the hydrogen load. As battery SOC has expressed in (7), the state of hydrogen (SOH) can be expressed as given in (9) [1].

$$\mathrm{SOH(t)} = \mathrm{SOH(0)} + \frac{1}{C_H}\int_0^t \dot{H}_m\left(P_H,\ t'\right) \cdot \mathrm{d\ t'} \tag{9}$$

where C_H is the gravimetric capacity of the hydrogen storage tank in kg, $\dot{H}_m$ is the hydrogen mass flow rate in kg/h, and P_H is the hydrogen pressure.

Battery storage has lower power capital cost and higher round-trip efficiency, whereas hydrogen storage has lower capital energy cost and longer storage duration.

Comparison of Battery Technologies

A cost comparison of five types of battery technologies has been made in Table 2. Lead-acid (LA) and lithium-ion (LI) batteries are popularly used in HEGS to store excess energy produced. Recently, advanced battery technologies like vanadium flow (VF) and zinc bromide (ZB) flow batteries are replacing LA and LI batteries. This is mainly because the life of flow batteries is not compromised even if the batteries are not charged for a long time. Moreover, flow batteries provide higher energy capacity than LA and LI batteries. Because of its high robustness, long durability and tolerance against high temperatures, nickel-iron (NI) is also becoming popular in HEGS for storing energy [52]. The annual SOC status of different battery technologies has been provided in Figure 9.

Table 2. Financial metrics of different battery technologies.

Battery Technology	Capital Cost ($ per kWh)	Replacement Cost ($ per kWh)	O&M Cost ($)	Life Span (Years)
LA battery	135	108	1.33 per year	5
LI battery	500	455	0	10
VF battery	535	465	0	25
ZB battery	800	735	0	25
NI battery	106	98	2.12	25

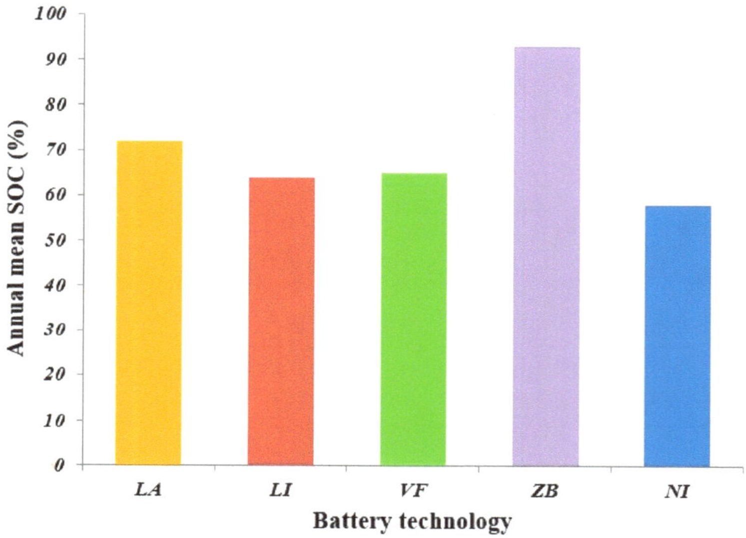

Figure 9. Annual mean SOC of different battery technologies.

3. Sizing and Economics of HEGS Components

For the effective function of HEGS, it is important to select the optimal sizing of the components to serve the energy demand of the selected load. This helps to achieve minimum COE after simulation. A number of optimal results are obtained after simulation, and different configurations of HEGS are expected to serve the load under consideration. The optimal solution is chosen based on optimum COE, NPC, and renewable fraction above 95% in HEGS.

3.1. Electric Load Demand Estimation

An academic building at SRM IST, Delhi-NCR campus, with 15 classrooms, has been considered a load. The details of the estimated load have been provided in Table 3. The load of one classroom consists of 6 LED lights, five ceiling fans, two air-conditioners, 1 LCD projector, 1 PA system (including microphone & speaker), one computer system and two plug points. The average electricity consumption of one classroom is estimated to be 26.58 kWh per day, and for 15 classrooms, it is approximately 400 kWh per day. The 24-h load profile and the yearly load profile have been presented in Figure 10a and 10b, respectively.

Table 3. Load estimation of 15 classrooms at SRM IST Delhi NCR campus.

Sr. No.	Load	Power (kW)	Quantity (Nos.)	Usage (h)	Total Load (kWh/Day)
1	LED light	0.024	6	6	0.864
2	Ceiling fans	0.020	5	6	0.6
3	Air conditioner (1.5 ton)	1.5	2	6	18
4	LCD Projector	0.28	1	6	1.680
5	PA system (microphone & speakers)	0.8	1	6	4.800
6	Computer system	0.1	1	6	0.6
7	Mobile charging point	0.003	2	6	0.036
	The total load for one classroom				26.58
	The total load for 15 classrooms				398.7 (~400)

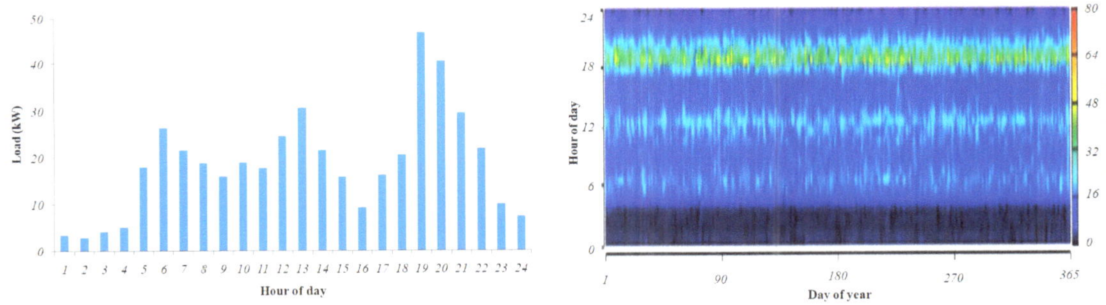

(a) (b)

Figure 10. (**a**) 24-h load profile and (**b**) yearly load profile of 15 classrooms at SRM IST Delhi-NCR campus.

3.2. *HOMER Pro Software*

Different software is available for finding the optimal HEGS configuration, like Hybrid2, RAPSim, and HYBRIDS. All these software is easy to use to provide long-term forecasts of HEGS performance. But the mentioned software does not offer economic analyses. On the other hand, hybrid Optimization of Multiple Electric Renewables (HOMER) software provides technical and economic analyses of HEGS [15].

In this study, the analysis has been done using HOMER software, which is developed by National Renewable Energy Laboratory, Golden, CO, USA. HOMER uses two optimization algorithms—the original grid search algorithm and the HOMER optimizer, to perform simulations. The optimized solutions are chosen based on the lowest NPC and lowest COE. Furthermore, the HOMER optimizer takes care of the maximum number of simulations per optimization, so the optimization process converges after a finite number of simulations.

It can be referred from Figure 11 that the geographical position of the selected location, weather conditions of the selected site, technical details of the components of the proposed HEGS and the load profile are the inputs to HOMER. Based on these inputs, the software performs simulations to obtain the optimal solutions. The outputs obtained from HOMER are the technical & economic data, number of estimated pollutants and sensitivity analysis, also known as what-if analysis, to understand the impact of one parameter on the other. The detailed flow chart of HOMER software has been provided as Figure A1 in Appendix B.

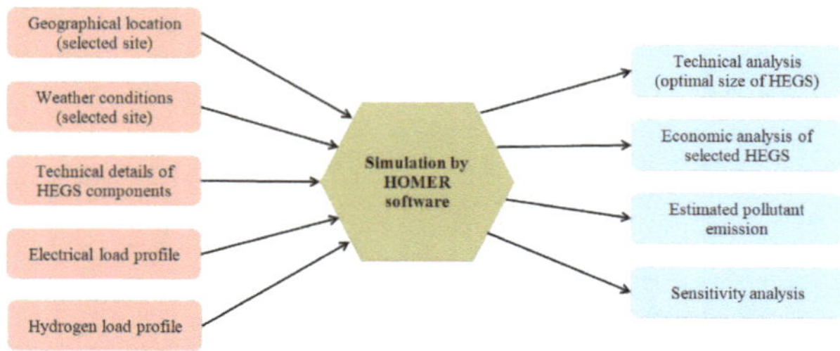

Figure 11. Block diagram process of HOMER software.

3.3. Cost Estimation of HEGS

The simulations performed by HOMER software can provide the optimal size of HEGS components to minimize the probability of loss of electricity supply and also minimize the NPC & COE. The HOMER performs technical and economic analysis by assuming that the cost of HEGS components increases at the same rate. The parameters considered to perform the calculation are annual interest rate, annual inflation rate and project lifetime. NPC is the summation of initial cost, replacement cost, cost of operation & maintenance (O&M), and cost of fuel, and from this, salvage value is subtracted. HOMER has used (10) to calculate NPC [53].

$$NPC = \sum_{N=1}^{t} r_d (C_{ini} + C_{rep} + C_{O\&M} + C_{fuel} - C_{sal}) \tag{10}$$

where, r_d (%) is the discount rate, N (years) is the project lifetime, C_{ini} ($) is the initial cost of the HEGS component, C_{rep} ($) is the replacement cost, $C_{O\&M}$ ($) is the operation and maintenance cost, C_{fuel} ($) is the fuel cost and C_{sal} ($) is the salvage value. If r is the annual interest rate, the discount rate can be calculated from (11) [54].

$$r_d = \frac{1}{(1+r)^N} \tag{11}$$

If r^o is the annual interest rate and f is the annual inflation rate, then the real annual interest rate, r, can be calculated from (12) [54].

$$r = \frac{r^0 - f}{1+f} \tag{12}$$

HOMER calculates the annualized cost, C_{ann}, by multiplying NPC with the capital recovery factor (CRF), as shown in (13) [54].

$$C_{ann} = CRF(r, N) \times NPC \tag{13}$$

CRF is a ratio used to calculate the present value of annual cash flows. HOMER calculates CRF as given in (14) [55].

$$CRF(r, N) = \frac{r(1+r)^N}{(1+r)^N - 1} \tag{14}$$

If E_{served} is the total electrical load served, HOMER calculates the levelized cost of energy, as shown in (15) [52].

$$COE = \frac{C_{ann}}{E_{served}} \tag{15}$$

HOMER uses (16) to calculate the levelized cost of hydrogen (COH) [1].

$$COH = \frac{C_{ann} + V_e(E_P + E_{G,sales})}{M_H} \quad (16)$$

where, V_e is the value of electricity, E_P is the primary electrical load, $E_{G,sales}$ is the total energy sold to the grid, and M_H is the total hydrogen production.

4. Results and Discussions

To perform simulation, HOMER takes the technical details of HEGS components, its initial cost, replacement cost, O&M cost, and lifespan of components as input. To calculate the economic parameters, discount rate, inflation rate and interest rate are also taken as input. The financial metrics provided as input to HOMER have been provided in Table 4. A total of 4607 solutions have been simulated by HOMER, out of which 48 feasible solutions have been obtained. Based on the selection criteria, four solutions have been selected for the analysis, which is presented in Section 4.1.

Table 4. Financial metrics of HEGS components.

Components	Initial Cost ($)	Replacement Cost ($)	O&M Cost ($)	Life Span
PV	$470 per kW	$470 per kW	$2.66 per year	25 years
DG	$665 per kW	$535 per kW	$0.027 per op. h	15,000 h
Converter	$195 per kW	$195 per kW	$4 per year	15 years
Battery	$800 per kWh	$735 per kWh	$10 per year	30 years
Electrolyser	$1500 per kW	$1500 per kW	$0.05 per year	20 years
Reformer	$3500 per kW	$3200 per kW	$200 per year	25 years
Hydrogen tank	$600 per kg	$600 per kg	$10 per year	25 years

Electricity generation from renewable sources is lowly reliable. This is due to uncertainty in weather conditions. In Figure 5, the renewable source of electricity generation is a solar PV array. The excess electricity generated from the PV array will charge the battery bank and be used by the electrolyser to generate hydrogen. In the proposed HEGS, DG serves as a captive power generation source when the PV array & battery bank cannot meet the load demand and the grid is under maintenance/repair work. Therefore, the inclusion of DG increases the reliability of HEGS. The composition of diesel has been 0.4% sulphur content, 88% carbon content, 43.2 MJ/kg of lower heating value and a density of 820 kg/m^3. Estimated emissions per litre of diesel consumption are 16.5 g of CO, 0.72 g of unburned hydrocarbon, 0.1 g of particulate matter and 15.5 g of NOx.

4.1. Optimization Results and Economic Analysis

The criteria for choosing the optimum solutions have been—(1) continuous supply of electricity to load, (2) optimum COE & NPC, (3) high value of the renewable fraction, and (4) low value of pollutant emission. Based on these criteria, four combinations have been chosen for the proposed HEGS—(1) PV/BT/Grid/ETL/RFM/HT (without DG), (2) PV/DG/BT/Grid/ETL/RFM/HT, (3) PV/DG/BT/ETL/RFM/HT (without Grid), and (4) PV/DG/Grid/ETL/RFM/HT (without battery). The simulation results for the proposed HEGS obtained from HOMER have been presented in Table 5. From a financial metrics point of view, comparisons have been made for COE ($/kWh), NPC ($), operating cost ($/year), initial cost ($), O&M cost ($/year), fuel cost ($/year). Combination 1 has the lowest NPC ($1,285,037) & least COE ($0.395 per kWh). The operating cost ($26,475 per year) is the lowest for combination 2. The lowest O&M cost ($21,955 per year) has been obtained for combination 3. The fuel cost ($107 per year) is the lowest for combinations 2 and 3. It can be observed from Table 5 that nominal battery capacity (7000 kWh) is the highest in combination 3. This provides more storage of electrical energy compared to other combinations. The lowest battery storage capacity is found in combination 2 (1000 kWh), as the grid is connected here to take care of any shortfall while serving the load demand. The comparison of total system cost for the selected HEGS combinations has been shown in Figure 12. A comparison of COE and the cost of hydrogen is shown in Figure 13.

Table 5. Optimization results obtained for proposed HEGS (Part—I).

Architecture											Financial Metrics					
HEGS Combination	**PV (kW)**	**DG (kW)**	**Battery (No. of Strings)**	**Battery Nominal Capacity (kWh)**	**Grid (kW)**	**Electrolyser (kW)**	**Reformer (kW)**	**Hydrogen Tank (kg)**	**Converter (kW)**	**Dispatch Strategy**	**COE ($/kWh)**	**NPC ($)**	**Operating Cost ($/Year)**	**Initial Cost ($)**	**O&M Cost ($/Year)**	**Fuel Cost ($/Year)**
1.	200		3	3000	100	100	100	100	38	CC	0.395	1,285,037	27,903	663,890	22,015	
2.	200	82	1	1000	100	100	100	100	40	LF	0.402	1,306,485	26,475	717,116	21,990	107
3.	200	82	7	7000		100	100	100	70	CC	0.408	1,327,521	26,942	727,756	21,955	107
4.	200	82			100	100	100	100	42	CC	0.444	1,442,017	32,584	716,655	28,692	2832

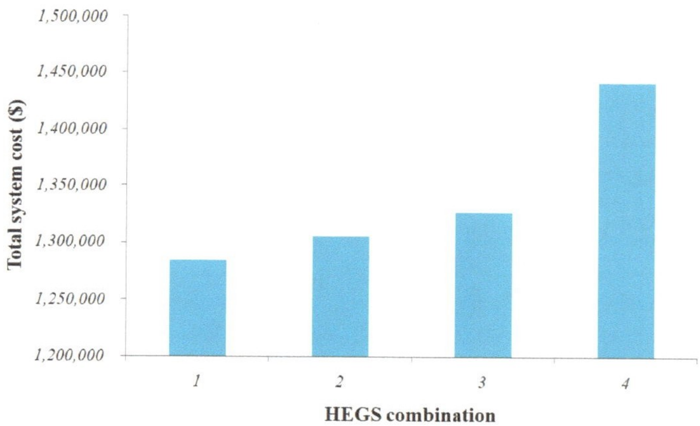

Figure 12. Comparison of total system cost for selected HEGS combination.

4.2. Technical Analysis

The summary of the technical report is provided in Table 6. From a technical point of view, combination 1, 3 & 4 follows the cycle charging (CC) dispatch strategy. In CC strategy, the DG or grid operates at full capacity to serve the load. The excess electricity produced in this case is utilized to charge the battery bank and serve the electrolyser. Combination 2 follows the load-following (LF) strategy. In the LF strategy, the DG or grid operates to generate enough electricity to meet the load demand. The battery bank is charged when a renewable source like a solar PV array generates electricity. The renewable fraction is found to be highest for combination 3, which is 99.8% and the lowest can be observed for combination 4, which is 44.2%. It has been estimated that in combination 4, energy purchased from the grid has been approximately 20% compared to 1% in the case of combinations 1 & 2 and 0% in the case of combination 3. The usage of DG is highest in the case of combination 4 compared to other combinations. It is also observed that the percentage of excess electricity generation (17%) is highest in combination four among all other combinations. This may be due to the absence of a battery bank in combination 4, and DG is used as a compensator.

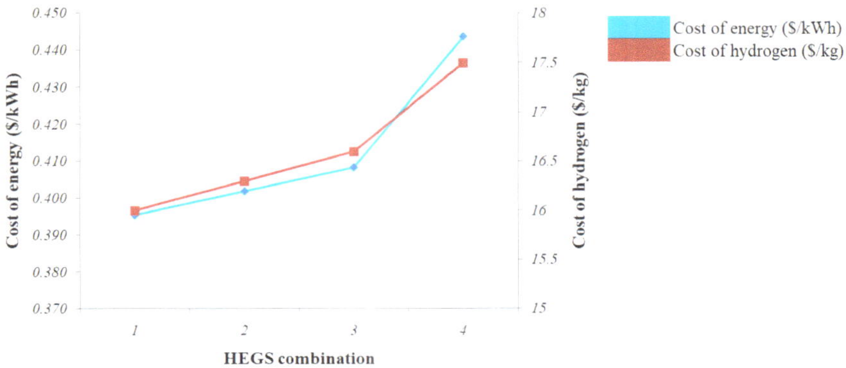

Figure 13. Comparison of cost of energy & cost of hydrogen for selected HEGS combination.

Table 6. Optimization results obtained for proposed HEGS (Part—II).

	Architecture											Electricity Generated (kWh/Year)				Grid	
HEGS Combination	**PV (kW)**	**DG (kW)**	**Battery (No. of Strings)**	**Battery Nominal Capacity (kWh)**	**Grid (kW)**	**Electrolyser (kW)**	**Reformer (kW)**	**Hydrogen Tank (kg)**	**Converter (kW)**	**Dispatch Strategy**	**Renewable Fraction (%)**	**DG**	**PV**	**Excess Electricity Production (%)**	**Excess Electricity Production (kWh/Year)**	**Energy Purchased (kWh)**	**Energy Sold (kWh)**
1.	200		3	3000	100	100	100	100	38	CC	97.7	0	296,363	3	9273	3285	0
2.	200	82	1	1000	100	100	100	100	40	LF	97.8	246	296,363	3	9255	2987	0
3.	200	82	7	7000		100	100	100	70	CC	99.8	246	296,363	3	9016	0	0
4.	200	82			100	100	100	100	42	CC	44.2	6533	296,363	17	62,542	74,976	0

The emission of CO_2 into the atmosphere is causing a planetary crisis. So, HEGS combination 3, which has a maximum renewable fraction (99.8%), has been selected as the optimal solution. This combination 3 provides a stand-alone solution for the proposed HEGS, which not only reduces the dependency on the grid but has no contribution towards CO_2 emissions from the grid. The absence of a grid in this combination has been compensated by increased battery nominal capacity (7000 kWh) compared to other HEGS combinations. The SOC of the battery bank is shown in Figure 14, and it is observed that for most days, SOC lies in the range of 80% to 100%. Therefore, the energy from the battery bank has mainly been used between the 0 to 11th hour and the 18th to 24th hours. Figure 15 shows the monthly average of electricity production from solar PV array and DG and the total electrical load served. The yearly power output from the solar PV array and DG is shown in Appendix B in Figures A2 and A3, respectively. The input power taken by the electrolyser has been shown in Figure A4 in Appendix B. Figure 16 shows the monthly average of hydrogen production from the electrolyser and the reformer, along with the total hydrogen load served. The yearly hydrogen output from the reformer is shown in Figure A5, and the yearly status of the hydrogen tank level is shown in Figure A6 in Appendix B.

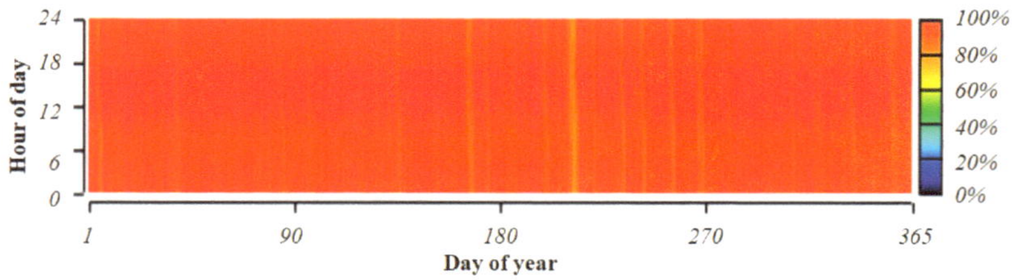

Figure 14. Yearly SOC status of battery bank of HEGS combination 3.

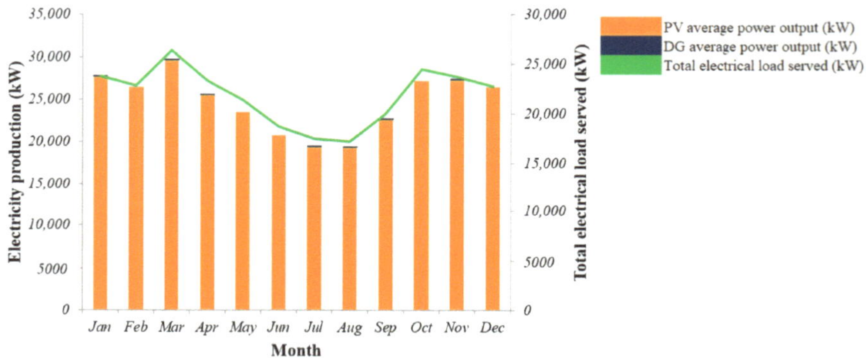

Figure 15. Monthly average electricity production by HEGS combination 3.

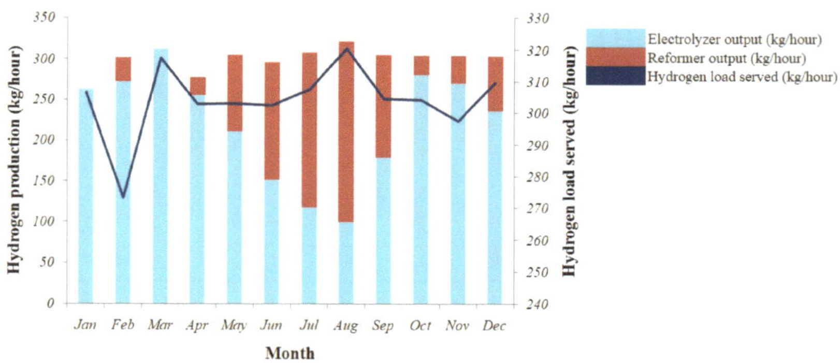

Figure 16. Monthly average hydrogen production by HEGS combination 3.

4.3. Time-Series Analysis

To check the energy balance for 24-h operation of combination 3 HEGS, a time-series analysis was done on the 2nd day of January. The electricity generators have been solar PV array & DG, while the battery charges after meeting the load demand and supplies electricity when needed. The electricity has been consumed by the AC load and the electrolyser. The contributions of all these mentioned HEGS components have been plotted in Figure 17. The observations from this plot have been provided in Table 7.

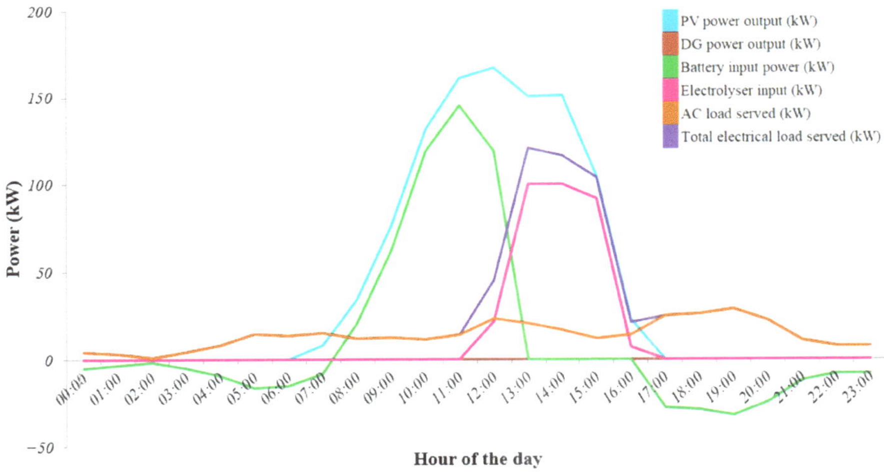

Figure 17. Time series analysis for 24 h operation of HEGS.

Table 7. Time series analysis for the 2nd day of January.

Time Duration	HEGS Component Contribution
0 to 6th hours	• Solar PV array has been unable to generate electricity without sunlight. • Energy served in the battery bank served the load.
6th to 17th hours	• Solar PV array generates electricity. • Battery bank charges till the 13th hour & it did not operate from the 13th to the 16th hour. It started to serve the load (along with the solar PV array) from the 16th hour. • The electrolyser consumed electricity from the 11th to 17th hour and AC primary load.
17th to 23rd hours	• As load demand reduces, it is met with energy stored in the battery bank

On this date, the DG was not required to be operated. The energy balance calculation, provided in Table 8, has been done for the 12th hour of the day. This validates that the total electricity generated equals the total consumed.

Table 8. Energy balance calculation from time series plot at 12th hour.

	HEGS Component	Power Generated/Consumed
Electricity generated	Solar PV array	A = 167 kW
	Diesel generator	B = 0 kW
Electricity served to load	Battery bank	C = 119 kW
Electricity consumed	Electrolyser	D = 21.86 kW
Inverter operation	Input power to the inverter	E (=A + B − C − D) =26.14 kW
	Output power to the inverter (Inverter efficiency = 90%)	F = 23.53 kW
Electricity consumed	AC load served	G = 23.53 kW (this value is equal to F)
	Total electrical load served	(=D + G) =45.39 kW

4.4. Pollutant Emissions

It is imperative to consider the pollutant emission before choosing an optimal solution for the proposed HEGS. The summary of pollutant emissions is provided in Table 9. The highest CO_2 emission has been observed from combination 4 (53,888 kg/year) and the lowest from combination 3 (247 kg/year). In HEGS combination 4, the energy is generated through a solar PV array, and if required, DG is operated to meet the load demand. In this combination, the battery bank is not a part of HEGS, so there is no provision for storing energy. On referring Table 6, it can be observed that DG has generated 6533 kWh per year of electricity, which is the highest value among all combinations. This means that DG has been operated a maximum number of times and consumed fuel worth $2832 annually. So, concerning emissions from DG, pollutant emission is highest in combination 4. In case of a shortfall of electricity generation from PV and DG, energy has to be purchased from the grid to fulfill the load demand. The grid generates electricity using fossil fuels, and hence, the emission of CO_2 is highest among the other HEGS combinations.

Table 9. Summary of pollutant emissions for proposed HEGS.

	Architecture												Emissions					
HEGS Combination	PV (kW)	DG (kW)	Battery (No. of Strings)	Battery Nominal Capacity (kWh)	Grid (kW)	Electrolyser (kW)	Reformer (kW)	Hydrogen Tank (kg)	Converter (kW)	Dispatch Strategy	Renewable Fraction (%)	CO_2 (kg/Year)	CO (kg/Year)	UHC (kg/Year)	PM (kg/Year)	SO_2 (kg/Year)	NOx (kg/Year)
1.	200		3	3000	100	100	100	100	38	CC	97.7	2076	0	0	0	0	4.4
2.	200	82	1	1000	100	100	100	100	40	LF	97.8	2134	1.55	0.0678	0.00942	8.68	5.46
3.	200	82	7	7000		100	100	100	70	CC	99.8	247	1.55	0.0678	0.00942	0.468	1.46
4.	200	82			100	100	100	100	42	CC	44.2	53,888	41	1.79	0.248	219	139

The grid is not a part of combination 3, so the emissions from grid operation have been avoided. It can also be observed from Table 9 that the nominal capacity of the battery bank (7000 kWh) is the highest among all combinations. So, this combination provides maximum provision for the storage of excess electricity. The DG has been operated least often, so CO_2 emissions are the lowest in this combination. Figure 18 presents the comparison of CO_2 emissions & renewable fractions for selected HEGS combinations.

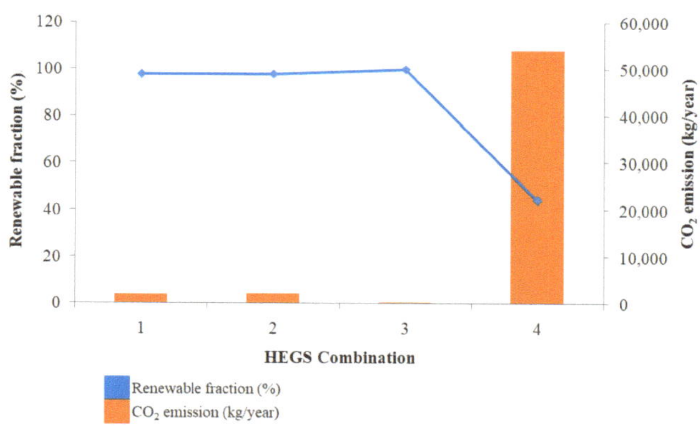

Figure 18. Comparison of CO_2 emissions & renewable fraction for selected HEGS combinations.

5. Sensitivity Analysis

A HEGS is a complex connection of different components. The optimum working of HEGS depends on different variables. Changes in some variables may bring huge changes in the working of HEGS. So, it is imperative to study the impact of such variables on the operation of HEGS. This what-if analysis is known as sensitivity analysis. The sensitivity analysis has been performed for combination 3. This combination has followed the CC dispatch strategy. The analysis has been performed between—(1) the size of the solar PV array and the number of battery strings and (2) the reformer and hydrogen tank capacity. The sensitivity variables are NPC, renewable fraction, and CO_2 emissions. The variation in COE has been marked in each of the analyses for a clear overview.

The optimization plot has been plotted between COE and renewable fraction, considering 25 years project lifetime. This plot is shown in Figure 19. The encircled points in the plot are the feasible solutions for that particular sensitivity case. It can be observed that many points lie in the region of high renewable fraction, which has the cost of energy between $0.4 to 0.6 per kWh.

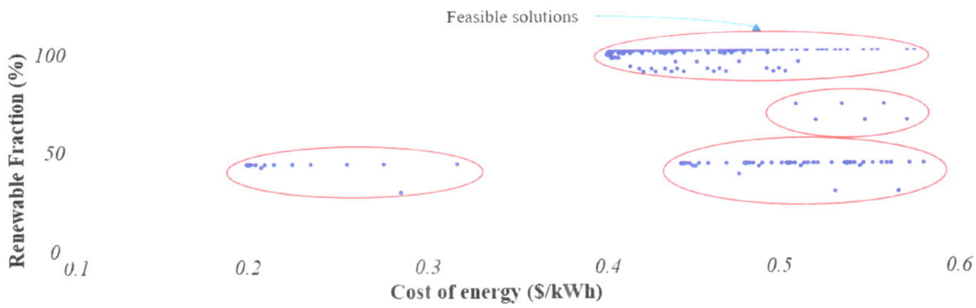

Figure 19. Optimization plot between COE and renewable fraction for 25 years project lifetime.

In Figure 20a, the impact of variation in NPC with COE values superimposed has been shown. The least COE has been obtained as $0.408 per kWh for seven battery strings, 200 kW solar PV array, 82 kW DG, 100 kW ETL, 100 kW RFM, and 100 kg HT with NPC of $1,327,521. In Figure 20b, the impact of variation in renewable fraction has been analyzed to understand the percentage of green energy production. It can be observed from this plot that the point of least COE lies within 90% to 100% of the renewable fraction. The impact of variation in CO_2 emissions has been shown in Figure 20c. It shows CO_2 emissions reductions in areas where the renewable fraction is above 90%. Figure 21 presents the impact of variation in NPC on the capacity reformer and hydrogen tank capacity.

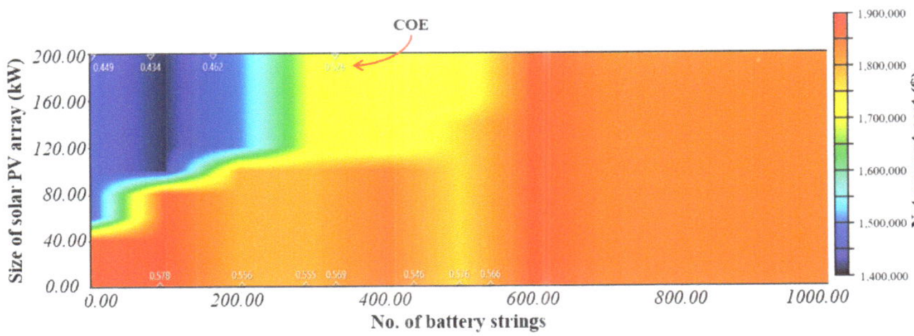

(**a**)

Figure 20. *Cont.*

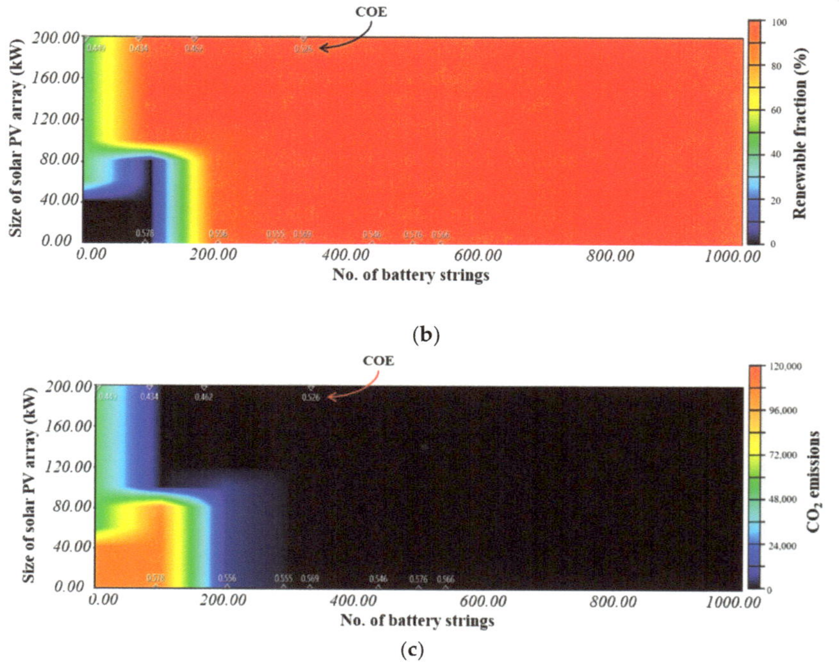

Figure 20. Sensitivity analysis between the size of solar PV array & number of battery strings with variation in (**a**) NPC, (**b**) renewable fraction, and (**c**) CO_2 emissions.

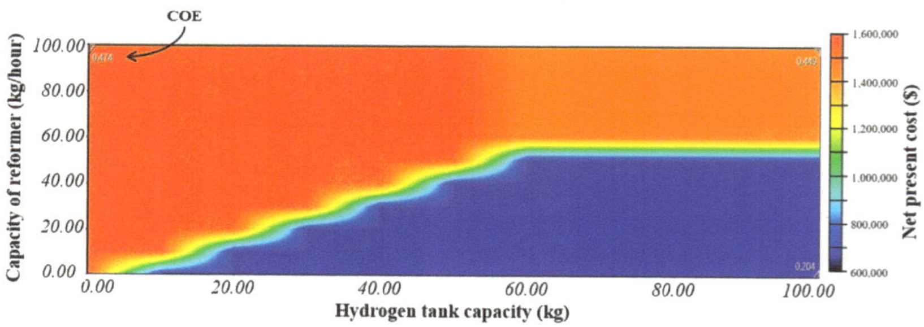

Figure 21. Sensitivity analysis for reformer and hydrogen tank capacity with variation in NPC.

6. Conclusions

In this study, the aim has been to compare the proposed HEGS solutions for on-grid and off-grid operations. The selected optimal solution has been taken as PV (200 kW)/DG (82 kW)/BT (7000 kWh) for electricity generation along with ETL (100 kW)/RFM (100 kW)/HT (100 kg) for green hydrogen production. The optimal solution has been chosen based on low O&M cost ($21,955 per year), a high renewable fraction (99.8%), and low CO_2 emissions (247 kg/year). The conclusions drawn from this work are mentioned below:

1. The electricity generation has been using 200 kW PV, 82 kW DG, 7000 kWh BT, and a converter of 70 kW, with a CC dispatch strategy.
2. The initial cost of HEGS has been estimated to be $727,756, and the operating cost is $26,942 per year.
3. The NPC has been estimated to be $1,327,521; the estimated COE will be $0.408 per kWh, and the cost of hydrogen will be $16.6 per kg.
4. The CO_2 emission will be reduced to 247 kg/year.

5. Sensitivity analysis has been performed in two parts—(a) the impact of variation in solar PV array size & number of battery strings has been studied with sensitivity variables such as NPC, renewable fraction & CO_2 emissions, and (b) the impact of variation in reformer capacity & hydrogen tank capacity, has been studied with NPC as sensitivity variable.

In future work, the government policies, subsidies provided by the government and tariffs regulated by the energy sector can be considered to improvise the analysis.

Author Contributions: Resources, G.F. and L.E.; Supervision, H.S. and V.B.; Writing—original draft, S.S.; Writing—review & editing, H.S., V.B. and P.K. All authors have read and agreed to the published version of the manuscript.

Funding: This research received no external funding.

Institutional Review Board Statement: Not applicable for studies not involving humans or animals.

Informed Consent Statement: Not applicable for studies not involving humans.

Data Availability Statement: The authors confirm that all data have been presented.

Conflicts of Interest: The authors declare no conflict of interest.

Appendix A

Table A1. Technical details of solar PV array, DG, and battery.

	Description	Value
Solar PV array	Type of panel	Flat plate
	Rated capacity (kWp)	1
	Capital cost ($/kW)	470
	Replacement cost ($/kW)	470
	O&M cost ($/year)	2.66
	Lifetime (years)	25
	Derating factor (%)	80
	Temperature coefficient (per °C)	−0.5
	Nominal operating cell temperature (°C)	47
	Efficiency (%)	0.13
Diesel generator	Fuel	Diesel
	Capital cost ($/kW)	665
	Replacement cost ($/kW)	535
	O&M cost ($/h)	0.027
	Fuel price ($/L)	1.14
	Lifetime (hours)	15,000
	CO (g/L/fuel)	16.5
	Unburned HC (g/L fuel)	0.72
	Particulates (g/L fuel)	0.1
	Fuel sulfur to PM (%)	2.2
	NOX (g/L fuel)	15.5
Battery	Nominal voltage (V)	600
	Nominal capacity (kWh)	1000
	Maximum capacity (Ah)	1670
	Round-trip efficiency (%)	90
	Maximum charge current (A)	1670
	Maximum discharge current (A)	5000

Appendix B

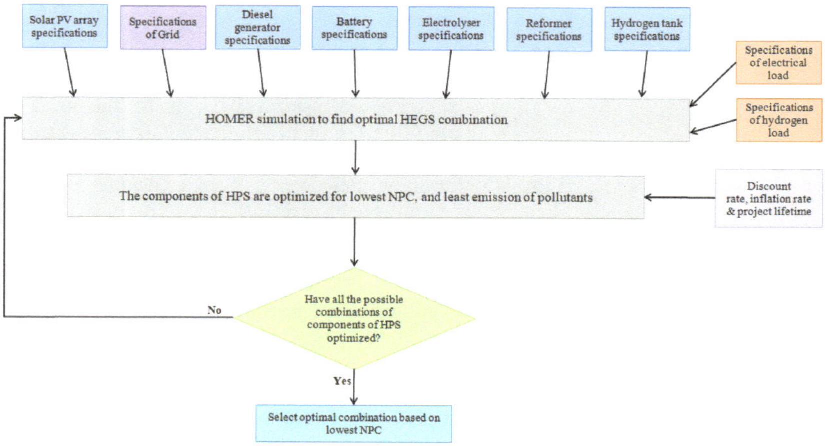

Figure A1. Flow chart of HOMER software.

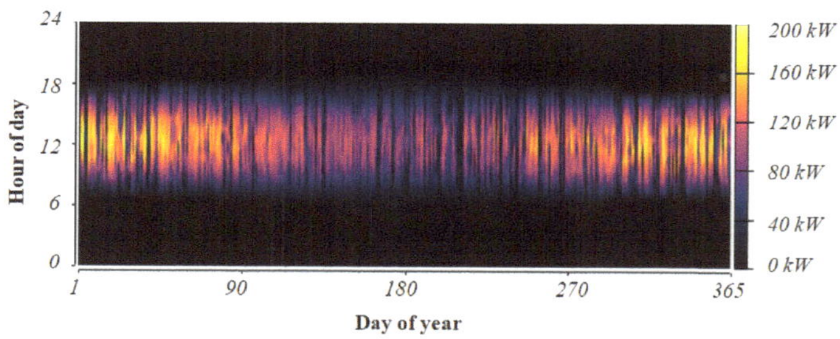

Figure A2. Power output from solar PV array in HEGS combination 3.

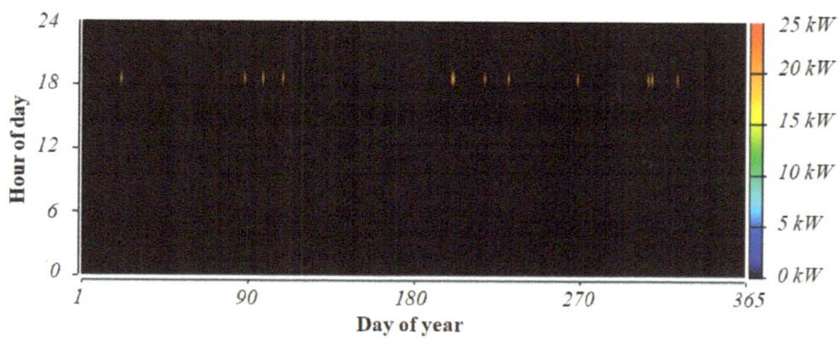

Figure A3. Power output from DG in HEGS combination 3.

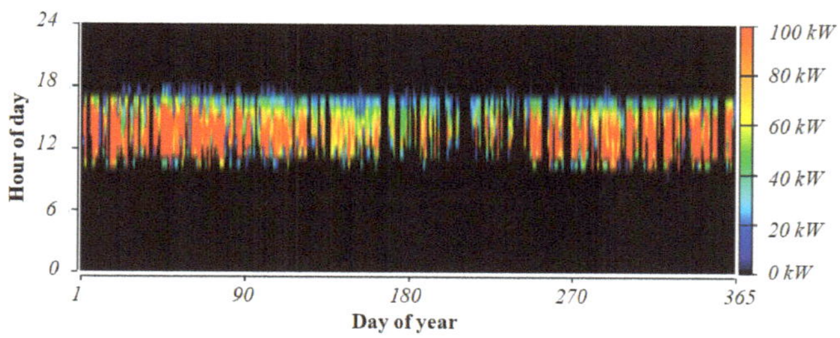

Figure A4. Electrolyser input power in HEGS combination 3.

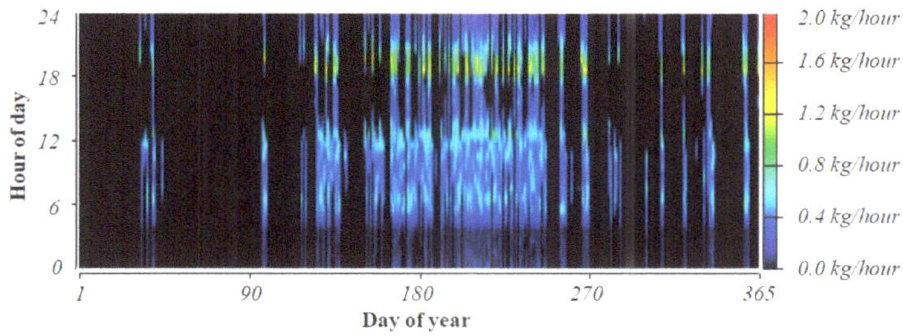

Figure A5. Hydrogen output from reformer in HEGS combination 3.

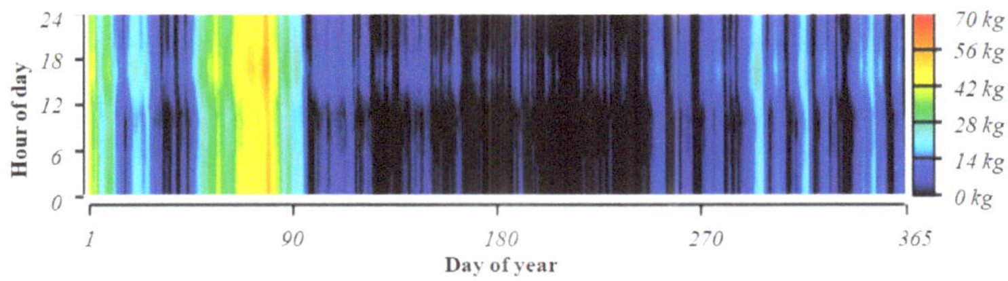

Figure A6. Hydrogen tank level in HEGS combination 3.

References

1. Abdin, Z.; Mérida, W. Hybrid energy systems for off-grid power supply and hydrogen production based on renewable energy: A techno-economic analysis. *Energy Convers. Manag.* **2019**, *196*, 1068–1079. [CrossRef]
2. Ekonomou, L.; Fotis, G.P.; Vita, V.; Mladenov, V. Distributed Generation Islanding Effect on Distribution Networks and End User Loads Using the Master-Slave Islanding Method. *J. Power Energy Eng.* **2016**, *4*, 1–24. [CrossRef]
3. Amini, A.; Abedi, M.; Nesari, E.; Daryadel, E.; Kolahi, M.; Mianabadi, H.; Fisher, J. The paris agreement's approach toward climate change loss and damage. *World Aff.* **2023**, *186*, 46–80. [CrossRef]
4. Shapiro, A.F.; Metcalf, G.E. The macroeconomic effects of a carbon tax to meet the U.S. Paris agreement target: The role of firm creation and technology adoption. *J. Public Econ.* **2023**, *218*, 104800. [CrossRef]
5. Sharma, H.; Pal, N.; Singh, Y.; Sadhu, P.K. Development and Simulation of Stand Alone Photovoltaic Model Using Matlab/Simulink. *Int. J. Power Electron. Drive Syst.* **2015**, *6*, 703. [CrossRef]
6. Hofrichter, A.; Rank, D.; Heberl, M.; Sterner, M. Determination of the optimal power ratio between electrolysis and renewable energy to investigate the effects on the hydrogen production costs. *Int. J. Hydrogen Energy* **2023**, *48*, 1651–1663. [CrossRef]
7. Fotis, G.; Dikeakos, C.; Zafeiropoulos, E.; Pappas, S.; Vita, V. Scalability and Replicability for Smart Grid Innovation Projects and the Improvement of Renewable Energy Sources Exploitation: The FLEXITRANSTORE Case. *Energies* **2022**, *15*, 4519. [CrossRef]
8. Peyvandi, M.; Hajinezhad, A.; Moosavian, S.F. Investigating the intensity of GHG emissions from electricity production in Iran using renewable sources. *Results Eng.* **2023**, *17*, 100819. [CrossRef]
9. Dogan, E.; Hodžić, S.; Šikić, T.F. Do energy and environmental taxes stimulate or inhibit renewable energy deployment in the European Union? *Renew. Energy* **2023**, *202*, 1138–1145. [CrossRef]
10. He, S.; Barati, B.; Hu, X.; Wang, S. Carbon migration of microalgae from cultivation towards biofuel production by hydrothermal technology: A review. *Fuel Process. Technol.* **2023**, *240*, 107563. [CrossRef]
11. Sambhi, S.; Sambhi, S.; Bhadoria, V.S. IoT based optimized and secured ecosystem for energy internet: The state of the art. In *Internet of Things in Business Transformation: Developing an Engineering and Business Strategy for Industry 5.0*; Scrivener Publishing LLC: Beverly, MA, USA, 2021; pp. 91–126. [CrossRef]
12. Mukhammadiev, M.M.; Urishev, B.U.; Uulu, A.A.; Almardanov, O.; Karimova, N.E.; Murodov, H. The role of renewable energy sources in providing the efficiency of the power system in the conditions of digital energy transformation. In *AIP Conference Proceedings*; AIP Publishing LLC: Melville, NY, USA, 2023; Volume 2552, p. 050019. [CrossRef]
13. Kreishan, M.Z.; Fotis, G.P.; Vita, V.; Ekonomou, L. Distributed Generation Islanding Effect on Distribution Networks and End User Loads Using the Load Sharing Islanding Method. *Energies* **2016**, *9*, 956. [CrossRef]
14. Adewumi, O.B.; Fotis, G.; Vita, V.; Nankoo, D.; Ekonomou, L. The Impact of Distributed Energy Storage on Distribution and Transmission Networks' Power Quality. *Appl. Sci.* **2022**, *12*, 6466. [CrossRef]
15. Sambhi, S.; Sharma, H.; Kumar, P.; Fotis, G.; Vita, V.; Ekonomou, L. Techno-Economic Optimization of an Off-Grid Hybrid Power Generation for SRM IST, Delhi-NCR Campus. *Energies* **2022**, *15*, 7880. [CrossRef]
16. Le, S.T.; Nguyen, T.N.; Linforth, S.; Ngo, T.D. Safety investigation of hydrogen energy storage systems using quantitative risk assessment. *Int. J. Hydrogen Energy* **2023**, *48*, 2861–2875. [CrossRef]

17. Bovo, A.; Poli, N.; Trovò, A.; Marini, G.; Guarnieri, M. Hydrogen energy storage system in a Multi-Technology Microgrid:technical features and performance. *Int. J. Hydrogen Energy* **2023**, *48*, 12072–12088. [CrossRef]
18. Walter, V.; Göransson, L.; Taljegard, M.; Öberg, S.; Odenberger, M. Low-cost hydrogen in the future European electricity system–Enabled by flexibility in time and space. *Appl. Energy* **2023**, *330*, 120315. [CrossRef]
19. National Green Hydrogen Mission. Available online: https://pib.gov.in/PressReleasePage.aspx?PRID=1888547 (accessed on 8 January 2023).
20. Assareh, E.; Dejdar, A.; Ershadi, A.; Jafarian, M.; Mansouri, M.; Roshani, A.S.; Azish, E.; Saedpanah, E.; Lee, M. Techno-economic analysis of combined cooling, heating, and power (CCHP) system integrated with multiple renewable energy sources and energy storage units. *Energy Build.* **2023**, *278*, 120315. [CrossRef]
21. Van, L.P.; Chi, K.D.; Duc, T.N. Review of hydrogen technologies based microgrid: Energy management systems, challenges and future recommendations. *Int. J. Hydrogen Energy* 2023; *in press*. [CrossRef]
22. Pang, K.Y.; Liew, P.Y.; Woon, K.S.; Ho, W.S.; Alwi, S.R.W.; Klemeš, J.J. Multi-period multi-objective optimisation model for multi-energy urban-industrial symbiosis with heat, cooling, power and hydrogen demands. *Energy* **2023**, *262*, 125201. [CrossRef]
23. Sambhi, S.; Sharma, H.; Bhadoria, V.; Kumar, P.; Chaurasia, R.; Chaurasia, G.S.; Fotis, G.; Vita, V.; Ekonomou, L.; Pavlatos, C. Economic Feasibility of a Renewable Integrated Hybrid Power Generation System for a Rural Village of Ladakh. *Energies* **2022**, *15*, 9126. [CrossRef]
24. Zafeiropoulou, M.; Mentis, I.; Sijakovic, N.; Terzic, A.; Fotis, G.; Maris, T.I.; Vita, V.; Zoulias, E.; Ristic, V.; Ekonomou, L. Forecasting Transmission and Distribution System Flexibility Needs for Severe Weather Condition Resilience and Outage Management. *Appl. Sci.* **2022**, *12*, 7334. [CrossRef]
25. Marocco, P.; Ferrero, D.; Lanzini, A.; Santarelli, M. The role of hydrogen in the optimal design of off-grid hybrid renewable energy systems. *J. Energy Storage* **2022**, *46*, 103893. [CrossRef]
26. Arevalo, P.; Benavides, D.; Lata-garcía, J.; Jurado, F. Energy control and size optimization of a hybrid system (photovoltaic-hidrokinetic) using various storage technologies. *Sustain. Cities Soc.* **2020**, *52*, 101773. [CrossRef]
27. Xia, T.; Rezaei, M.; Dampage, U.; Alharbi, S.; Nasif, O.; Borowski, P.; Mohamed, M. Techno-Economic Assessment of a Grid-Independent Hybrid Power Plant for Co-Supplying a Remote Micro-Community with Electricity and Hydrogen. *Processes* **2021**, *9*, 1375. [CrossRef]
28. Wu, Z.; Zhu, P.; Yao, J.; Tan, P.; Xu, H.; Chen, B.; Yang, F.; Zhang, Z.; Ni, M. Thermo-economic modeling and analysis of an NG-fueled SOFC-WGS-TSA-PEMFC hybrid energy conversion system for stationary electricity power generation. *Energy* **2020**, *192*, 116613. [CrossRef]
29. Temiz, M.; Javani, N. Design and analysis of a combined floating photovoltaic system for electricity and hydrogen production. *Int. J. Hydrogen Energy* **2020**, *45*, 3457–3469. [CrossRef]
30. Nasser, M.; Megahed, T.F.; Ookawara, S.; Hassan, H. Performance evaluation of PV panels/wind turbines hybrid system for green hydrogen generation and storage: Energy, exergy, economic, and enviroeconomic. *Energy Convers. Manag.* **2022**, *267*, 115870. [CrossRef]
31. Kamel, A.A.; Rezk, H.; Abdelkareem, M.A. Enhancing the operation of fuel cell-photovoltaic-battery-supercapacitor renewable system through a hybrid energy management strategy. *Int. J. Hydrogen Energy* **2021**, *46*, 6061–6075. [CrossRef]
32. Ghenai, C.; Salameh, T.; Merabet, A. Technico-economic analysis of off grid solar PV/Fuel cell energy system for residential community in desert region. *Int. J. Hydrogen Energy* **2020**, *45*, 11460–11470. [CrossRef]
33. Jahangir, M.H.; Cheraghi, R. Economic and environmental assessment of solar-wind-biomass hybrid renewable energy system supplying rural settlement load. *Sustain. Energy Technol. Assess.* **2020**, *42*, 100895. [CrossRef]
34. Rad, M.A.V.; Ghasempour, R.; Rahdan, P.; Mousavi, S.; Arastounia, M. Techno-economic analysis of a hybrid power system based on the cost-effective hydrogen production method for rural electrification, a case study in Iran. *Energy* **2020**, *190*, 116421. [CrossRef]
35. Cao, Y.; Dhahad, H.A.; Togun, H.; Anqi, A.E.; Farouk, N.; Farhang, B. A novel hybrid biomass-solar driven triple combined power cycle integrated with hydrogen production: Multi-objective optimization based on power cost and CO_2 emission. *Energy Convers. Manag.* **2021**, *234*, 113910. [CrossRef]
36. Rezaei, M.; Naghdi-Khozani, N.; Jafari, N. Wind energy utilization for hydrogen production in an underdeveloped country: An economic investigation. *Renew. Energy* **2020**, *147*, 1044–1057. [CrossRef]
37. Khosravi, A.; Sanna Syri, M.E.H.; Assad, M.M. Thermodynamic and economic analysis of a hybrid ocean thermal energy conversion/photovoltaic system with hydrogen-based energy storage system. *Energy* **2019**, *172*, 304–319. [CrossRef]
38. Singh, S.; Chauhan, P.; Aftab, M.A.; Ali, I.; Hussain, S.M.S.; Ustun, T.S. Cost Optimization of a Stand-Alone Hybrid Energy System with Fuel Cell and PV. *Energies* **2020**, *13*, 1295. [CrossRef]
39. Almutairi, K.; Dehshiri, S.S.H.; Mostafaeipour, A.; Jahangiri, M.; Techato, K. Technical, economic, carbon footprint assessment, and prioritizing stations for hydrogen production using wind energy: A case study. *Energy Strat. Rev.* **2021**, *36*, 100684. [CrossRef]
40. Google Maps. Available online: http://maps.google.com/ (accessed on 10 December 2022).
41. NASA Power—Data Access Viewer. Available online: https://power.larc.nasa.gov/data-access-viewer/ (accessed on 10 ecember 2022).
42. Solar Resource Glossary. Available online: https://www.nrel.gov/grid/solar-resource/solar-glossary.html (accessed on 10 December 2022).

43. Direct Solar Irradiation. Available online: https://www.sciencedirect.com/topics/engineering/direct-normal-irradiation (accessed on 10 December 2022).
44. Sharma, H.; Pal, N.; Kumar, P.; Yadav, A. A control strategy of hybrid solar-wind energy generation system. *Arch. Electr. Eng.* **2017**, *66*, 241–251. [CrossRef]
45. Kumar, P.; Pal, N.; Sharma, H. Techno-economic analysis of solar photo-voltaic/diesel generator hybrid system using different energy storage technologies for isolated islands of India. *J. Energy Storage* **2021**, *41*, 102965. [CrossRef]
46. Kumar, P.; Pal, N.; Sharma, H. Optimization and techno-economic analysis of a solar photo-voltaic/biomass/diesel/battery hybrid off-grid power generation system for rural remote electrification in eastern India. *Energy* **2022**, *247*, 123560. [CrossRef]
47. Kumar, P.; Pal, N.; Sharma, H. Performance analysis and evaluation of 10 kWp solar photovoltaic array for remote islands of Andaman and Nicobar. *Sustain. Energy Technol. Assess.* **2020**, *42*, 100889. [CrossRef]
48. Technology Brief: Analysis of Current-Day Commercial Electrolyzers. Available online: https://www.nrel.gov/docs/fy04osti/36705.pdf (accessed on 15 November 2022).
49. Sorrenti, I.; Zheng, Y.; Singlitico, A.; You, S. Low-carbon and cost-efficient hydrogen optimisation through a grid-connected electrolyser: The case of GreenLab skive. *Renew. Sustain. Energy Rev.* **2023**, *171*, 113033. [CrossRef]
50. Chen, Z.; Yiliang, X.; Hongxia, Z.; Yujie, G.; Xiongwen, Z. Optimal design and performance assessment for a solar powered electricity, heating and hydrogen integrated energy system. *Energy* **2023**, *262*, 125453. [CrossRef]
51. Sijakovic, N.; Terzic, A.; Fotis, G.; Mentis, I.; Zafeiropoulou, M.; Maris, T.I.; Zoulias, E.; Elias, C.; Ristic, V.; Vita, V. Active System Management Approach for Flexibility Services to the Greek Transmission and Distribution System. *Energies* **2022**, *15*, 6134. [CrossRef]
52. Sambhi, S.; Sharma, H.; Bhadoria, V.; Kumar, P.; Chaurasia, R.; Fotis, G.; Vita, V. Technical and Economic Analysis of Solar PV/Diesel Generator Smart Hybrid Power Plant Using Different Battery Storage Technologies for SRM IST, Delhi-NCR Campus. *Sustainability* **2023**, *15*, 3666. [CrossRef]
53. Fotis, G.; Vita, V.; Maris, T.I. Risks in the European Transmission System and a Novel Restoration Strategy for a Power System after a Major Blackout. *Appl. Sci.* **2022**, *13*, 83. [CrossRef]
54. Vita, V.; Fotis, G.; Pavlatos, C.; Mladenov, V. A New Restoration Strategy in Microgrids after a Blackout with Priority in Critical Loads. *Sustainability* **2023**, *15*, 1974. [CrossRef]
55. Wang, Z.; Fang, G.; Wen, X.; Tan, Q.; Zhang, P.; Liu, Z. Coordinated operation of conventional hydropower plants as hybrid pumped storage hydropower with wind and photovoltaic plants. *Energy Convers. Manag.* **2023**, *277*, 116654. [CrossRef]

designs

MDPI

Article

Control over Grid Reactive Power by Using a Powerful Regenerative Controlled-Speed Synchronous Motor Drive

Aleksandr S. Maklakov [1,*], Aleksandr A. Nikolaev [2] and Tatyana A. Lisovskaya [3]

1 Project Activity Center, Scientific Activity Sector, Moscow Polytechnic University, 107023 Moscow, Russia
2 Research and Development Department, Nosov Magnitogorsk State Technical University, 455000 Magnitogorsk, Russia; alexniko@inbox.ru
3 Department of Automation and Control, Moscow Polytechnic University, 38 Bolshaya Semyonovskaya Street, 107023 Moscow, Russia; sosedkova92@mail.ru
* Correspondence: a.s.maklakov@mospolytech.ru

Abstract: The authors propose a technique for reactive power compensation using a powerful regenerative controlled-speed synchronous motor drive (SMD) based on a three-level (3L) neutral point clamped (NPC) active front-end rectifier (AFE) and a voltage source inverter (VSI). The review of technical solutions for reactive power compensation showed that the limitations on the transmitted reactive power in the system under consideration still have not been studied. The paper provides a mathematical description and proposes synthesis-friendly block diagrams of the mathematical 3L-NPC-AFE-VSI and SMD models. The developed models allow defining the instantaneous values of the total 3L-NPC-AFE power consumed from the grid depending on the SMD load diagram. It is noted that the 3L-NPC-AFE-VSI-SMD system is designed without considering the opportunities for reactive power generation. It was determined that the limit value of reactive power generated by a 3L-NPC-AFE depends on the DC link voltage, the grid current consumption and the modulation index. The possibility of reactive power compensation by the SMD system through a 3L-NPC-AFE was experimentally tested on the main drive of a metal plate hot rolling mill. The analysis of the results obtained showed that during the breakdown, an SMD can generate reactive power equal to 16% of the total rated power using a 3L-NPC-AFE at a rated DC link voltage and without overcurrent. It was shown that generating reactive power is expedient in low-load SMD operation modes or at idle. Research in this area is promising due to the widespread use of high-power SMD based on a 3L-NPC-AFE-VSI and the tightening of requirements for energy saving and efficiency and supply voltage quality. The proposed reactive power control technique can be used as part of an industrial smart grid.

Keywords: power converters; regenerative electric drive; control systems; reactive power compensation; power quality

Citation: Maklakov, A.S.; Nikolaev, A.A.; Lisovskaya, T.A. Control over Grid Reactive Power by Using a Powerful Regenerative Controlled-Speed Synchronous Motor Drive. *Designs* **2023**, *7*, 62. https://doi.org/10.3390/designs7030062

Academic Editors: Surender Reddy Salkuti and Dibin Zhu

Received: 11 April 2023
Revised: 26 April 2023
Accepted: 4 May 2023
Published: 8 May 2023

1. Introduction

The increase in the global demand for energy resources resulted in the emergence of new engineering systems based on power conversion equipment that can control large active and reactive power flows producing minimum adverse impacts on the environment and featuring high efficiency. Reactive power compensation technologies allow improving the energy quality and efficiency of various electrical complexes and systems. Most reactive power problems in the grid can be mitigated or solved with passive inductive or capacitive filters, static thyristors (STC) or synchronous reactive power compensators (STATCOM). However, this is expensive and may cause additional EMC problems.

Medium-voltage AC regenerative electric drives (AC RED), based on three-level neutral point clamped (3L-NPC) active front ends (AFE) and voltage source inverters (VSI), have similar opportunities to STATCOMs. These systems were widely introduced in metal plate rolling mills, excavators, electric trains, etc. [1–3]. Modern AC REDs have a number of advantages over the previous generation electric drives featuring thyristor rectifier modules, namely: (1) the capacity to regenerate the power to the grid in the generating modes of the electric drive with minimum distortion; (2) the capacity to adjust the cos(φ); and (3) the improved harmonic composition of the power consumed from the grid due to the use of specialized AFE pulse width modulation (PWM) algorithms [4,5]. Today, no full understanding has been formed of the reactive power compensation possibilities using high-power AC REDs with an AFE considering the possible limitations and associated problems. However, such modes are used ubiquitously and prove their efficiency.

Gonzalo Alonso Orcajo et al. [6–8] demonstrated the possibility of reactive power compensation in the grid via the modernization of several main electric drives of hot rolling mill stands. New electric drives feature three-level NPC AFE converters facilitating the electric motors' consumption of active power to maintain the required rolling process and compensate for some of the reactive power consumed by neighboring devices. In terms of their practical significance, the solutions presented in the articles are straightforward and can be recommended for any other similar systems. However, these works do not cover in detail the issues of determining the possible limitations for reactive power compensation under the maximum current and voltage in the DC bus of the frequency converter. These problems require a deeper theoretical analysis to spread the suggested solutions to a wider class of objects.

In the article [9] by Pandit et al., an excavator power system was reviewed, with the power coefficient in the point of common connection (PCC) selected to provide the maximum reduction of voltage fluctuations on the high-voltage bushing of the excavator. To get the desired cos(φ), the high-power AC RED system used an AFE to compensate for the reactive power consumption from the auxiliary loads of the excavator. Note that this was achieved when the voltage in the DC link was two times higher than the voltage in the secondary winding of the transformer. This solution poses some questions about the operating conditions of the AC motor because the high voltage in the DC link has an adverse effect on the electromagnetic processes on the VSI side (low modulation index, higher harmonics, in-phase currents/voltages, etc.).

Brovanov et al. [10] provide a detailed analysis of electromagnetic processes occurring in the power circuit of a three-level converter with the NPC topology in autonomous electric power generation systems. As mentioned before, during the reactive power generation or consumption mode operation, the loads on the semiconductor converter modules become even more unequally distributed while the voltage in the DC link becomes more unbalanced, and the quality of converter power decreases insignificantly. However, these factors do not affect the converter operation if there is proper cooling provided, proper DC link voltage rectification algorithms are used, and proper sine filter parameters are selected.

This paper studies the possibilities and defines limitations in reactive power control using a powerful regenerative controlled-speed synchronous motor drive (SMD) based on a 3L-NPC-AFE. The paper has the following structure. Section 1 reviews the current research and formulates the study problem and objective. Section 2 covers the SMD model development for considering the load diagram impact on reactive power compensation modes. Section 3 analyzes vector diagrams of AFE operation modes. Section 4 gives the mathematical description of a 3L-NPC-AFE-VSI and analyzes the reactive power compensation mode. Section 5 provides the experimental study results. Finally, general conclusions are formulated.

2. Developing a Mathematical Model of the SMD System

The SMD's mathematical model is required to calculate the current consumption when operating in compliance with a process load diagram. The obtained data are used to form a disturbing effect in the DC voltage maintaining the circuit of the 3L-NPC-AFE control system.

All things considered, 3L-NPC-AFE-VSI-SMD systems are widely applied in the low-speed drives of metallurgical equipment characterized by increased requirements to the response, overload capacity and speed control quality, in wind turbines and traction drives, as well as in high-performance grinding units in the mining, cement and chemical industries. A simplified diagram of the SMD system is shown in Figure 1.

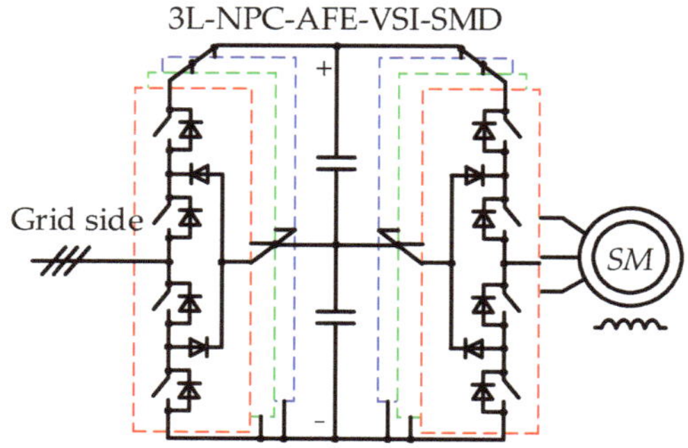

Figure 1. SMD system diagram.

The decisive SMD advantages are the lack of power, overload capacity and current slew rate limitations, as well as the possibility of a significant improvement in dynamics by changing the synchronous motor design. SMDs are developed, as a rule, using custom-made high-torque synchronous motors. A high rate of electromagnetic use is decisive for developing their control systems, which predetermined high requirements for the response of the internal stator current control loops [11].

The equivalent circuit of a synchronous machine is derived from a generalized AC machine; the stator and rotor windings are fed with AC and DC, respectively. In the steady state, the rotor rotates simultaneously with the field in the air gap. A diagram of the SM model for mathematical derivations is presented in Figure 2.

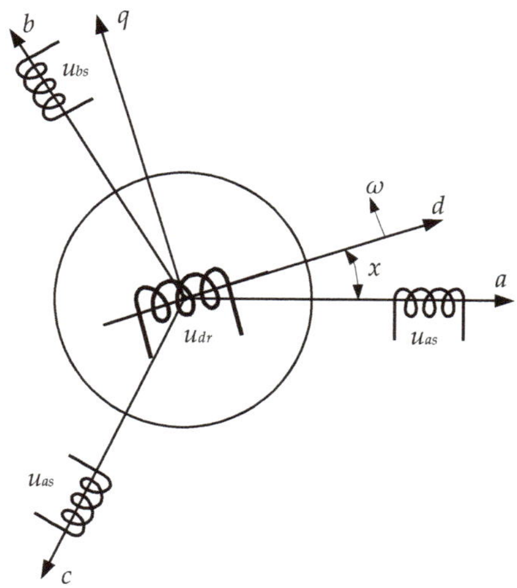

Figure 2. Three-phase SM diagram.

As a rule, rotors have a single excitation winding along the longitudinal axis. In some cases, one observes the use of compensated SMs containing an additional transverse excitation winding. In a frequency-controlled drive, this winding is not required; moreover, it has an adverse effect, making it difficult to boost transients. Therefore, it is further assumed that the rotor has a single longitudinal excitation winding. Furthermore, the SM is assumed to be laminated, so the impact of eddy currents in the massive core parts can be neglected. The voltage equilibrium equation for the stator windings is as follows:

$$\|u_s\| = R_s\|i_s\| + \frac{d}{dt}\|\psi_s\| \tag{1}$$

where $\|u_s\| = \begin{Vmatrix} u_{sa} \\ u_{sb} \\ u_{sc} \end{Vmatrix}$ is the stator voltage; $\|i_s\| = \begin{Vmatrix} i_{sa} \\ i_{sb} \\ i_{sc} \end{Vmatrix}$ is the stator current.

The stator flux linkages are as follows:

$$\|\psi_s\| = \|L'_{ss}\|\|i_s\| + \|M'_{sr}\|i_f \tag{2}$$

where i_f is the excitation winding current.

The stator inductance matrix is expressed as follows:

$$\|L'_{ss}\| = \begin{Vmatrix} L_{aa} & L_{ab} & L_{ac} \\ L_{ba} & L_{bb} & L_{bc} \\ L_{ca} & L_{cb} & L_{cc} \end{Vmatrix} = \begin{Vmatrix} L_{ave} + L_m\cos 2x & -M_{ave} + L_m\cos\left(2x - \frac{2\pi}{3}\right) & -M_{ave} + L_m\cos\left(2x + \frac{2\pi}{3}\right) \\ -M_{ave} + L_m\cos\left(2x - \frac{2\pi}{3}\right) & L_{ave} + L_m\cos\left(2x + \frac{2\pi}{3}\right) & -M_{ave} + L_m\cos 2x \\ -M_{ave} + L_m\cos\left(2x + \frac{2\pi}{3}\right) & -M_{ave} + L_m\cos 2x & L_{ave} + L_m\cos\left(2x - \frac{2\pi}{3}\right) \end{Vmatrix} \tag{3}$$

where L_{ave} and M_{ave} are the average values of the self-inductance of the stator phase and the mutual inductance between two stator phases.

The matrix of mutual inductances of the stator phase and the excitation winding is as follows:

$$\|M_{sr}\| = \begin{Vmatrix} L_{af} \\ L_{bf} \\ L_{cf} \end{Vmatrix} = \begin{Vmatrix} M_f\cos x \\ M_f\cos\left(x - \frac{2\pi}{3}\right) \\ M_f\cos\left(x + \frac{2\pi}{3}\right) \end{Vmatrix} \tag{4}$$

The flux linkage of the excitation winding is as follows:

$$\|\psi_f\| = \begin{Vmatrix} M_f\cos x & M_f\cos\left(x - \frac{2\pi}{3}\right) & M_f\cos\left(x + \frac{2\pi}{3}\right) \end{Vmatrix}\|i_s\| + L_{ff}i_f \tag{5}$$

where L_{ff} is the self-inductance of the excitation winding; M_f is the maximum mutual inductance between the excitation winding and the stator phase.

Due to the synchronous (magnetic and electric) machine rotor asymmetry, equations with constant coefficients can only be obtained when transformations are conducted in the $dq0$ reference frame, the axes of which are rigidly connected to the rotor. The matrix of this transformation for the stator equation is as follows:

$$\|A_s\| = \frac{2}{3}\begin{Vmatrix} \cos x & \cos\left(x - \frac{2\pi}{3}\right) & \cos\left(2x + \frac{2\pi}{3}\right) \\ -\sin x & -\sin\left(x - \frac{2\pi}{3}\right) & -\sin\left(x + \frac{2\pi}{3}\right) \\ \frac{1}{2} & \frac{1}{2} & \frac{1}{2} \end{Vmatrix} \tag{6}$$

The formulas for transforming stator currents into the $dq0$ axis components obtained from the equation $\|i_{dq0}\| = A_s\|i_s\|$, have the following form:

$$\begin{cases} i_d = \frac{2}{3}\left(i_{sa}\cos x + i_{sb}\cos\left(x - \frac{2\pi}{3}\right) + i_{sc}\cos\left(x + \frac{2\pi}{3}\right)\right); \\ i_q = -\frac{2}{3}\left(i_{sa}\sin x + i_{sb}\sin\left(x - \frac{2\pi}{3}\right) + i_{sc}\sin\left(x + \frac{2\pi}{3}\right)\right); \\ i_0 = \frac{1}{3}\left(i_{sa} + i_{sb} + i_{sc}\right). \end{cases} \tag{7}$$

Both parts of Equation (1) are multiplied by the *As* matrix on the left to obtain, in the expanded form, the following:

$$\begin{aligned} u_d &= i_d R_s + \tfrac{d\psi_d}{dt} + \omega\psi_q; \\ u_q &= i_q R_s + \tfrac{d\psi_q}{dt} + \omega\psi_d; \\ u_0 &= i_0 R_s + \tfrac{d\psi_{s0}}{dt}. \end{aligned} \tag{8}$$

A similar operation conducted with Equation (2) gives the following equations for the stator flux linkages:

$$\begin{aligned} \psi_d &= i_d L_d + i_f M_f; \\ \psi_q &= i_q L_q; \\ \psi_{s0} &= i_0 L_0, \end{aligned} \tag{9}$$

where $L_d = L_{ave} + M_{ave} + \frac{3}{2}L_m;\ L_q = L_{ave} + M_{ave} - \frac{3}{2}L_m;\ L_0 = L_{ave} - 2M_{ave}$.

Considering the first Equation (7), the excitation winding flux linkage Equation (5) can be represented as follows:

$$\psi_f = \frac{3}{2} i_d M_f + i_f L_{ff}. \tag{10}$$

The rotor winding equation is as follows:

$$u_f = i_f R_f + \frac{d\psi_f}{dt}. \tag{11}$$

The SM electromagnetic torque is expressed as follows:

$$M = \frac{3}{2} Z_p (\psi_d i_q - \psi_q i_d). \tag{12}$$

The rotor motion equation is as follows:

$$M - M_s = \frac{J}{Z_p}\frac{d\omega}{dt}, \tag{13}$$

where J is mass moment of inertia; Z_p is a pair of poles.

Equations (8)–(13) form a complete system of equations for a synchronous machine. It can be transformed into a form suitable for the synthesis of an automated control system. Considering Equations (10) and (11), the following is obtained:

$$u_f = i_f R_f + L_{ff}\frac{di_f}{dt} + \frac{3}{2}M_f\frac{di_d}{dt}, \tag{14}$$

from which, $\frac{di_f}{dt} = \frac{1}{L_{ff}}\left(u_f + i_f R_f - \frac{3}{2}M_f\frac{di_d}{dt}\right)$.

From Equations (8) and (9), we obtain the following:

$$u_d = i_d R_s + L_d\frac{di_d}{dt} - \omega L_q i_q + M_f\frac{di_f}{dt}. \tag{15}$$

Substituting Equation (14) into Equation (15) and considering that *d/dt* = *p*, we obtain the following:

$$u_d = R_s(1 + pT'_{sd})i_d + K_f U_f - K_f R_f i_f - \omega\psi_q, \tag{16}$$

where $T'_{sd} = \frac{L'_d}{R_s};\ L'_d = L_d - \frac{3M_f^2}{2L_{ff}};\ K_f = \frac{M_f}{L_{ff}}$.

From Equations (8) and (9), we obtain the following:

$$u_q = R_s(1 + pT_{sq})i_q + \omega\psi_d; \tag{17}$$

where $T_{sq} = \frac{L_q}{R_s}$.

From Equations (9) and (14), we obtain the following:

$$u_f = R_f(1 + pT''_{d0})i_f + K_d p\psi_d, \quad (18)$$

where $T''_{d0} = \frac{L'_f}{R_f}$; $L'_f = L_{ff} - \frac{3M_f^2}{2L_d}$; $K_d = \frac{3M_f}{2L_d}$.

The vector–matrix system of equations is synthesized based on the SM block diagram with the main controlled coordinates i_d, i_q and ψ_f.

From Equation (11), considering Equation (10), we obtain the following:

$$u_f = \frac{1 + pT'_{do}}{T'_{do}}\psi_f - \frac{3}{2}K_f R_f i_d, \quad (19)$$

where $T'_{do} = \frac{L_{ff}}{R_f}$.

In this case, the longitudinal stator flux linkage can be represented as follows:

$$\psi_d = i_d L'_d + K_f \psi_f. \quad (20)$$

The block diagram with the main controlled coordinates i_d, i_q and ψ_f was built using Equations (9), (12), (13) and (16)–(20) and is shown in Figure 3. This diagram is convenient for the synthesis of a vector control system.

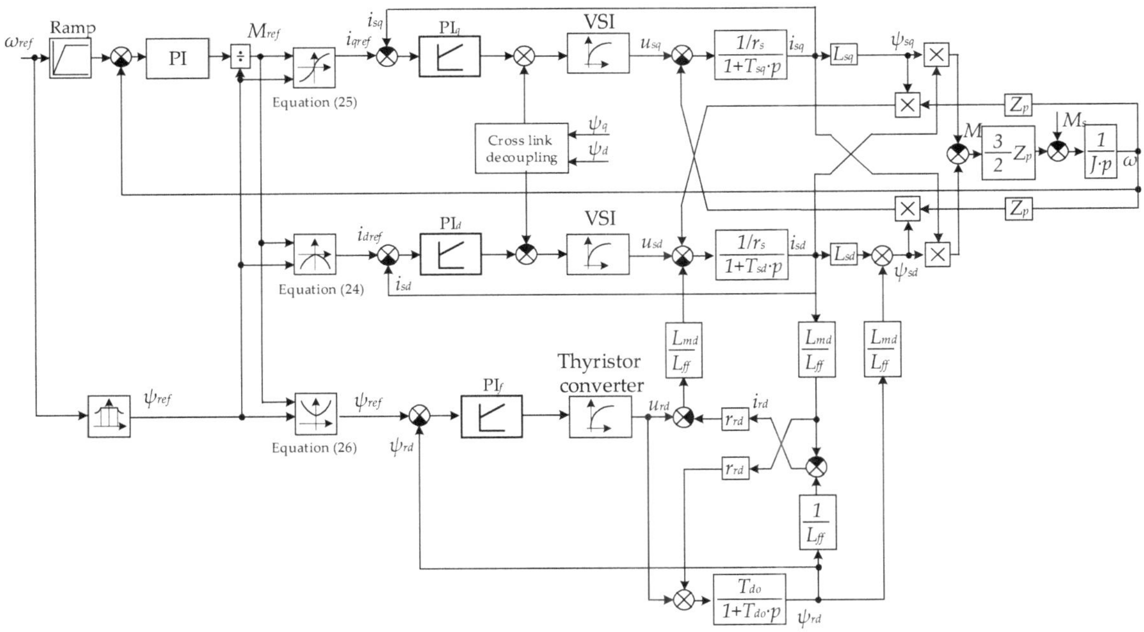

Figure 3. Block diagram of the VSI-SMD vector control system.

The SM control algorithm in the reference frame associated with the rotor and the main conditions for implementing it can be formulated if we assume that the stator windings are powered by a VSI operating in the AC power source mode.

The requirements for the optimal use of the SM power-to-size ratio and the installed capacity of the 3L-NPC-AFE-VSI necessitate meeting the following conditions:

- Providing the SM operation with a rated stator flux linkage within the entire load change and speed control range (except for the field weakening mode with two-zone control):

$$\psi_{sref} \sim \psi_{srate} \tag{21}$$

- Providing the SM operation with a power factor equal to (or close to) unity:

$$Q_s = \frac{3}{2}(\psi_d i_d + \psi_q i_q) = 0. \tag{22}$$

Based on Equations (21) and (22), the synchronous motor's electromagnetic torque is determined as follows:

$$M = \frac{3}{2} Z_p \psi_{srate} i_s. \tag{23}$$

When conditions (21)–(23) are fulfilled, the stator's full flux linkage vector is always perpendicular to the $\widetilde{I}_s$ vector, i.e., $Re(\widetilde{\psi}_s^* \widetilde{I}_s) = 0$. Since the SM's electromagnetic torque is always proportional to the stator and rotor flux linkages, this solves the problem of optimal use of its power-to-size ratio. For vector control of a synchronous motor using Equation (23), the following dependencies were obtained:

$$i_d = -\frac{L_q}{\psi_s \sqrt{1 + \frac{M^2}{\left(\frac{3}{2} Z_p \psi_s\right)^2} \cdot \frac{L_q^2}{\psi_s^2}}} \cdot \frac{M^2}{\left(\frac{3}{2} Z_p \psi_s\right)^2} \tag{24}$$

$$i_q = -\frac{L_q}{\sqrt{1 + \frac{M^2}{\left(\frac{3}{2} Z_p \psi_s\right)^2} \cdot \frac{L_q^2}{\psi_s^2}}} \cdot \frac{M}{\frac{3}{2} Z_p \psi_s} \tag{25}$$

$$\psi_f = -\frac{1}{\sqrt{1 + \frac{M^2}{\left(\frac{3}{2} Z_p \psi_s\right)^2} \cdot \frac{L_q^2}{\psi_s^2}}} \cdot \left[L_{ff} I_{f0} + \frac{L_d' L_q}{K_f \psi_s} \cdot \frac{M^2}{\left(\frac{3}{2} Z_p \psi_s\right)^2} \right] \tag{26}$$

where I_{f0} is the excitation current according to the no-load characteristic.

Figure 1 shows the block diagram of a closed-loop control system. The controllers are synthesized in the $dq0$ coordinate system: the speed control loop with a PI controller is an outer control loop, and the control loops of the stator current components i_d and i_q and the rotor winding flux linkage ψ_f are the inner control loops. The three aforementioned loops in the aggregate make a multidimensional torque loop. The drive signals at the input of the control loops of the longitudinal and transverse components of the stator currents and the flux linkage of the excitation winding are implemented according to function Equations (24)–(26). With a high response of the internal control loops, the function equations compensate for the nonlinearities of the SM's control object in the reference frame selected.

Figure 1 shows the compensators of the two main crosslinks for rotational EMF. When choosing the stator current components i_d and i_q and the excitation winding flux linkage ψ_f as the controlled variables, only these internal crosslinks require compensation. The crosslinks between the excitation and longitudinal component circuits are very weak. The compensation of the main SM crosslinks and nonlinearities is achieved by almost full decoupling of the i_d, i_q and ψ_f control loops. Controllers are implemented as active filters. To eliminate the static control error, current PI controllers are used in the $dq0$ reference frame.

Thus, the 3L-NPC-AFE-VSI-SMD is developed on the basis of the conditions stipulated for the optimal use of the SM and frequency converter powers; thus, the reactive power generation mode will be limited by active power consumption. Therefore, the 3L-NPC-AFE operation opportunities and limitations for the reactive power generation in various drive operation modes should be analyzed in more detail.

3. AFE Power Description

The possibilities of reactive power compensation within the system of an AFE and the grid can be conveniently reviewed using a single-phase equivalent circuit shown in Figure 4a. The system is shown as the EMF of the AFE E_{AFE} connected over the reduced inductive reactance X_L at the terminal point of the secondary transformer winding EMF (grid) E_g. The difference between E_{AFE} and E_g corresponds to X_L, the voltage drop for which is U_L. Further, review of the AFE operation using vector diagrams in Figure 4b–d features marked limitations for the maximum current I_{gmax} and AFE EMF E_{AFEmax}. Accounting for these limitations allowed obtaining the acceptable AFE operation range, which is crosshatched with black lines in Figure 3.

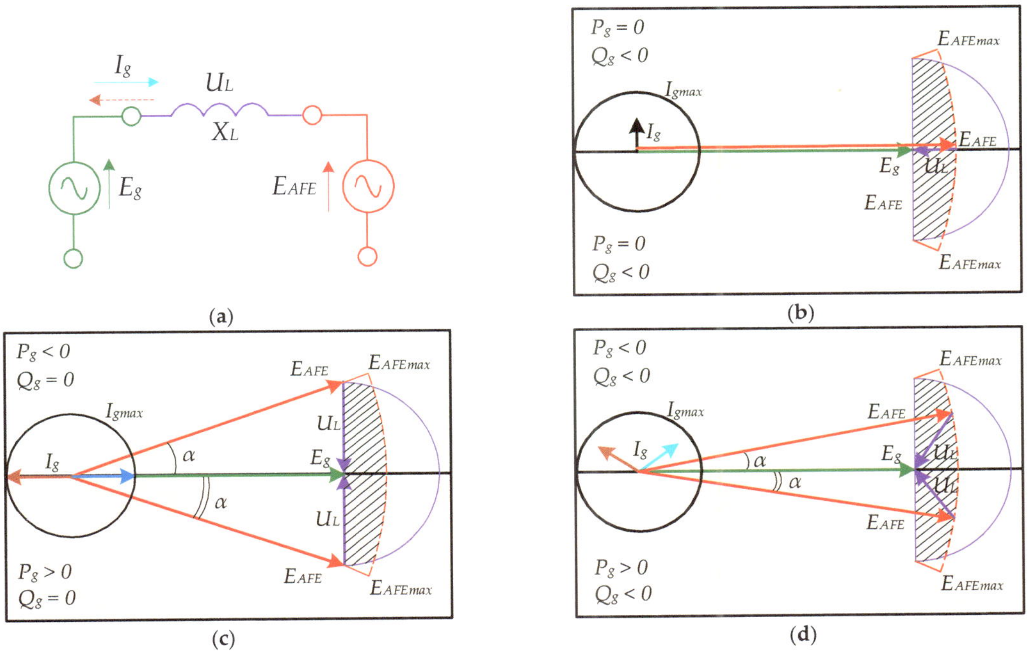

Figure 4. The specifics of AFE operation during reactive power compensation. (**a**) Single-phase equivalent circuit of an AFE. (**b**) Reactive power generation mode without active power consumption. (**c**) Reactive power generation mode with active power consumption. (**d**) Reactive power generation mode with active power generation.

Figure 4b shows the positioning of the vectors during the AFE operation in the reactive power generation mode (similar to a static VAR compensator). This mode is possible during idling operation. Note that in this mode, the maximum reactive current value is severely limited as the ceilings of E_{AFE} are also limited by E_{AFEmax}. It is possible to increase the generated reactive power through the additional charging of the capacitors in the DC link.

In the lower region of Figure 4c, the AFE operates in the rectifier mode. Current I_g consumed from the grid is in phase with grid voltage E_g, which results in a single power coefficient at the converter input. This mode requires adjusting the amplitude and phase of the E_{AFE} converter voltage relative to E_g so that the voltage drop over the reduced inductance U_L is perpendicular to E_g. In the upper region of Figure 4c, the AFE operates in the inverter mode. Vector I_g is opposite in phase to vector E_g, which also results in a single power coefficient at the converter input. For this mode, the E_{AFE} voltage phase must outrun voltage E_g by angle α. In the reviewed modes, the AFE can operate at maximum current I_{gmax}. They can be used to determine the possible reactive power generation region.

In the lower region of Figure 4d, the AFE operates in the rectifier mode, and in the upper region of Figure 4d, it operates in the inverter mode with reactive power generation. For them, the voltage drop U_L is not perpendicular to E_g, which results in current I_g outrunning or lagging behind E_g. The analysis of the diagram shows that the generated reactive power can also be increased using the additional charging of the capacitors in the DC link. However, it is necessary to account for the consumed active power to prevent the tripping of the maximum permissible AFE current.

The analysis of the vector diagrams of AFE operation in Figure 4 shows that it is possible to compensate for the reactive power both in the motoring and inverter operation modes.

Taking into account an equivalent active resistance R at the common coupling point, the equations for the calculation of active, reactive and apparent power can be composed as follows:

$$P_g = \frac{E_g \cdot E_{AFE} \cdot (R \cdot \cos(\alpha) + X_L \cdot \sin(\alpha)) - E_g^2 \cdot R}{R^2 + X_L^2}, \tag{27}$$

$$Q_g = \frac{E_g \cdot E_{AFE} \cdot (X_L \cdot \cos(\alpha) + R \cdot \sin(\alpha)) - E_g^2 \cdot X_L}{R^2 + X_L^2}, \tag{28}$$

$$S_g = \sqrt{\frac{E_g^2 \cdot \left(E_{AFE}^2 - 2 \cdot E_g \cdot E_{AFE} \cdot \cos(\alpha) + E_g^2\right)}{R^2 + X_L^2}}. \tag{29}$$

where P_g, Q_g and S_g are the dependencies of the active, reactive and apparent powers of the grid and α is the shift angle between E_g and E_{AFE}.

When AFEs operate in the PWM mode, the maximum amplitude of the continuous (useful) component of the phase voltage E_{AFEmax} is determined by the U_{dc} value and the PWM algorithm. Controlled AFE operation requires the voltage in the U_{dcmin} DC link to be at such a minimum level that allows for blocking of the forward diodes. Thus, the DC link voltage should be higher than the grid voltage amplitude value, considering the voltage drop across the active inductive input impedance.

It is known that six-step control is the best option for the DC link voltage. Currently, for most AFEs as part of an SMD, PWM with selected harmonic elimination (SHE) is used to provide the best quality of converted voltage under the low semiconductor module's switching frequency [11–15]. Along with good electromagnetic compatibility, PWM with SHE facilitates the use of the DC link voltage at the six-step modulation level ($K_{PWM} \approx 2/\pi$) [16–18]. The E_{AFE} vector length is controlled using the modulation index m, which is calculated as follows:

$$m = \frac{\pi \cdot E_{AFE}}{2 \cdot U_{dc}} \tag{30}$$

To define the limit E_{AFEmax} value, the following equation is used:

$$E_{AFE\max} = U_{dc} \cdot K_{PWM} = \frac{2}{\pi} \cdot U_{dc} \tag{31}$$

At different m values, the converted power quality parameters change. The relevant studies in the area have shown that the higher the m value, the lower the THD of the converter output voltage [19–23]. In practice, the m value close to 1 (if we describe a range from 0 to $4/\pi$) is chosen while providing a margin for maintaining stable AFE and VSI operation in dynamic modes (a voltage drop in the DC link during shock application of the electric drive load) [24–27].

Considering the aforementioned Equations (30) and (31), the active, reactive and apparent powers can be determined as follows:

$$P_g = \frac{E_g \cdot m \cdot U_{dc} \cdot \frac{2}{\pi} \cdot (R \cdot \cos(\alpha) + X_L \cdot \sin(\alpha)) - E_g^2 \cdot R}{R^2 + X_L^2}, \tag{32}$$

$$Q_g = \frac{E_g \cdot m \cdot U_{dc} \cdot \frac{2}{\pi} \cdot (X_L \cdot \cos(\alpha) + R \cdot \sin(\alpha)) - E_g^2 \cdot X_L}{R^2 + X_L^2}, \quad (33)$$

$$S_g = \sqrt{\frac{E_g^2 \cdot \left(\left(m \cdot U_{dc} \cdot \frac{2}{\pi}\right)^2 - 2 \cdot E_g \cdot m \cdot U_{dc} \cdot \frac{2}{\pi} \cdot \cos(\alpha) + E_g^2 \right)}{R^2 + X_L^2}}. \quad (34)$$

Based on Equations (32)–(34), the limit values of the active, reactive and apparent powers at the set values of the grid voltage and equivalent active inductive impedance on the AFE side can be controlled by the modulation index, the DC link voltage and the shift angle α. The problem of generating reactive power in the system under consideration is solved using an AFE control system.

4. Synthesis of a 3L-NPC-AFE-VSI Control System

Figure 5 shows a 3L-NPC-AFE-VSI circuit diagram representing 24 fully controlled semiconductor modules T_1–T_{24}, 24 forward diodes D_1–D_{24}, 12 clamped diodes D_{c1}–D_{c12}; $S_{abc1AFE}$–$S_{abc4AFE}$ and $S_{abc1VSI}$–$S_{abc4VSI}$ are the states of the 3L-NPC-AFE-VSI semiconductor modules at each phase rack a, b and c, taking logical values 0 or 1; u_{dc1}, u_{dc2} are the instantaneous voltage values on the equivalent capacities of the DC link; u_{abcAFE}, u_{abcVSI} and i_{abcAFE}, i_{abcVSI} are the phase voltages and currents of the 3L-NPC-AFE-VSI; C_{dc1}, C_{dc2} are the equivalent DC link capacitance values.

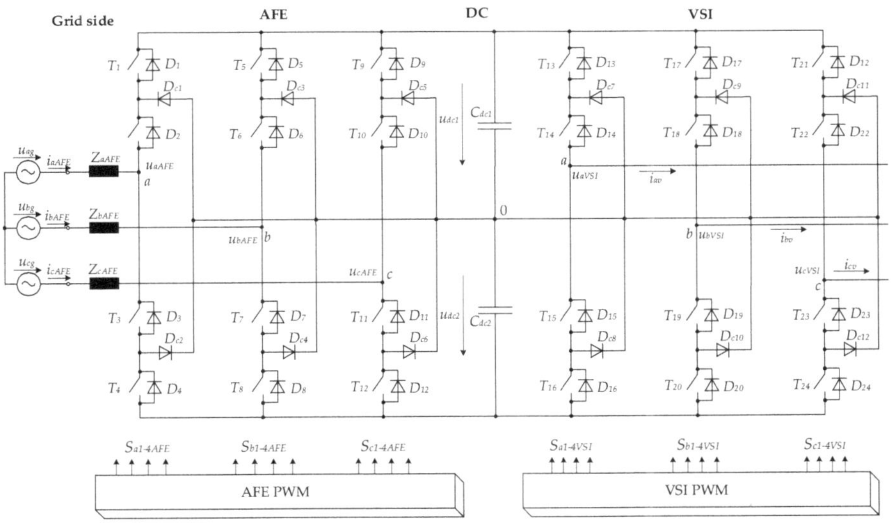

Figure 5. Block diagram of a 3L-NPC-AFE-VSI.

When building a mathematical model, discrete logical functions f_{abc} describing the states of modules in each 3L-NPC-AFE-VSI phase rack should be considered:

$$f_{abcAFE} = \begin{cases} 1, & \rightarrow (S_{abc1AFE} \text{ and } S_{abc2AFE}) = 1 \text{ and } (S_{abc3AFE} \text{ and } S_{abc4AFE}) = 0 \\ 0, & \rightarrow (S_{abc2AFE} \text{ and } S_{abc3AFE}) = 1 \text{ and } (S_{abc1AFE} \text{ or } S_{abc4AFE}) = 0 \\ -1, & \rightarrow (S_{abc3AFE} \text{ and } S_{abc4AFE}) = 1 \text{ and } (S_{abc1AFE} \text{ and } S_{abc2AFE}) = 0 \end{cases}, \quad (35)$$

$$f_{abcVSI} = \begin{cases} 1, & \rightarrow (S_{abc1VSI} \text{ and } S_{abc2VSI}) = 1 \text{ and } (S_{abc3VSI} \text{ and } S_{abc4VSI}) = 0 \\ 0, & \rightarrow (S_{abc2VSI} \text{ and } S_{abc3VSI}) = 1 \text{ and } (S_{abc1VSI} \text{ or } S_{abc4VSI}) = 0 \\ -1, & \rightarrow (S_{abc3VSI} \text{ and } S_{abc4VSI}) = 1 \text{ and } (S_{abc1VSI} \text{ and } S_{abc2VSI}) = 0 \end{cases} \quad (36)$$

Equations (35) and (36) are optimized to the state of the switching functions F_{abc} generating logic signals (0 or 1) that determine the voltage of the DC link connected to each 3L-NPC-AFE-VSI phase:

$$\begin{aligned} F_{abc1AFE} &= \frac{f_{abcAFE}(f_{abcAFE}+1)}{2}; \\ F_{abc2AFE} &= \frac{f_{abcAFE}(f_{abcAFE}-1)}{2}; \\ F_{abc1VSI} &= \frac{f_{abcVSI}(f_{abcVSI}+1)}{2}; \\ F_{abc2VSI} &= \frac{f_{abcVSI}(f_{abcVSI}-1)}{2}, \end{aligned} \tag{37}$$

A system of equations describing electromagnetic processes was built based on Equations (35)–(37) and the voltage and current equilibrium equations obtained with Kirchhoff's laws, as well as the power balance in a three-phase system, as follows:

$$\begin{cases} u_{abcAFE} = u_{dc1} \cdot D_{abc1AFE} + u_{dc2} \cdot D_{abc2AFE} + i_{abcAFE} \cdot R_{abcAFE} + L_{abcAFE} \cdot \frac{di_{abcAFE}}{dt} \\ u_{abcVSI} = u_{dc1} \cdot D_{abc1VSI} + u_{dc2} \cdot D_{abc2VSI} \\ u_{dc1} = \frac{(F_{a1AFE} \cdot i_{aAFE} + F_{b1AFE} \cdot i_{bAFE} + F_{c1AFE} \cdot i_{cAFE}) - (F_{a1VSI} \cdot i_{aVSI} + F_{b1} \cdot i_{bVSI} + F_{c1} \cdot i_{cVSI})}{C_{dc1} \cdot p} \\ u_{dc2} = \frac{(F_{a2} \cdot i_{aAFE} + F_{b2} \cdot i_{bAFE} + F_{c2} \cdot i_{cAFE}) - (F_{a2VSI} \cdot i_{aVSI} + F_{b2VSI} \cdot i_{bVSI} + F_{c2VSI} \cdot i_{cVSI})}{C_{dc2} \cdot p} \\ D_{abc1AFE} = F_{abc1AFE} - \frac{1}{3} \cdot \sum_{n=a}^{b.c} F_{n1AFE}; \\ D_{abc2AFE} = F_{abc2AFE} - \frac{1}{3} \cdot \sum_{n=a}^{b.c} F_{n2AFE}; \\ D_{abc1VSI} = F_{abc1VSI} - \frac{1}{3} \cdot \sum_{n=a}^{b.c} F_{n1VSI}; \\ D_{abc2VSI} = F_{abc2VSI} - \frac{1}{3} \cdot \sum_{n=a}^{b.c} F_{n2VSI}. \\ P_g = u_{ag} \cdot i_{aAFE} + u_{bg} \cdot i_{bAFE} + u_{cg} \cdot i_{cAFE} \\ Q_g = \frac{1}{\sqrt{3}} \left(\left(u_{bg} - u_{cg} \right) \cdot i_{aAFE} + (u_{cg} - u_{ag}) \cdot i_{bAFE} + \left(u_{ag} - u_{bg} \right) \cdot i_{cAFE} \right) \end{cases} \tag{38}$$

where u_{abcg}, u_{abcAFE} and u_{abcVSI} are the phase voltages; i_{abcAFE} and i_{abcVSI} are the phase currents; $F_{abc12AFE}$ and $F_{abc12VSI}$ are the switching functions in the $dq0$ reference frame; p is the instantaneous active power; q is the instantaneous reactive power.

The system of Equation (38) considers the discrete nature of generating the 3L-NPC-AFE-VSI output voltages and the voltage imbalance in the DC link. Figure 6 shows the block diagram of the developed logical–mathematical model.

The AFE control system is further synthesized in a rotating reference frame $dq0$ directed along the grid vector. After the $abc/dq0$ coordinate transformation, the system of Equations (38) takes the following form:

$$\begin{cases} u_{dg} = u_{dAFE} + i_{dAFE} \cdot R_{dAFE} + L_{dAFE} \cdot \frac{di_{dAFE}}{dt} - \omega_g \cdot L_{dAFE} \cdot i_{qAFE} \\ 0 = u_{qAFE} + i_{qAFE} \cdot R_{qAFE} + L_{qAFE} \cdot \frac{di_{qAFE}}{dt} + \omega_g \cdot L_{qAFE} \cdot i_{dAFE} \\ u_{dqAFE} = u_{dc1} \cdot F_{dq1AFE} + u_{dc2} \cdot F_{dq2AFE} \\ u_{dqVSI} = u_{dc1} \cdot F_{dq1VSI} + u_{dc2} \cdot F_{dq2VSI} \\ u_{dc1} = \frac{3}{2} \cdot \frac{\left(F_{d1AFE} \cdot i_{dAFE} + F_{q1AFE} \cdot i_{qAFE}\right) - \left(F_{d1VSI} \cdot i_{dVSI} + F_{q1VSI} \cdot i_{qVSI}\right)}{C_{dc1} \cdot p} \\ u_{dc2} = \frac{3}{2} \cdot \frac{\left(F_{d2AFE} \cdot i_{dAFE} + F_{q2AFE} \cdot i_{qAFE}\right) - \left(F_{d2VSI} \cdot i_{dVSI} + F_{q2VSI} \cdot i_{qVSI}\right)}{C_{dc2} \cdot p} \\ P_g = \frac{3}{2} \cdot u_{dg} \cdot i_{dAFE} \\ Q_g = -\frac{3}{2} \cdot u_{dg} \cdot i_{qAFE} \\ S_g = \frac{3}{2} \cdot u_{dg} \cdot \sqrt{i_{dAFE}^2 + i_{qAFE}^2} \end{cases} \tag{39}$$

where u_{dqg}, u_{dqAFE} and u_{dqVSI} are the grid phase voltages; i_{dqAFE} and i_{dqVSI} are the phase currents; R_{dqAFE}, L_{dqAFE} are the active inductive resistances on the AFE side; ω_g is the angular grid voltage frequency; $F_{dq12AFE}$ and $F_{dq12VSI}$ are the switching functions.

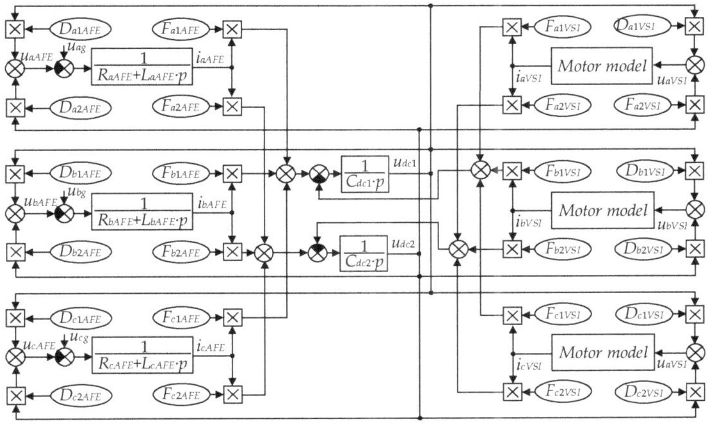

Figure 6. AFE voltage-oriented control block diagram in the *abc* reference frame.

Figure 7 shows the block diagram of the 3L-NPC-AFE-VSI's logical–mathematical model in the *dq*0 reference frame.

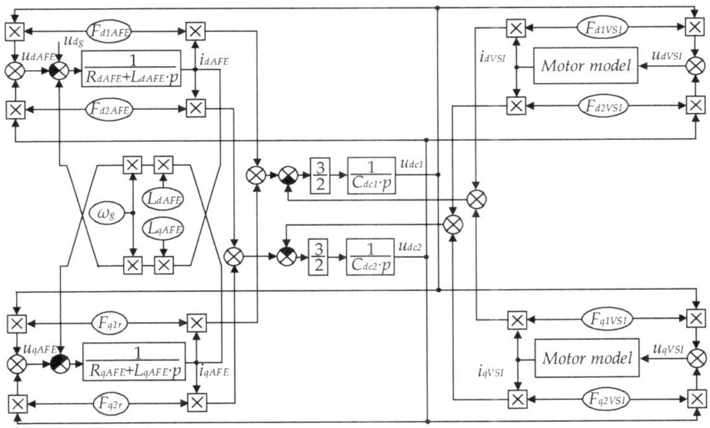

Figure 7. AFE voltage-oriented control block diagram in the *dq*0 rotating reference frame.

The system of Equation (39) shows that the reactive power consumed is fully determined by the reactive current i_{qAFE}. Emerging i_{qAFE} will cause a sharp increase in the voltage drop across the inductive resistance and, therefore, the need to increase the output voltage E_{AFE} by the voltage drop value $X_L i_q$. The values of the maximum generated reactive power and current can be determined from the system of Equations (39) under the condition $i_d = 0$ (STATCOM mode) as follows:

$$\begin{aligned} Q_{g\max} &= -\frac{3}{2}\cdot\left(\frac{u_{dg}\cdot m\cdot U_{dc}\cdot\frac{2}{\pi}-u_{dg}^2}{X_L}\right) \\ i_{qAFE\max} &= -\frac{2\cdot Q_{g\max}}{3\cdot u_{dg}} \end{aligned} \tag{40}$$

The Equations system (39) shows that with an increase in the i_{dAFE} current, the maximum reactive power will decrease. This is because the maximum apparent power consumption of the semiconductor converter is limited by the maximum active power consumption of the SMD ($S_{gmax} \approx P_{gmax}$). The reactive power limit is defined as follows:

$$Q_g \leq \sqrt{S_{g\max}^2 - P_g^2}. \tag{41}$$

This limitation in the AFE control system is implemented using the dependent current limitation according to the following formula:

$$i_q \leq \sqrt{i_{dAFE\max}^2 - i_{dAFE}^2}. \tag{42}$$

Figure 8 shows an AFE control system for reactive power compensation comprising two control loops. The inner current circuit is implemented in the $dq0$ reference frame, oriented to the grid phase voltage (voltage-oriented control (VOC)) [28–30]. The i_{dAFE} current provides DC voltage control. The i_{qAFE} current is used to control reactive power. The circuit includes the following items: input AFE phase current i_{abcmeg} and voltage u_{abcmeg}; grid angle θ; references i_{dqref} and measured i_{dqmeg} phase currents in the $dq0$ coordinates; references $u_{dqAFEref}$ and measured instantaneous phase voltage u_{dqmeg} in the $dq0$ coordinates; reference u_{dcref} and measured u_{dcmeg} DC voltage; and selective harmonic elimination pulse width modulation (SHEPWM). The AFE control system calculates the modulation index m and the shift angle α as follows:

$$m = \frac{\pi \cdot \sqrt{u_{dAFE}^2 + u_{qAFE}^2}}{2 \cdot U_{dc}}, \tag{43}$$

$$\alpha = arctg\left(\frac{u_{qAFE}}{u_{dAFE}}\right). \tag{44}$$

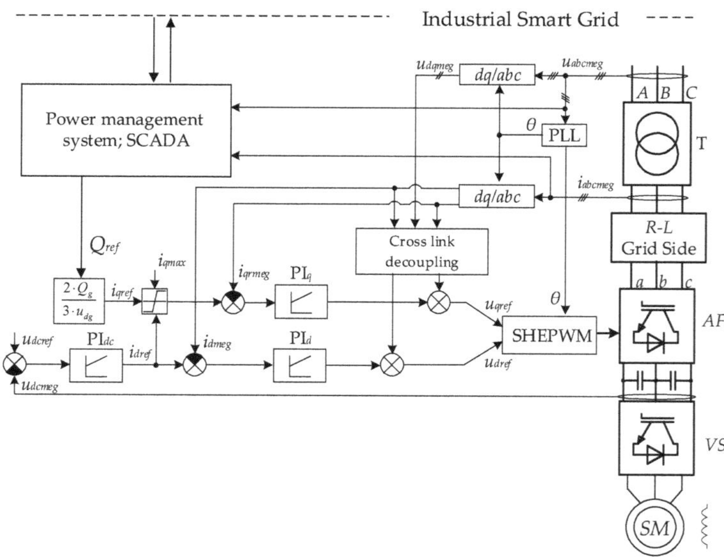

Figure 8. AFE voltage-oriented control block diagram.

Of great promise is AFE control with dependent current limitation to compensate reactive power in a network as part of an industrial smart grid. When implementing reactive power compensation modes, the precalculation or real-time monitoring of reactive power in the grid is required to produce a command for the generation or consumption of reactive power Q_{ref} at the local AFE control level.

5. Experimental Studies

5.1. Laboratory Studies

Validation of mathematical models was carried out at a power conversion equipment laboratory. A photo and the equipment composition are presented in Figure 9. The specifications of the laboratory prototype and the testing system are presented in Table 1.

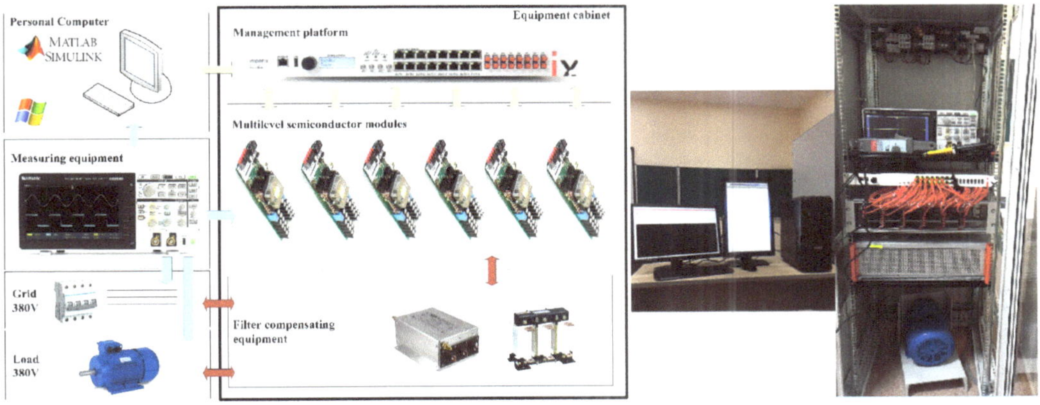

Figure 9. The laboratory equipment.

Table 1. Equipment parameters.

Power modules	1 × P924F33 Vincotech; reverse voltage IGBT, 600 V; permissible continuous current IGBT, 30 A; admissible continuous current of the reverse diode, 27 A; maximum switching frequency IGBT, 50 kHz; voltage drop, 1.5–2 V
Capacitors	2 × 517 μF Panasonic EEU-EE2W470S (two batteries of 11 × 47 μF each); maximum voltage, 450 V
Control drivers	4 × Avago ACPL-P345
Current sensor	1 × LEM HLSR 20-P/SP33; nominal range, ±20 A, 450 kHz; instrument error, ±1%
Voltage sensor	1 × resistive voltage divider + Avago ACPL-C87B; bandwidth, 25 kHz; instrument error, ±0.1%
FPGA	1 × Xilinx XC9536XL-10VQG44C, 10 ns; 36 microelements
Microcontrollers	2 × Microchip PIC24F04KA201, 16 bit, 16 MHz; 9 × 10-bit ADCs; sampling rate, 500 ksps

Direct access to the control signal of the semiconductor modules was gained through optic fiber inputs. Then, the galvanically isolated voltage probes of the DC and AC links at the converter input were connected to analog outputs. Overvoltage, overcurrent and overheating protection were built into each converter's NPC cells, and the converters were programmed by the researchers using a built-in microcontroller and a complex programmable logic device (CPLD). The mezzanine module was designed to transmit signals from the optic inputs and the analog outputs between the CPLD and the BoomBox control platform, including a digital signal processor (DSP) and a field-programmable gate array (FPGA). The converter control tasks were distributed between the DSP and the FPGA. The FPGA controlled the correct distribution of the IGBT control impulses, and the DSP processed the converter control system program. The computing core of the DSP was a 32-bit TMS320C28346 processor manufactured by Texas Instruments with a clock rate of 300 MHz and featuring 256 MB NOR flash memory, 2 MB RAM and 300 MIPS (with a floating point).

To take an oscillogram, Windows^{TM}-based graphic software of the BoomBox control platform was used. The probes were connected to the BoomBox control platform using the plug-and-play method and received directly from the control platform. The output signal was transmitted over a screened twisted pair of probes integrated into RJ45 cables. The program code for the control platform was developed in the MATLAB/Simulink. A fragment of the developed model is shown in Figure 10.

The results of experimental studies of the AFE reactive power generation using the laboratory equipment are presented below. Figure 11 shows an experiment with three reference reactive power generations of 9 kVAr when active power consumption on the load side was from 0 to 8 kW. The AFE control system provided the set with the maximum apparent power of 10 kVA. The measured DC link voltage was constant, 600 V. As noted earlier, if power consumption from the load side increases, the AFE control system reduces the reactive current reference and, as a result, the generated reactive power.

The next experiment with the active power regenerative mode from 0 to 8 kW is shown in Figure 12. As in the previous experiment, the level of generated reactive power decreased with an increase in the regenerative active power, and the apparent power in the system was controlled by the AFE at the reference value of 10 kVA.

The modulation index in both experiments varied from 1.1 at the maximum reactive power generation of 10 kVAR to 1.0 at the minimum 2 kVAR. Thus, the results of the studies performed using the laboratory equipment confirmed the adequacy of the obtained constraints (41) and (42) when using an AFE as a reactive power compensator.

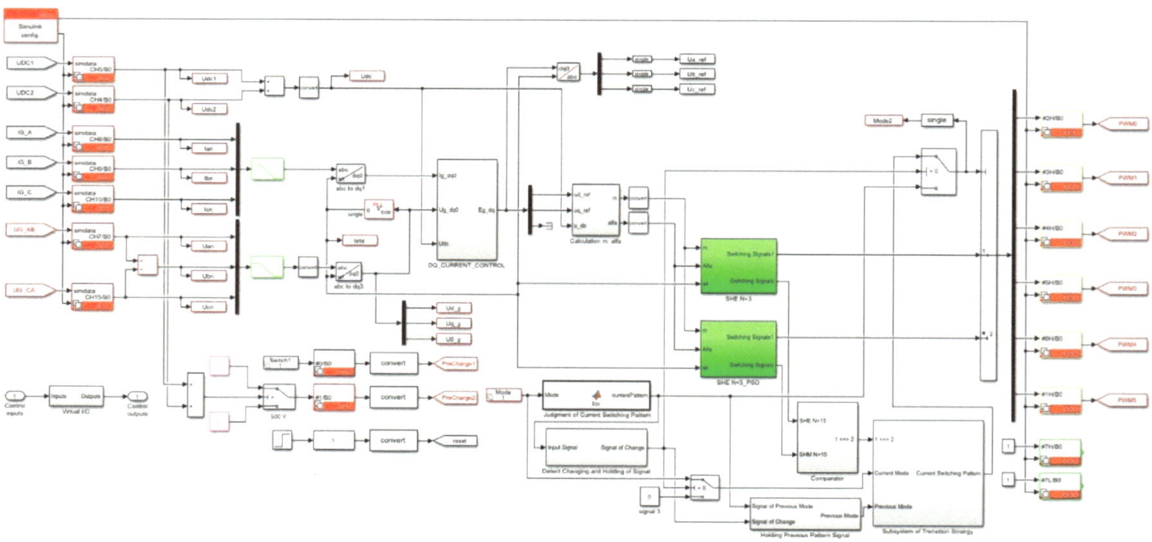

Figure 10. Part of the mathematical model of the object (label * is default data port).

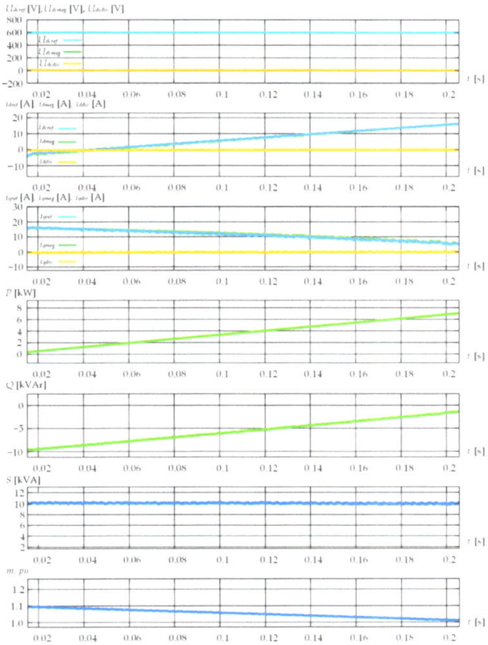

Figure 11. AFE reactive power generation mode with active power consumption.

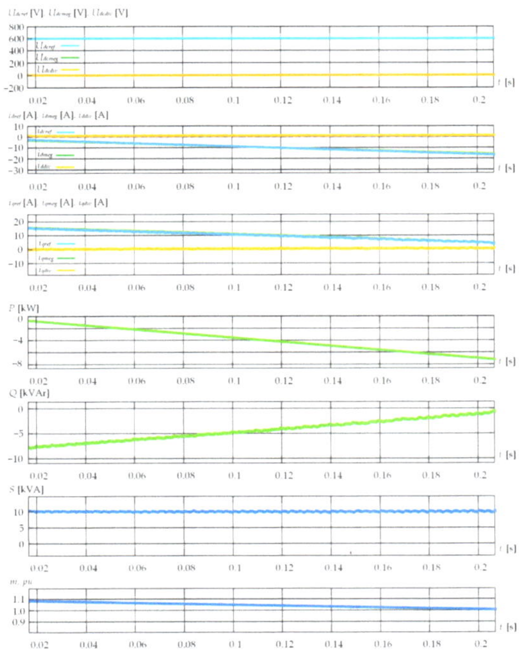

Figure 12. AFE reactive power generation mode with active power regeneration.

5.2. Industrial Experiment

The main drive of a plate mill rolling stand was investigated as the experimental object to verify the adequacy of the developed logical–mathematical models and the proposed reactive power compensation technique. An industrial drive is a system in which a DYMMZ 8025-8V synchronous motor with a rated power of 12 MW of each working roll is connected to three parallel-connected converters based on an AFE (PWM with the SHE algorithm) and a VSI (SVPWM algorithm). The object's electrical circuit is shown in Figure 13 while the basic specifications of the motor are provided in Table 2, and those of the converter in Table 3.

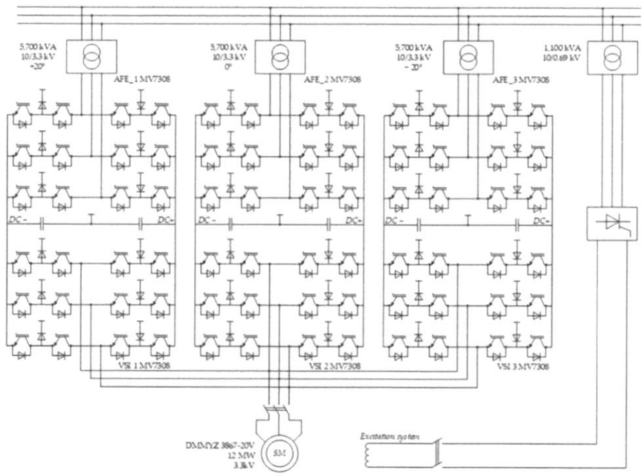

Figure 13. Powerful regenerative controlled-speed synchronous motor drive power diagram.

Table 2. Initial motor specifications.

Parameter	Designation	Value
Rated power	P_2, MW	12
Rated stator current	I_s, A	3600
Rated stator voltage	U_s, kV	3.3
Rated frequency	f_s, Hz	10
Stator's active resistance	R_s, Ohm	0.167
Rotor's active resistance	R_r, Ohm	0.066
Inductive reactance in the D-axis	X_d, Ohm	0.12
Inductive reactance in the Q-axis	X_q, Ohm	0.113
Overload capacity		3

Table 3. Initial converter specifications.

Parameter	Designation	Value
Maximum active power	P_{AFE}, MW	12
Rated current	I_{AFE}, A	2400
Rated voltage	U_{AFE}, kV	3.3
Rated frequency	f_g, Hz	50
DC voltage	U_{dc}, V	5020
Modulation index	$p.u.$	0.85
AFE input inductance	mH	7.5
DC link capacity	μF	6170
Overload capacity		1.2

Based on the initial object parameters, the maximum reactive power of 80 kVAR was found using Equation (40) with a modulation index of 1. When increasing the modulation index to 1.1, the maximum reactive power would increase sharply to 2.5 MVAR. Generating reactive power above these values without increasing the DC link voltage becomes problematic. An increased modulation index may cause unstable operation of the AFE due to voltage drops in the DC link when an abruptly variable load emerges on the SMD side. Increasing the DC link voltage also deteriorates the operating conditions of the SMD and semiconductor modules. Thus, to generate reactive power, the SMD's idle mode and roughing-down conditions were used. Figure 14 shows the SMD operation load diagram for mastering the reactive power generation mode. The modulation index was set to 1.1 where the total generated reactive power was 6 MVAR. Figure 15 shows the active, reactive and apparent power measurement results.

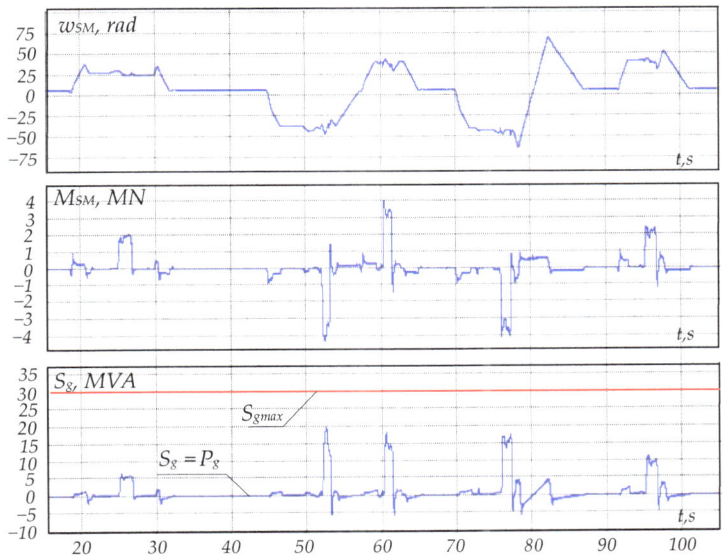

Figure 14. SMD operation load diagram during the roughing-down.

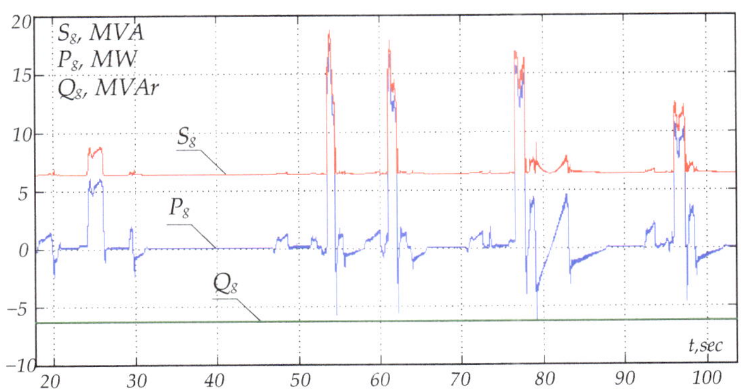

Figure 15. Reactive power generation by a 3L-NPC-AFE at 15% of the rated SMD power during the roughing-down.

6. Conclusions

This paper reviews the possibilities of reactive power compensation and power quality improvement in the grid using a high-power 3L-NPC-AFE-VSI-SMD. The analysis of the obtained results led to the following conclusions:

1. The review of previous studies revealed underestimated limitations when implementing reactive power compensation modes using a powerful regenerative controlled-speed synchronous motor drive based on a 3L-NPC-AFE-VSI;
2. Industrial 3L-NPC-AFE-VSI-SMD systems are developed without considering the possibility of reactive power compensation in the grid. Active power control is a priority; furthermore, it significantly limits the reactive power generation range;
3. Increasing the DC link voltage and/or the 3L-NPC-AFE modulation index allows for the increasing and controlling of the compensated reactive power. This paper proposes formulas for determining the maximum reactive power and current;
4. Experimental studies at a rolling mill showed that with the recommended modulation index and DC link voltage values, it was virtually impossible to generate reactive power. It was proposed to increase the modulation index to 1.1 during low-load SDM operation modes and at idle, which allowed generating reactive power at a level of 15% of the rated active power;
5. The proposed 3L-NPC-AFE reactive power control technique with a dependent current limitation can be used as part of an industrial smart grid to reduce the reactive power at the boundary.

Author Contributions: Conceptualization, A.S.M.; methodology, A.S.M.; software, T.A.L.; validation, A.A.N.; formal analysis, A.S.M. and A.A.N.; investigation, A.S.M. and T.A.L.; data curation, A.S.M. and T.A.L.; writing—review and editing, A.S.M.; visualization, A.S.M. All authors have read and agreed to the published version of the manuscript.

Funding: This work was financially supported by Moscow Polytechnic University under a grant named after Pyotr Kapitsa.

Institutional Review Board Statement: Not applicable.

Informed Consent Statement: Not applicable.

Data Availability Statement: Not applicable.

Conflicts of Interest: The authors declare no conflict of interest.

References

1. Abu-Rub, H.; Bayhan, S.; Moinoddin, S.; Malinowski, M.; Guzinski, J. Medium-Voltage Drives: Challenges and existing technology. *IEEE Power Electron. Mag.* **2016**, *3*, 29–41. [CrossRef]
2. Rajesh, D.; Ravikumar, D.; Bharadwaj, S.K.; Vastav, B.K.S. Design and control of digital DC drives in steel rolling mills. In Proceedings of the International Conference on Inventive Computation Technologies (ICICT), Coimbatore, India, 26–27 August 2016. [CrossRef]
3. Jing, T.; Maklakov, A.S. A Review of Voltage Source Converters for Energy Applications. In Proceedings of the International Ural Conference on Green Energy, Chelyabinsk, Russia, 4–6 October 2018. [CrossRef]
4. Steczek, M.; Jefimowski, W.; Szeląg, A. Application of Grasshopper Optimization Algorithm for Selective Harmonics Elimination in Low-Frequency Voltage Source Inverter. *Energies* **2020**, *13*, 6426. [CrossRef]
5. Al-Hitmi, M.; Ahmad, S.; Iqbal, A.; Padmanaban, S.; Ashraf, I. Selective Harmonic Elimination in a Wide Modulation Range Using Modified Newton–Raphson and Pattern Generation Methods for a Multilevel Inverter. *Energies* **2018**, *11*, 458. [CrossRef]
6. Orcajo, G.A.; Diez, J.R.; Cano, J.M.; Norniella, J.G.; Pedrayes González, J.F.; Rojas, C.H. Coordinated Management of Electrical Energy in a Steelworks and a Wind Farm. *IEEE Trans. Ind. Appl.* **2022**, *58*, 5488–5502. [CrossRef]
7. Alonso Orcajo, G.; Rodríguez, D.J.; Cano, J.M.; Norniella, J.G.; Ardura, G.P.; Llera, T.R.; Cifrián, R.D. Retrofit of a Hot Rolling Mill Plant with Three-Level Active Front End Drives. *IEEE Trans. Ind. Appl.* **2018**, *54*, 2964–2974. [CrossRef]
8. Orcajo, G.A.; Rodríguez, J.; Ardura, P.; Cano, J.M.; Norniella, J.G.; Llera, R.; Cifrián, D. Dynamic Estimation of Electrical Demand in Hot Rolling Mills. *IEEE Trans. Ind. Appl.* **2016**, *52*, 2714–2723. [CrossRef]
9. Pandit, P.; Mazumdar, J.; May, T.; Koellner, W.G. Real-Time Power Quality Measurements from a Conventional AC Dragline. *IEEE Trans. Ind. Appl.* **2010**, *46*, 1755–1763. [CrossRef]
10. Dybko, M.A.; Brovanov, S.V.; Kharitonov, S.A. Mathematical Simulation Technique for Power Systems based on Diode-Clamped Multilevel VSC. In Proceedings of the Eurocon 2013, Zagreb, Croatia, 1–4 July 2013; pp. 941–948. [CrossRef]

11. Slezhanovskii, O.V.; Datskovsky, L.K.; Kuznetsov, I.S.; Lebedev, E.D.; Tarasenko, L.M. *Systems of Subordinate Regulation of AC Electric Drives with Valve Converters*; Energoatomizdat: Moscow, Russia, 1983.
12. Cheng, J.; Xu, T.; Chen, D.; Chen, G. Dynamic and steady state response analysis of selective harmonic elimination in high power inverters. *IEEE Access* **2021**, *9*, 75588–75598. [CrossRef]
13. Steczek, M.; Chudzik, P.; Szeląg, A. Combination of SHE- and SHM-PWM techniques for VSI DC-link current harmonics control in railway applications. *IEEE Trans. Ind. Electron.* **2017**, *64*, 7666–7678. [CrossRef]
14. Khattak, F.A.; Rehman, H.U. Improved Selective Harmonics Elimination Strategy for Multilevel Inverters with Optimal DC Values. In Proceedings of the International Conference on Emerging Power Technologies, Topi, Pakistan, 1–6 April 2021. [CrossRef]
15. Patil, S.D.; Kadu, A.; Dhabe, P. Improved Control Strategy for Harmonic Mitigation in Multilevel Inverter. In Proceedings of the 5th International Conference on Intelligent Computing and Control Systems, Madurai, India, 6–8 May 2021. [CrossRef]
16. Moeini, A.; Dabbaghjamanesh, M.; Kimball, J.W.; Zhang, J. Artificial neural networks for asymmetric selective harmonic current mitigation-PWM in active power filters to meet power quality standards. *IEEE Trans. Ind. Appl.* **2020**. [CrossRef]
17. de Carvalho Alves, M.D.; Leão e Silva Aquino, R.N.A. Digital implementation of SHE-PWM modulation on FPGA for a multilevel inverter. In Proceedings of the Simposio Brasileiro de Sistemas Eletricos (SBSE), Niteroi, Brazil, 1–6 May 2018. [CrossRef]
18. Chen, J.; Shen, Y.; Chen, J.; Bai, H.; Gong, C.; Wang, F. Evaluation on the Autoconfigured Multipulse AC/DC Rectifiers and Their Application in More Electric Aircrafts. *IEEE Trans. Transp. Electrif.* **2020**, *6*, 1721–1739. [CrossRef]
19. Padmanaban, S.; Dhanamjayulu, C.; Khan, B. Artificial neural network and Newton Raphson (ANN-NR) algorithm based selective harmonic elimination in cascaded multilevel inverter for PV applications. *IEEE Access* **2021**, *9*, 75058–75070. [CrossRef]
20. Radionov, A.A.; Gasiyarov, V.R.; Maklakov, A.S. Hybrid PWM on the basis of SVPWM and SHEPWM for VSI as part of 3L-BtB-NPC converter. In Proceedings of the IECON 43rd Annual Conference of the IEEE Industrial Electronics Society, Beijing, China, 29 October–1 November 2017; pp. 1232–1236. [CrossRef]
21. Nikolaev, A.A.; Gilemov, I.G.; Denisevich, A.S. Analysis of Influence of Frequency Converters with Active Rectifiers on the Power Quality in Internal Power Supply Systems of Industrial Enterprises. In Proceedings of the International Conference on Industrial Engineering, Applications and Manufacturing, Moscow, Russia, 1–4 May 2018.
22. Wu, B.; Narimani, M. *High-Power Converters and AC Drives*, 2nd ed.; Wiley-IEEE Press: Hoboken, NJ, USA, 2017.
23. Nikolaev, A.; Maklakov, A.; Bulanov, M.; Gilemov, I.; Denisevich, A.; Afanasev, M. Current Electromagnetic Compatibility Problems of High-Power Industrial Electric Drives with Active Front-End Rectifiers Connected to a 6–35 kV Power Grid: A Comprehensive Overview. *Energies* **2023**, *16*, 293. [CrossRef]
24. Holmes, D.G.; Lipo, T.A. *Pulse Width Modulation for Power Converters: Principles and Practice*, 1st ed.; Wiley-IEEE Press: Hoboken, NJ, USA, 2003.
25. Maklakov, A.S.; Jing, T.; Radionov, A.A.; Gasiyarov, V.R.; Lisovskaya, T.A. Finding the best programmable PWM pattern for three-level active front-ends at 18-pulse connection. *Machines* **2021**, *9*, 127. [CrossRef]
26. Nikolaev, A.A.; Gilemov, I.G.; Bulanov, M.V.; Kosmatov, V.I. Providing Electromagnetic Compatibility of High-Power Frequency Converters with Active Rectifiers at Internal Power Supply System of Cherepovets Steel Mill. In Proceedings of the International Scientific-Technical Conference Alternating Current Electric Drives, Ekaterinburg, Russia, 1–8 May 2021.
27. Marquez Alcaide, A.; Leon, J.I.; Laguna, M.; Gonzalez-Rodriguez, F.; Portillo, R.; Zafra-Ratia, E.; Vazquez, S.; Franquelo, L.G.; Bayhan, S.; Abu-Rub, H. Real-Time Selective Harmonic Mitigation Technique for Power Converters Based on the Exchange Market Algorithm. *Energies* **2020**, *13*, 1659. [CrossRef]
28. Martínez, C.; Rohten, J.; Garbarino, M.; Andreu, M.; Silva, J.; Baier, C.; Morales, R. Operating Region for AFE Configuration under Variable Frequency Grid. In Proceedings of the IEEE International Conference on Automation/XXIV Congress of the Chilean Association of Automatic Control (ICA-ACCA), Valparaíso, Chile, 22–26 March 2021; pp. 1–6. [CrossRef]
29. Espinosa, E.E.; Melin, P.E.; Garcés, H.O.; Baier, C.R.; Espinoza, J.R. Multicell AFE Rectifier Managed by Finite Control Set–Model Predictive Control. *IEEE Access* **2021**, *9*, 137782–137792. [CrossRef]
30. Chowdhury, M.M.R.; Sütő, Z. Validation of a real-time simulation model of a three-phase Active-Front-End (AFE) rectifier. In Proceedings of the 8th International Youth Conference on Energy (IYCE), Eger, Hungary, 6–9 July 2022; pp. 1–5. [CrossRef]

designs

MDPI

Review

Recent Advances, Development, and Impact of Using Phase Change Materials as Thermal Energy Storage in Different Solar Energy Systems: A Review

Farhan Lafta Rashid [1,*], Mudhar A. Al-Obaidi [2,3], Anmar Dulaimi [4,5,*], Haitham Y. Bahlol [3] and Ala Hasan [6,*]

1 Petroleum Engineering Department, College of Engineering, University of Karbala, Karbala 56001, Iraq
2 Technical Institute of Baquba, Middle Technical University, Baquba 32001, Iraq; dr.mudhar.alaubedy@mtu.edu.iq
3 Technical Instructor Training Institute, Middle Technical University, Baghdad 10074, Iraq; haitham_yaqout@mtu.edu.iq
4 College of Engineering, University of Warith Al-Anbiyaa, Karbala 56001, Iraq
5 School of Civil Engineering and Built Environment, Liverpool John Moores University, Liverpool L3 2ET, UK
6 VTT Technical Research Centre of Finland Ltd., P.O. Box 1000, FI-02044 Espoo, Finland
* Correspondence: farhan.lefta@uokerbala.edu.iq (F.L.R.); a.f.dulaimi@uowa.edu.iq (A.D.); ala.hasan@vtt.fi (A.H.)

Abstract: The efficient utilization of solar energy technology is significantly enhanced by the application of energy storage, which plays an essential role. Nowadays, a wide variety of applications deal with energy storage. Due to the intermittent nature of solar radiation, phase change materials are excellent options for use in several types of solar energy systems. This overview of the relevant literature thoroughly discusses the applications of phase change materials, including solar collectors, solar stills, solar ponds, solar air heaters, and solar chimneys. Despite the complexity of their availability and high costs, phase change materials are utilized in the majority of solar energy techniques because of the considerable technical improvements they provide. While numerous studies have investigated the progress of phase change materials used in solar energy applications such as photovoltaic systems, it is vital to understand the conceptual knowledge of employing phase change materials in various types of solar thermal energy systems. Investigations into the use of phase change materials in solar applications for the purpose of storing thermal energy are still being carried out to upgrade the overall performance. This paper briefly reviews recently published studies between 2016 and 2023 that utilized phase change materials as thermal energy storage in different solar energy systems by collecting more than 74 examples from the open literature. This study focuses on demonstrating the maturity of phase change materials and their integration into solar energy applications. Based on the findings, proposals for new research projects are made.

Keywords: phase change materials (PCMs); solar energy; energy storage; solar energy technology; energy storage system

Citation: Rashid, F.L.; Al-Obaidi, M.A.; Dulaimi, A.; Bahlol, H.Y.; Hasan, A. Recent Advances, Development, and Impact of Using Phase Change Materials as Thermal Energy Storage in Different Solar Energy Systems: A Review. *Designs* **2023**, *7*, 66. https://doi.org/10.3390/designs7030066

Academic Editors: Surender Reddy Salkuti and Dibin Zhu

Received: 12 April 2023
Revised: 4 May 2023
Accepted: 15 May 2023
Published: 17 May 2023

1. Introduction

Even though solar energy has been used passively since prehistoric days, its technical application emerged in 2013 [1]. The development of solar thermal collector equipment from 1877 to the current times has improved human living in terms of comfort and economics [2]. Solar energy is a time-dependent renewable source of energy that is inexhaustible and eco-friendly. Among the most often used applications of solar energy is for thermal purposes. On the other hand, thermal energy storage is an efficient way to combat the sporadic and diffuse character of solar energy [3]. Techniques for energy storage can work towards closing the gap between supply and demand for energy. Among the most feasible methods for storing solar energy involves the utilization of specific organic and inorganic substances, which are referred to as phase change materials (PCMs), which enable the latent heat of fusion to be harnessed [4]. To improve the thermal performance of solar heating systems, PCMs can be used as an effective tool. PCMs can effectively store additional thermal energy during the day through fusion and release it at sunset via solidification at a consistent temperature, which leads to a higher storage efficiency.

PCM has a number of benefits. Specifically, its high heat storage capabilities compared to other heat storage materials lead to a decrease in the size of the thermal storage facility [5]. The isothermal phase change feature of PCMs, wherein heat is charged and discharged at a virtually fixed temperature (phase change temperature) matching the phase transition, seems to be another significant benefit. As a consequence, they are perfect for applications involving space heating where thermal comfort can be attained by reducing temperature swings [6]. However, PCMs have a number of demerits such as PCM systems necessitating a long operational time to recuperate the installation cost [7]. Conventional PCMs also have other drawbacks such as low thermal stability, thermal cycling, and low thermal conductivity [8,9]. In addition, the super-cooling effect lowers the PCM's efficiency, leading to inadequate heat recovery, making it impossible to fix the PCM system without damaging it. To improve PCMs, several investigators have attempted to add various compounds; unfortunately, this has led to lower performance [10,11]. Prior to installation, the PCM's extremely poor thermal conductivity must be increased since it hinders heat transfer during the solid–liquid transition [12]. Phase segregation may happen in PCM because it may contain many elements, which would undermine its long-term stability [13]. Building envelope organic PCMs have a substantial impact on fire safety. In using organic PCMs in building envelopes, scientists have advised using fire retardants to increase fire safety [14,15].

Despite that there has been a number of attempts to review the implementation of PCMs in different solar energy systems, scholars focused on discussing the concepts of these materials with outlining the influential parameters towards its implications in the solar energy industry. For instance, Nazir et al. (2019) [16] focused on the influences of thermos-physical properties of PCMs such as the melting point, and density and thermal conductivity of the organic, and inorganic materials on their implication in thermal energy storage applications. Additionally, Sikiru et al. (2022) [17] carried out a comprehensive review that provided a thorough examination of the recent advances and impact of PCMs on solar energy systems. Specifically, the authors described how photovoltaic (PV) systems could be improved in terms of performance by integrating heat storage devices into the PV cell energy harvesting. Additionally, it covered the techniques for configuring thermal storage, evaluation of PCM type and hybrid performance, PCM entry requirements, and property testing, as well as advancements in heat and mass transfer fluids for solar energy harvesting and thermal storage mechanisms and their uses. However, both Nazir et al. (2019) [16] and Sikiru et al. (2022) [17] have not covered the potential applications of PCMs in other solar energy fields such as solar air heater, a solar pond, solar collector, solar still, and a solar chimney.

This study discusses the applications of PCMs in solar collectors, solar still, solar ponds, solar air heaters, and solar chimneys. Thoroughly, a variety of technical, research, and development approaches pertaining to PCMs are addressed. Additionally, a full grasp of the characteristics of PCM is outlined to be improved and utilized in a range of applications. The findings of the current investigation may guide future research since they will help scientists understand the numerous advancements made in PCM, which still require further advancement.

2. PCMs in Different Solar Energy Systems

The use of PCMs in solar thermal facilities is meant to boost the intensity of the heat that is stored as a result of the conversion of solar energy. The system elements in which PCMs are deployed are frequently identified in the literature to be traditional thermal energy storage devices, for example water tanks [18–20] or systems with a separate tank filled just with PCM [21]. The next sections illustrate the evaluation of the conducted studies on the utilization of PCMs in solar collector systems, solar stills systems, solar bond systems, solar air heater systems, and solar chimney systems. Several aspects will be discussed for each solar energy system including the configuration, studied parameters and highlighting of the most interesting results.

2.1. Solar-Collector-Related Studies

A solar collector system is a technology that harnesses solar power to produce heat or electricity using sun radiation. There are two main types of solar collectors: photovoltaic (PV) (to generate electricity) and thermal (to heat water or air) collectors. A flat plate or a set of tubes that absorb sunlight and transmit the heat to a fluid or air that is pumped through the system make up the fundamental structure of a thermal solar collector [22]. The amount of sunlight available, the angle and orientation of the collector, the type of fluid utilized, and the temperature of the air or water heated, and the system's insulation all have an impact on how well a solar collector performs. By choosing the right collector design and size, maximizing the orientation and tilt of the collectors to maximize solar radiation, using premium building materials, and limiting heat losses through adequate insulation, enhancing the effectiveness of a solar collector system is attainable [23]. To guarantee optimum efficiency and the system's longevity, proper system maintenance, including routine collector cleaning of the panels to remove dust or debris is essential [24]. Solar collector systems can offer a clean and sustainable energy source for a number of uses, including electricity generation, hot water production, and residential and commercial heating, with proper design and implementation. Overall, solar collectors can decrease carbon emissions and energy expenses while also raising a property's value.

This section discusses the recent studies that focused on the utilization of PCMs as thermal energy storage in solar collectors while highlighting the benefits of PCMs, which undoubtedly have influenced their analysis and conclusions.

A relatively new proposal for an evacuated tube collector/storage (ETC/S) system including a PCM was proposed in 2016 by Feliński and Sekret [25]. Experiments were carried out to investigate the impact of incorporating commercial-grade paraffin as a PCM in evacuated tubes with heat pipes, with the aim of enhancing the thermal performance of ETC/S systems. The melting temperatures of the hydrocarbons that make up paraffin wax typically range from 47 to 68 °C. The precise melting point of paraffin wax might change depending on the unique makeup of the paraffin wax. Furthermore, the total amount of useable heat as a function of the presence of PCM in ETC/s. By allowing the recovery of heat that had been stored throughout the discharge cycle, the employment of paraffin has augmented the solar thermal operational time. Clearly, it was possible to raise the heating temperature of the medium to a practical level (approximately 45 °C). The findings also demonstrated that the paraffin-combined ETC/S system produced a total amount of useable heat that raised by 45–79% if compared to an ETC (Figure 1). These findings were reported throughout the discharge cycle considering the mass flow rate of the heating fluid.

In 2016, a numerical analysis carried out using an Artificial Neural Networks model was provided by Serale et al. [26]. These simulations were carried out to evaluate the code and get a rough idea of how well the unique notion performed. To assess the efficacy of the slurry phase change materials (PCS) (this is a suspended material of micro-encapsulated phase change material (n-eicosane) in water or in a carrier fluid of water and glycol), and contrast it with that of a conformist solar thermal collector, several (dynamic) boundary conditions including the PCM concentration, location, and orientation were used based on based a flat-plate solar thermal collector. In this regard, solar energy extraction was utilized as a function of the use of slurry phase change materials. The results of the simulations have demonstrated the resilience of the model and its suitability for use in a recent investigation. Additionally, it was demonstrated that the use of PCS as a heat transfer fluid might boost solar energy extraction to varying degrees, depending on the region. Cold regions can get the biggest benefits.

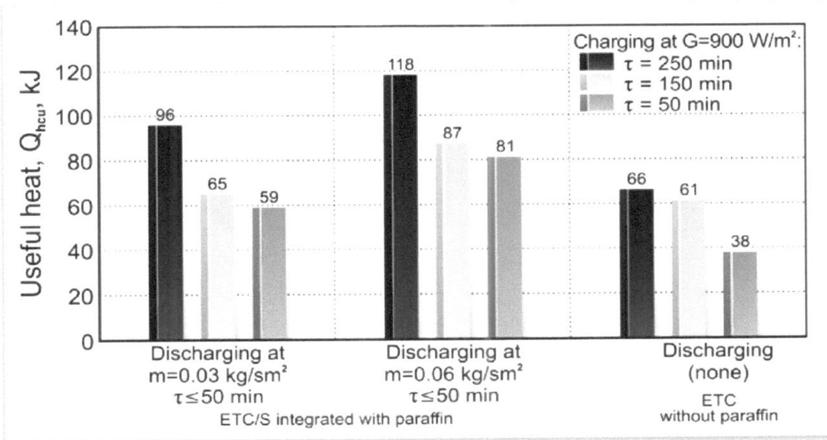

Figure 1. Useful heat is received by the heat collection unit throughout the charge and discharge cycles [25].

In 2017, Li and Zhai [27] presented a solar collector designed for mid-temperature operations. The PCM in this system, which has a melting temperature of 119 °C and is composed of erythritol and expanding graphite, was put within the aluminum pipes before being located inside the evacuated tubes. To conduct heat from PCM to fluid, the heat pipes were occupied in PCM. Based on thermos-physical property tests, the composite PCM with 3 wt% expanded graphite was selected as a good material for heat storage. It should be noted that heat storage was utilized as a function of tube diameter in this study. The thermal performance of this system was then investigated using both the experimental outfit and the computational model. The exposure tests were performed in August, and the analysis revealed that for mid-temperature consumption, the storage performance extended to 40.17%, according to Figure 2. The experimental data also served to verify the numerical model. For heat storage, a phase transition material composed of 3 weight percent expanded graphite and 97 weight percent erythritol is employed. Latent heat, thermal conductivity, and melting point are all 118.69 °C, 312.2 J/g, and 2.4 W/m K, respectively. In comparison to pure erythritol, the addition of expanded graphite improves thermal conductivity by 241.4%.

A hybrid slurry made of microencapsulated phase change materials (MPCMs) of paraffin@melamine resin mixed in an ethanol/water mixture-based multi-walled carbon nanotube (MWCNT) nanofluid was developed and testified by Wang et al. (2019) [28]. Temperature changes in PCM in the cases of charging and discharging were identified as a function of concentration. The study found that when PCM melted and the surface of MPCM exhibited high reflectivity, the temperature rise of the hybrid MPCM-MWCNT slurry was lower than that of pure MWCNT nanofluid in the base fluid (an ethanol/water mixture). However, the final temperatures of the hybrid slurries were higher than those of the MWCNT nanofluid during the thermal discharging phase and increased with the amount of MPCM present. The hybrid slurry comprising 15 wt% MPCMs had a comparable terminal temperature that was approximately 4.6 °C greater than the base fluid, as illustrated in Figure 3.

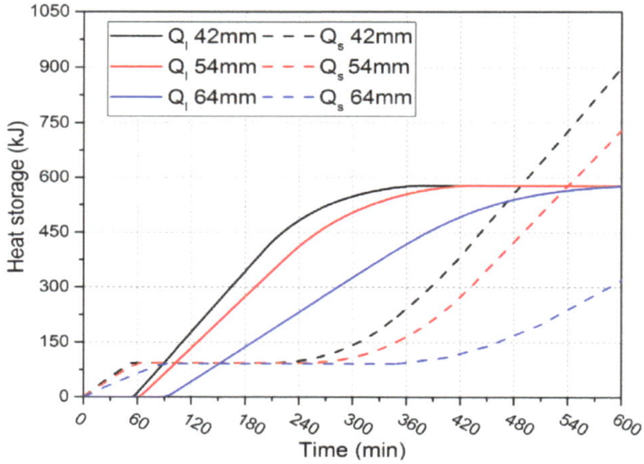

Figure 2. Changes in heat storage with several storage tube diameters [27].

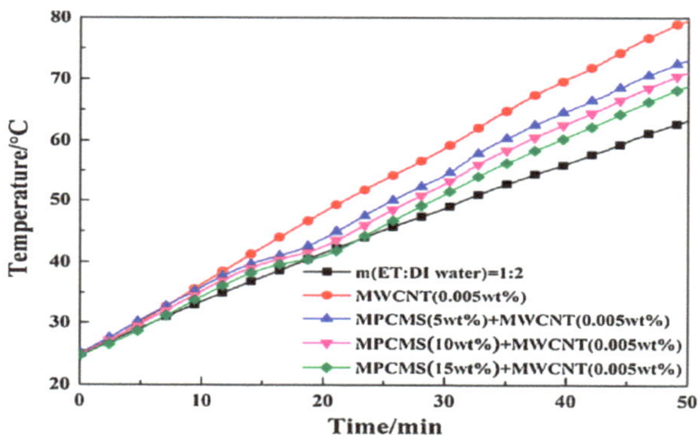

(a)

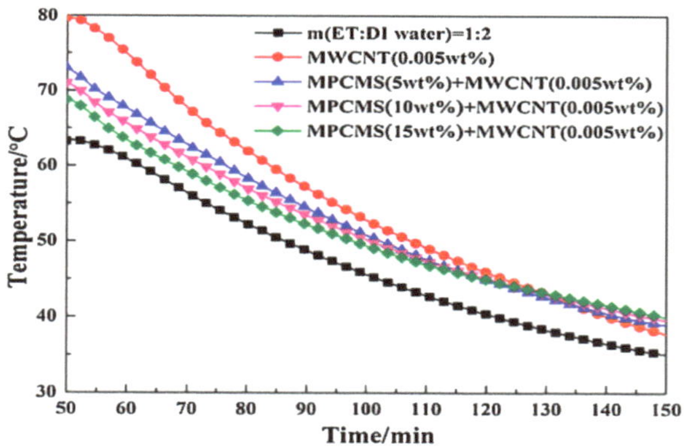

(b)

Figure 3. Temperature changes in MPCM-MWCNT slurries at two stages of (**a**) charging and (**b**) discharging with various concentrations [28].

To overcome the phenomenon of overheating and freezing, a brand-new dual-PCM for flat plate solar collectors was suggested by Wang et al. (2019) [29]. Spherical particles were used to encapsulate the PCM, which comprised 20% high-purity graphite and 80% natural grease. The solar collector contained two layers of PCMs, one with 70 °C of phase change temperature (PCT) and the other with 15 °C of PCT. In other words, the PCM liquefies and absorbs heat when exposed to high temperatures, such as when the water temperature raises the PCM's high melting temperature above 70 °C. This helps prevent the working fluid from becoming overheated. When the low-melting PCM reaches a temperature of less than 15 °C in a low-temperature situation, the PCM solidifies and releases heat to prevent the working fluid from freezing. Furthermore, this study focused on considering the thermal efficiency of the solar collector as a function of inlet collector temperature. They were positioned in the dual-PCM collector's area beneath the absorber plate. According to the test results, high-temperature conditions can prolong the time it takes for the absorber plate temperature to rise from 60 to 78 °C by 1.6 h. Further, the dual-PCM collector can be employed to prevent freezing and overheating. Figure 4 shows that the performance of the dual-PCM collector was elevated by 24.1% and 19.6%, respectively, when low-melting PCM was positioned beneath high-melting PCM and the contrary situation.

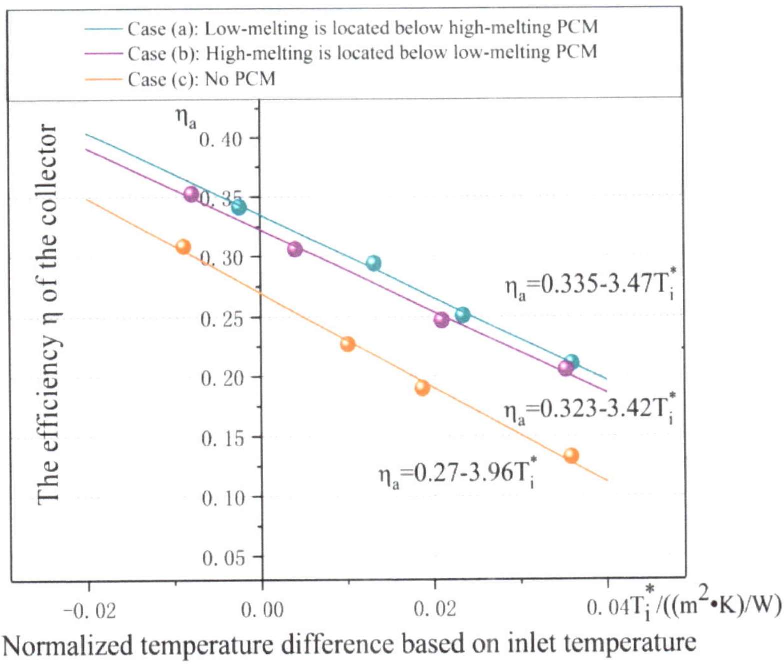

Figure 4. Proficiencies of the PCM collector and conservative collector [29] ($T_i^* = \frac{(t_i - t_a)}{I}$ is the normalized temperature difference based on the inlet temperature ((m^2·K)/W); t_a is the ambient temperature (°C), I is the solar radiation intensity.)

In situ, polymerization was used by Gao et al. (2020) [30] to create the microcapsule, which had an octadecane core and a melamine-formaldehyde (MF) resin shell that had been amended with a reduced form of commercial graphene oxide (rGO) and oleic acid-coated Fe_3O_4 magnetic nanoparticles in the shell (OA-MNs). In this regard, it is vital to admit that the thermal conductivity of PCM was incorporated as a function of PCM type. According to the experimental findings, introducing rGO significantly improved the thermal conductivity of magnetic PCMs while having no effect on their phase change activity, as shown in Figure 5. In the meantime, the MF resin shell that functioned as a protective shield significantly increased the thermal stability of octadecane in rGO-MPCMs. The photo-thermal conversion performance was further improved by the rGO-MPCMs slurry's superior optical absorption and thermal storage capabilities compared to MPCMs slurry and deionized water. The rGO-MPCMs in the slurry showed remarkable stability and renewability in the presence of an external magnetic field.

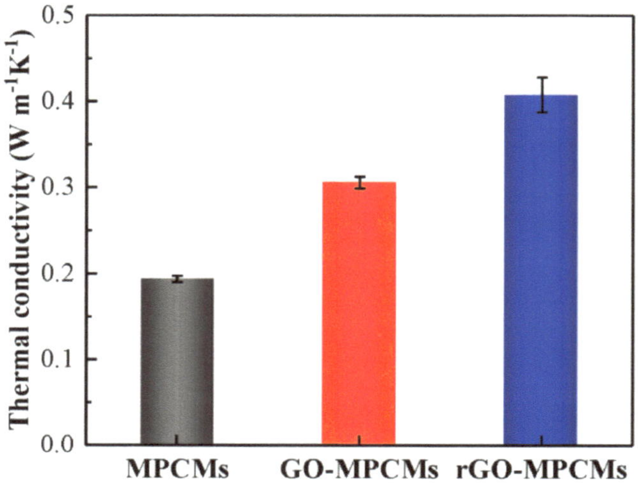

Figure 5. Thermal conductivities of bulk MPCMs, GO-MPCMs and rGO-MPCMs [30].

A detailed comparative examination of a conventional flat plate solar collector (FPSC) and a prototype with an equivalent thermal storage system via PCM was testified by Palacio et al. in 2020 [31]. Experimental and quantitative methods were used to study the devices' thermal performance. An experimental setup measured the temperature of the glass cover, absorber plate, and PCM, in addition to the temperature of the ambient air and the water at the inlet and outflow. Palacio et al. in 2020 [31] introduced the thermal collector efficiency as a function of the contract conduction between the absorber and PCM. With this approach, experimental findings were assessed in order to determine how well both collectors performed under identical operating circumstances. Additionally, several trial scenarios were taken into account to enhance the prototype's thermal efficiency. Analysis was using on two different PCMs (PCM1: Ecopetrol Semirefinada of 60 °C melting point, and PCM2: Rubitherm RT-47 of 41 °C fusion temperature), a contact thermal conduction increase, and collector disposition. According to this study, the contact conduction between the absorber and PCM and the selecting of the PCM are crucial components in enhancing the collector efficacy in comparison to traditional FPSC. Tilting the collector did not significantly improve the latitude of the laboratory investigation.

An innovative evacuated tube solar collector (ETSC) featuring a nano-PCM with fins was proposed and investigated by Elarem et al. 2021 [32]. The effectiveness was examined in relation to the adding of copper (Cu) nanoparticles to paraffin wax. Two PCMs were prepared and tested (PCM1: paraffin wax + 1% nano-Cu, and PCM2: paraffin wax + 2% nano-Cu). The simulation of heat transport through the energy storage process was carried out using ANSYS Fluent (Note: ANSYS is a software that provides a range of simulation capabilities, including structural analysis, fluid dynamics, electromagnetics, and system-level simulations). Last but not least, a set of comparative simulations were run to see how well the ETSC performs with and without the solar parabolic trough reflector (SPTR). In this study, the melting of PCM was identified as a function of fin thickness. It has been concluded that the paraffin phase shift heat transmission in the ETSC was significantly affected by the addition of fins. The PCM melts more rapidly as a result of lowering the fin thickness. Additionally, adding 1% Cu to the PCM was the appropriate mass concentration for raising the heat transfer fluid output temperature by 2 K, as shown in Figure 6. Additionally, it was found that a final flow rate of 0.003 kg/s was necessary for the PCM's entire mass to melt. In reality, Nano-PCM can achieve a greater temperature, although briefly. The heat transfer coefficient between the copper tube and the HTF is improved by including Cu nanoparticles into the PCM. The reason for this is because the Nano-PCM filled Evacuated Tube Solar Collector (ETSC) shows improved energy storage during charging, allowing for a longer duration of hot HTF supply during discharging when solar radiation is at its peak.

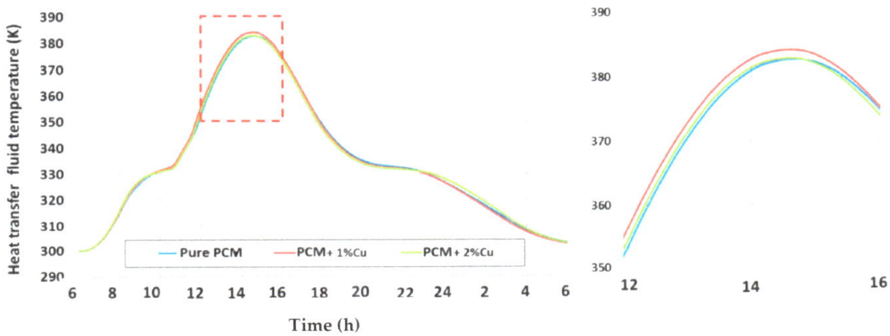

Figure 6. Influence of adding Cu nanoparticles on the heat transfer fluid temperature [32].

The performance of a heat pipe evacuated tube collector (HPETC) with a new approach, which included either one or both evacuated tubes (ET) and two separate storage tanks that incorporated nano- or micro-enhanced PCMs, was analyzed by Alshukri et al. (2021) [33]. The two separate tanks were filled with improved paraffin wax in the suggested designs, whereas ET was occupied with improved medical paraffin wax (grade-A). These two forms of paraffin wax were improved with either ZnO micro-particles or 5 weight percent CuO nanoparticles (MP-ZnO). In this regard, the collector efficiency was expressed as a function of micro or nanoparticle type. The outcomes demonstrated that the combination of improved paraffin wax with NP-CuO and MP-ZnO yields an efficacy enhancement range of 36.8–50% and 25.43–41.4% correspondingly in both ET and the isolated tanks. In contrast to a standard reference collector without PCM, the combination of improved paraffin wax with NP-CuO and MP-ZnO in the only ET increased efficacy with a range of 33.8–45.7% and 23.8–26.7%, respectively, as illustrated in Figure 7.

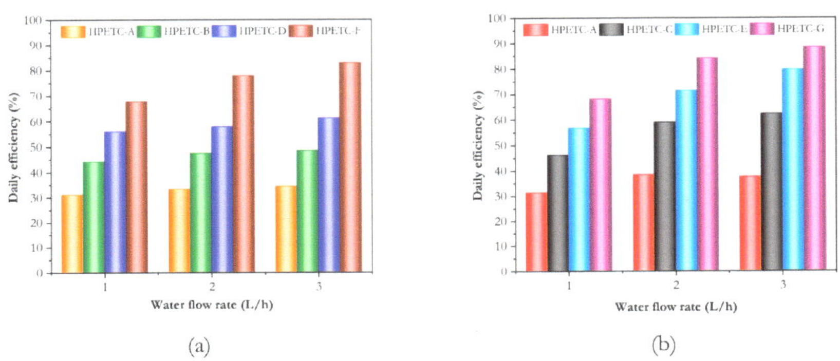

Figure 7. The daily thermal efficiency at three various flow rates for (**a**) tests 1–3 and (**b**) tests 4–6 [33].

To overcome the difficulty of solar collectors to immediately store solar energy, Sadeghi et al. (2022) [34] effectively utilized a shape-stabilized phase change material (SSPCM) in a tankless direct-absorption evacuated tube solar collector. The SSPCM utilized in this study consisted of sodium acetate trihydrate (SAT), carboxymethyl-cellulose (CMC), sodium phosphate monohydrate (SPM), and expandable graphite. In this regard, expandable graphite was used to provide the PCM with a shape-stability feature in addition to improving its thermal conductivity. Sadeghi et al. (2022) [34] demonstrated the collector heat gain as a function of flow rate. After determining the appropriate constituent concentrations, the suggested salt hydrate PCM was built utilizing a range of relevant component concentrations and tested through a number of cycles to ensure that it could consistently store heat. The solar system was further refilled in the stagnation mode and then released at flow rates of 10, 27, and 40 (liter per hour) LPH (without water flow). According to the results, the collector-storage system improved thermal efficiency in the stagnation mode from 66% to 82%. Additionally, increasing the flow rate from 10 to 27 LPH did not significantly diminish the heat gain of the collector; however, a decrease in the discharge efficiency occurred after employing a flow rate of 40 LPH.

Yeh et al. (2022) [35] explored a storage-integrated solar thermal collector that had a small footprint and used a compatible phase change material (PCM) for storage. In this study, sodium acetate trihydrate (SAT), a salt hydrate-based composite PCM, was utilized to create a shape-stabilized PCM (SSPCMs) due to its low leakage and excellent thermal conductivity properties. When the built SSPCMs underwent experimental testing, a numerical model was designed to compare the results with the data and estimate the developed SSPCMs' effective thermal characteristics. Furthermore, a model was used to investigate the (dis)charging behavior of a PCM in a storage-integrated solar thermal collector consisting of an evacuated tube and a double-spiral coil heat exchanger. In this regard, the discharging time was illustrated as a function of geometric optimization. With a minimal outlet temperature of 55 °C, a geometrical optimization has been performed to obtain a 2.6-fold longer discharging time. Furthermore, the circulating water is established to be supportive in rising the PCM charging rate by 9% by transporting heat from the surface of the tube to its center.

Nekoonam and Ghasempour (2022) [36] examined the thermal responses of a solar system with thermal energy storage. To address the PCM's low thermal conductivity for energy storage and increase the system's charging and storage efficiencies, a novel technique incorporating the use of sensible-latent materials (encapsulated spherical capsules of PCM and synthetic oil) as the ideal setup with two different options was used (option 1: outlet temperature of the unit with no temperature limitation; option 2: outlet temperature of the unit with threshold restriction). Charging time was considered in this study as a function of the Heat Transfer Fluid (HTF) flow rate. The findings indicated that a 5% rise (reduction) in the input HTF temperature reduced (increased) the charging time by 25% (32%) and enhanced (decreased) the total stored energy by 20% (by 7%). Moreover, a 50% change in the HTF flow rate causes a 12% gain or a 25% drop in charging time. For the first and second optimization scenarios, the best configuration stores 4940 and 4725 kWh of energy during the charge, which is approximately 14% more than the single PCM/sensible investigated situations.

Wen et al. (2022) [37] conducted a detailed performance assessment of an inventive concentrating system made up of a photovoltaic/thermal (PV/T) module with PCM and a solar thermal (ST) collector with thermoelectric generators (TEG) connected in a series. In this technology, the PV/T module produces low-temperature thermal energy in addition to photoelectric power generation, and the ST module and TEG on ST provide the secondary heated water and secondary thermoelectric output. Furthermore, The PV cell's operational temperature can be lowered by the PCM during the day, and the platform can avoid freezing in the winter by releasing heat at night. The electrical and thermal performance were demonstrated in this study as a function of the Fresnel lens. Theoretical analyses revealed that the PV/T-ST system, which consists of a sequence of the ST and T modules, accomplishes well in terms of thermal and overall performance. Throughout the course of the whole day of activity, the water tank in the PV/T-ST system has a maximum temperature that is approximately 5 °C higher than the PV/T-PV/T system. The electrical and thermal performances using the 1.1-m2 Fresnel lens are also 10.65% and 65.22%, respectively, at 12:30. TEG converts some thermal energy into electrical energy over the day, increasing the quantity of high-grade energy in the system by approximately 321.53 kJ.

By using molecular dynamics modelling, Hatamleh et al. (2022) [38] examined the thermal behavior of glass in the vicinity of PCM (decane). In their research, temperature and density profiles were used to examine the atomic behavior of the structure. Furthermore, heat flow and thermal conductivity were used to examine the simulated structure's thermal behavior. The quickly altering of PCM was expressed as a function of the design of the structure. According to the numerical findings, fluid particle density and temperature maximums are 0.09643 atoms/cm^3 and 1067.51 K, respectively. The structure exhibits a thermal conductivity of 0.74 W/m K and a heat flow of 650.27 W/m^2. Overall, incorporating phase change materials in the design can prevent energy loss, lower energy expenses, and facilitate rapid phase changes in the materials, positively impacting the structure's performance.

Algarni et al. (2020) [39] explored whether a novel evacuated tube solar collector (ETSC) concept that incorporates nano-enhanced PCM (Ne-PCM) (0.33 wt% nano-copper with paraffin wax composite) as a latent heat thermal energy storage system could improve the performance of a conservative evacuated tube collector (ETC). This innovative technology combines the U-pipe of the evacuated tube solar collector (ETSC) with the Ne-PCM to form a single unit (ETSC/Ne-PCM) that can generate heat even when solar intensity is limited or inaccessible. By heat accumulation and the thermal insulation of the evacuated tubes, the new system functions as a thermal promoter, extending the time it takes to produce hot water and reducing temperature swings. Additionally, the effectiveness of FTSC was articulated as a function of the weight percent of Cu/PCM. It was discovered that adding 0.33 weight percent of copper/PCM composite to the ETSC increased its effectiveness by 32%. Moreover, the ETSC/Ne-PCM system with 0.08 L/min can deliver hot water up to 50 °C for roughly 2 h longer than standard ETC systems.

Karami et al. (2023) [40] examined the thermal efficiency of CuO nanofluid, CuO/Al_2O_3 binary nanofluid, and CuO nanofluid mixed with micro-encapsulated phase transition material in direct absorption solar evacuated tube collectors (MPCM). To simulate the MPCM slurry, two models—single-phase and mixture—were employed. The single-phase model is advised to shorten calculation time owing to the negligible distinction between the two-phase model's results. The coefficient of performance was justified as a function of the volume % of nanoparticles. It was discovered that 0.06% Al_2O_3 + 0.002% CuO/water nanofluid and 0.06% Al_2O_3 + 0.002% CuO/(ethylene glycol) EG nanofluid, which are equivalent to 28.34 K and 36.64 K, respectively, provide the largest changes in outlet temperature. In both water- and EG-based nanofluids, the collector efficiency utilizing binary nanofluids is greater than that of using CuO nanofluids. According to the observations, the addition of MPCM and CuO nanofluid boosts the coefficient of performance by 4.53% and reduces heat loss by 5.84%. Across all working fluids, the combination of MPCM and CuO nanofluid simultaneously had the maximum accuracy.

A PCM of Tritriacontane paraffin is directly incorporated within the heat pipe evacuated tube solar collector (HPETC) and has copper (Cu) porous metal implanted in was developed by Pawar and Sobhansarbandi (2023) [41]. The performance of the suggested system was contrasted against that of a traditional HPETC system under comparable circumstances in order to make comparisons. Taking into account the distinction of fin temperature, output water temperature, energy, exergy efficacy, and cost analysis, each system's thermal performance was evaluated. The maximum temperature increase was identified as a function of PCM and Cu porous that fill the evacuated tube. The results demonstrated that the evacuated tube filled with PCM + Cu porous metal was able to reach a maximal temperature increase of approximately 21 °C when solar radiation was at its highest amount. The difference in fin temperature between the ordinary evacuated tube and the evacuated tube filled with PCM + Cu porous metal was observed to be 36.1 °C after sunset. Additionally, on 9–11 April, at 10 PM, the highest water outlet temperature differences of 8 °C, 10 °C, and 11 °C, respectively, were discovered between the two designs. The conventional system only recorded a daily energy efficiency maximum of 36.91%, but the suggested scheme showed a maximum energy efficiency of 85.64%.

Studies on phase transition materials for solar collectors are shown in Table 1. Specifically, Table 1 introduces details of configurations, type of study, studied parameters and highlights the most important findings for those studies approximately solar collectors.

Table 1 can show that the ETSC's efficacy was found to rise by 32% with the use of 0.33 weight percent of copper/PCM composite. The prototype with a PCM raised the night outlet temperature by 2 °C and improved the greatest gathered energy by 630 Wh, if compared to the collector without a PCM. By using this collector-storage system, the thermal efficiency during stagnation mode increased from 66% to 82%.

2.2. Solar Still-Related Studies

A solar still system is a device that evaporates and condenses water to purify it using sun energy. A solar still's basic structure consists of a shallow basin or trough with a transparent cover, typically made of glass or plastic, that lets sunlight in and warms the water inside. Water vapor rises and condenses on the bottom of the lid as it evaporates, gathering there to be collected in a different container [42]. A solar still's effectiveness is affected by a number of variables, including the quantity of sunshine it receives, the surface area of the basin, the material used for the cover, and the temperature and humidity of the surrounding air [43]. For maximum effectiveness and system longevity, the system must be properly maintained, which includes routinely cleaning the basin and cover. In places where access to clean water is scarce or non-existent, solar still systems can offer a low-cost, long-term alternative for producing drinkable water.

This section attempts to provide a comprehensive review of recent studies that encountered the implications of PCMs in solar stills with a critical analysis of the overall performance.

Using a new shape-stabilized phase change material (SSPCM), Cheng et al. (2019) [44] provided an original experiment as well as a simulation model to assess the efficiency of solar still. SSPCM of 5 wt% expanded graphite was specifically used. High thermal conductivity, a steady form, and strong sun absorption were all characteristics of the SSPCM. The SSPCM employed in the investigation had a thermal conductivity of 1.50 W/m K and a solar absorption of 0.94, making it a suitable replacement for the traditional metal absorber plate. Next, a test for solar still with PCM (SSPCM) was conducted. According to the trial findings, as depicted in Figure 8, the SSPCM's daily productivity was 3.41 L/m^2, 43.3% greater than that of the traditional solar still without PCM. The modeling findings demonstrated that when the thermal conductivity of SSPCM improved from 0.2 to 4.0 W/m K, the daily production rose from 42% to 53% compared to that of traditional solar still. The proportion of boosting productivity augmented from 21.5% to 57.5% as the melting temperature of SSPCM rose from 34 to 50 °C.

Vigneswaran et al. (2019) [45] conducted an investigation to appraise the thermal efficiency of three passive solar stills of the same size. The first still was a conventional solar still (Still 1), the second was a still with a single PCM (paraffin wax) (Still II), and the third was a still with two PCMs (Still III). The two PCMs utilized in Still III had different temperature ranges for their phase change but similar capacities for latent heat storage. PCM1 and PCM2 had a temperature range for a phase change of 58.03 °C to 64.5 °C and 53.05 °C to 62 °C, respectively. The tests were achieved in Chennai, India from February to May. PCM1 was designed to discharge stored latent heat energy once the solar radiation began to diminish, while PCM2 was intended to discharge heat when PCM1 had almost discharged all the stored energy. The researchers analyzed the thermal performance of the three stills in terms of hourly yield day^{-1} and exergy performance. The yields of Stills I, II, and III were 3.680, 4.020, and 4.400 L/m^2/day, respectively.

Table 1. Outline of studies on phase change material for solar collectors.

Authors (Year) [Reference]	Configuration/Composition	Type of Study	Studied Parameters	Highlighted Results/Findings
Feliński and Sekret (2016) [25]	A system that combines an evacuated tube collector with a PCM storage unit.	Experimental	Effect of using PCM.	The paraffin-integrated ETC/S system produced a total improvement of 45–79% in usable heat.
Serale et al. (2016) [26]	Location, orientation, and PCM concentration were implemented to appraise the efficacy of the PCS-based technology.	Experimental	Effect of PCS as a heat transfer fluid.	Depending on the climate, the use of PCS as a heat transfer fluid can boost solar energy extraction to varying degrees.
Li and Zhai (2017) [27]	Before being inserted into the evacuated tubes, the aluminum pipes were filled with PCM made of erythritol and expanding graphite.	Numerical and experimental	Impact of using PCM.	Using PCM led to enhancing the storage efficacy by 40.17% for mid-temperature utilization.
Wang et al. (2019) [28]	Hybrid slurries containing MPCMs dispersed in ethanol/water-based multi-walled carbon nanotube (MWCNT) nanofluid are used.	Experimental	Impact of hybrid MPCM-MWCNT slurry.	Compared to the base fluid, a combination of ethanol and water, the temperature increase in the hybrid MPCM-MWCNT slurry was less.
Wang et al. (2019) [29]	A novel dual PCM in flat plate solar collector.	Experimental	Impact of A novel dual PCM.	When low-melting PCM was positioned beneath high-melting PCM and when the conditions were reversed, the dual-PCM collector's performance rose by 24.1% and 19.6%, respectively.
Gao et al. (2020) [30]	Microcapsule, with melamine-formaldehyde resin shell and octadecane core.	Experimental	Effect of adding rGO.	The inclusion of rGO significantly increased the thermal conductivity of microcapsules containing magnetic phase change material.
Palacio et al. (2020) [31]	A traditional flat plate solar collector and a corresponding prototype equipped with a thermal storage system using PCM.	Experimental	Effect of using PCM.	In comparison to the collector without a PCM, the prototype with a PCM raised the night outlet temperature by 2 °C and improved the maximum gathered energy by 630 Wh.
Elarem et al. (2021) [32]	Evacuated tube solar collector integrating a nano-PCM with fins.	Numerical	Effect of using fins and nanoparticles on the PCM.	The inclusion of fins has a major influence on the paraffin phase transition heat transfer in the ETSC. The PCM melts more rapidly as the fin thickness decreases. Additionally, it was discovered that 1% Cu fed to the PCM was the optimal mass concentration for a 2 K rise in the HTF outlet temperature.

Table 1. *Cont.*

Authors (Year) [Reference]	Configuration/Composition	Type of Study	Studied Parameters	Highlighted Results/Findings
Alshukri et al. (2021) [33]	A new method of integrating nano- or micro-enhanced PCMs was applied to an evacuated tube collector, which included one or both evacuated tubes and two separate storage tanks.	Experimental	Effect of improved paraffin wax with NP-CuO and MP-ZnO.	For an evacuated tube, the addition of improved paraffin wax with NP-CuO and MP-ZnO increased efficiency by a range of 33.8–45.7% and 23.8–26.7%, respectively.
Sadeghi et al. (2022) [34]	A tankless direct-absorption evacuated tube solar collector was successfully applied with a shape-stabilized PCM.	Experimental	Effect of shape-stabilized PCM.	This collector-storage system increased the thermal performance in the stagnation mode from 66% to 82%.
Yeh et al. (2022) [35]	Storage-combined solar thermal collector and its compatible PCM.	Experimental and numerical	Impact of PCM.	The PCM charging rate is observed to increase by 9% thanks to the circulating water's ability to transfer heat from the tube's surface to its interior.
Nekoonam and Ghasempour (2022) [36]	Solar system with thermal energy storage.	Numerical	Effect of configuration.	The optimum configuration retains at approximately 14% more energy during charging than the single PCM/sensible cases, at 4940 and 4725 kWh for the first and second optimization scenarios, respectively.
Wen et al. (2022) [37]	Photovoltaic/thermal module with PCM and a solar thermal collector with thermoelectric generators connected in series.	Experimental and numerical	Effect of configuration.	Thermal and overall performance are both strong points of the PV/T-ST system, which integrates the ST and PV/T modules in series.
Hatamleh et al. (2022) [38]	Molecular dynamics simulation was used to investigate the behavior of glass in the presence of decane PCM.	Numerical	Effect of using PCM.	While fast changing the phase of the materials, phase change materials can limit energy loss, lower energy expenditures, and improve a building's aesthetic.
Algarni et al. (2020) [39]	Evacuated tube solar collector integrating nano-enhanced PCM.	Experimental	Impact of copper/PCM composite.	The ETSC was shown to be 32% more effective when 0.33 weight percent of copper/PCM composite was added. Furthermore, due to its particular mass flow rate of 0.08 L/min, the ETSC/Ne-PCM can provide hot water for up to 2 h longer than conventional ETC systems, with temperatures reaching up to 50 °C.

Table 1. *Cont.*

Authors (Year) [Reference]	Configuration/Composition	Type of Study	Studied Parameters	Highlighted Results/Findings
Karami et al. (2023) [40]	Using CuO nanofluid, CuO/Al_2O_3 binary nanofluid, and CuO nanofluid coupled to microencapsulated PCM, direct absorption solar evacuated tube collectors are used.	Numerical	The influence of water and ethylene glycol and nanofluid volume fraction.	The effectiveness of the collector is increased by 4.53% and the energy loss is decreased by 5.84% when MPCM and CuO nanofluid are combined.
Pawar and Sobhansarbandi (2023) [41]	A PCM embedded with copper porous metal which is incorporated within the HPETC.	Experimental	Impact of a PCM embedded with copper porous metal.	The evacuated tube filled with PCM + Cu porous metal had a rise in maximum temperature of almost 21 °C.

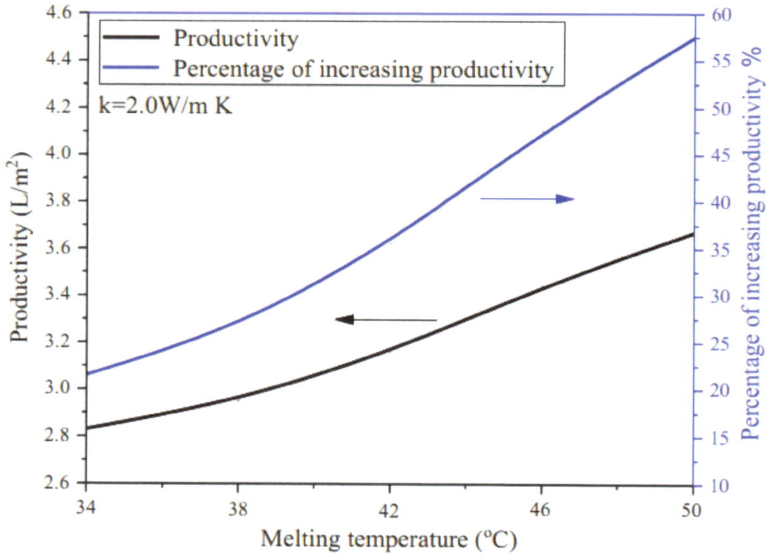

Figure 8. The daily output and the increase in daily output of the shape-stabilized PCM (SSPCM) with various melting temperatures [44].

Abu-Arabi et al. (2020) [46] conducted a study to evaluate the efficiency of adapted solar still under different scenarios. The solar still was modified and tested in three options: SC connected to an external collector (SCC), solar still with glass cooling (SC), and SCC with PCM (SCCP). In this regard, three PCMs utilized: sodium thiosulfate pentahydrate, sodium acetate trihydrate, and paraffin wax. The results showed that the productivity of the SC, SCC, and SCCP systems (with SAT as PCM) increased from 0.9 to 3.4 mL/min, 2.35 to 10 mL/min, and 3 to 11.9 mL/min, respectively. This is specifically conducted cncreased from 200 to 700 W/m^2. The adding of an external collector and PCM improved productivity by 2.4 fold. An increase in the coolant mass flow rate from 0 to 10 kg/s has caused an increase in productivity for all systems. Moreover, an increase in the hot water circulation rate of the external collector has also increased productivity for the SCC and SCCP systems. The best PCM was the one with the greatest melting point and latent heat of fusion, and the ideal mass ratio of basin water to PCM was 2:1.

Incorporating hollow cylindrical pin fins was proposed by Yousef et al. (2019) [47] to improve the heat transfer properties of a PCM storage unit used in a solar still system. These fins functioned as thermal conductivity enhancers (TCEs). The study considered three scenarios: Case 1, a conservative solar still without PCM; Case 2, a solar still with PCM; and Case 3, a solar still with a pin-finned heat sink embedded in the PCM. The efficiency of each scenario was assessed and compared through experimental evaluation under the same climatic conditions. The findings elucidated that the existence of PCM had a negative impact on daytime freshwater productivity, but overall freshwater yield increased significantly. Moreover, the solar still with pin-finned PCM (Case 3) demonstrated superior thermal efficiency compared to both the conservative still (Case 1) and the solar still with only PCM (Case 2). The total daily water productivity was found to be 17% and 7% higher in Case 3 than in Cases 1 and 2, respectively. Additionally, the study showed that the inclusion of PCM in Cases 2 and 3 extended the productivity of distilled water after sunset by roughly 5 and 7 h, respectively.

A unique PCM tube shape was created by Elashmawy et al. (2020) [48] for usage within a tubular solar still with a parabolic solar concentrator. Twelve aluminum tubes with paraffin wax within each serve as a PCM. To improve the thermal qualities, copper rods are placed inside the paraffin wax. The tubular solar still is exposed to stronger sun radiation thanks to a parabolic concentrator solar tracking device. The experimental work was carried out in Ha'il City, Saudi Arabia, which is located at 27.64° N and 41.75° E and is 965 m above sea level. After being coated black, the 12 aluminum tubes were put within the trough. The productivity and device performance in freshwater has significantly improved, with comparatively low production costs. The productivity and performance of the created device are improved by 40.51% and 38.25%, respectively, when PCM tubes are used. When utilizing a device with PCM tubes, the yield, efficiency, and cost per liter are 5.55 L/m^2 day, 44.1%, and 0.00782 USD, respectively. In contrast, using a device without PCM tubes yields 3.95 L/m^2 day, 31.9% efficiency, and a cost of $0.0163 USD per liter.

Kabeel et al. (2020) [49] conducted experiments to elevate the efficiency of tubular solar still during both daytime and nighttime. They achieved this by adding enhancements to the design and construction, specifically by constructing closed copper tubes filled with PCM (paraffin wax), which suited the basin of the still. The enhancements led to a noteworthy rise in the production of distilled water, where conventional tubular stills produced 4.1–4.31 L/m^2/day, while utilizing closed copper tubes filled with PCM increased productivity to 8.82–9.05 L/m^2/day during June–August 2019, resulting in a 110–115.1% increase in accumulated productivity under similar conditions. Moreover, daily efficiency also improved, with conventional tubular stills achieving between 32.9% and 33.8%, whereas the utilization of closed copper tubes filled with PCM increased daily performance to 70.9–72.7%, representing a 114.4–115.5% increase in daily performance. Overall, the outcomes demonstrated that the use of copper tubes filled with PCM is a promising option for achieving optimal performance in tubular solar stills.

Solar-driven-spray flash evaporation (SFE) desalination unit improved using latent heat fluids was reported by Chen et al. in 2021 [50]. The addition of microencapsulated MPCM (TH-ME 58 (SiO_2-paraffin)) in a NaCl aqueous solution improves the working fluid's thermal energy storage and SFE process. The simulation results indicate that the use of MPCM in droplets can delay the temperature drop during the flash evaporation process due to their higher effective heat capacity. By adding 0.18 mass fraction of MPCM to the flash evaporation fluid, the water production ratio increased by 23.1%, and the specific energy consumption reduced by 18.3%. This is clearly depicted in Figure 9.

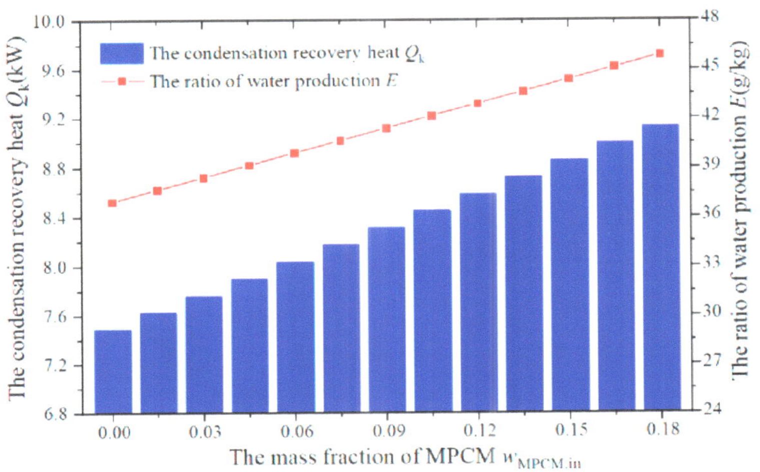

(a)

Figure 9. *Cont.*

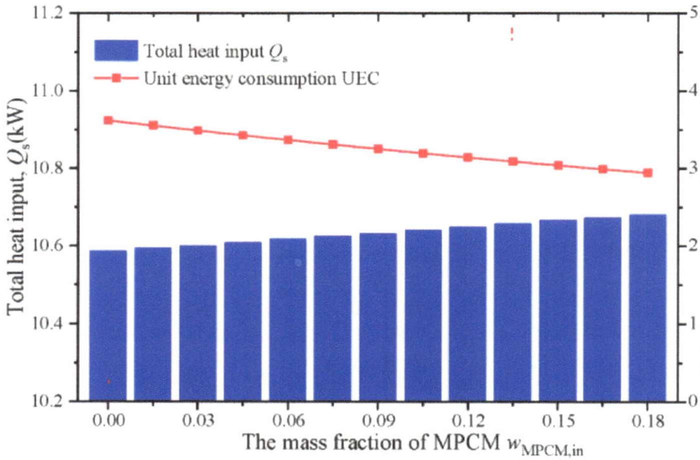

(**b**)

Figure 9. SFE-condensation efficiency for different added solutions of different mass fractions of MPCM (**a**) the ratio of water production; (**b**) unit energy consumption [50].

Jahanpanah et al. (2021) [51] studied the impact of low-temperature PCM on the overall efficiency and productivity and basin temperature in a single-slope solar still. A commercialized inorganic salt hydrate + water + Additive PCM (Product number: PGSCR Co. PCM28/315) of 28 °C melting point, low temperature, and latent heat of fusion of 225 kJ kg^{-1} was chosen for this objective as a cutting-edge method in the industry. Investigations were conducted on three cases: (1) still without PCM, (2) still with 3 kg of PCM, and (3) still with 6 kg of PCM. The operative temperature, hourly and collective productivities, and desalination effectiveness were all observed and computed throughout each experiment. It was discovered that the addition of 6 kg of PCM can raise desalination efficiency from 28.13% to 36.42% while improving production overall by 30.3% (Figure 10). The outcomes demonstrate that the performance of the solar stills might be effectively enhanced by the use of low-temperature PCMs.

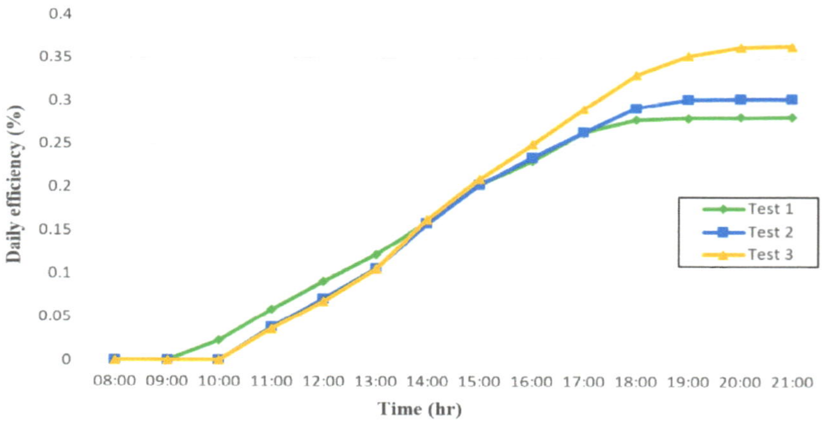

Figure 10. Daily desalination efficiency of the unit for three test cases [51].

In a study conducted by Gnanavel et al. (2021) [52], the efficiency of solar still was enhanced by incorporating a PCM. The experimental setup consisted of an aluminum basin with and without the phase change material, and tests were conducted from 7 a.m. to 6 p.m. to determine hourly and cumulative productivity. Trimethylolethane and paraffin C18 were used as PCMs, and the results indicated that paraffin C18 resulted in greater productivity. CFD analysis was performed to validate these findings.

The inclined and traditional solar stills were coupled by Hansen et al. (2022) [53] to increase the daily distillation yield as a whole. Both stills store latent heat through the use of paraffin wax. In the course of the day, the PCM stores thermal energy and releases heat as needed. The solar still's productive yield throughout the night is provided by the thermal energy storage of latent heat. The solar still is connected with external add-ons such as flat, steeped, and fin-shaped absorbers to improve distillation. The experiment was conducted outside, and daily production was assessed. This study determined that fin-shaped absorbers were more productive, and it was combined with PCM to produce a supreme amount of 5.62 L/day of distillate. The inclined still with the fin-shaped absorber was shown to be more creative when paired with basin still, with a production increase of 74.5%, and when coupled to PCM and basin still, with an increase in productivity of 87.96%.

Ajdari and Ameri (2022) [54] constructed an inclined stepped solar still with baffles to study the impact of the input brine flow rate, CuO/graphene oxide (GO) nanocomposite weight percent, a volume ratio of individual nanoparticles in the nanocomposite, existence of paraffin wax as a PCM, and distillate yield on the glass, brine, and bottom temperatures. As a consequence, the freshwater yield increased by 48.12 and 81.59%, respectively, with 0.03 wt% CuO and GO. Freshwater volume rose by 81.59%, thanks to a nanocomposite with a volume ratio of 30/70 for CuO/GO. Additionally, it was discovered that reducing the brine flow rate from 30 to 8 Lh^{-1} enhanced the productivity of freshwater while raising it to 68 Lh^{-1} decreased the distillate yield. Last but not least, adding paraffin wax as PCM to the solar still's stages increased distillate production by 32.8%. In conclusion, it is shown that the CuO/GO nanocomposite is a promising alternative to be used in solar stills to elevate their productivity.

Abdullah et al. (2022) [55] carried out an experimental investigation to enhance the efficiency of the Tray Solar Still (TSS). The rate of heat transfer between the TSS absorber and the saline water has improved. The effectiveness of TSSs with flat and finned absorber forms was therefore examined. Conventional solar stills (CSS), TSS, and finned TSS were the SSs that were examined (FTSS). The finned absorbers were covered with jute wick materials, allowing water from the wick feed to gently rise through the porous material. PCMs of paraffin wax combined with copper oxide nanoparticles have been used to examine the FTSS efficiency for further improving TSS efficiency. The water basin has also been heated using three electric heaters. To occupy the same space as the FTSS and function as a power supply for electric heaters, a photovoltaic cell was constructed adjacent to the back wall of the FTSS. According to the results, using heaters and PCMs with nanoparticles increased FTSS productivity by 166 and 136% as contrasted to that of the CSS, respectively. Moreover, employing electrical heaters and PCMs with nanoparticles simultaneously increased the FTSS's daily water productivity by 196% over the CSS.

By adding an internal sidewall reflector, hollow circular fins, and a nanoparticle-PCM, Tuly et al. (2022) [56] improved the efficiency of actively modified double-slope solar distiller units. One standard still and four modified stills with the specified connections were utilized in the experiment. According to the findings, the solar still modified with a reflector, fins, and nano-PCM produced a max of 1853 mL per day. The system's most efficient adaptation was discovered to be the employment of a paraffin wax-Al_2O_3 nanoparticle mixture (paraffin wax and 3 wt% Al_2O_3), which offered nearly twice as much cumulative productivity as the pure PCM (just paraffin) scenario. In the situations of pure PCM and nano-PCM, distillate accumulation was superior to the conventional scenario by 61.36% and 92%, respectively. In addition, the combined effect of the circular fin, reflector, and nano-PCM resulted in achieving the highest energy efficiency of 21.56%.

A combined double-slope solar still (DSSS) and PCM-TES w(ere proposed by Afolabi et al. in 2023 [57]. Using vacuum mold-filling processes, PCM was microencapsulated in an epoxy resin composite. Specifically, they used 1.5 wt% of 40 nm sized Zn-nanoparticle and 5 kg paraffin PCM. To determine the impact of TES on productivity, the collected data from the traditional DSSS and DSSS-TES have been evaluated. The DSSS-TES system attained average daily temperatures of 65.2 °C, 77.5 °C, 82.4 °C, 79.5 °C, and 68.4 °C for the glass cover, humid air, salt water, still basin absorber, and TES cavity, respectively. With 7.5 L of potable water every day and a 3 h expansion of the working period, DSSS-TES has produced more. Moreover, rates of condensation and evaporation surged by 105% along with the rise in production. The system's integration of TES has decreased heat losses, and a microencapsulated insulator has stopped PCM nanocomposite leakages. Desalinated water has not been discovered to contain any metals, microbes, or organic pollutants.

Bacha et al. (2023) [58] introduced an inventive design for tubular solar distillers that enhances productivity while minimizing costs. This was accomplished by replacing the flat absorber with a convex one covered in black jute wick material and incorporating an energy storage reservoir containing paraffin wax beneath the convex absorber. An oil heat exchanger was also included with a solar parabolic concentrator serving as an additional heat source. The effectiveness of this new design was assessed through the creation of two tubular solar distillers and conducting tests in Al-Kharj, Saudi Arabia. The first distiller was a standard tubular solar distiller used as a control, while the second was a modified convex tubular solar distiller (MCVTSD). The outcomes demonstrated the superior productivity of the MCVTSD, which achieved 13.58 $L/m^2/day$. Furthermore, the pioneering configuration has upgraded thermal and exergy efficiencies, and overall productivity by 39.32%, and 4.99%, 233.25%, respectively, compared to the conventional tubular solar distiller.

The evaluation of the studies described in Table 2 shows that the joint effects of circular fin, reflector, and nano-PCM result in a maximum energy efficacy of 21.56%. The overall freshwater yield still significantly increases as a result of PCM, which also has a detrimental effect on daytime freshwater productivity. Moreover, productivity was boosted by 2.4 fold with the inclusion of PCM and an external collector.

2.3. Solar Pond-Related Studies

A solar pond system is a device that employs a sizable, shallow pond to gather and store solar energy for use in heating and electricity production, among other things. A solar pond's basic structure is made up of a number of salt concentration-varying layers that together form a gradient that traps solar energy and warms the water at the pond's bottom. The heated water can subsequently be used for a variety of purposes, including the production of energy through a turbine, heating of enclosed spaces, and industrial activities. The depth of the pond, the gradient in salinity, the amount of solar radiation, and the surrounding temperature all have an impact on how well a solar pond performs [59]. By choosing the right pond shape and size, maximizing the salinity gradient to optimize thermal stratification, and properly insulating the pond, one can increase the efficiency of a solar pond system [60]. For the system to operate at its best and last as long as possible, it must be properly maintained, which includes monitoring the salinity gradient and routinely clearing the pond of sediment and other waste. Particularly in regions with plenty of sunlight and restricted access to conventional energy sources, solar pond systems can offer a low-cost, environmentally friendly solution for producing thermal energy in a range of applications.

The current section provides a valuable summary of the associated research on PCMs in solar pond systems.

Table 2. Outline of studies on phase change material for solar still.

Authors (Year) [Reference]	Configuration	Type of Study	Studied Parameters	Highlighted Results
Cheng et al. (2019) [44]	Solar still with using a new shape-stabilized phase change material (SSPCM).	Experimental	Effect of SSPCM in solar still.	The solar still utilizing SSPCM demonstrated a daily output of 3.41 L/m^2, indicating a productivity increase of 43.3% compared to the standard solar still without SSPCM.
Vigneswaran et al. (2019) [45]	Solar stills with three different configurations were tested: a conventional solar still (Still 1), a solar still with a single phase change material (Still II), and a solar still with two PCMs (Solar III).	Experimental	Impact of using PCM.	The yield of Still I was 3.680 $L/m^2/day$, while Still II and III produced 4.02 and 4.4 $L/m^2/day$, respectively. The productivity of Still III was 4.40 $L/m^2/day$, representing a 19.6% and 9.5% increase in yield compared to Still I and II, respectively.
Abu-Arabi et al. (2020) [46]	Solar still, solar still linked to a third-party collector, and solar still with glass cooling with PCM. The three PCMs—sodium thiosulfate pentahydrate, paraffin wax, and sodium acetate trihydrate—were all different.	Experimental	Effect of external collector and PCM.	The productivity was boosted by 2.4 fold with the inclusion of the external collector and PCM.
Yousef et al. (2019) [47]	Hollow cylindrical pin fins are included in PCM.	Experimental	Effect of using hollow cylindrical pin fins incorporated into PCM.	The overall freshwater yield still significantly increases when PCM is present, which has a negative impact on daytime freshwater productivity. The cumulative daily yield of distilled water in Case 3 is 17% and 7% higher than that of Case 1 and Case 2, respectively.
Elashmawy et al. (2020) [48]	The PCM used in the solar still consists of paraffin wax filled inside 12 aluminum tubes.	Experimental	Using paraffin wax in 12-aluminum tubes.	The productivity and efficiency of the created device are improved by using PCM tubes by 40.51% and 38.25%, respectively.
Kabeel et al. (2020) [49]	Closed copper tubes filled with PCM were assembled and fixed in the basin of tubular solar still.	Experimental	Impact of closed copper tubes filled with PCM.	The daily performance of traditional tubular still varies between 32.9 and 33.8%, but when closed copper tubes filled with PCM are used, the daily performance increases to 70.9 to 72.7% with an increase of 114.4 to 115.5%.
Chen et al. (2021) [50]	NaCl aqueous solution with the inclusion of PCM in microcapsules.	Experimental	Impact of using MPCM.	Due to their greater effective heat capacity, droplets with MPCM may experience a slower temperature drop during the flash evaporation process. The inclusion of MPCM with a mass fraction of 0.18 into the flash evaporation solution resulted in a 23.1% increase in the water production ratio.
Jahanpanah et al. (2021) [51]	PCM enhanced single-slope solar still.	Experimental	Impact of using PCM in a single-slope solar still.	The addition of 6 kg of PCM might boost output by 30.3% overall and boost desalination efficiency from 28.13% to 36.42%.

Table 2. *Cont.*

Authors (Year) [Reference]	Configuration	Type of Study	Studied Parameters	Highlighted Results
Gnanavel et al. (2021) [52]	Solar still uses phase change materials.	Experimental	Impact of using paraffin 18.	Paraffin C18 material produces more productivity. Paraffin C18 material produces 209.5% more productivity than the case without using paraffin C18.
Hansen et al. (2022) [53]	Paraffin wax is used in the inclined and conventional solar still.	Experimental	Impact of the fin-formed absorber and PCM.	The inclined still with a fin-shaped absorber was particularly productive, with productivity increases of 74.5% when paired with a basin still and 87.96% when combined with PCM and a basin still.
Ajdari and Ameri (2022) [54]	The water production ratio experienced a 23.1% increase when MPCM with a mass fraction of 0.18 was added to the flash evaporation solution.stepped solar still with baffles.	Experimental	Effect of using a nanocomposite in solar still.	The volume of the freshwater was enhanced by 81.59% using a nanocomposite with a volume ratio of 30/70 for CuO/GO.
Abdullah et al. (2022) [55]	PCMs with nanoparticles are used in solar still.	Experimental	Effect of PCMs mixed with copper oxide-nanoparticles.	Using heaters and PCMs with nanoparticles increased the FTSS productivity by 166 and 136% as contrasted to that of the CSS, respectively.
Tuly et al. (2022) [56]	By combining the effects of an internal sidewall reflector, hollow circular fins, and a phase change material mixed with nanoparticles, active modified double-slope solar distiller units are created nano-PCM.	Experimental	Effect of internal sidewall reflector, hollow circular fins, and nanoparticle PCM.	The mutual effect of the circular fin, reflector, and nano-PCM results in the greatest energy efficiency of 21.56%. The mutual effect of the circular fin, reflector, and nano-PCM results in the greatest productivity efficiency of 21.56%.
Afolabi et al. (2023) [57]	Double-slope solar is still integrated with PCM-TES.	Experimental	PCM.	The condensation and evaporation index rose by 105% along with the rise in production. The system's integration of TES has minimized heat losses, and a microencapsulated insulator has stopped leaks from the PCM nanocomposite.
Bacha et al. (2023) [58]	Use a convex absorber rather than a flat one, cover it with black jute wick material, and use a paraffin wax-filled energy storage reservoir beneath the convex absorber.	Experimental	The impact of utilizing a convex absorber, wick materials, and PCM reservoir in conjunction with a solar parabolic concentrator.	The MCVTSD's high productivity was 13.58 $L/m^2/day$. In this case, MCVTSD's novel design increased productivity and thermal and energy efficiency in comparison to CTSD by 233.25%, 39.32%, and 4.99%, respectively.

Sarathkumar et al. (2017) [61] added Al_2O_3 nanoparticles to PCM to enhance the performance of the solar pond. Al_2O_3 nanoparticles have strong thermal conductivity, and PCM has good heat absorption ability. The time required for solar pond charging and discharging can be greatly decreased, and the rate of heat transfer can be boosted, by the addition of Al_2O_3 nanoparticles. The copper tube contains a mixture of PCM and Al_2O_3 nanoparticles. This encapsulation, which consists of PCM mixed with nanoparticles, is positioned in the pond's LCZ (lower convective zone). In the LCZ, a heat exchanger extracts heat. Power plants can be provided with useful steam to produce electricity. The most usable heat that may be produced is 167 W. Just until evening 6.00 PM could the solar pond without nanoparticles supply water at a warmer temperature. Yet, the solar pond containing Al_2O_3 nanoparticles was capable of providing hot water until 7.30 PM in the evening. As illustrated in Figure 11, the solar pond containing Al_2O_3 particles was still capable to produce hot water for 1.5 h after sunset.

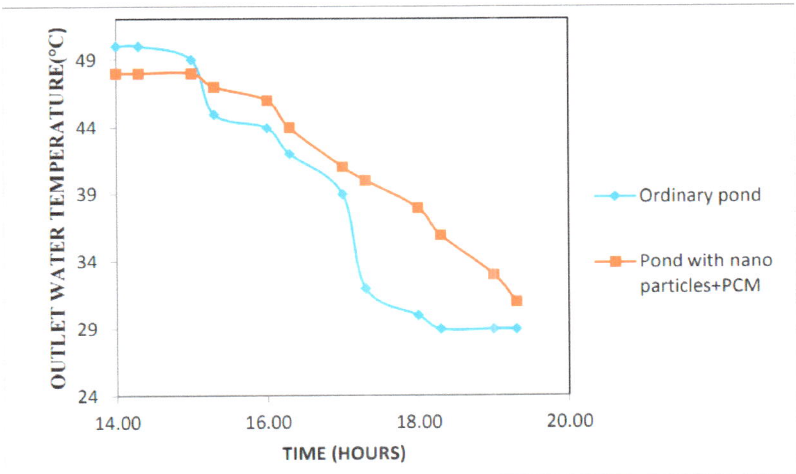

Figure 11. Evaluation of outlet water temperature of two different ponds [61].

Amirifard et al. (2018) [62] combined a solar pond with latent heat storage to improve performance stability. They analyzed three different configurations: two parallel and one series layout and presented the results in terms of temperature changes, thermal energy, and process efficiency. The conducted study evaluated the performance of four PCMs including the n-pentacosane, n-hexacosane, n-docosane and n-tetracoscane. The series layout showed a 6.1% increase in average efficiency of discharging time compared to the pond without PCM, while the parallel layout showed a 5.4% increase.

Experimentally, a salt-gradient solar pond (SGSP) subjected to a solar simulator was provided for thermal evaluation by Ines et al. (2019) [63]. Environmental chambers were used for the experiments. To compare the integration and absence of ultra violet (UV) filters, two experimental prototype solar ponds were constructed in Italy. Additionally, illustrative findings demonstrating the impact of the simulator spectrum on an SGSP's temperature distribution and the consequences of thermal performance have been examined. Due to its higher potential to store latent heat, the PCM has been utilized in this study as a material that can absorb and release significant amounts of energy during solidification and melting. The outcome demonstrated that the solar pond with PCM (Plexiglas Poly (methyl methacrylate)) is capable of serving as a hot water source for home presentations by simultaneously gathering and storing solar energy. As demonstrated in Figure 12, the difference in temperature has decreased due to the pond's highest temperature being lower with PCM while the lowest temperature was roughly identical in both ponds.

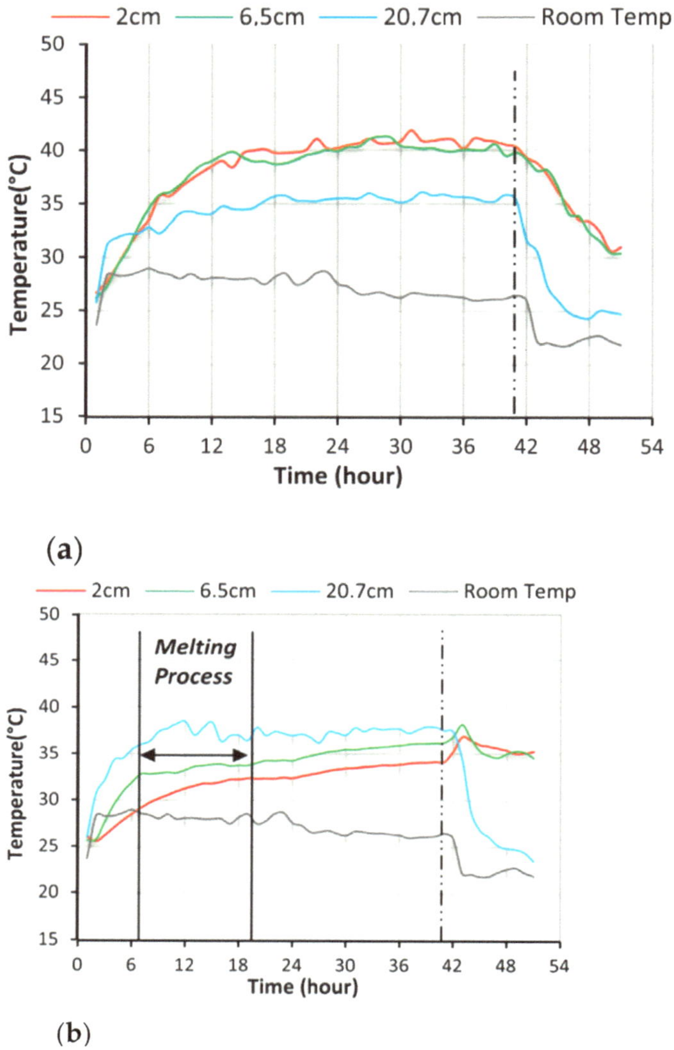

Figure 12. Temperature fluctuations of SGSP without and with PCM throughout the operational time [63] (**a**) without phase change materials, (**b**) with phase change materials.

A one-dimensional model for transient heat transfer utilizing PCM was proposed by Beik et al. (2019) [64]. The convection, conduction, and radiation heat exchange with the ambient was used to determine the UCZ temperature. Paraffin Wax was utilized as PCM in the LCZ to extend the effects. According to numerical findings, a sizable amount of energy is wasted in this configuration due to reflection, walls, convection, and evaporation. Only 6% of the available solar radiation was stored, which can be improved with better insulation. Additionally, this study establishes optimal guidelines for the depth-to-radius ratio of the system, beyond which heat loss can be neglected. Moreover, the application of PCM in the solar pond yielded more stable temperatures during heat extraction, as indicated by both experimental and numerical findings.

Wang et al. (2020) [65] evaluated the effect of incorporating composite PCM heat storage capsules into the heat storage layer of a salt-gradient solar pond on its thermal efficiency, using both numerical and experimental methods. The solar pond with PCM capsules (paraffin wax 50 and paraffin wax 60) exhibited a smaller temperature range during the phase change process compared to the conventional solar pond, but not during the non-phase transition process. Moreover, the inclusion of PCM phase change units has reduced the flow rate of the heat storage zone. The high latent heat of the PCM has a more pronounced suppressive influence on the flow. Consequently, adding PCM to the solar pond has an optimistic influence on preserving a fixed temperature and steadiness within a certain temperature range.

Rghif et al. (2021) [66] conducted a numerical investigation to determine the impact of a layer of PCM (paraffin wax) and the Dufour effect on the thermal efficiency of a salt-gradient solar pond (SGSP). The analysis considered two cases, the first case with an SGSP without a PCM layer, and the second case with an SGSP having a PCM layer located at its bottom. The findings of the simulation showed that the SGSP with a PCM layer had a lower temperature and thermal efficiency and higher heat losses across the saline water-free surface compared to the SGSP without a PCM layer. In Case 1, elevating the dimensionless Dufour coefficient from 0 to 0.8 resulted in a drop in the storage zone's dimensionless temperature from 0.894 to 0.764 and a drop in thermal storage efficacy from 37.41% to 31.85%. Similar to Case 1, Case 2 saw a reduction in the dimensionless temperature of the storage zone from 0.800 to 0.688 and a drop in the thermal storage efficacy from 33.32% to 28.66% for the same variation spectrum in the dimensionless Dufour coefficient.

A pond combined with paraffin wax PCM (RT35 HC) and a conventional one were planned, constructed, and tested by Colarossi et al. (2022) [67] to compare their respective performances. The average temperature in the pond with the RT35 HC is, according to the results, roughly 4 °C cooler over 10 days (equivalent to a 5.7% decrease). After that, a second experiment was run to compare two melting temperatures (RT35 HC and RT44 HC). According to the findings, a greater melting temperature makes it possible to smooth out temperature peaks. Due to the release of latent heat, a lower melting temperature ensures a greater and more stable temperature at night, making the application suitable in cases where a fixed temperature need at the outlet fluid is present.

In the LCZ of a small solar pond, Colarossi and Principi (2022) [68] demonstrated an innovative application of PCMs. Aluminum cylinders that have been stacked on the bottom contained paraffin wax (RT-35 HC). Both with and without the PCM, the solution's deep temperature is being tracked. With a laser shadowgraph approach, the stability of the pond is examined in order to see how thermal convection affects the surfaces. Findings indicate that after a 6 h heating cycle, the LCZ of the solar pond with PCM is approximately 3 °C cooler than the reference example. The shadowgraph study demonstrates that the reference case's increased thermal convection damages the interface to the point of failure. The solar pond is more stable, according to PCM testing.

Wang et al. (2022) [69] investigated the effect of adding steel wires to paraffin in the composite PCM on heat transfer performance. The study found that adding 10 wt% of steel wires significantly improved the heat transfer coefficient, with a greatest value of 18.17 W m^{-2} $°C^{-1}$ and an increase of 18.84% during melting, and a maximum value of 21.59 W m^{-2} $°C^{-1}$ and an increase of 20.88% during freezing. In addition, an experimental study was conducted in a solar pond with 20 PCM capsule tubes added. According to the study, the addition of RT 50 and RT 60 paraffin waxes to the solar pond reduced day-night temperature differences by 2.87 °C and 2.53 °C, respectively, compared to the conditions without any PCM. Moreover, the inclusion of RT 50 and RT 60 resulted in exergy increments of 9.34% and 14.13%, respectively. This implies that adding PCM to the solar pond improved the quality of the thermal energy produced.

In 2022, Reza et al. [70] carried out a study to compare the thermal and salinity characteristics of two pilot salt-gradient solar ponds with and without a PCM made of paraffin wax. They looked at three distinct heat extraction techniques and discovered that instability can occur at the upper and lower salinity gradient limits. The findings demonstrated that the PCM-equipped salt-gradient solar pond outperformed the pond without the PCM in terms of average outlet temperature for the internal heat exchanger, thermal and salinity stability, and temperature drop during heat extraction. The minimum and maximum temperatures of the pond were controlled by PCM, which increased evaporation (8.4%) and enhanced thermal efficiency during rapid or prolonged heat extraction. The research additionally indicated that while heat extraction from the non-convective zone (NCZ) can affect the stability of the NCZ, heat extraction from the convective layer is preferable to that from the NCZ.

In summary, it can be stated that the maximum increase in outlet water temperature from the solar pond was 6.5% for using PCM than the case without using PCM. This is already indicated by Beik et al. (2019) [64] and demonstrated in Figure 13.

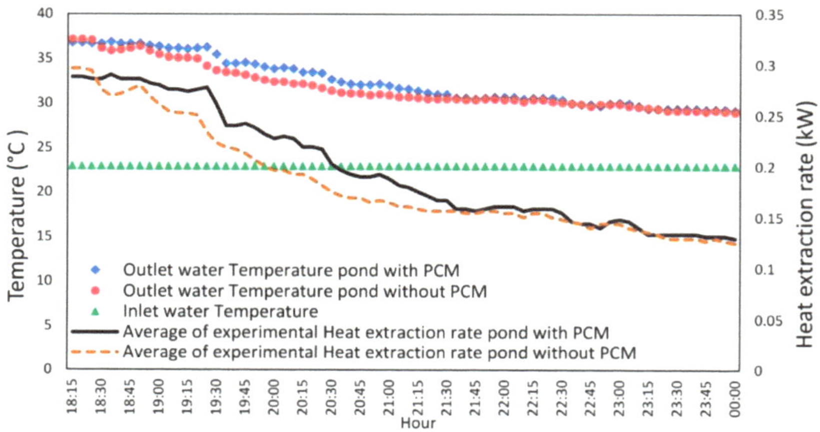

Figure 13. Heat extraction of the ponds with and without PCM and ponds' temperature against operational time [64].

A review of the research included in Table 3 demonstrates that adding PCM units to the solar pond improved its stability and capacity to sustain a constant temperature. Moreover, utilizing the PCM in the solar pond results in a temperature that is more consistently stable while the heat is being extracted. Additionally, the average efficiency of discharging time for the series arrangement reveals a 6.1% increase in the pond without phase change material (PCM). Moreover, this increase for the parallel layout is roughly 5.4%.

2.4. Solar Air Heater-Related Studies

In order to heat air for applications such as room heating and ventilation, a solar air heater system employs solar energy. A flat plate or a group of tubes serving as the basic building blocks of a solar air heater are used to absorb sunlight and transmit heat to the air being circulated within the system. The heated air can then be utilized in place of or as an addition to conventional heating systems. The amount of sunshine available, the direction and tilt of the collector, the airflow rate, and the surrounding temperature all have an impact on how well a solar air heater performs [71]. By choosing the right collector type and size, adjusting the collectors' orientation and tilt to maximize solar radiation, and reducing heat losses through adequate insulation, it is possible to increase the efficiency of a solar air heater system. For the system to operate at its best and last as long as possible, proper maintenance is required, including routine air filter replacement and collector cleaning [72]. In a number of applications, solar air heater systems can offer an affordable, environmentally friendly alternative for space heating and ventilation, especially in regions with ample sunlight and high heating demands.

Table 3. Outline of studies on phase change material for the solar pond.

Authors (Year) [Reference]	Configuration	Type of Study	Studied Parameters	Highlighted Results
Sarathkumar et al. (2017) [61]	Al_2O_3 nanoparticles with PCM to enhance the performance of solar ponds.	Experimental	Impact of Al_2O_3 with PCM in the solar pond.	Up to 7.30 p.m. in the evening, the solar pond with Al_2O_3 nanoparticles was able to heat the water.
Amirifard et al. (2018) [62]	Solar pond with latent heat storage.	Experimental and numerical	Effect of using PCM.	The average efficiency of discharge time for the series arrangement is 6.1% higher in the pond without PCM. Moreover, this improvement for the parallel arrangement is roughly 5.4%.
Ines et al. (2019) [63]	Solar pond with PCM.	Experimental	Impact of PCM in the solar pond.	While the lowest temperature was roughly comparable in both ponds, the highest temperature of the pond decreased with PCM, reducing the temperature differential.
Beik et al. (2019) [64]	In the LCZ of the solar pond, paraffin wax as PCM was used.	Experimental and numerical	Effect of using PCM in the solar pond.	While extracting heat from the solar pond, using the PCM results in more consistently stable temperatures.
Wang et al. (2020) [65]	Incorporating the composite PCM heat storage capsules into the salt-gradient solar pond's heat storage layer.	Experimental and numerical	Impact of adding PCM units to the solar pond.	The solar pond's stability and ability to maintain a steady temperature were both improved with the addition of PCM units.
Rghif et al. (2021) [66]	The layer of a PCM in a salt-gradient solar pond (SGSP).	Numerical	Impact of using PCM.	In contrast to the SGSP without the PCM layer, the temperature, thermal efficiency, and amount of heat losses over the saline water-free surface are all inferior.
Colarossi et al. (2022) [67]	Solar pond integrated with PCM (RT35 HC) and a traditional one.	Experimental	Impact of PCM.	A greater melting temperature permits moderately high peaks of temperature. Yet, a lower melting point ensures a greater and steadier nighttime temperature.
Colarossi and Principi (2022) [68]	PCMs in the LCZ of a small solar pond.	Experimental	Phase change material.	After a 6 h heating cycle, the LCZ of the solar pond with PCM is approximately 3 °C cooler than the baseline example.
Wang et al. (2022) [69]	Composite PCM of paraffin and steel wires applied in the solar pond.	Experimental	Effect of PCM.	By incorporating PCM, solar ponds can produce heat that is of higher quality and with a wider range of applications.
Reza et al. (2022) [70]	Using Paraffin Wax (PW) as the phase transition material in two pilot salt-gradient solar ponds and not in them (PCM).	Experimental and analytical	Effect of PCM.	Higher evaporation (8.4%) in the SPP was caused by the use of PCM to regulate the maximum and minimum temperatures of the pond.

There are profound advantages and a few limitations and shortcomings of the current utilization of PCMs as thermal energy storage in solar air heaters. This section covers a broad range of topics related to PCMs, including their thermal properties, applications, and challenges.

Research on the thermal efficiency of a v-corrugated plate solar air heater and a flat plate solar air heater with and without PCM (paraffin wax) at various mass flow rates of 0.062, 0.028, and 0.009 kg/s was undertaken by Kabeel et al. (2016) [73]. They also looked into the impact of changing the PCM's thickness beneath the absorber plate. According to the experimental results, the flat plate solar air heater with the same mass flow rate of 0.062 kg/s had a temperature of 1–5.5 °C for 2.5 h after sunset while the v-corrugated plate solar air heater's outlet temperature was 1.5–7.2 °C greater than the surrounding temperature when PCM was used. Furthermore, the outcomes presented in Figure 14 indicate that the v-corrugated plate solar air heater equipped with PCM had a 12% higher daily performance than the one without PCM. Moreover, when the mass flow rate was 0.062 kg/s, the v-corrugated plate solar air heater with PCM showed a 15% and 21.3% improvement in daily performance compared to the flat plate solar air heater with and without PCM, respectively.

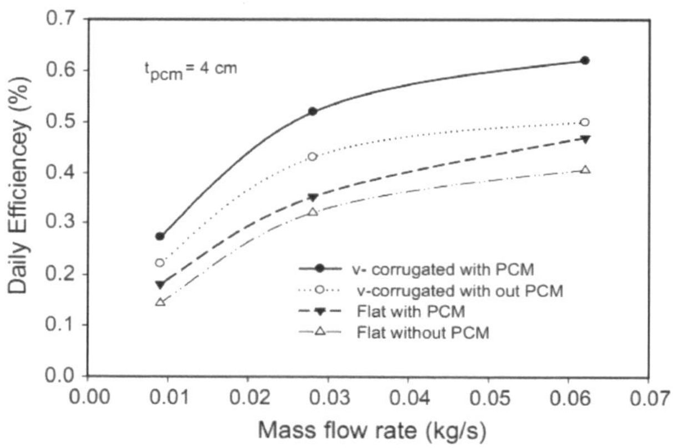

Figure 14. An assessment of daily performance between v-corrugated and flat plate solar air heaters vs. the mass flow rates with and without PCM [73].

El Khadraoui et al. (2017) [74] developed an indirect forced convection solar dryer using PCM (paraffin wax). A solar air panel heats the drying agent directly, while a drying chamber and a solar energy accumulator (a solar air collector with a PCM cavity) are also part of the system. The objective of this study was to investigate the feasibility of using a solar air heater with PCM to store solar energy during the day and release it at night. The solar dryer was tested both with and without PCM while under no load. The solar energy accumulator's daily energy effectiveness was estimated to be 33.9%, compared to its daily energy efficiency of 8.5%. The findings show that after employing the solar energy accumulator, the drying chamber's temperature stayed 4–16 °C above the surrounding air throughout the night. Moreover, when using the solar dryer with PCM, the relative humidity in the drying chamber was 17–34.5% lower than the ambient relative humidity.

A double-pass solar air heater was studied by Salih et al. (2019) [75] using a series of rectangular capsules filled with PCM made of paraffin wax. The innovative system was tested during the charge/discharge process using an indoor projector simulator, and the three-dimensional forced convection turbulent flow in the double-pass solar heater was solved using a model based on a SIMPLE solver. As shown in Figure 15, the results revealed that raising the air flow rate caused the melting duration to take longer and the paraffin melting temperature to drop. For different sun intensities of 825 W/m^2, 725 W/m^2, and 625 W/m^2 at the same airflow speed of 0.6 kg/min, it was discovered that the heater's best discharging period and air temperature increase were 3 h with 17.95–3 °C, 2 h with 14–3 °C, and 1.25 h with 11–2.5 °C.

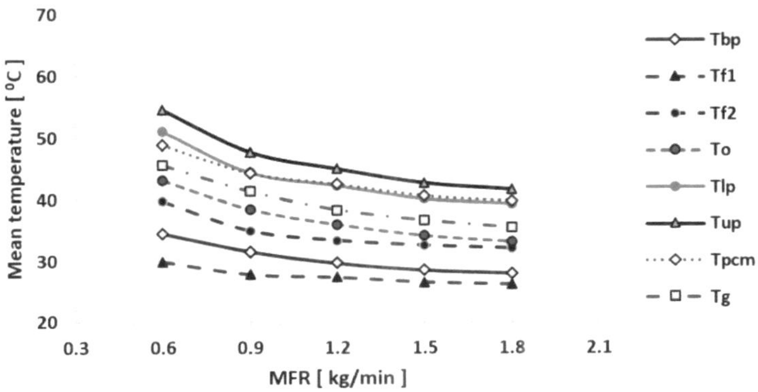

Figure 15. The mean temperature in the DP-SAH with PCM versus air MFR for various solar irradiance of 825 W/m^2 [75].

The thermal efficiency of solar air heaters was studied in experiments by Abuşka et al. (2019) [76] to determine the effects of employing PCM and honeycomb as an internal fin structure in the PCM panel. A total of 6 different air mass flow rates, ranging from 0.008 to 0.048 kg/s, were used during the tests. A total of 2 distinct configurations of the heat storage material were tested. The Type I heater (which utilized PCM with a honeycomb core) and the Type II heater (which used solely PCM). The third heater (Type III), which served as a reference, had a flat absorber plate without PCM. The findings revealed that for mass flow rates of 0.032, 0.04, and 0.048 kg/s, the Type I and II heaters with PCM's average daytime thermal performance improved from 8.4% to 9.0%, while for mass flow rates of 0.008, 0.016, and 0.024 kg/s, it declined from 0.8% to 8.0%. The heater without honeycomb was more effective by 0.1% to 4.4% per day when compared to the heaters with PCM, according to the comparison. The Type I and II heaters with PCM were more effective, ranging from 2.6% to 22.3%, when the daily thermal performance was taken into account, especially at high mass flow rates.

The study conducted by Raj et al. (2019) [77] aimed to improve the thermal efficiency of the double-pass solar air heater system (DPSAHS) by incorporating PCM with thermal lag. Additionally, they investigated the impact of metallic macro-encapsulation on the heat storage and recovery of the PCM coupled with the DPSAHS. The study investigates the impact of encapsulation geometry on the storage and discharge of organic paraffin wax PCM. The research found that rectangular and cylindrical macro-encapsulates equipped with DPSAHS had an average encapsulation performance of 47.2% and 67%, respectively. The study also compared the performance of DPSAHS with and without PCM storage. The analysis showed that the DPSAHS with PCM storage improved operational time at a slightly higher construction cost.

SunilRaj et al. (2020) [78] conducted an experiment on a solar air heater with a V-trough solar thermal collector integrated with PCM (paraffin wax) blended with Al_2O_3 nanomaterials at a fill ratio of 0.90, and a drying chamber. The experiment aimed to test the efficiency of the system in drying 0.5 kg capsicum, and the results demonstrated a significant enhancement in moisture removal. The researchers determined that the design of the system is uncomplicated and exhibits excellent thermal efficiency. The research was carried out under sunny conditions in Bengaluru (12.96° N, 77.63° E) and demonstrated that the daily thermal efficiency of the air heater with thermal storage system ranged from 12% to 65%.

In order to assess the temperature responsiveness and competence of a standard solar air heater (SAH) with SAH fitted with a PCM constructed of paraffin packs, Ameri et al. (2021) [79] conducted an experiment. The experiment looked at how the use of PCM, its melting temperature, and the layout of slabs of PCM with various melting points affected the thermal responsiveness of SAHs. On a Board, four distinct combinations of 2 kinds of paraffin wax with various melting temperatures between 40 °C and 50 °C were fitted. In Kerman, Iran, the experiment was carried out at two separate air mass flow rates. The findings demonstrated that utilizing PCM increased the SAH's daily thermal efficiency from 53.1% to 62.6%. Using PCM with a greater melting point temperature, SAHs' outlet temperatures rose by over 5 °C, and their daily efficiency increased by approximately 3%. (Figure 16). The experiment also showed that the configuration with evenly distributed PCM types with various melting points had the best daily thermal efficiency and produced an enhanced thermal efficiency of approximately 5%.

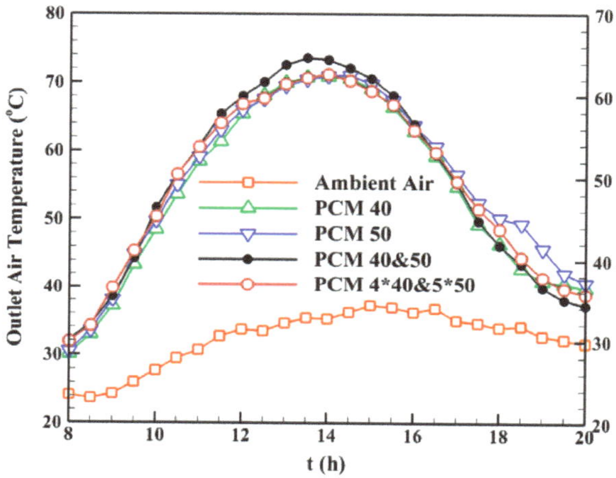

(a)

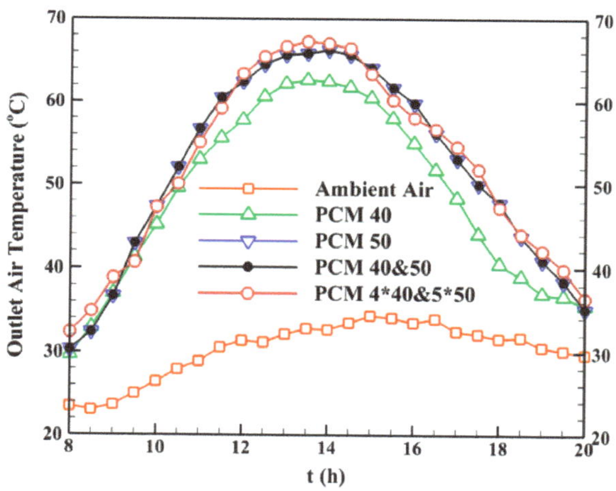

(b)

Figure 16. The variation of outlet temperatures in different scenarios for air mass flow rates of (**a**) 0.006 kg/s and (**b**) 0.01 kg/s, data acquired on 20 July 2019, and 21 July 2019 [79].

Mahdi and colleagues (2021) conducted numerical simulations to investigate the melting/solidification process in a metal foam containing PCM for both sequential and simultaneous operational modes. Specifically, Mahdi et al. (2021) [80] used organic phase change material paraffin wax type RT70HC (RUBITHERM). The composite was included in a rectangular compound that heated a room as a space heater after being cooled by air flowing through a central channel. As a result of quicker charging rates than mode II, the results showed that modes I (8 h charging and 8 h discharging separately) and III (2 h charge and 14 h simultaneous charging-discharging) both reached full melting (2 h charging and 2 h discharging separately, repeated for 16 h). The temperature distribution in Mode III was more uniform, leading in consistent heat transfer between the PCM and cooling fluid. Increasing the porosity of the metal foam had a negative impact on the rate of liquid development, resulting in complete PCM melting in 6.5 h for 90% porosity and 78% melting in 8 h for 95% porosity. Moreover, by increasing the air flow rate from 0.01 kg/s to 0.03 kg/s, the final mean temperature of the PCM was reduced from 69.9 °C to 66.4 °C.

Madhulatha et al. (2021) [81] utilized ANSYS FLUENT to optimize TESS design through numerical analysis. The researchers evaluated three diverse PCMs, including paraffin wax, n-octadecane, and calcium chloride hexahydrate ($CaCl_2{\cdot}6H_2O$), and examined inline, staggered, and circular arrangements of tubes in the TESS. They altered the air intake temperature to TESS from 40 °C to 125 °C and adjusted the mass flow rate of air velocity at the inlet from 3.5 to 5.5 m s^{-1} in steps of 0.5 m s^{-1}. The results introduced the average tube wall temperature, average surface heat transfer coefficient, average PCM temperature in tubes, and average heat absorbed by PCM. When comparing staggered and inline grid arrangements, tubes in staggered arrangements were found to increase heat absorbed by PCM by 10.45% and 48.03%, respectively, with paraffin wax and $CaCl_2{\cdot}6H_2O$ as PCMs, while there was an insignificant 0.09% improvement with n-octadecane.

Dinesh et al. (2022) [82] carried out a study to test the efficacy of a baffled solar-based air heater (SAH-BP) with organic PCM backup. The investigation was carried out over two similar solar days in February 2021 and interacted with operating modes, namely baffled SAH without organic PCM (SAH-BP) and baffled SAH with organic PCM. The organic PCM was stored underneath the absorber, and the baffle plates were connected over the absorber surface in an alternate sequential manner. The trials were kept at a constant flow rate of 0.18 kg/s and the organic PCM utilized was regular commercial-grade paraffin. The findings demonstrated that the baffled SAH's energy efficiency was increased by 11.25% by adding organic PCM.

The energy consumption (EC) of a conference hall in Riyadh that had a timetable of occupancy from 8 a.m. to 12 p.m. was evaluated by Abu-Hamdeh et al. in 2022 [83]. The installation of an economizer and heat recovery equipment allowed for the development of software that changes the active and inactive hours in accordance with the mechanical ventilation timetable. In the first method, the injection of cold air into the indoor area in the summer was adjusted in the range of 3–12 (i.e., ACH = 3, 6, 9 and 12) by constructing programming on the air change (ACH) parameter. The findings showed that the economizer's effectiveness was more apparent with a lower ACH value. The economizer's performance was lower than expected, despite the improvement in ACH, leading to a maximum decrease of 3.07% in EC. In the second method, a program was developed to limit the use of heat recovery to the temperature difference between the interior and exterior of the building. By using this method, EC was decreased by 34% in the summer and 95% in the winter. Yet, the yearly EC drop was capped at 35% because of the significant amount of cooling load. In the third procedure, HS29 PCM was introduced to the structure, and it was found that while it occasionally performed poorly during the winter, it performed admirably throughout the summer. The approach utilized has ultimately decreased annual EC by 5.6%.

An experiment was carried out by Palacio and colleagues (2022) [84] to investigate the effectiveness of a double-channel solar air heater (SAH) with and without a PCM used as a thermal energy storage system. In order to include macro-encapsulated PCM

(Rubitherm RT 45) in the absorber plate, one of two similar prototypes was adjusted. The three weather conditions that were tested in the study were clear skies with strong solar radiation, partly cloudy skies, and clear skies with a large decrease in solar radiation at noon. The experiment demonstrated that the double-channel solar air heater (SAH) with a PCM as a thermal energy storage system had maximum outlet temperatures of 82 °C, 62 °C, and 79.5 °C for the three cases studied, respectively. In contrast, the SAH without PCM consistently had an outlet temperature that was 10 °C lower than the SAH with PCM. The PCM enabled the system to maintain heating during day irradiation transients and extend it after dusk without affecting thermal efficiency.

Chaatouf et al. (2022) [85] conducted a study on a solar air heater (SAH) integrated with a PCM storage unit to optimize its efficiency for drying apricots. The SAH was analyzed under real meteorological conditions in Eastern Morocco using the computational fluid dynamics (CFD) method and a C++ UDF (User-Defined Function) subroutine written in the C++ programming language ((Note: C++ UDF is a type of function used in CFD, such as ANSYS Fluent, to extend the functionality of the software beyond the built-in capabilities). To meet the drying needs of apricots, the SAH was subjected to testing with natural and forced convection and optimization based on different tilts and air velocities. Results showed that without PCM, the SAH produced high temperatures above 70 °C that could negatively affect product quality. However, with a PCM layer (paraffin RT RT58), the temperature decreased throughout the day and increased after sunset, with an optimal combination of 1 m/s air velocity and 4 cm PCM thickness for forced convection. For natural convection, a 60° tilt was found to be appropriate for enhancing the mass flow rate, which significantly affected the outlet temperature of SAH and the melting process of PCM. In the drying chamber, these two ideal instances were further examined.

Verma et al. (2022) [86] used a transient computational iterative solution procedure to predict the thermal efficiency of a solar air heater with PCM (paraffin wax RT42) using differential airflow arrangements. The finite volume method was employed to solve the governing differential equations. The researchers used the first-order upwind advection scheme to stimulate air temperatures. The findings demonstrated that the PCM integrated into the solar air heater enabled large thermal energy storage, instantaneous and prolonged thermal backups, and air heat gain as corresponding to the daytime and changeable air flow rate configurations. The highest thermal performance was 63% at 0.05 kg/s total mass flow rate. The airflow rate settings that can deliver an outlet air temperature of at least 303 K during the night were also found by the researchers.

Sharol et al. (2022) [87] conducted an experimental analysis to investigate the effect of incorporating a thermal energy storage material inside the tube of a double-pass solar air heater with a cross-matrix absorber (DPSAH-CMA) on its thermal efficiency. The researchers tested the SAH under different operating conditions both indoors and outdoors. Results showed that the DPSAH-CMA with PCM (paraffin wax) was able to buffer thermal energy more effectively, with a rate of -0.81 °C min^{-1}, compared to the one without PCM at a mass flow rate of 0.005 kg/s. The PCM also led to an increase in the cumulative heat rate of the DPSAH-CMA, with an extreme value of 1548.54 W at the greatest solar radiation of 900 W m^{-2}. The maximum storage performance of the DPSAH-CMA with PCM was at the matching greatest mass flow rate and was inversely related to solar radiation values. When comparing the performance of the two SAHs under outdoor conditions, the DPSAH-CMA with PCM attained the greatest thermal performance, outperforming the DPSAH-CMA without PCM by 17%. Moreover, the DPSAH-CMA's exergy efficiency was found to be 23% with PCM and 15% without PCM, as per Sharol et al. (2022) [87].

Unidirectional flow, PCM-rod (ZN-110H), and spiral tubes were used in an experiment by Luo et al. (2002) [88] to increase the thermal performance of an evacuated tube solar air heater. Using an experimental test platform, the thermal efficiency of the SAH-PCM was assessed. ANSYS software was utilized to examine the temperature and flow fields in the vacuum tube, and an environmental assessment was performed. The numerical study identified the ideal PCM-rod size and estimated the spiral tube's heat augmentation coefficient to be 1.1. The results of the experimental testing showed that the SAH-PCM could effectively prevent outlet temperature swings, albeit at the expense of a 31.5 °C lower average outlet temperature. The rate of heat loss was decreased by 58.5%, and PCM's thermal efficiency could increase to 844.9%. Moreover, the operation hours at night were extended by over 3 h. These findings demonstrate that phase change materials and spiral tubes, as well as unidirectional flow structural design, can enhance the thermal efficiency of the air heater.

Farzan et al. (2022) [89] developed a double-pass solar air heater with a perforated absorber plate and integrated PCM. Three scenarios were tested over 24 h at two mass flow rates. Results showed the use of PCM and PCM/EMM increased total heat gain and energy efficiency by approximately 45.88% and 48.19%, respectively, compared to traditional SAHs. The SAH with PCM/EMM had a higher exhaust temperature in the evening, reaching 76% daily performance, but reduced exergy efficiency due to lower exhaust air temperature.

A form of construction material that uses PCMs in a ventilation system to control indoor temperatures and save energy is known as an interlayer ventilated phase change material (IVPC) system was developed in 2023 by Fan et al. [90] (Figure 17). A layer of PCM is often sandwiched between two layers of construction material, such as drywall or plasterboard. A ventilation system is subsequently built underneath the PCM layer, and the PCM layer is perforated to facilitate airflow. The PCM layer becomes heated by solar light during the day, melting and absorbing heat. The PCM solidifies at night when the temperature decreases and releases heat. Air is circulated through the perforations by the ventilation system behind the PCM layer, which enables the heat to be dispersed uniformly throughout the structure. This facilitates temperature control indoors and lessens the demand for HVAC systems. Buildings with a lot of thermal mass, such as concrete or masonry structures, benefit greatly from interlayer ventilated PCM systems. Utilizing PCMs can increase the thermal mass of the building, enhancing its capacity to absorb and store heat. The ventilation system also aids in keeping the PCM layer from overheating and keeps the building's temperature constant [90].

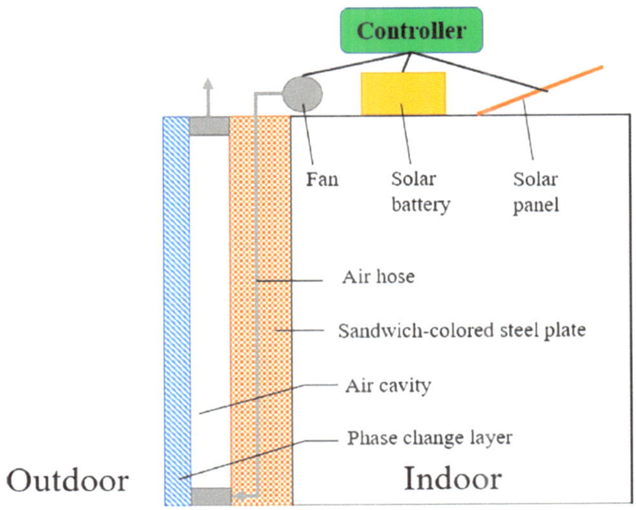

Figure 17. The basic structure of an interlayer ventilated phase change material [90].

Interlayer ventilated phase change material (IVPC), which can be used in conjunction with a solar air collector to boost thermal performance, was proposed by He et al. (2022) [91]. The performance of the IVPCs (PCM24 and EGPCM) was evaluated by trials, and the findings indicated that the ideal solar hot air supply temperature and speed should be between 35 and 45 °C and 3 to 4 m/s, respectively, when the IVPC finished the latent heat storage process. Five hours could pass while the latent heat release time was preserved. The impact of linking the IVPC with a solar air collection system (IVPSS) on heating temperature changes in a rural building located in a cold region of China was also simulated by the researchers using TRNSYS (Transient System Simulation Tool). TRNSYS is a software package for simulating the behavior of energy systems. In comparison to buildings without the IVPSS, the IVPSS was able to keep the inside temperature of the building within the range of 8 to 16 °C, a rise of 5 to 8 °C. This shows that the IVPC can efficiently utilize solar energy to change peak loads, decrease temperature swings in rooms, and enhance indoor thermal comfort, offering a clean energy alternative for heating farms in cold climates.

Brahma et al. (2023) [92] deliberate a new solar air heater (SAH) with the ability to incorporate PCM at the Department of Energy, Tezpur University. The performance of a solar air heater (SAH) integrated with different PCMs, namely acetamide, stearic acid, and paraffin wax, was evaluated and compared with a conventional SAH during both charging and discharging periods. The absorbed, useful, and lost heat were estimated for all three PCMs, and it was observed that paraffin wax exhibited the highest values for absorbed heat (36.56 kW), useful heat (14.13 kW), stored heat (21.59 kW), and efficiency (52.83%). Stearic acid (52.31%) followed closely behind, while acetamide (47.25%) had the lowest efficiency. The performance of the SAH with acetamide was indicated to be 15.09% higher than that of the SAH without acetamide, while the efficiency of the SAH with paraffin wax was 8.18% higher than that of the SAH without PCM, and the performance of the SAH with acetamide was 6.67% higher than that of the SAH without PCM.

A recent study by SunilRaj and Eswaramoorthy (2023) [93] evaluated the performance of a forced convection V-trough solar air heater (FCVTSAH) that incorporated a latent thermal energy storage system. The FCVTSAH system included a three-pass solar air heater, an absorber unit containing paraffin wax as a phase change material, an air blower, and a cylindrical drying unit. The authors used dry air temperatures ranging from 35 °C to 68 °C to assess the FCVTSAH's thermodynamic capability for drying agricultural items. The experimental tests were conducted at coordinates (12.96° N and 77.63° E) from 20 April to 22 April , 2020. The study reported that the energy performance of the FCVTSAH ranged from 2% to 50.5% on average, while its exergy performance ranged from 2.4% to 25%. The energy and exergy efficiencies of the latent heat storage unit were between 20.3% and 24.5% and 8.9% and 12.8%, respectively. Additionally, the drying unit's energy efficiency ranged from 20.5% to 74.3%.

A list of studies on phase change material for solar air heaters is shown in Table 4. The analysis of the trials in Table 4 can demonstrate that the storage efficiency of DPSAH-CMA with PCM was proportionate to solar radiation values and was optimum at higher mass flow rates. On a daily basis, the air heater's thermodynamic performance varies between 12% and 65% when equipped with a thermal storage system. When using PCM such as paraffin wax, n-octadecane, and $CaCl_2{\cdot}6H_2O$ and tubes arranged in a circular pattern, heat absorption by the PCM was enhanced by 9.17%, 1.55%, and 1.08%, respectively, compared to tubes in a staggered design.

Table 4. Outline of studies on phase change material for solar air heater.

Authors (Year) [Reference]	Configuration	Type of Study	Studied Parameters	Highlighted Results
Kabeel et al. (2016) [73]	Varying the thickness of PCM below the absorber plate.	Experimental	Altering the thickness of PCM underneath the absorber plate and its impact.	The outlet temperature of the v-corrugated plate solar air heater with PCM was higher than that of the flat plate solar air heater, ranging from 1.5–7.2 °C above ambient air temperature during the 3.5 h after sunset, compared to 1–5.5 °C for the flat plate solar air heater with a mass flow rate of 0.062 kg/s.
El Khadraoui et al. (2017) [74]	Solar dryer using PCM.	Experimental	Impact of PCM in SAH.	It was discovered that the relative humidity in the drying chamber of the solar dyer with PCM was 17 to 34.5% lower than the relative humidity in the immediate vicinity.
Salih et al. (2019) [75]	Double-pass solar air heater using multiple rectangular capsules filled with paraffin wax based on a PCM.	Experimental and numerical	Air flow rate and PCM.	The increased airflow rate causes the melting process to take longer to complete and lowers the paraffin's melting temperature.
Abuşka et al. (2019) [76]	The PCM panel of SAH utilizes honeycomb as the internal fin structure along with PCM.	Experimental	Impact of PCM and honeycomb.	At mass flow rates of 0.032, 0.04, and 0.048 kg/s, the Type I-II heaters showed an average daytime thermal performance increase from 8.4% to 9.0%. Conversely, at mass flow rates of 0.008, 0.016, and 0.024 kg/s, the average daytime thermal performance decreased from 0.8% to 8.0%.
Raj et al. (2019) [77]	Double-pass solar air heater system with PCM.	Experimental	The effect of the geometry of encapsulation used for storage.	For DPSAHS fitted with rectangular and cylindrical macro-encapsulates, average encapsulate efficiencies of 47.2% and 67%, respectively, were found.
SunilRaj et al. (2020) [78]	PCMs mixed Al_2O_3 nanomaterials are inserted in a V-shaped solar thermal collector that makes up the SAH.	Experimental	Thermal energy storage.	The daily thermal efficiency range of an air heater with thermal storage system is 12% to 65%.
Ameri et al. (2021) [79]	The traditional solar air heater and the solar air heater with paraffin phase change material (PCM).	Experimental	Using PCMs in SAHs.	The output temperature of SAHs increases by nearly 5 °C when PCMs with higher melting point temperatures are used, and the daily performance of SAHs decreases by approximately 3%.

Table 4. *Cont.*

Authors (Year) [Reference]	Configuration	Type of Study	Studied Parameters	Highlighted Results
Mahdi et al. (2021) [80]	SAH with metal foam saturated with PCM.	Numerical	The effect of the metal foam and the variation in coolant flow rate on the solidification performance.	The porosity and liquid development rate are negatively related. When the air flow rate is increased from 0.01 kg/s to 0.03 kg/s, the end means PCM temperature decreased from 69.9 °C to 66.4 °C.
Madhulatha et al. (2021) [81]	The SAH was evaluated using three distinct PCMs, namely paraffin wax, n-octadecane, and calcium chloride hexahydrate ($CaCl_2{\cdot}6H_2O$).	Numerical	Type and arrangement of PCM.	When using paraffin wax, n-octadecane, and $CaCl_2{\cdot}6H_2O$ as the PCM, tubes in a circular layout boosted PCM's ability to absorb heat by 9.17%, 1.55%, and 1.08%, respectively, relative to tubes in a staggered design.
Dinesh et al. (2022) [82]	Baffled solar-based air heater with organic PCM.	Experimental	Impact of organic PCM in a baffled SAH.	The integration of organic PCM into the confused SAH improved its energy-efficiency performance.
Abu-Hamdeh et al. (2022) [83]	Solar air heater equipped with PCM.	Experimental	Impact of using PCM.	The solar air heater's yearly EC was eventually decreased by 5.6% by using PCM.
Palacio et al. (2022) [84]	Double-channel solar air heater (SAH) with and without PCM.	Experimental	PCM in the SAH.	The SAH with PCM's highest outlet temperatures in the three cases were, respectively, 62 °C, 82 °C, and 79.5 °C.
Chaatouf et al. (2022) [85]	SAH with a heat storage unit made of PCM block.	Numerical	The SAH's performance can be affected by the type of PCM used and its thickness.	The temperature is changed by adding a PCM layer so that it drops during the day and rises at night.
Verma et al. (2022) [86]	PCM embedded parallel flow in solar air heater.	Numerical	Impact of PCM embedded parallel flow in solar air heater.	With a total mass flow rate of 0.05 kg/s, the highest heat efficiency of roughly 63% is attained.

Table 4. *Cont.*

Authors (Year) [Reference]	Configuration	Type of Study	Studied Parameters	Highlighted Results
Sharol et al. (2022) [87]	A double-pass solar air heater with a cross-matrix absorber and thermal energy storage material inside the tube.	Experimental	The impact of incorporating a thermal energy storage material inside the tube of a double-pass solar air heater.	The greatest DPSAH-CMA-with-PCM storage efficiency was achieved at greater mass flow rates and was inversely correlated with solar radiation levels.
Luo et al. (2002) [88]	A solar air heater with evacuated tubes, featuring unidirectional flow, a PCM-rod, and a spiral tube.	Experimental	The structural design, spiral tube, and the use of PCM.	The use of phase change materials, spiral tubes, and unidirectional flow in the air heater's structural design can enhance its thermal performance.
Farzan et al. (2022) [89]	Novel double-pass SAH design using integrated PCM and a perforated absorber plate.	Experimental	Effect of PCM/EMM.	The innovative SAH employing PCM/EMM has a daily efficiency of 76%.
He et al. (2022) [91]	An interlayer ventilated PCM component coupled with SAH.	Experimental and numerical	Impact of PCM coupled with SAH.	The ideal solar hot air supply temperature and speed are 35–45 °C and 3–4 m/s, respectively, if IVPC successfully stores latent heat.
Brahma et al. (2023) [92]	Novel solar air heater with the provision of integrating PCM.	Analytical	Impact of using acetamide in SAH.	It was discovered that SAH with acetamide was 15.09% more effective than SAH without acetamide; likewise, SAH with paraffin wax was revealed to be 8.18% more effective; and acetamide was discovered to be 6.67% more effective than SAH without PCM.
SunilRaj and Eswaramoorthy (2023) [93]	V-trough solar air heater with latent thermal energy storage unit.	Experimental	Effect of latent thermal energy storage unit.	Latent heat storage units' respective energy and exergy efficiencies range from 20.3% to 24.5% and 8.9% to 12.8%.

The following Equation (1) represents the thermal energy balance required to characterize the operation of solar air collectors [94]:

$$M_C C_P \frac{dT_{p,ave}}{dT} + mC_P(T_{out} - T_{in}) = \eta_0 I A_C - U_C(T_{p,ave} - T_e)A_C \quad (1)$$

M_C is the mass of the collector, m is the mass flow rate, C_P is air-specific heat, $T_{p,ave}$ is the average plate temperature, T is the temperature, T_{out}, T_{in} are outlet and inlet temperatures, respectively, η_0 is the optical yield, I is the solar radiation, A_C is the surface area of the collector, U_C is the energy loss coefficient, and finally T_e is the temperature of the environment. The energy loss coefficient takes into account all collector losses, including those at the top, sides, and bottom. Temperature and emissivity of the absorbent bed, as well as wind convection on the top cover, have a significantly bigger impact on these losses than any other element.

2.5. Solar Chimney-Related Studies

A solar chimney system is a piece of equipment that powers natural ventilation in buildings using solar power. A solar chimney's fundamental structure is composed of a tall, vertical shaft with a glass surface that lets light in to warm the air inside. Vents are situated at the base of the chimney, and when the air inside the chimney warms up, it rises and pulls cooler air in. A flow of air produced by this natural convection can be utilized to cool and ventilate the building [95]. A solar chimney's efficiency is affected by a number of variables, such as the chimney's height and direction, the size and location of the vents, and the amount of solar radiation available. A solar chimney system's effectiveness can be increased by using high-quality building materials, choosing the right chimney shape and size, and orienting and positioning the chimney to catch as many sun rays as possible. For the system to operate at its best and last as long as possible, proper maintenance is required, including cleaning the vents and checking the efficiency of the chimney [96]. Notably in hot and dry locations, solar chimney systems can offer an affordable, environmentally friendly alternative for natural ventilation and cooling in buildings.

The current section presents a comprehensive review of existing research on PCMs on the solar chimney providing a critical analysis of the findings and discussing potential limitations or gaps in the research.

Li et al. (2017) [97] developed a numerical model to investigate the influence of different parameters on the thermal efficiency of a solar chimney coupled to PCM (paraffin RT 42). The model was validated, and the parameters studied included air flow rate, melting/freezing time, and temperature difference between the inlet and outlet. The study found that the phase change temperature of the PCM had a significant impact on the thermal efficiency of the solar chimney, with wider phase change temperature differences leading to earlier complete melting. However, this also caused fluctuations in room air temperature, which may result in thermal discomfort. The specific heat of the PCM affected the contribution of sensible heat to the melting/freezing process. The research findings indicate that enhancing the thermal conductivity of the PCM resulted in a rise in the air temperature difference and average mass flow rate from 1.5 to 2 °C, and from 0.033 to 0.038 kg/s, respectively, as the thermal conductivity increased from 0.2 to 0.6 W/m °C.

Bin et al. (2017) [98] investigated the impact of PCM position in the hybrid wall on the chimney effect, temperature, air velocity, and mass flow rate. As shown in Figure 18, a greater temperature difference was observed between the heating surface and the air gap than between PCM and the absorber. The conductivity of $Na_2CO_3 \cdot 10H_2O$ PCM and chimney effect during charging time caused convection that led to an immediate increase in the heat surface temperature of PCM, followed by a brief period of fixed temperature (melting temperature) if the density of radiating heat was sufficient. However, in the absence of steady temperature, the heat surface temperature of PCM declined rapidly during the discharging time (solidification temperature).

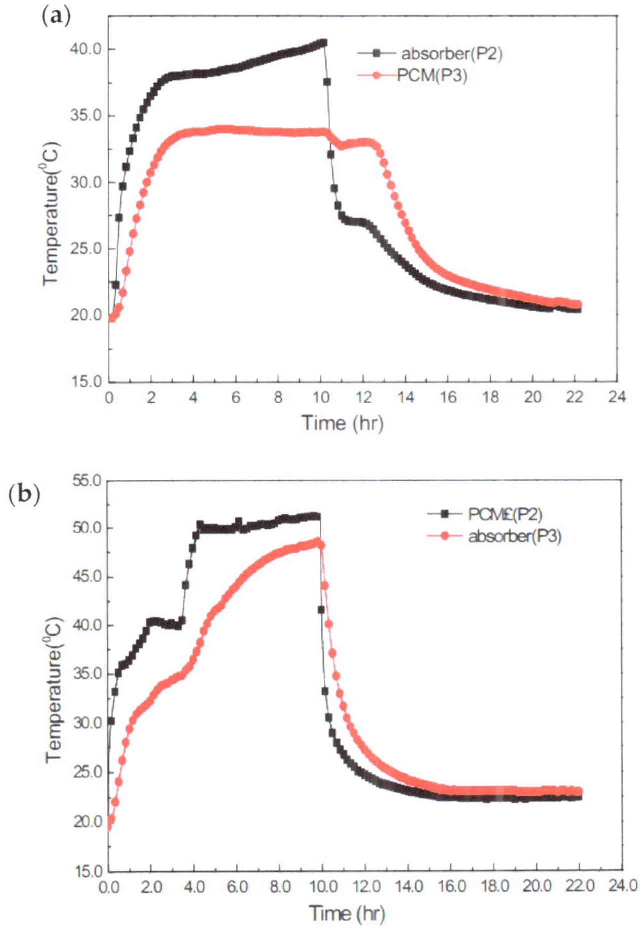

Figure 18. The heated surface temperature development of PCM and absorber [98]. (**a**) the heated surface temperature development of PCM and absorber when PCM was behind of the absorber at the height of 750 mm, (**b**) the heated surface temperature development of PCM and absorber when PCM was in front of the absorber at the height of 750 mm.

In order to improve natural ventilation and decrease heat gain admittance, Thantong et al. (2018) [99] published an evaluation of the thermal performance of a new arrangement of solar chimneys incorporating a phase change material (SC-PCM) (RT 42 paraffin wax). A 0.05 m air gap separates the two walls of this SC-PCM, which is made up of two walls. Three layers make up the exterior wall that faces the sun; the first is a cement board panel that has been painted black and is only 0.01 m thick. A paraffin-containing tank (0.015 m × 0.8 m × 1.28 m) is the second stratum. A 0.005 m thick zinc sheet makes up the third layer. The inner wall is made of thin, 0.07 m thick concrete. The SC-PCM was incorporated into the south facade of a modest house with a volume of 4.05 m^3 and walls that were 10 cm thick from concrete. Another home model with the same volume was used to contrast thermal efficiencies to a straightforward concrete wall (SW). Figure 19 illustrates the experimental findings that the interior temperature of the solar chimney with PCM room was lower than that of the room with a single concrete wall.

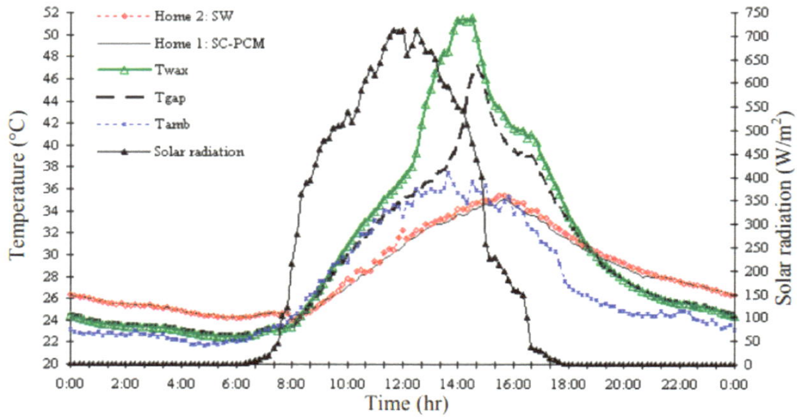

Figure 19. Hourly change in temperature of indoor air of the two houses, wax and air gap of the SC-PCM, ambient temperature and solar radiation on the vertical plane (5 February 2017) [99].

In their 2018 study, Fadaei et al. investigated the impact of latent heat storage (LHS) on a solar chimney pilot, conducting two sets of experiments—one with paraffin wax (C20) PCM and one without. Using a 3 m chimney height and 3 m collector diameter on the University of Tehran campus, they measured temperature and velocity parameters to evaluate the system's efficiency. Results showed that the maximum air velocity was 2 m/s for the system with PCM thermal storage, compared to 1.9 m/s for the conventional solar chimney. The maximum absorber surface temperature was also 3 °C higher for the PCM system than the conventional system. The LHS system increased the average mass flow rate of the pilot by approximately 8.33%, as shown in Figure 20.

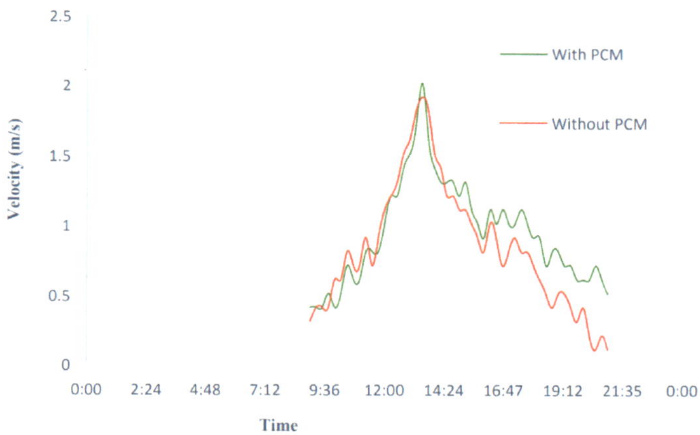

Figure 20. The velocity of the fluid, with and without PCM on August 22nd [100].

Dordelly et al. (2019) [101] explored the effect of adding a PCM (Rubitherm RT44) on the efficiency of two solar chimney prototypes in a laboratory setting, in order to make it a practical option for year-round use. Solar chimneys are known to provide continuous ventilation and improve indoor air quality. The prototype used in this study was made of 2 cm plywood plates with a thermal conductivity of 0.15 W/m K and a size of 3.50 × 1.00 × 0.30 m. After a 6 h charging period, the solar chimney achieved a mean ventilation rate of over 70 m^3/h, with a relatively low gain of 550 W/m^2 from 7 halogen lamps directed towards an effective collector area of 3.00 m^2. The findings showed that integrating PCM improved the ventilation rate and slowed the rate of decrease during the discharge phase (6 h) when the halogen lamps did not deliver any energy to the solar chimney, as demonstrated in Figure 21.

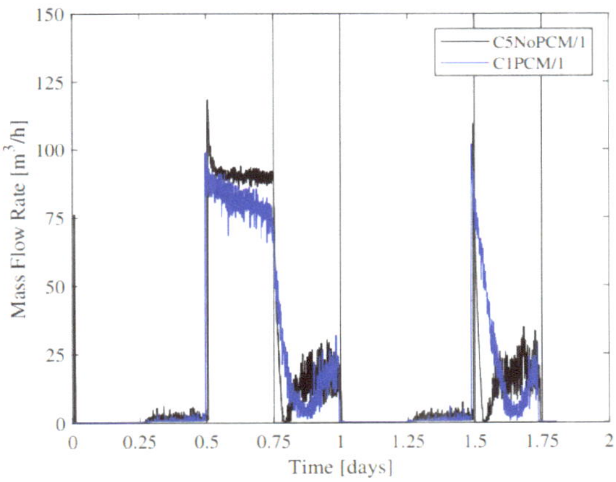

Figure 21. Mass flow rate results under similar operating conditions for C5NoPCM-1 and C1PCM-1 [101].

Tiji et al. (2020) [102] explored the efficiency of a solar chimney with PCM storage (Sodium sulphate decahydrate ($Na_2SO_4 \cdot 10H_2O$)) with and without fins linked to the absorber plate. The aim was to investigate the velocity and temperature distributions, in addition to the airflow rate inside a room, with and without the PCM and fins, and then compare the results to a non-PCM chimney. The researchers simulated the system using CFD simulation for an entire winter day. The findings indicated that the utilization of PCM has improved the temperature uniformity inside the room but did not reach thermal comfort conditions, with a mean temperature of 14.68 °C. To address this, fins were added to the absorber plate, resulting in a 20% improvement in the mean room temperature in comparison to the non-finned case. The results are illustrated in Figure 22.

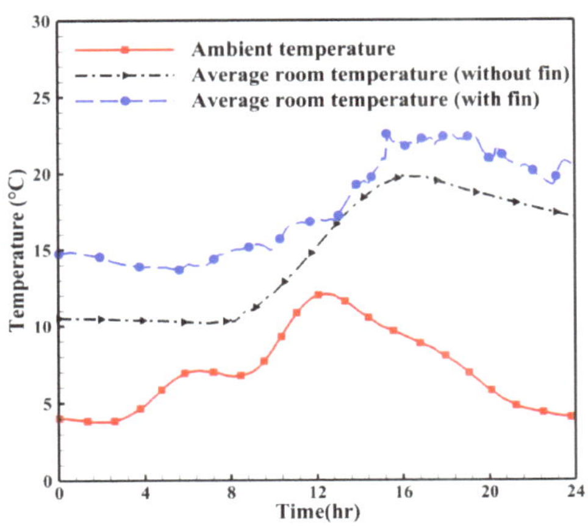

Figure 22. The room's mean temperature variation for a whole day with PCM in two cases, with and without fin [102].

Phase change capsules (PCC) of paraffin wax are stacked over various sieve beds to create porous layers of heat storage in a new method of phase change heat storage for solar heating reported by Chen and Chen (2020) [103]. The flow of heated air in the system is propelled by the buoyancy force produced by the solar chimney. The Brinkman–Forchheimer extended Darcy model and double energy equations are used to investigate heat transport and flow in the porous thermal storage layer. The model takes local thermal non-equilibrium (LTNE) in the porous beds into account. Additionally, the impacts of the materials, porosity, and particle size of PCC in the porous thermal storage as well as the features of the flow channel on the thermal storage efficiency are investigated using the k- turbulence model in conjunction with the aforementioned equations. According to the findings, the thermal storage duration of the PCC layer is 7% to 25% longer than that of the Tuff and Basalt layers, as illustrated in Figure 23.

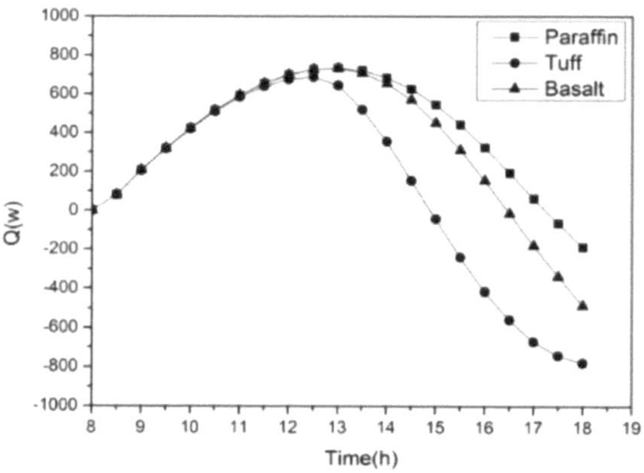

Figure 23. Variation of heat storage capacity with time in the system with different particle materials [103].

Ashouri and Hakkaki-Fard (2021) [104] aimed to improve the performance of an inclined rooftop solar chimney (SC) that integrates PCM and a photovoltaic module (SC-PCM-PV) system. The study examined the efficiency of the system by employing finned absorbers and composite PCMs, and created a 3D-CFD model using the finite volume method. Various design parameters, such as the number and thickness of fins, PCM type (paraffin RT-50, SAT/CF, Pa/CF), and PCM mass were investigated to assess their impact on the length of natural ventilation, ventilation capacity, and power output. This research marks the first time such an investigation has been conducted. The findings demonstrated that raising PCM mass does not considerably lengthen breathing duration but did decrease the ventilation capacity for finless instances. On the contrary, hand, using a finned absorber increases ventilation capacity and duration by 7.7% and 17.6%, respectively. Furthermore, copper-foam-based composite PCMs lengthen the time that the SC-PCM-PV system is ventilated.

Nateghi and Jahangir (2022) [105] conducted a study using Energy Plus to simulate the thermal performance of a house under three different climates. The study included three modes of operation: without solar chimney (SC), with SC, and with SC combined with a layer of PCM. In this regard, the authors used three types of PCMs including the paraffin RT42, RT35, and 25 RT. The study found that incorporating PCM in solar chimneys resulted in lower thermal comfort in hot-arid climates for both heating and ventilation modes. However, in hot-humid and cold semi-arid climates, PCM improved the performance of the SC in providing indoor thermal comfort. The study revealed that the average thermal comfort indices (PMV) improved in hot-humid and cold semi-arid cities by using PCM. The cooling mode of the cold semi-arid climate (Tehran) benefited the most from using PCM.

Li et al. (2022) [106] examined how the inclination angle and heat flux impact the thermal efficiency of a solar chimney with PCM. They varied the heat fluxes (400, 500, and 600 W/m^2) and inclination angles (30°, 45°, and 60°), and found that the melting times of the PCM were longer at lower inclination angles and heat fluxes. The inclination angle had a greater effect on the heat transfer than the heat flux. The solar chimney inclined at 45° had the highest air flow rate, and peak air velocities were 0.27–0.37 m/s during the charging process, and 0.06–0.1 m/s during discharge.

Long et al. (2022) [107] explored a solar chimney-enhanced active hybrid evaporative cooling (SCEAHE) system that incorporated phase change material (PCM) using a verified numerical model. The integration of PCM in the SCEAHE system improved its thermal inertia. According to the modeling results, the absorber surface's maximum temperature with PCM was 78.8 °C, which was 16.2% lower than its maximum temperature without PCM. The PCM was charged and discharged for roughly 9.5 h (05:30–15:00) and 14.5 h, respectively. Approximately 5 h (15:30–20:30) were spent with the PCM entirely melted, and the solidification of the PCM occurred from 02:00 to 7:15, reaching a maximum of 24% at 05:30. While the highest air flow rate reduced by 17.8% to 209.5 m^3/h during the day, the integration of PCM raised the SCEAHE air flow rate by 50% at night. For the SCEAHE with and without PCM, the output air temperatures were from 24.8 to 26.5 °C and from 24.4 to 27.2 °C, respectively.

Overall, it was earlier suggested that the significant variables governing the flow in a solar chimney were ρ; A; V; q''; cp; β; hc; g. In a subsequent work of [108], certain primitive variables were grouped together and presented in Equation (2):

$$\rho AV\frac{V^2}{2} = f^n\left(\rho, g, \frac{q''A_r}{C_P}, \beta, h_C\right) \tag{2}$$

Additionally, the dimensionless relationship was found to be [109]:

$$\begin{gathered}\prod 1 = f^n\left(\prod 2\right) \text{ or} \\ \frac{\frac{1}{2}m\cdot V^2}{\rho h_C^{\frac{7}{2}} g^{\frac{3}{2}}} = f^n\left(\frac{q''A_r\beta}{\rho C_P h_C^{\frac{5}{2}} g^{\frac{1}{2}}}\right)\end{gathered} \tag{3}$$

According to the Buckingham Pi theorem,

$$s = n - b \tag{4}$$

In this case, one dimensionless variable is desired with four basic dimensions [mass (M), length (L), time (t), and temperature (θ)] involved, or s = 1 and b = 4. Hence, only five quantities are essential.

Equation (1) involves six quantities: $\rho A\frac{V^3}{2}$; $\rho, g, \frac{q''A_r}{C_P}, \beta$ and h_C.

Thus, two quantities are selected from among the six to merge into one new term.

The chimney effect, dependent on g and hc, is among the driving mechanisms in the solar chimney system. Therefore, $g.hc$ is chosen to be a new variable. Therefore, the pertinent variables are reduced to be [110]:

$$\rho AV\frac{V^2}{2} = f^n\left(\rho, \frac{q''A_r}{C_P}, \beta, g.h_C\right) \tag{5}$$

A is flow area, ρ is density, V is flow velocity, g is gravitational acceleration, q'' is isolation, A_r is roof area, C_P is specific heat, β is coefficient of volumetric thermal expansion, h_C is chimney height, m is mass flow rate.

The studies related to the usage of phase change material in solar chimneys are displayed in Table 5. The evaluation of the studies in Table 4 shows that the SCEAHE system's maximum absorber surface temperature with PCM was 78.8 °C, which was 16.2% lower than it was without PCM. For the studied circumstances, the solar chimney with a 45° inclination also had the highest air flow rate. Moreover, using PCM in solar chimneys in both SC functioning modes results in unhappiness in hot arid climates (ventilation and heating).

Table 5. Outline of studies on phase change material for solar chimney.

Authors (Year) [Reference]	Configuration	Type of Study	Studied Parameters	Highlighted Results
Li et al. (2017) [97]	Solar chimney integrated with PCM.	Numerical	Phase change temperature.	The air temperature differential increased from 1.5 to 2 °C and the average mass flow rate increased from 0.033 to 0.038 kg/s as the thermal conductivity of the PCM increased from 0.2 to 0.6 W/m °C, respectively.
Bin et al. (2017) [98]	PCM in the hybrid wall of the solar chimney.	Experimental	The effect of the position of PCM in the hybrid wall.	Without a period of steady temperature, the heat surface temperature of the PCM will rapidly decline during the discharging time (solidification temperature).
Thantong et al. (2018) [99]	A new arrangement of solar chimneys integrating a PCM.	Experimental	Impact of integrating solar chimney with PCM.	Integrating a solar chimney with a PCM is highly effective since it can increase airflow and significantly lower heat gain through the south wall.
Fadaei et al. (2018) [100]	Latent heat storage in a solar chimney pilot.	Experimental	Latent heat storage.	With the SC with PCM and the traditional solar chimney (CSC), the highest absorber surface temperatures are 72 °C and 69 °C, respectively.
Dordelly et al. (2019) [101]	Integrating a Phase Change Material (PCM) in two different laboratory solar chimney prototypes.	Experimental	PCM.	When there is no energy coming from the halogen lamps to the solar chimney (6 h discharge), PCM integration offers a faster ventilation rate and a slower drop.
Tiji et al. (2020) [102]	PCM-enhanced solar chimney with and without considering fins connected to the absorber plate.	Numerical	Impact of using PCM and fins.	Utilizing fines leads to a 20% increase in the room's mean temperature for the PCM-based solar chimney system in comparison to the non-finned instance.
Chen and Chen (2020) [103]	Novel phase change heat storage for solar heating.	Numerical	Effect of the porosity and particle size in the porous layer.	Enhanced thermal storage efficiency is possible at the system's ideal ratio of flow channel width to bed thickness, which depends on the porosity and particle size in the porous layer.
Ashouri and Hakkaki-Fard (2021) [104]	A solar chimney system integrated with a PCM and photovoltaic module, installed on an inclined rooftop.	Experimental	Impact of using paraffin/copper foam (Pa/CF) as the PCM.	The utilization of paraffin/copper foam (Pa/CF) as the PCM leads to a power output of 5.54 kWh and ventilation duration of 22 h and 30 min. These values are approximately 16% and 101% higher, respectively, than those obtained with the SC-PV system.

Table 5. *Cont.*

Authors (Year) [Reference]	Configuration	Type of Study	Studied Parameters	Highlighted Results
Nateghi and Jahangir (2022) [105]	There are three distinct Energy Plus modes: without SC, with SC, and with SC plus a layer of PCM.	Experimental	Effect of combining PCM in solar chimneys.	In hot-arid climates, combining PCM in solar chimneys leads to dissatisfaction in both SC operation modes (ventilation and heating).
Li et al. (2022) [106]	Inclination angle and heat flow on the integrated PCM-based thermal performance of a solar chimney.	Experimental	The inclination angle of SC integrated with PCM.	The solar chimney with a 45° generated the greatest air flow rate.
Long et al. (2022) [107]	PCM-based SCEAHE system.	Experimental	Effect of using PCM.	The SCEAHE system's maximum absorber surface temperature with PCM was 78.8 °C, which is 16.2% lower than without PCM.

3. Critical Evaluation of Utilized PCMs in the Covered Solar Energy Systems

This study discussed the implications of PCMs in different types of solar energy systems including the solar still, solar ponds, solar air heaters, and solar chimneys. The types and thermos-physical characteristics of the OCMs used in solar energy systems are shown in Table 6. Table 6 depicts the melting temperature, latent heat, and thermal conductivity as they play important roles in the utilization of PCMs. As a critical evaluation of the covered studies, it can be stated that the PCMs enable to absorption and store of heat, which efficiently contributes to a reduction in heat loss and an increase in the solar collector system's energy efficiency. Compared to conventional materials, PCMs are able to store more heat which results in more stable energy output. Furthermore, the utilization of PCMs in solar energy systems has resulted in a smaller overall system due to less storage space being needed. System longevity can also be increased the ability of PCMs to lessen thermal stress on the solar collector system. More importantly, the PCMs can lessen energy expenses by minimizing the need for backup energy sources at times of poor solar energy availability. However, the addition of PCMs to the solar collector system may result in an increase in the system's initial cost, which could make it less affordable for some consumers. Additionally, PCMs can only store thermal energy within a certain temperature range, they are less effective at extremely hot or cold temperatures. The longevity of the system is quite related to the number of heat cycles that PCMs can withstand before degrading. Notably, the routine maintenance for the solar collector system is vital to guarantee optimal performance, which in turn would increase the operational cost. There is also a necessity for additional space for energy storage, and therefore the overall system design may be affected. Beyond the above demerits, the PCMs have low thermal conductivity coefficients, which reduce the heat storage and release during melting and crystallization operations and thus exhibit relatively large volume changes during phase transitions.

Table 6 shows the existence of several other effective PCMs with competent thermal conductivity larger than 1 W/m K and substantial latent heat, in contrast to a significant number of PCMs with poor thermal conductivity. In order to protect the PCM from moisture and other environmental variables, it is crucial to use these materials for future research and optimization. By using particular methods such as fins and heat exchangers, the heat exchange between the PCM and the environment should also be increased. The addition of thermally conductive fillers, such as carbon nanotubes, graphene, or metal particles, can also boost the thermal conductivity and latent heat of PCMs. Another crucial issue that needs to be taken into account is the PCMs' melting temperature. In general, PCMs with higher melting temperatures have more energy to store. This implies that they have greater energy storage capacity per unit of mass or volume, which may be helpful in particular applications. Additionally, PCMs with higher melting points typically have better thermal stability, which means they can hold onto their properties for a longer time. In fact, a PCM's latent heat directly correlates to how much energy it can store or release during a phase shift. In comparison to PCMs with lower latent heats, those with higher latent heats can store or release more energy per unit of mass or volume. A balance between the desired thermal properties and other factors, most importantly the energy storage capacity, must be found. Beyond this, it should be acknowledged that the higher thermal conductivity values may come with tradeoffs, such as reduced energy storage capacity or increased costs. In this regard, the choice of whether to raise or lower a PCM's melting temperature should be made based on the application's needs. This is because PCMs with high melting temperatures are likely to have low thermal conductivity. Finally, since PCMs' molecular structure directly affects their thermal conductivity, future research should focus on determining what that structure should be. For instance, altering the PCM chains' length and branching might enhance the passage of heat between molecules.

Table 6. Overview of types and thermos-physical characteristics of PCMs used in solar applications.

Type of PCM	Range of Melting Temperature (°C)	Range of Latent Heat Capacity (J/g)	Thermal Conductivity (W/m K)	Authors (Year) [Reference]
Commercial-grade paraffin wax	47–68	160–220	≤0.2 *	Feliński and Sekret (2016) [25]
n-eicosane micro-encapsulated PCM) suspended in water or a mixture of water and glycol	36–38	195	0.23	Serale et al. (2016) [26]
Erythritol and expanding graphite (0–4 wt.%)	130.45–129.56	320.7–308.8	0.703–2.674	Li and Zhai (2017) [27]
Microencapsulated phase change material of paraffin@melamine resin mixed in an ethanol/water mixture	42.5–49.8	166	-	Wang et al. (2019) [28]
PCM was encapsulated in spherical particles and made of 20% high-purity graphite and 80% natural grease	15–70	190–210	3–5	Wang et al. (2019) [29]
A microcapsule of an octadecane core and a melamine-formaldehyde resin shell with graphene oxide and oleic acid-coated Fe_3O_4 magnetic nanoparticles in the shell (rGO-MPCMs)	25.82–28.34	73.05	0.4076	Gao et al. (2020) [30]
PCM1: Ecopetrol Semirefinada PCM, and PCM2: Rubitherm RT-47 PCM	60, 41–46	210, 160	0.9, 0.2	Palacio et al. (2020) [31]
PCM1: Paraffin wax + 1% nano-Cu, and PCM2: paraffin wax + 2% nano-Cu	59.57, 59.14	160.3, 172.2	0.196, 0.226	Elarem et al. (2021) [32]
PCM1: Medical paraffin and PCM2: grade A paraffin waxes with ZnO micro-particles or 5 wt% CuO nanoparticles	45, 64	190, 268	0.21, 0.211	Alshukri et al. (2021) [33]
Shape-stabilized PCM of sodium phosphate monohydrate, sodium acetate trihydrate, expandable graphite and carboxymethylcellulose	55–57	193.62	-	Sadeghi et al. (2022) [34]
Shape-stabilized PCM of sodium acetate trihydrate and salt hydrate	57	218.5	1.5	Yeh et al. (2022) [35]
Encapsulated spherical capsules and a synthetic oil	280–380	75–266	0.5–0.92	Nekoonam and Ghasempour (2022) [36]
Decane PCM and glass (amorphous silica)	-	-	0.74	Hatamleh et al. (2022) [38]
Composite of paraffin wax with 0.33 wt% nano-copper	51.2	227	0.31	Algarni et al. (2020) [39]
CuO/water nanofluid, CuO/ethylene glycol nanofluid, binary ($CuO + Al_2O_3$)/water nanofluid, binary ($CuO + Al_2O_3$)/EG nanofluid in two different volume fractions, and CuO/water nanofluid	36.85	192.66	0.749	Karami et al. (2023) [40]

Table 6. *Cont.*

Type of PCM	Range of Melting Temperature (°C)	Range of Latent Heat Capacity (J/g)	Thermal Conductivity (W/m K)	Authors (Year) [Reference]
Tritriacontane paraffin PCM and copper metal foam	72	256	0.21	Pawar and Sobhansarbandi (2023) [41]
Shape-stabilised PCM of 5 wt% expanded graphite	46	195	1.5	Cheng et al. (2019) [44]
PCM1: paraffin wax and PCM2: two PCMs of different phase change temperature ranges	58.03–64.5, 53.05–62	-	-	Vigneswaran et al. (2019) [45]
Three PCMs: Sodium thiosulfate pentahydrate, sodium acetate trihydrate, and paraffin wax	48.5, 58, 56–58	208.8, 265, 226	0.16–0.37, 1.44, 0.2–0.3	Abu-Arabi et al. (2020) [46]
PCM1: and PCM2: hollow cylindrical pin fins embedded in the PCM	56–58, 60.85	-	-	Yousef et al. (2019) [47]
12 aluminum tubes with paraffin wax with copper rods	-	226	-	Elashmawy et al. (2020) [48]
Copper tubes filled with paraffin were as PCM	56	-	0.24	Kabeel et al. (2020) [49]
TH-ME 58 (SiO_2-paraffin))	46.06–62.49	153.76	0.287	Chen et al. (2021) [50]
A commercialized inorganic salt hydrate + water + Additive PCM (Product number: PGSCR Co. PCM28/315)	28	225	0.6 ± 15%	Jahanpanah et al. (2021) [51]
Paraffin wax combined with copper oxide nanoparticles (975 g PCM + 25 g CuO)	53	187	0.28	Abdullah et al. (2022) [55]
Paraffin wax and 3 wt% Al_2O_3	56–58	226	0.24	Tuly et al. (2022) [56]
0.5–2.5 wt% of 40 nm sized Zn-nanoparticle and 5 kg paraffin PCM	36–42	-	0.3398–0.4001	Afolabi et al. (2023) [57]
Paraffin wax	54.5	190	0.24	Bacha et al. (2023) [58]
4 PCMs including the n-pentacosane, n-hexacosane, n-docosane and n-tetracoscane.	54, 56, 44, 51	238, 257, 249, 255	0.21, 0.21, 0.21, 0.21	Amirifard et al. (2018) [62]
Paraffin wax	46	210	-	Beik et al. (2019) [64]
Paraffin wax 50 and paraffin wax 60	48–50, 58–60	171.1, 213.8	0.2, 0.2	Wang et al. (2020) [65]
Paraffin wax	41–44	255	0.2	Rghif et al. (2021) [66]

Table 6. *Cont.*

Type of PCM	Range of Melting Temperature (°C)	Range of Latent Heat Capacity (J/g)	Thermal Conductivity (W/m K)	Authors (Year) [Reference]
Paraffin waxes of RT35 HC and RT44 HC	34–36, 41–44	240, 250	0.2, 0.2	Colarossi et al. (2022) [67]
RT 50 and RT 60 paraffin waxes	less than 55	171.1, 213.8	0.215, 0.212	Wang et al. (2022) [69]
Paraffin wax	46	210	-	Reza et al. (2022) [70]
Paraffin wax	54	190	0.21	Kabeel et al. (2016) [73]
Paraffin wax	56–60	214.4	0.21	El Khadraoui et al. (2017) [74]
Paraffin wax	38–43	174	0.2	Salih et al. (2019) [75]
Paraffin wax	58–60	180.8	0.241	Raj et al. (2019) [77]
Paraffin wax type RT70HC (RUBITHERM)	69	260	0.2	Mahdi et al. (2021) [80]
Three PCMs of paraffin wax, n-octadecane, and calcium chloride hexahydrate	64, 30, 30	268.336, 245, 170	0.21, 0.358, 1.09	Madhulatha et al. (2021) [81]
HS29 PCM	26–29	190	1.05	Abu-Hamdeh et al. (2022) [83]
Rubitherm RT 45	41–46	160	0.2	Palacio et al. (2022) [84]
Paraffin RT58	58	180	0.2	Chaatouf et al. (2022) [85]
Paraffin wax RT42	38–43	-	0.2	Verma et al. (2022) [86]
Paraffin wax	58–60	214.4	0.2	Sharol et al. (2022) [87]
PCM-rod (ZN-110H)	54.4	147.7	0.9	Luo et al. (2002) [88]
Interlayer ventilated phase change material of PCM24 and EGPCM	29.81, 26.32	216.05–214.23, 172.36–164.89	0.22, 0.81	He et al. (2022) [91]
Acetamide, stearic acid, and paraffin wax	112.8, 63.4, 65.8	210.21, 200.7, 198.7	0.592, 0.18, 0.24	Brahma et al. (2023) [92]

Table 6. *Cont.*

Type of PCM	Range of Melting Temperature (°C)	Range of Latent Heat Capacity (J/g)	Thermal Conductivity (W/m K)	Authors (Year) [Reference]
Paraffin RT 42	38–43	174	0.2	Li et al. (2017) [97]; Thantong et al. (2018) [99]
paraffin wax (C_20)	44–46	189	0.21	Fadaei et al. (2018) [100]
Rubitherm RT44	41–44	250	0.15	Dordelly et al. (2019) [101]
Sodium sulphate decahydrate ($Na_2SO_4 \cdot 10H_2O$)	33.85	126	-	Tiji et al. (2020) [102]
Paraffin wax	28.85–30.85	-	0.36	Chen and Chen (2020) [103]
Paraffin RT-50, SAT/CF, Pa/CF	50, 56.75, 42.67	160, 271, 270.7	0.2, 3.4, 6.2	Ashouri and Hakkaki-Fard (2021) [104]
Three PCMs of paraffin RT42, RT35, and 25 RT	38–43, 34–36, 22–26	174, 157, 148	0.2, 0.2, 0.19	Nateghi and Jahangir (2022) [105]; Li et al. (2022) [106]; Long et al. (2022) [107]

*: Ukrainczyk et al. 2010 [111].

4. Conclusions

Solar energy is an attractive alternative that helps to reduce reliance on non-renewable energy sources. Among renewable solar energy applications, domestic solar water heating systems (DSWHS) are widely used. However, these systems face challenges, the primary one being the inability to provide sufficient hot water on demand due to reduced solar radiation during nighttime or cloudy days. This review study focused on the current implications of PCMs in different solar thermal energy systems. The following conclusions are made after reviewing more than 90 relevant studies and analyzing their results:

1. Solar collector: Adding fins significantly affects the ETSC's ability to transfer heat from the paraffin during its phase transition. It was discovered that as the fins' thickness decreases, the PCM melts more quickly. Additionally, it was found that adding 1% Cu to the PCM was the ideal mass concentration for raising the HTF output temperature by 2 K.
2. Solar collector: The storage efficiency for mid-temperature use was increased by more than 40% as a result of using PCM.
3. Solar still: Due to their larger effective heat capacities, droplets with MPCM may experience a slower temperature drop during the flash evaporation process.
4. Solar still: The SSPCM-equipped solar still entails an increase in productivity of less than 44% beyond the productivity of the standard solar still.
5. Solar bond: A solar pond containing Al_2O_3 nanoparticles was capable of providing hot water till 7:30 p.m. in the evening.
6. Solar bond: The temperature difference decreased as a result of the PCM-induced fall in the pond's maximum temperature, even though both ponds' minimum temperatures were roughly similar.
7. Solar air heater: Using PCM in the solar air heater eventually reduced annual energy consumption.
8. Solar air heater: Employing PCMs with greater melting point temperatures causes SAH outlet temperatures to rise by nearly 5 °C and their daily performance to drop insignificantly (by approximately 3%).
9. Solar air heater: The increased airflow velocity causes the melting time to take longer to complete and lowers the paraffin's melting temperature.
10. Chimney: When fins are used in the PCM-based solar chimney system, the mean temperature of the room is 20% higher than in the non-finned scenario.
11. Chimney: Without a period of steady temperature, the heat surface temperature of the PCM will rapidly decline throughout the discharging time (solidification temperature).
12. Chimney: When paraffin/copper foam (Pa/CF) is used as the PCM instead of the SC-PV system, the power output (5.54 kWh) and ventilation time (22 h and 30 min), respectively, are the highest.

5. Recommendation for Future Work

Despite the many benefits of solar energy, it is mainly characterized by its features of discontinuities and instability. To address these shortcomings, numerous researchers have combined solar energy applications with thermal energy storage technology. In recent years, latent heat storage has drawn a lot of interest. Some advantages of the material include a high energy density and nearly isothermal operation during both melting and solidification. For many years, PCM preparation and heat exchanger design were the principal topics of latent heat storage research [112]. Certain elements with high thermal conductivity are mixed into PCMs to create a new composite material because many PCMs have low thermal conductivity, which negatively impacts heat charge and discharge. The following recommendations are given for additional research:

1. In the future, the solar pond could additionally utilize other organic PCMs for capsule manufacturing, such as fatty acids, polyglycerol, and capric acids.

2. To assure large and beneficial social consequences, more research is required on the development of efficient and affordable PCMs with reduced ageing effects for solar thermal energy storage applications.
3. It is required to enhance the properties of PCM such as the conductivity parameter for better results.
4. Advanced research in the materials science of PCMs can enable innovative engineering solutions to integrate PCMs into solar thermal systems, aligning with the clean energy roadmap.
5. PCMs can have better thermal stability, less leakage, and longer durability if they are perfectly encapsulated.
6. The total effectiveness and stability of the solar energy system can be increased by combining PCM technology with other energy storage technologies, such as batteries or super-capacitors.

Author Contributions: Conceptualization, F.L.R. and M.A.A.-O.; methodology, F.L.R. and A.H.; software, H.Y.B.; validation, F.L.R., H.Y.B. and A.H.; formal analysis, M.A.A.-O. and H.Y.B.; investigation, F.L.R.; resources, A.D. and A.H.; data curation, F.L.R., A.D. and M.A.A.-O.; writing—original draft preparation, F.L.R., M.A.A.-O., A.D., H.Y.B. and A.H.; writing—review and editing, F.L.R., M.A.A.-O., A.D., H.Y.B. and A.H.; visualization, F.L.R.; supervision, F.L.R. and A.H.; project administration, F.L.R. and A.D.; funding acquisition, A.H. and A.D. All authors have read and agreed to the published version of the manuscript.

Funding: This research received no external funding.

Data Availability Statement: Not applicable.

Acknowledgments: The financial support of Kerbala University and University of Warith Al-Anbiyaa Universities in Iraq is gratefully acknowledged.

Conflicts of Interest: The authors declare no conflict of interest.

Abbreviations

Symbol	Definition
CSS	Conventional solar still
DSSS	Double-slope solar still
ETSC	Evacuated tube solar collector
ETC/S	Evacuated tube collector/storage
FPSC	Flat plate solar collector
HPETC	Heat pipe evacuated tube collector
HTF	Heat transfer fluid
LHTES	Latent heat thermal energy storage
MCVTSD	Modified convex tubular solar distiller
MWCNT	Multi-walled carbon nanotube
PCMs	Phase change materials
PCS	Slurry phase change materials
SPTR	Solar parabolic trough reflector
SSPCM	Shape-stabilized phase change material
TESS	Thermal energy storage system

References

1. Perlin, J. *Let It Shine: The 6000-Year Story of Solar Energy*; New World Library: Novato, CA, USA, 2013.
2. Daniels, F.; Duffie, J. *Solar Energy Research*; University of Wisconsin Press: Madison, WI, USA, 1955.
3. Bhushan, B.; Singh, R. A review on methodology of artificial roughness used in duct of solar air heaters. *Energy* **2010**, *35*, 202–212. [CrossRef]
4. Zhao, D.L.; Li, Y.; Dai, Y.J.; Wang, R.Z. Optimal study of a solar air heating system with pebble bed energy storage. *Energy Convers. Manag.* **2011**, *52*, 2392–2400. [CrossRef]
5. Ghoneim, A.; Klein, S.; Duffie, J. Analysis of collector storage building walls using phase change materials. *Sol. Energy* **1991**, *47*, 237–242. [CrossRef]

6. Pasupathy, A.; Athanasius, L.; Velraj, R.; Seeniraj, R. Experimental investigation and numerical simulation analysis on the thermal performance of a building roof incorporating phase change material (PCM) for thermal management. *Appl. Thermal Eng.* **2008**, *28*, 556–565. [CrossRef]
7. Bland, A.; Khzouz, M.; Statheros, T.; Gkanas, E. PCMs for residential building applications: A short review focused on disadvantages and proposals for future development. *Buildings* **2017**, *7*, 78. [CrossRef]
8. Al-Yasiri, Q.; Szabó, M. Paraffin as a phase change material to improve building performance: An overview of applications and thermal conductivity enhancement techniques. *Renew. Energy Environ. Sustain.* **2021**, *6*, 38. [CrossRef]
9. Khedher, N.B.; Bantan, R.A.; Kolsi, L.; Omri, M. Performance investigation of a vertically configured LHTES via the combination of nano-enhanced PCM and fins: Experimental and numerical approaches. *Int. Commun. Heat Mass Transf.* **2022**, *137*, 106246. [CrossRef]
10. Fan, Y.; Zhang, X.; Wang, X.; Li, J.; Zhu, Q. Super-cooling prevention of microencapsulated phase change material. *Thermochim. Acta* **2004**, *413*, 1–6. [CrossRef]
11. Yamagishi, Y.; Sugeno, T.; Ishige, T.; Takeuchi, H.; Pyatenko, A.T. An evaluation of microencapsulated PCM for use in cold energy transportation medium. In Proceedings of the 31st Intersociety Energy Conversion Engineering Conference, Washington, DC, USA, 11–16 August 1996; pp. 2077–2083.
12. Jin, Y.; Wan, Q.; Ding, Y. PCMs heat transfer performance enhancement with expanded graphite and its thermal stability. *Proc. Eng.* **2015**, *102*, 1877–1884. [CrossRef]
13. Letcher, T.M.; Law, R.; Reay, D. *Storing Energy: With Special Reference to Renewable Energy Sources*; Elsevier: Oxford, UK, 2016.
14. Sittisart, P.; Farid, M.M. Fire retardants for phase change materials. *Appl. Energy* **2011**, *88*, 3140–3145. [CrossRef]
15. Song, G.; Ma, S.; Tang, G.; Yin, Z.; Wang, X. Preparation and characterization of flame retardant form-stable phase change materials composed by EPDM, paraffin and nano magnesium hydroxide. *Energy* **2010**, *35*, 2179–2183. [CrossRef]
16. Nazir, H.; Batool, M.; Osorio, F.J.B.; Isaza-Ruiz, M.; Xu, X.; Vignarooban, K.; Phelan, P.; Kannan, A.M. Recent developments in phase change materials for energy storage applications: A review. *Int. J. Heat Mass Transf.* **2019**, *129*, 491–523. [CrossRef]
17. Sikiru, S.; Oladosu, T.L.; Amosa, T.I.; Kolawole, S.Y.; Soleimani, H. Recent advances and impact of phase change materials on solar energy: A comprehensive review. *J. Energy Storage* **2022**, *53*, 105200. [CrossRef]
18. Cabeza, L.F.; Ibanez, M.; Sole, C.; Roca, J.; Nogues, M. Experimentation with a water tank including a PCM module. *Sol. Energy Mater. Sol. Cells* **2006**, *90*, 1273–1282. [CrossRef]
19. Chaurasia, P.B.L. Phase change material in solar water heater storage system. In Proceedings of the 8th International Conference on Thermal Energy Storage, Stuttgart, Germany, 28 August–1 September 2000.
20. Canbazoglu, S.; Sahinaslan, A.; Ekmekyapar, A.; Aksoy, Y.G.; Akarsu, F. Enhancement of solar thermal energy storage performance using sodium thiosulfate pentahydrate of a conventional solar water-heating system. *Energy Build.* **2005**, *37*, 235–242. [CrossRef]
21. Haillot, D.; Franquet, E.; Gibout, S.; Bedecarrats, J.P. Optimization of solar DHW system including PCM media. *Appl. Energy* **2013**, *109*, 470e5. [CrossRef]
22. Hussain, F.; Othman, M.Y.H.; Sopian, K.; Yatim, B.; Ruslan, H.; Othman, H. Design development and performance evaluation of photovoltaic/thermal (PV/T) air base solar collector. *Renew. Sustain. Energy Rev.* **2013**, *25*, 431–441. [CrossRef]
23. Rashidov, Y.K.; Ramankulov, S.A. Improving the Efficiency of Flat Solar Collectors in Heat Supply Systems. *Cent. Asian J. Theor. Appl. Sci.* **2021**, *2*, 152–159.
24. Alshukri, M.J.; Hussein, A.K.; Eidan, A.A.; Alsabery, A.I. A review on applications and techniques of improving the performance of heat pipe-solar collector systems. *Solar Energy* **2022**, *236*, 417–433. [CrossRef]
25. Feliński, P.; Sekret, R. Experimental study of evacuated tube collector/storage system containing paraffin as a PCM. *Energy* **2016**, *114*, 1063–1072. [CrossRef]
26. Serale, G.; Goia, F.; Perino, M. Numerical model and simulation of a solar thermal collector with slurry Phase Change Material (PCM) as the heat transfer fluid. *Solar Energy* **2016**, *134*, 429–444. [CrossRef]
27. Li, B.; Zhai, X. Experimental investigation and theoretical analysis on a mid-temperature solar collector/storage system with composite PCM. *Appl. Therm. Eng.* **2017**, *124*, 34–43. [CrossRef]
28. Wang, Z.; Qu, J.; Zhang, R.; Han, X.; Wu, J. Photo-thermal performance evaluation on MWCNTs-dispersed microencapsulated PCM slurries for direct absorption solar collectors. *J. Energy Storage* **2019**, *26*, 100793. [CrossRef]
29. Wang, D.; Liu, H.; Liu, Y.; Xu, T.; Wang, Y.; Du, H.; Wang, X.; Liu, J. Frost and High-temperature resistance performance of a novel dual-phase change material flat plate solar collector. *Sol. Energy Mater. Sol. Cells* **2019**, *201*, 110086. [CrossRef]
30. Gao, G.; Zhang, T.; Jiao, S.; Guo, C. Preparation of reduced graphene oxide modified magnetic phase change microcapsules and their application in direct absorption solar collector. *Sol. Energy Mater. Sol. Cells* **2020**, *216*, 110695. [CrossRef]
31. Palacio, M.; Rincón, A.; Carmona, M. Experimental comparative analysis of a flat plate solar collector with and without PCM. *Solar Energy* **2020**, *206*, 708–721. [CrossRef]
32. Elarem, R.; Alqahtani, T.; Mellouli, S.; Aich, W.; Ben Khedher, N.; Kolsi, L.; Jemni, A. Numerical study of an Evacuated Tube Solar Collector incorporating a Nano-PCM as a latent heat storage system. *Case Stud. Therm. Eng.* **2021**, *24*, 100859. [CrossRef]
33. Alshukri, M.J.; Eidan, A.A.; Najim, S.I. The influence of integrated Micro-ZnO and Nano-CuO particles/paraffin wax as a thermal booster on the performance of heat pipe evacuated solar tube collector. *J. Energy Storage* **2021**, *37*, 102506. [CrossRef]
34. Sadeghi, G.; Mehrali, M.; Shahi, M.; Brem, G.; Mahmoudi, A. Experimental analysis of Shape-Stabilized PCM applied to a Direct-Absorption evacuated tube solar collector exploiting sodium acetate trihydrate and graphite. *Energy Convers. Manag.* **2022**, *269*, 116176. [CrossRef]

35. Yeh, C.-Y.; Boonk, K.; Sadeghi, G.; Mehrali, M.; Shahi, M.; Brem, G.; Mahmoudi, A. Experimental and numerical analysis of thermal performance of shape stabilized PCM in a solar thermal collector. *Case Stud. Therm. Eng.* **2022**, *30*, 101706. [CrossRef]
36. Nekoonam, S.; Ghasempour, R. Modeling and optimization of a thermal energy storage unit with cascaded PCM capsules in connection to a solar collector. *Sustain. Energy Technol. Assess.* **2022**, *52*, 102197. [CrossRef]
37. Wen, X.; Ji, J.; Li, Z.; Song, Z. Performance analysis of a concentrated system with series photovoltaic/thermal module and solar thermal collector integrated with PCM and TEG. *Energy* **2022**, *249*, 123777. [CrossRef]
38. Hatamleh, R.I.; Abu-Hamdeh, N.H.; Khoshaim, A.; Alzahrani, M.A. Using phase change material (PCM) to improve the solar energy capacity of glass in solar collectors by enhancing their thermal performance via developed MD approach. *Eng. Anal. Bound. Elem.* **2022**, *143*, 163–169. [CrossRef]
39. Algarni, S.; Mellouli, S.; Alqahtani, T.; Almutairi, K.; Khan, A.; Anqi, A. Experimental investigation of an evacuated tube solar collector incorporating nano-enhanced PCM as a thermal booster. *Appl. Therm. Eng.* **2020**, *180*, 115831. [CrossRef]
40. Karami, M.; Shahini, N.; Behabadi, M.A.A. Numerical investigation of double-walled direct absorption evacuated tube solar collector using microencapsulated PCM and nanofluid. *J. Mol. Liq.* **2023**, *377*, 121560. [CrossRef]
41. Pawar, V.R.; Sobhansarbandi, S. Heat transfer enhancement of a PCM-porous metal based heat pipe evacuated tube solar collector: An experimental study. *Sol. Energy* **2023**, *251*, 106–118. [CrossRef]
42. Rashid, F.L.; Al-Obaidi, M.A.; Hussein, A.K.; Akkurt, N.; Ali, B.; Younis, O. Floating solar stills and floating solar-driven membranes: Recent advances and overview of designs, performance and modern combinations. *Solar Energy* **2022**, *247*, 355–372. [CrossRef]
43. Hussein, A.K.; Rashid, F.L.; Abed, A.M.; Al-Khaleel, M.; Togun, H.; Ali, B.; Akkurt, N.; Malekshah, E.H.; Biswal, U.; Al-Obaidi, M.A.; et al. Inverted Solar Stills: A Comprehensive Review of Designs, Mathematical Models, Performance, and Modern Combinations. *Sustainability* **2022**, *14*, 13766. [CrossRef]
44. Cheng, W.-L.; Huo, Y.-K.; Nian, Y.-L. Performance of solar still using shape-stabilized PCM: Experimental and theoretical investigation. *Desalination* **2019**, *455*, 89–99. [CrossRef]
45. Vigneswaran, V.; Kumaresan, G.; Dinakar, B.; Kamal, K.K.; Velraj, R. Augmenting the productivity of solar still using multiple PCMs as heat energy storage. *J. Energy Storage* **2019**, *26*, 101019. [CrossRef]
46. Abu-Arabi, M.; Al-Harahsheh, M.; Ahmad, M.; Mousa, H. Theoretical modeling of a glass-cooled solar still incorporating PCM and coupled to flat plate solar collector. *J. Energy Storage* **2020**, *29*, 101372. [CrossRef]
47. Yousef, M.S.; Hassan, H.; Kodama, S.; Sekiguchi, H. An experimental study on the performance of single slope solar still integrated with a PCM-based pin-finned heat sink. *Energy Procedia* **2019**, *156*, 100–104. [CrossRef]
48. Elashmawy, M.; Alhadri, M.; Ahmed, M.M. Enhancing tubular solar still performance using novel PCM-tubes. *Desalination* **2021**, *500*, 114880. [CrossRef]
49. Kabeel, A.; Abdelgaied, M.; Harby, K.; Eisa, A. Augmentation of diurnal and nocturnal distillate of modified tubular solar still having copper tubes filled with PCM in the basin. *J. Energy Storage* **2020**, *32*, 101992. [CrossRef]
50. Chen, Q.; Xu, G.; Xia, P. The performance of a solar-driven spray flash evaporation desalination system enhanced by microencapsulated phase change material. *Case Stud. Therm. Eng.* **2021**, *27*, 101267. [CrossRef]
51. Jahanpanah, M.; Sadatinejad, S.J.; Kasaeian, A.; Jahangir, M.H.; Sarrafha, H. Experimental investigation of the effects of low-temperature phase change material on single-slope solar still. *Desalination* **2021**, *499*, 114799. [CrossRef]
52. Gnanavel, C.; Saravanan, R.; Chandrasekaran, M. CFD analysis of solar still with PCM. *Mater. Today Proc.* **2021**, *37*, 694–700. [CrossRef]
53. Hansen, R.S.; Mary, M.B.Q.; Subramanian, S.S.; Raj, J.A.; Gnanaraj, S.J.P.; Appadurai, M. Utilization of PCM in inclined and single basin solar stills to improve the daily productivity. *Mater. Today Proc.* **2022**, *62*, 967–972. [CrossRef]
54. Ajdari, H.; Ameri, A. Performance assessment of an inclined stepped solar still integrated with PCM and CuO/GO nanocomposite as a nanofluid. *J. Build. Eng.* **2022**, *49*, 104090. [CrossRef]
55. Abdullah, A.; Omara, Z.; Essa, F.A.; Alqsair, U.F.; Aljaghtham, M.; Mansir, I.B.; Shanmugan, S.; Alawee, W.H. Enhancing trays solar still performance using wick finned absorber, nano- enhanced PCM. *Alex. Eng. J.* **2022**, *61*, 12417–12430. [CrossRef]
56. Tuly, S.; Ayon, A.; Hassan, R.; Das, B.K.; Khan, R.; Sarker, M. Performance investigation of active double slope solar stills incorporating internal sidewall reflector, hollow circular fins, and nanoparticle-mixed phase change material. *J. Energy Storage* **2022**, *55*, 105660. [CrossRef]
57. Afolabi, L.O.; Enweremadu, C.C.; Kareem, M.; Arogundade, A.I.; Irshad, K.; Islam, S.; Oladosu, K.; Elfaghi, A.M.; Didane, D.H. Experimental investigation of double slope solar still integrated with PCM nanoadditives microencapsulated thermal energy storage. *Desalination* **2023**, *553*, 116477. [CrossRef]
58. Ben Bacha, H.; Abdullah, A.; Kabeel, A.E.; Abdelgaied, M. Design and development of a tubular solar distiller using a convex absorber, wick materials, and PCM reservoir combined with a solar parabolic concentrator. *J. Energy Storage* **2023**, *62*, 106897. [CrossRef]
59. Shahid, M.I.; Asim, M.; Farhan, M.; Sheikh, M.F.; Ashraf, M.U.; Arshad, H.; Alghamdi, A.; SAlshahrani, A.; Bahaddad, A.A.; Almarhabi, K.A. Design and performance analysis of salinity gradient solar pond under different climatic and soil conditions. *PLoS ONE* **2023**, *18*, e0279311. [CrossRef]
60. Kumar, A.; Singh, K.; Verma, S.; Das, R. Inverse prediction and optimization analysis of a solar pond powering a thermoelectric generator. *Solar Energy* **2018**, *169*, 658–672. [CrossRef]
61. Sarathkumar, P.; Sivaram, A.; Rajavel, R.; Kumar, R.P.; Krishnakumar, S. Experimental Investigations on The Performance of A Solar Pond by using Encapsulated Pcm with Nanoparticles. *Mater. Today Proc.* **2017**, *4*, 2314–2322. [CrossRef]

62. Amirifard, M.; Kasaeian, A.; Amidpour, M. Integration of a solar pond with a latent heat storage system. *Renew. Energy* **2018**, *125*, 682–693. [CrossRef]
63. Ines, M.; Paolo, P.; Roberto, F.; Mohamed, S. Experimental studies on the effect of using phase change material in a salinity-gradient solar pond under a solar simulator. *Solar Energy* **2019**, *186*, 335–346. [CrossRef]
64. Beik, A.J.G.; Assari, M.R.; Tabrizi, H.B. Transient modeling for the prediction of the temperature distribution with phase change material in a salt-gradient solar pond and comparison with experimental data. *J. Energy Storage* **2019**, *26*, 101011. [CrossRef]
65. Wang, H.; Xiaomeng, M.; Liugang, Z.; Zhang, X.; Mei, Y.; Zhang, A. Numerical and experimental study of effect of paraffin phase change heat storage capsules on the thermal performance of the solar pond. *Energy Explor. Exploit.* **2020**, *39*, 1010–1023. [CrossRef]
66. Rghif, Y.; Zeghmati, B.; Bahraoui, F. Modeling the influences of a phase change material and the Dufour effect on thermal performance of a salt gradient solar pond. *Int. J. Therm. Sci.* **2021**, *166*, 106979. [CrossRef]
67. Colarossi, D.; Pezzuto, M.; Principi, P. Effect of PCM melting temperature on solar ponds performance: Design and experimental investigation. *Solar Energy* **2022**, *242*, 225–233. [CrossRef]
68. Colarossi, D.; Principi, P. Experimental investigation and optical visualization of a salt gradient solar pond integrated with PCM. *Sol. Energy Mater. Sol. Cells* **2022**, *234*, 111425. [CrossRef]
69. Wang, H.; Zhang, C.Y.; Zhang, L.G. Effect of steel-wires and paraffin composite phase change materials on the heat exchange and exergetic performance of salt gradient solar pond. *Energy Rep.* **2022**, *8*, 5678–5687. [CrossRef]
70. Assari, M.R.; Beik, A.J.G.; Eydi, R.; Tabrizi, H.B. Thermal-salinity performance and stability analysis of the pilot salt-gradient solar ponds with phase change material. *Sustain. Energy Technol. Assess.* **2022**, *53*, 102396. [CrossRef]
71. Tchinda, R. A review of the mathematical models for predicting solar air heaters systems. *Renew. Sustain. Energy Rev.* **2009**, *13*, 1734–1759. [CrossRef]
72. ElGamal, R.; Kishk, S.; Al-Rejaie, S.; ElMasry, G. Incorporation of a solar tracking system for enhancing the performance of solar air heaters in drying apple slices. *Renew. Energy* **2021**, *167*, 676–684. [CrossRef]
73. Kabeel, A.; Khalil, A.; Shalaby, S.; Zayed, M. Experimental investigation of thermal performance of flat and v-corrugated plate solar air heaters with and without PCM as thermal energy storage. *Energy Convers. Manag.* **2016**, *113*, 264–272. [CrossRef]
74. El Khadraoui, A.; Bouadila, S.; Kooli, S.; Farhat, A.; Guizani, A. Thermal behavior of indirect solar dryer: Nocturnal usage of solar air collector with PCM. *J. Clean. Prod.* **2017**, *148*, 37–48. [CrossRef]
75. Salih, S.M.; Jalil, J.M.; Najim, S.E. Experimental and numerical analysis of double-pass solar air heater utilizing multiple capsules PCM. *Renew. Energy* **2019**, *143*, 1053–1066. [CrossRef]
76. Abuşka, M.; Şevik, S.; Kayapunar, A. A comparative investigation of the effect of honeycomb core on the latent heat storage with PCM in solar air heater. *Appl. Therm. Eng.* **2019**, *148*, 684–693. [CrossRef]
77. Raj, A.; Srinivas, M.; Jayaraj, S. A cost-effective method to improve the performance of solar air heaters using discrete macro-encapsulated PCM capsules for drying applications. *Appl. Therm. Eng.* **2019**, *146*, 910–920. [CrossRef]
78. SunilRaj, B.; Eswaramoorthy, M. Experimental study on hybrid natural circulation type solar air heater with paraffin wax based thermal storage. *Mater. Today Proc.* **2020**, *23*, 49–52. [CrossRef]
79. Ameri, M.; Sardari, R.; Farzan, H. Thermal performance of a V-Corrugated serpentine solar air heater with integrated PCM: A comparative experimental study. *Renew. Energy* **2021**, *171*, 391–400. [CrossRef]
80. Mahdi, J.M.; Mohammed, H.I.; Talebizadehsardari, P.; Ghalambaz, M.; Majdi, H.S.; Khan, A.; Yaïci, W.; Giddings, D. Simultaneous and consecutive charging and discharging of a PCM-based domestic air heater with metal foam. *Appl. Therm. Eng.* **2021**, *197*, 117408. [CrossRef]
81. Madhulatha, G.; Kumar, M.M.J.; Sateesh, P. Optimization of tube arrangement and phase change material for enhanced performance of solar air heater- A numerical analysis. *J. Energy Storage* **2021**, *41*, 102876. [CrossRef]
82. Dinesh, S.; Saminathan, R.; Patil, M.M.; Baviskar, P.R.; Hadidi, H.; Vignesh, S.; Kumar, P.M. Investigating the single pass baffled solar air heater (SAH) with an organic PCM (OPCM). *Mater. Today Proc.* **2022**, *62*, 5245–5249. [CrossRef]
83. Abu-Hamdeh, N.H.; Alsulami, R.A.; Hatamleh, R.I. A case study in the field of building sustainability energy: Performance enhancement of solar air heater equipped with PCM: A trade-off between energy consumption and absorbed energy. *J. Build. Eng.* **2022**, *48*, 103903. [CrossRef]
84. Palacio, M.; Ramírez, C.; Carmona, M.; Cortés, C. Effect of phase-change materials in the performance of a solar air heater. *Sol. Energy* **2022**, *247*, 385–396. [CrossRef]
85. Chaatouf, D.; Ghiaus, A.-G.; Amraqui, S. Optimization of a solar air heater using a phase change material for drying applications. *J. Energy Storage* **2022**, *55*, 105513. [CrossRef]
86. Verma, G.; Singh, S.; Chander, S.; Dhiman, P. Numerical investigation on transient thermal performance predictions of phase change material embedded solar air heater. *J. Energy Storage* **2022**, *47*, 103619. [CrossRef]
87. Sharol, A.; Razak, A.; Majid, Z.; Azmi, M.; Tarminzi, M.; Ming, Y.; Zakaria, Z.; Harun, M.; Fazlizan, A.; Sopian, K. Effect of thermal energy storage material on the performance of double-pass solar air heater with cross-matrix absorber. *J. Energy Storage* **2022**, *51*, 104494. [CrossRef]
88. Luo, Q.; Li, B.; Wang, Z.; Su, S.; Xiao, H. A study of unidirectional spiral tube for air evacuation in a solar heater with phase-change material. *J. Build. Eng.* **2022**, *46*, 103659. [CrossRef]
89. Farzan, H.; Zaim, E.H.; Amiri, T. Performance investigation on a new solar air heater using phase change material/expanded metal mesh composite as heat storage unit: An experimental study. *J. Energy Storage* **2022**, *47*, 103602. [CrossRef]

90. Fan, Z.; Zhao, Y.; Shi, Y.; Liu, X.; Jiang, D. Thermal performance evaluation of a novel building wall for lightweight building containing phase change materials and interlayer ventilation: An experimental study. *Energy Build.* **2023**, *278*, 112677. [CrossRef]
91. He, Z.; Ma, H.; Lu, S.; Sun, Y. Research on the thermal performance of interlayer ventilated PCM component coupled with solar air collector. *Energy Build.* **2022**, *255*, 111698. [CrossRef]
92. Brahma, B.; Shukla, A.K.; Baruah, D.C. Design and performance analysis of solar air heater with phase change materials. *J. Energy Storage* **2023**, *61*, 106809. [CrossRef]
93. SunilRaj, B.; Eswaramoorthy, M. Performance study on forced convection mode V-trough solar air heater with thermal storage: An experimental approach. *Mater. Today Proc.* **2023**, *72*, 1664–1672. [CrossRef]
94. Bayrak, F.; Oztop, H.F.; Hepbasli, A. Energy and exergy analyses of porous baffles inserted solar air heaters for building applications. *Energy Build.* **2013**, *5*, 338–345. [CrossRef]
95. Zhai, X.Q.; Song, Z.P.; Wang, R.Z. A review for the applications of solar chimneys in buildings. *Renew. Sustain. Energy Rev.* **2011**, *15*, 3757–3767. [CrossRef]
96. Tariq, R.; Torres-Aguilar, C.E.; Xamán, J.; Zavala-Guillén, I.; Bassam, A.; Ricalde, L.J.; Carvente, O. Digital twin models for optimization and global projection of building-integrated solar chimney. *Build. Environ.* **2022**, *213*, 108807. [CrossRef]
97. Yongcai, L.; Shuli, L.; Jun, L. Effects of various parameters of a PCM on thermal performance of a solar chimney. *Appl. Therm. Eng.* **2017**, *127*, 1119–1131. [CrossRef]
98. Liu, B.; Wang, M.; Wang, Q.; Mao, S.; Rachid, B. Effect of the Position of the Phase Change Material (PCM $Na_2CO_3.10H_2O$) on the Solar Chimney Effect. *Energy Procedia* **2017**, *139*, 462–467. [CrossRef]
99. Thantong, P.; Khedari, J.; Chantawong, P. Investigation of thermal performance by applying a solar chimney with PCM towards the natural ventilation of model house under Climate of Thailand. *Mater. Today Proc.* **2018**, *5*, 14862–14867. [CrossRef]
100. Fadaei, N.; Kasaeian, A.; Akbarzadeh, A.; Hashemabadi, S.H. Experimental investigation of solar chimney with phase change material (PCM). *Renew. Energy* **2018**, *123*, 26–35. [CrossRef]
101. Dordelly, J.C.F.; El Mankibi, M.; Roccamena, L.; Remion, G.; Landa, J.A. Experimental analysis of a PCM integrated solar chimney under laboratory conditions. *Solar Energy* **2019**, *188*, 1332–1348. [CrossRef]
102. Tiji, M.E.; Eisapour, M.; Yousefzadeh, R.; Azadian, M.; Talebizadehsardari, P. A numerical study of a PCM-based passive solar chimney with a finned absorber. *J. Build. Eng.* **2020**, *32*, 101516. [CrossRef]
103. Chen, W. Analysis of heat transfer and flow in the solar chimney with the sieve-plate thermal storage beds packed with phase change capsules. *Renew. Energy* **2020**, *157*, 491–501. [CrossRef]
104. Ashouri, M.; Hakkaki-Fard, A. Improving the performance of the finned absorber inclined rooftop solar chimney combined with composite PCM and PV module. *Solar Energy* **2021**, *228*, 562–574. [CrossRef]
105. Nateghi, S.; Jahangir, M.H. Performance evaluation of solar chimneys in providing the thermal comfort range of the building using phase change materials. *Clean. Mater.* **2022**, *5*, 100120. [CrossRef]
106. Li, W.; Li, Z.; Xie, L.; Li, Y.; Long, T.; Huang, S.; Lu, J.; Wang, Z. Evaluation of the thermal performance of an inclined solar chimney integrated with a phase change material. *Energy Build.* **2022**, *270*, 112288. [CrossRef]
107. Long, T.; Li, W.; Lv, Y.; Li, Y.; Liu, S.; Lu, J.; Huang, S.; Zhang, Y. Benefits of integrating phase-change material with solar chimney and earth-to-air heat exchanger system for passive ventilation and cooling in summer. *J. Energy Storage* **2022**, *48*, 104037. [CrossRef]
108. Koonsrisuk, A.; Chitsomboon, T. A single dimensionless variable for solar chimney power plant modeling. *Solar Energy* **2009**, *83*, 2136–2143. [CrossRef]
109. Koonsrisuk, A.; Chitsomboon, T. Dynamic similarity in solar chimney modeling. *Solar Energy* **2007**, *81*, 1439–1446. [CrossRef]
110. Koonsrisuk, A.; Chitsomboon, T. Partial geometric similarity for solar chimney power plant modeling. *Solar Energy* **2009**, *83*, 1611–1618. [CrossRef]
111. Ukrainczyk, N.; Kurajica, S.; Šipušić, J. Thermophysical comparison of five commercial paraffin waxes as latent heat storage materials. *Chem. Biochem. Eng. Q.* **2010**, *24*, 129–137.
112. Pielichowska, K.; Pielichowski, K. Phase change materials for thermal energy storage. *Prog. Mater. Sci.* **2014**, *65*, 67–123. [CrossRef]

designs

Review

Electricity Pricing and Its Role in Modern Smart Energy System Design: A Review

Jiaqi Liu, Hongji Hu, Samson S. Yu * and Hieu Trinh

School of Engineering, Deakin University, 75 Pigdons Rd, Waurn Ponds, Geelong, VIC 3216, Australia; ljiaqi@deakin.edu.au (J.L.); hj.hu@ieee.org (H.H.); hieu.trinh@deakin.edu.au (H.T.)
* Correspondence: samson.yu@deakin.edu.au

Abstract: Energy is the foundation for human survival and socio-economic development, and electricity is a key form of energy. Electricity prices are a key factor affecting the interests of various stakeholders in the electricity market, playing a significant role in the sustainable development of energy and the environment. As the number of distributed energy resources (DERs) increases, today's power systems no longer rely on a vertical market model and fixed electricity pricing scheme but instead depend on power dispatch and dynamic pricing to match supply and demand. This can help prevent significant fluctuations in supply–load imbalance and maintain system stability. Modern power grids have evolved by integrating information, communication, and intelligent control technologies with traditional power systems, giving rise to the concept of smart electric grids. Choosing an appropriate pricing scheme to manage large-scale DERs and controllable loads in today's power grid become very important. However, the existing literature lacks a comprehensive review of electricity pricing in power systems and its transformative impact on shaping the energy landscape. To fill this void, this paper provides a survey on the developments, methods, and frameworks related to electricity pricing and energy trading. The review mainly considers the development of pricing in a centralized power grid, peer-to-peer (P2P) and microgrid-to-microgrid (M2M) energy trading and sharing, and various pricing methods. The review will cover the pricing schemes in modern power systems, particularly with respect to renewable energy sources (RESs) and batteries, as well as controllable load applications, and the impact of pricing schemes based on demand-side ancillary services (DSAS) for grid frequency support. Lastly, this review article describes the current frameworks and limitations of electricity pricing in the current energy market, as well as future research directions. This review should offer a great overview and deep insights into today's electricity market and how pricing methods will drive and facilitate the future establishment of smart energy systems.

Keywords: demand-side ancillary services; electricity market; electricity pricing; energy sharing; peer-to-peer energy trading

Citation: Liu, J.; Hu, H.; Yu, S.S.; Trinh, H. Electricity Pricing and Its Role in Modern Smart Energy System Design: A Review. *Designs* **2023**, *7*, 76. https://doi.org/10.3390/designs7030076

Academic Editors: Surender Reddy Salkuti and Dibin Zhu

Received: 23 May 2023
Revised: 14 June 2023
Accepted: 15 June 2023
Published: 16 June 2023

1. Introduction

Throughout most of the 20th century, traditional electricity markets were dominated by vertically integrated monopoly utilities, with consumers paying a fixed price based on usage. However, the escalating electricity demand has heightened the significance of optimizing power system management, minimizing costs, and mitigating environmental pollution. Researchers have developed new frameworks and algorithms to improve the planning and operation of electricity markets over the years. A single organization no longer manages the power market but involves various stakeholders, including producers, distributors, and consumers, who can exchange real-time information [1]. Consumers now have the capability to monitor their real-time electricity requirements and actively engage in the dispatch of power within the system.

In the past decade, DERs have increased significantly, changing the way energy is produced, transmitted, and consumed. It has increased the number of prosumers who can both buy and sell energy, resulting in a more decentralized and open power system [2]. With the integration of renewable energy sources (RESs) into the grid and the transition toward a low-carbon energy model, adopting new methodologies for pricing energy and enhancing the flexibility and efficiency of the energy market have become imperative [3]. Demand-side management is an important approach, with demand response (DR) being a change based on consumer behavior. As time and electricity prices change, users can alter their electricity consumption from normal patterns or are encouraged to reduce consumption and receive rewards when prices are high, or system reliability is threatened [4]. DR has the advantages of peak shaving, higher RESs utilization, rapid response, and cost reduction [5].

In recent years, the establishment of local energy markets has become essential to facilitate the direct trading of renewable energy generation among prosumers, eliminating the need for intermediaries [6]. As a result, P2P energy trading has become a way for prosumers to participate in electricity markets. P2P allows energy trading between prosumers and energy sharing, increasing their benefits. P2P energy trading provides users with greater flexibility, increases the utilization of RES, and improves the reliability and stability of the power system [7]. There are also some common dynamic pricing schemes in the power system, including time-of-use pricing, real-time pricing, critical peak pricing, demand-side bidding, interruptible load compensation pricing, and direct load control [8]. Dynamic pricing tracks future price fluctuations based on overall consumer demand, promoting more energy conservation as higher prices reflect lower consumer demand [9]. In general, the pricing of the electricity system can be summarized into two methods: One approach is to implement dynamic pricing, where the price of energy changes is based on factors such as demand, the availability of renewable energy, and time of day. This would incentivize consumers to shift their energy usage to times when renewable sources are abundant, reducing stress on the grid and lowering overall energy costs. Another method is to use blockchain technology to create a P2P energy trading system, where individuals and businesses can buy and sell excess energy directly to each other without intermediaries. This would make the market more decentralized and increase competition, leading to lower prices and more efficient use of renewable energy.

Existing literature reviews on power system electricity prices have predominantly focused on aspects such as electricity price prediction algorithms, strategies, and the development and status of specific countries. For example, in [10], the authors introduced the theory of electricity pricing based on economics and the development history of electricity prices in China. The main discussion is about price-based DR strategies for smart grids and the current state of research. In [11], the authors assessed the main driving factors of EU household electricity prices and developed a dynamic panel data model to promote the use of renewable energy under liberal policies. In [12], the historical demand and supply curves of the German day-ahead market were analyzed, demonstrating that the integration of renewable energy greatly reduces electricity prices. These review articles mainly focus on a country's electricity price strategy and current situation. The review article in [13] examines state-of-the-art electricity price forecasting algorithms, compares two prediction models, and provides datasets for evaluating future electricity price prediction research. In [14], the authors used neural basis expansion analysis techniques to study electricity price predictions for a wide range of years in the electricity market and found that the prediction accuracy of this technique is higher than that of the original prediction model. In [15], the authors conducted a comparative analysis of MG energy management on different campuses of different universities, trying to address the uncertainties introduced by RESs while considering ESSs technologies. In [16], the authors reviewed probabilistic electricity price forecasting and further expanded on it, and they proposed related usage methods and measures. However, these articles focus more on reviewing algorithms for electricity price prediction in power systems. In [17], the authors provided a comprehensive review of P2P energy sharing and trading, including system configurations and pricing mechanisms. The author analyzes the benefits of P2P energy trading, which include the potential to reduce energy costs, increase the uptake of renewable energy sources, and enhance energy security. However, the success of these systems also relies on the development of effective pricing mechanisms that balance the interests of producers, consumers, and distribution network operators. In [18], the role of energy pricing policies in electricity markets is reviewed, mainly analyzing aspects such as power supply, possibilities, efficiency, and price structures. In [19], the authors evaluated pilot test results using a meta-analysis method, analyzing the impact of various incentive measures (including demographic changes and participants voluntarily joining) on household electricity consumption behavior. These articles mainly review pricing policies and mechanisms in the electricity market.

Considering the above, current review articles still lack a comprehensive review of pricing mechanisms within the context of smart power grids that integrate distributed energy resources (DERs), loads, and demand-side auxiliary services. Considering this research gap in the literature, this review article focuses on P2P and M2M energy trading (and energy sharing) pricing methods. This study will encompass pricing in renewable energy sources (RESs), energy storage systems (ESSs), applications for controllable loads, and the provision of demand-side ancillary services (DSAS), exploring their impact on the pricing scheme within the grid. The main contributions of this paper are as follows:

1. Introduction of the development of power grid pricing in chronological order, from 1989 to the present, and changes in the framework of the electricity market so that readers can clearly understand the development of the electricity market;
2. A comprehensive review of electricity pricing enables researchers to better identify current research gaps. These include pricing (and energy sharing)-enabled control for modern distributed power grids with P2P and M2M energy trading; pricing in RESs, ESSs, and controllable loads applications; and the impact of DSAS provision on power pricing schemes;
3. A comprehensive summary of the current pricing framework and limitations of retail and wholesale markets, with suggestions for future electricity markets and research directions in this field.

The rest of this paper is organized as follows. Section 2 provides a brief overview of the development of electricity pricing and the main components of electricity pricing. Section 3 describes the pricing schemes of P2P and M2M energy trading. Section 4 reviews the pricing in power grids with RESs using batteries and controllable load applications. Section 5 presents DSAS provision and its impact on pricing schemes. Section 6 provides the current pricing framework for retail and wholesale markets and its limitations. Section 7 gives suggestions for future research directions. Finally, conclusions are drawn in Section 8.

2. Development of Pricing in the Power Grid

2.1. Brief History

The continuous changes in the power industry have driven reforms in energy pricing within the distribution system, improving efficiency while reducing wholesale electricity costs. Pricing is one of the essential factors in marketing strategies, benefiting stakeholders, maintaining the balance of the power system, and enabling the creation of new regional markets using more RESs. This provides more options for electricity access while reducing carbon emissions [20]. The electricity price reform will also be a great challenge for grid companies. The opening of the electricity sales market requires grid companies to have high market competitiveness [21]. In the evolving market environment, power grid companies have transitioned from a model focused solely on selling electricity to a profit model centered around electricity transmission and distribution.

The earliest analysis of electricity prices and power market reforms can be found in [22], which reveals the impact of regulatory reforms on retail prices for industrial customers and the ratio of industrial to residential prices from 1986 to 1994. The retail electricity market expanded, industrial electricity prices fell, but the price difference between industrial and residential users widened, indicating that industrial users benefited from the reforms [23]. It was also observed that unbundling electric generation and introducing wholesale electricity markets did not necessarily lower electricity prices. From 1995 to 2002, European countries proposed liberalizing markets to provide reliable and secure electricity supplies, but electricity prices in each market model showed an upward trend despite increased competition [24]. This trend was partly due to the significant investments needed to upgrade infrastructure and meet environmental regulations. Additionally, market concentration remained high in some countries, limiting the benefits of competition. The liberalization process also faced opposition from trade unions and some political groups who feared job losses and the loss of national control over energy resources. It was not until 2007 that energy reforms impacted household energy prices [25]. Prior to this, the state-owned electricity company had a monopoly on the energy market and set prices according to governmental regulations. However, the market was opened for competition, and private companies were allowed to enter the market. This led to a decrease in electricity prices for industrial and commercial customers, as the new competition forced the state-owned electricity company to lower prices in order to remain competitive. However, it took several years for this to impact household energy prices as the reforms were implemented gradually. Although unbundling electric generation was not proven to have a significant effect on household electricity prices, the freedom to choose electricity suppliers was valued by consumers, leading to price reductions. By 2011, the large-scale integration of RESs into the grid can lead to changes in electricity prices, but it is not always the case that consumers profit from using RESs [26]. In some cases, electricity prices may decrease due to an oversupply of electricity from RESs during high production, but this may not necessarily translate into lower retail prices for consumers. Additionally, some utility companies may charge higher rates to customers who use RESs due to the cost of implementing and maintaining the necessary infrastructure to integrate them into the grid. Therefore, while integrating RESs can lead to changes in electricity prices, the impact on consumers may vary depending on the specific circumstances. In 2014, there was a notable discrepancy in the power system where household electricity prices generally increased despite a decline in wholesale electricity prices [27]. This divergence can be attributed to the lack of a simultaneous competitive marketing scheme in the industry and the government's regulation of retail electricity prices, which prevented market forces from influencing them. Moreover, the lack of competition in the power system created a need for additional incentives for electricity retailers to lower prices and attract customers. This was exacerbated by the fact that many households could not switch electricity providers due to contractual obligations or a lack of awareness of alternative options. In recent years, researchers have proposed various new electricity price policies and mechanisms to adjust electricity market prices because there has been a push toward increased competition in the electricity market and the deregulation of retail prices. This led to more household pricing options and increased pressure on retailers to provide competitive prices in order to retain customers. In [28], the authors suggest a combined electricity price and capacity addition tariff as an analytical approach for evaluating the optimal pricing mechanism for wind and solar energy. The study concludes that the price of RESs will vary greatly depending on the location and type of RESs. A dynamic distributed solar energy pricing model is proposed in [29], considering unit cost, profit, and taxation. This incentive mechanism has resulted in a significant drop in electricity prices from 2017 to 2020. The authors in [30] conclude that different regional grids need to provide different price mechanisms. In today's electricity system, we can find that wholesale electricity prices can vary widely from hour to hour, and peaks often occur during peak hours due to the high cost of generating electricity. Almost all end users are currently charged some fixed rate retail electricity price [31]. However, this will cause users to consume more electricity during peak hours, introduce high costs to retailers, and negatively impact the grid. Retailers will therefore want to better allocate electricity at different times of the day by implementing proper demand-side management. Figure 1 shows important events during the development of electricity pricing.

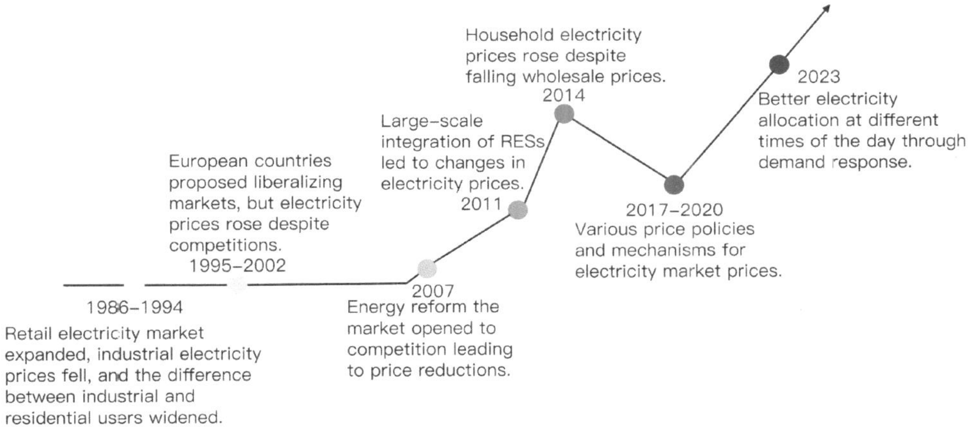

Figure 1. Timeline of key historical events of electricity pricing.

2.2. Electricity Price Structure

The electricity price paid by users encompasses not only the cost of energy consumption but also incorporates the power supply fee and service fees associated with energy usage. The components that contribute to the electricity price are shown in Figure 2.

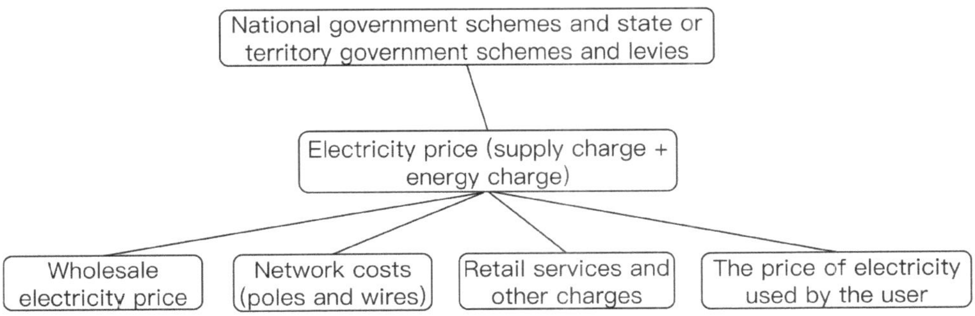

Figure 2. Components of electricity price.

A brief description of the main components of electricity price is as follows [32]:

1. Wholesale electricity price: The wholesale electricity price is simply the cost incurred when preparing to transmit electricity to the user. It includes the cost of generating electricity, distributing it, and the cost of operating and maintaining transmission infrastructure. No matter what kind of energy is used (coal and RESs), it will enter the wholesale market once generated. Wholesale electricity prices are determined by supply and demand in the wholesale electricity market, which is where energy producers and retailers buy and sell electricity. Retailers buy electricity from the wholesale market, where prices are generally set every 30 min but fluctuate based on supply and demand. The factors that influence wholesale electricity prices include fuel prices, weather, electricity demand, transmission capacity, and renewable energy generation.
2. Network costs (poles and wires): The network cost is the cost of the transmission and distribution network when transmitting energy required by the user to the user's place of consumption. It includes the cost of building, operating, maintaining, and upgrading the infrastructure necessary to transmit and distribute electricity or gas. Energy suppliers typically charge network costs to cover using the energy grid in

order to transport energy to customers. Network costs are regulated by regulatory bodies in each country to ensure that they are fair and transparent and that energy companies do not overcharge customers. These costs are usually calculated based on the amount of energy used by customers, as well as the distance between the energy source and the user's location. Some factors that can affect network costs include the age and condition of the infrastructure, the level of demand and consumption, the cost of raw materials and labor, and the level of investment in new technologies and renewable energy sources. Reducing network costs is often a key priority for energy suppliers, as it can help them remain competitive and provide cost-effective energy solutions to their customers.

3. National government schemes and state or territory government schemes and levies: Many national and state governments have environmental programs, such as Australia's Large Scale Renewable Energy Target and Small Renewable Energy Plan. The cost of these government programs can also affect electricity prices for consumers. However, it is important to note that while these programs may increase electricity prices in the short term, they are ultimately expected to lead to greater energy efficiency and a transition toward RESs, which may ultimately reduce prices in the long term. Furthermore, the cost of environmental damage caused by non-renewable sources is often not factored into electricity prices, so these programs may lead to cost savings in the long run by reducing the impact of pollution on public health and other resources. Overall, government environmental programs can have a complex and nuanced impact on electricity prices, but they are ultimately a necessary step toward a more sustainable and equitable energy future.
4. Retail services and other charges: Retail service is the bridge between energy retailers (such as ActewAGL) for consumers and the wholesale market. Energy retailers buy electricity for consumers from wholesale markets; arrange their meters, bills, and connections; and ensure that consumers can better manage their energy throughout the process. Retail services also offer customers a wide range of energy plans and options depending on their needs and preferences. They advise on energy efficiency, RESs, and the latest technologies that can help customers reduce their energy usage and bills. The retail services team manages customer accounts, billing, customer service, and support. They provide customers with information on their energy usage, tariffs, and billing and help them understand their consumption patterns and habits. Retail services also offer various payment options and plans to help customers manage their energy bills more effectively. In summary, retail services play a crucial role in ensuring the smooth and efficient operation of energy markets and in providing customers with high-quality services and support. Their expertise and knowledge help customers make informed decisions and take action to reduce their energy usage and bills while also contributing to a sustainable energy future.

2.3. The Evolution of the Electricity Market

The framework of the electricity market has also changed with the development and progress of society. For most of the 20th century, the electricity market has operated within the framework of a monopoly model, as shown in Figure 3a. In this model, a single entity integrates generation, transmission, and distribution to form vertical marketing [33]. The monopoly model controls the timing and price strategy of electricity flow in the grid and has strict rules for transferring electricity. This model has been criticized for its lack of competition, which can lead to higher prices and lower-quality service. It also discourages innovation and investment in the sector, as the sole entity may have different incentives to invest in new technologies or improve existing infrastructure. Additionally, consumers may not be able to choose their electricity provider or negotiate prices, resulting in limited options for them. Many countries have moved toward a more competitive market model, such as a liberalized or deregulated one. In these models, multiple companies are allowed to participate in the market, which can lead to increased competition, innovation, and investment. Consumers may have more options and are able to negotiate prices, leading to improved overall service. Then, the electricity market framework of the single-buyer model emerges, as shown in Figure 3b. Independent power producers and existing traditional power generation companies have generated competition on the power generation side. Purchasing agencies can purchase electricity from power generation companies or independent power producers [34]. However, distribution companies do not have the right to choose power producers, and they can only follow the policy of the framework. The main purpose of the single-buyer model is to encourage private participation in the electricity market. However, its disadvantage is that consumers still need to sign long-term contracts and are subject to regulated prices. Table 1 compares the advantages and disadvantages of the monopoly and single-buyer models [33,34].

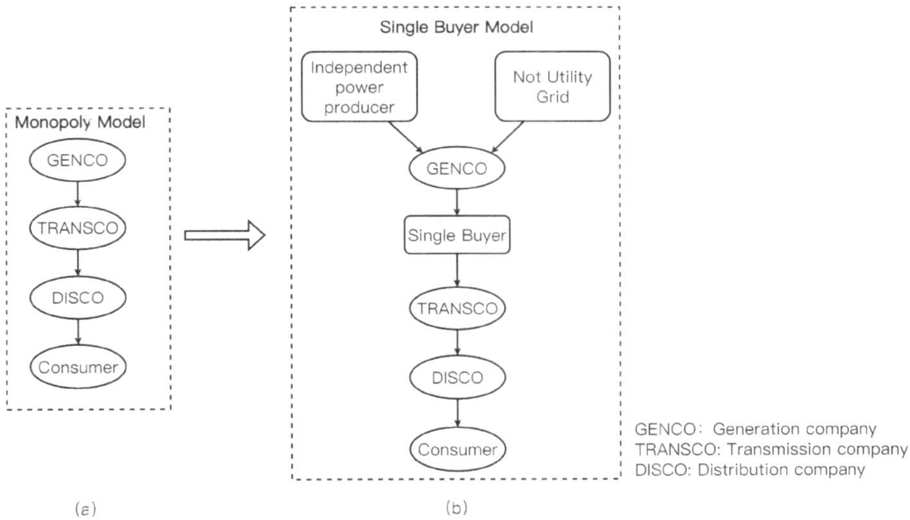

Figure 3. (**a**) Monopoly model; (**b**) single-buyer model.

Table 1. Advantages and disadvantages of the monopoly and single-buyer models.

Electricity Markets Model	Advantages	Disadvantages
Monopoly model	1. Efficient use of resources: The monopoly model enables a single firm to coordinate all production activities, ensuring efficient resource use and lower production costs and prices. 2. Economies of scale: The monopoly model allows the firm to operate at a larger scale, resulting in production efficiency and lower costs.	1. Lack of competition: The monopoly model has been criticized for its lack of competition, which can lead to higher prices and lower-quality service. 2. Barriers to entry: Monopolies can create barriers to entry that can limit competition, preventing new entrants from entering the market and stifling innovation. 3. Limited consumer choice: Consumers have limited choice, as only one provider is in the market. This can lead to a lack of product diversity and options for consumers. 4. Lack of innovation: Monopolies may not have the motivation to innovate, given that there is no competition. This could stall technological advancements in the industry.

Table 1. *Cont.*

Electricity Markets Model	Advantages	Disadvantages
Single-buyer model	1. Reduced transaction costs: In a single-buyer model, electricity trading occurs between a single seller and a single buyer. This reduces transaction costs for both parties as the need to negotiate multiple contracts is eliminated. 2. Promotes stability: The single-buyer model can promote stability in the power market by ensuring a large and consistent demand for electricity. This can help ensure that power plants are running at optimal levels, which can reduce prices and increase efficiency. 3. Encourages investment: The single-buyer model can also encourage investment in the power sector by providing investors with a stable and predictable revenue stream. This is particularly attractive to investors who may hesitate to invest in a more volatile market.	1. Risk of inefficiency: The single-buyer model can also lead to inefficiencies in the power sector. This is because the buyer may have different incentives than multiple buyers to negotiate the best prices, leading to higher prices. 2. Political interference: The single-buyer model can also be vulnerable to political interference, which can lead to prioritizing certain projects over others, even if they are not the most efficient or cost-effective. This can lead to a suboptimal allocation of resources in the power sector. 3. Other limits: Consumers still need to sign long-term contracts and are subjected to regulated prices.

The current electricity system market consists of a wholesale market model and a retail market model, as shown in Figures 4 and 5. In the wholesale market model, there is competition between generators and distributors [35]. Consumers can choose to buy electricity directly from retailers who buy electricity directly from the electricity generation company, or they can choose to buy electricity from distributors who buy electricity from multiple generators and then sell it to consumers. This model allows more competition, which can lead to lower prices for consumers. However, there are also concerns about the potential for market manipulation by large distributors and generators. In some cases, these companies may engage in anti-competitive behavior, such as price-fixing or limiting the electricity supply to drive up prices. Many countries have implemented regulations to ensure fair competition in the wholesale electricity market. These regulations may include measures that prevent market manipulation, increase transparency, and promote the development of renewable energy sources. In the retail market model, prosumers can obtain electricity from retailers or between prosumers and RESs (producers) [36]. Retailers purchase electricity from wholesale markets and sell it to end users, including prosumers. These retailers often offer fixed or variable rate plans for residential or commercial customers, considering factors such as demand, seasonality, and market prices. Prosumers can also choose to obtain electricity from RESs. They can enter into agreements with RESs to purchase a set amount of electricity directly, eliminating the need for an intermediary retailer. In some cases, prosumers may even generate and supply excess electricity back to the grid, receiving compensation from the retailer or network operator. This model gives customers the power to choose the seller, and its advantage is that it is more economical and efficient for consumers to purchase electricity. The downside is that transmission and distribution costs can increase during peak periods and in remote locations. Overall, the wholesale market model can effectively provide consumers with affordable electricity while promoting competition and innovation in the energy industry. However, it is important to balance the benefits of competition with the need for regulatory oversight to ensure fair and equitable market practices. The retail market model allows flexibility and provides choices for prosumers, who can select the most cost-effective and sustainable source of electricity for their needs.

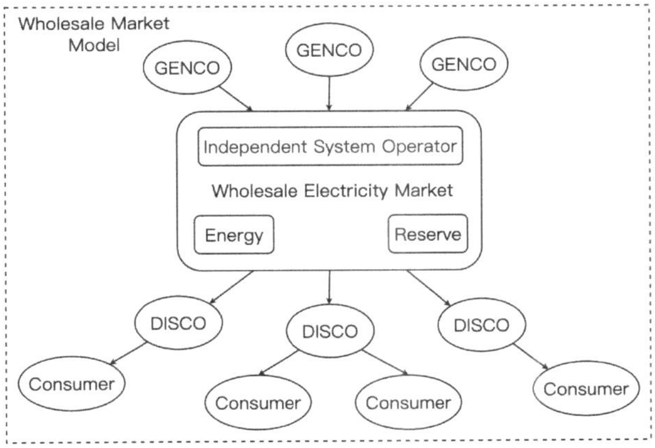

Figure 4. Wholesale market model.

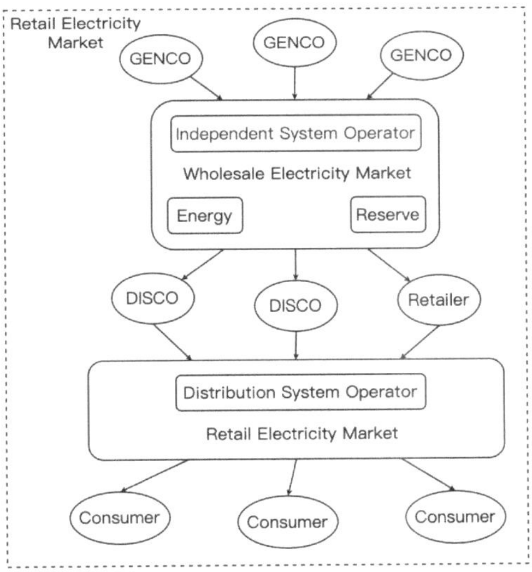

Figure 5. Retail electricity model.

3. P2P and M2M Energy Trading and Pricing in the Electricity Market

In recent years, the rapid expansion of distributed energy resources (DERs) has brought a significant transformation in energy systems, revolutionizing both the production and consumption of energy. Traditional consumers are gradually becoming prosumers who can store excess energy in storage devices, feed it back into the grid, or sell it to other prosumers. Direct energy trading between consumers and prosumers is called P2P energy trading, which can also be considered energy sharing (essentially free energy trading). With the advent of P2P energy trading, prosumers can exchange energy without intermediaries, increasing their benefits. A simplified P2P energy trading model diagram is shown in Figure 6 [37], featuring main participants such as power companies, energy sharing coordinators, consumers, and prosumers. Consumers can only purchase electricity, while prosumers can both buy electricity from or sell to other consumers or prosumers. Energy-sharing coordinators act as a platform for consumers and prosumers in negotiating energy transactions [38].

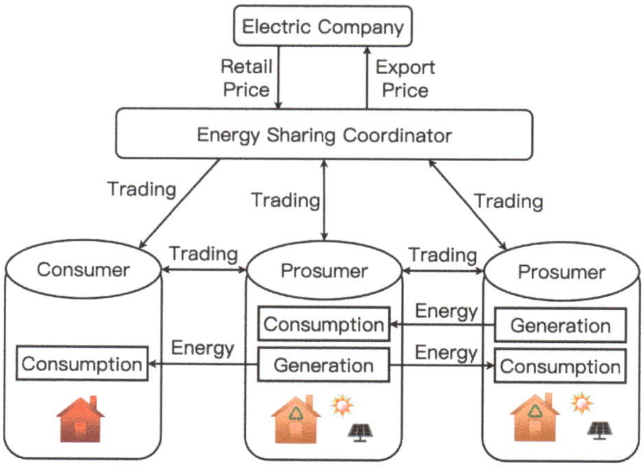

Figure 6. P2P energy trading model.

3.1. Pricing Mechanism of P2P and M2M Electricity Trading

P2P and M2M energy trading and sharing pricing mechanisms include auction theory, bargaining, and game mechanisms based on dynamic pricing models. Auction theory involves a competitive bidding process where buyers and sellers compete to determine the market price. This approach is suitable for P2P and M2M energy trading, where participants bid for the price they are willing to pay for energy. Bargaining is another pricing mechanism that allows buyers and sellers to negotiate a mutually acceptable price [39]. This approach is suitable for situations where participants have specific energy requirements and can negotiate a price that fits their needs. Game mechanisms based on dynamic pricing models involve setting prices that are dynamically based on market demand and supply. This approach is suitable when the market is volatile and depends on weather conditions, time of day, and geographical location. Moreover, for P2P energy sharing in MGs, pricing mechanisms need to consider factors, such as battery aging, the depreciation of RESs, energy transmission losses, and the wear and tear of related facilities, to avoid inaccurate initial pricing [40]. Table 2 summarizes the methods related to pricing mechanisms for P2P and M2M energy trading and sharing within five years.

Table 2. Pricing mechanisms for P2P and M2M energy trading and sharing.

Pricing Mechanism	Brief Description	Reference
Marginal pricing	Considering the transaction preference in the distribution network, a carbon-aware distribution location marginal pricing scheme for the distribution system operator's operation service pricing scheme is proposed to guide the P2P trading between producers and consumers.	[41]
	A new pricing mechanism (including preference, uncertainty, and local marginal price) for P2P energy trading is proposed. These pricing strategies can overcome current market uncertainty issues and increase the reliability of energy trading.	[42]

Table 2. *Cont.*

Pricing Mechanism	Brief Description	Reference
Dynamic pricing	A dynamic pricing mechanism for P2P energy trading is designed to enable the efficient trading of on-site energy and contribute to decarbonization and grid security goals—design dynamic price policies using multi-agent reinforcement learning with an open-source economic simulation framework built by Salesforce Research.	[43]
	A P2P energy trading mechanism for electric vehicles and solar power businesses is proposed. The mechanism is based on a dynamic pricing mechanism developed for storage energy prices.	[44]
	A framework for P2P trading prices considering dynamic retail electricity prices is proposed, through which prosumers can automatically generate bids and participate in auctions in P2P markets.	[45]
Three-part tariff price	This paper deals with pricing schemes for utilities and electricity consumers. The utility determines the time-varying price of non-renewable energy and the buy-back price of RESs while maximizing its profit. Electricity consumers decide whether to become prosumers according to the pricing policy.	[46]
Game theory	Based on the cooperative game of bargaining, a pricing mechanism for M2M energy trading is proposed. Additionally, a fast alternation method is introduced to solve the energy management problem of MGs in a decentralized manner. The results show that MGs can actively trade with each other and gain economic benefits.	[47]
	An efficient game theory-based approach is proposed to determine the electricity price mechanism in direct P2P electricity markets. Additionally, the participant's reserved price for electricity trading is considered to ensure that the participant's electricity transaction price is the best profit price.	[48]
	This paper considers two trading scenarios (smart homes and electric vehicles) and proposes a generalized Nash bargaining model to obtain optimal pricing strategies, effectively reducing overall costs.	[49]
Hybrid pricing	A power price mechanism framework is designed based on local households, and its optimization goal is to make the power distribution of adjacent families more balanced and stable. The framework utilizes real-time data from smart meters to monitor the power consumption of individual households and adjust the price accordingly.	[50]

Table 2. *Cont.*

Pricing Mechanism	Brief Description	Reference
Decentralized energy trading mechanism	A decentralized P2P energy trading mechanism in a distributed manner is proposed to maximize social welfare. The aim is to create a more efficient and dynamic energy market driven by market forces and consumer demand. The decentralized nature of the system also provides greater resilience and stability, as it reduces the risk of power outages and blackouts.	[51]
Feed-in tariff	By studying the techno-economic performance of renewable energy depreciation in energy-sharing systems, new energy management strategies have been proposed to improve the relative capacity of batteries.	[52]

By reviewing the literature related to the pricing mechanism of P2P and M2M energy trading, we can see that when the pricing mechanism for P2P energy trading and sharing appears in the power system, it is expected to revolutionize the way energy is bought and sold, enabling consumers to trade excess energy with each other via a decentralized platform. The pricing mechanism for P2P energy trading and sharing will be based on a market system, with consumers that are able to set prices for the energy they generate and consume. This will enable them to obtain the best value for their energy resources, and they are incentivized using RESs. The pricing mechanism will also consider factors such as the time of day and seasonal variations, as well as the level of demand in the market. This will ensure that consumers can sell their energy when prices are high and buy when prices are low. Ultimately, introducing P2P energy trading and sharing will create a more efficient and sustainable energy system, with consumers able to take control of their energy usage and make informed decisions about their energy needs. This will also enable the transition toward a more decentralized and decentralized energy system, with consumers becoming active participants in the energy market. In summary, P2P and M2M energy trading and sharing pricing mechanisms depend on various factors, including the market structure, participants' requirements, and market volatility. Dynamic pricing models and auction theory can help optimize the trading process while ensuring fair prices for all participants.

3.2. Recent Development of P2P and M2M Energy Trading and Sharing

In recent years, researchers' studies on P2P and M2M energy trading mechanisms in the power system have mainly focused on the following: energy trading platforms (allowing storage and trading), blockchain (equivalent to a distributed database), simulation experiments (to verify P2P energy trading mechanism), game theory (used to simulate the behavior of participants), optimization mechanism, algorithm, pricing mechanism, market design, and market regulation. These research areas aim to develop efficient, reliable, and scalable P2P and M2M energy trading mechanisms that can improve the overall performance of the electricity market and satisfy the needs of various stakeholders, including producers, consumers, and grid operators [53]. Some of the specific research topics that have emerged in recent years include peer-to-peer energy trading platforms based on blockchain technology; simulation experiments to evaluate the performance of different trading mechanisms; game-theoretic models to capture the strategic interactions among market participants; optimization algorithms to optimize the allocation and pricing of energy resources; pricing mechanisms that reflect the real-time value of electricity; market designs that balance the competing objectives of efficiency, fairness, and sustainability; and regulatory frameworks that ensure the security, privacy, and interoperability of energy markets. Table 3 summarizes the methods related to P2P and M2M energy trading mechanisms in the past five years.

Table 3. Methods of P2P and M2M energy trading and sharing.

Methods	Brief Description	Reference
Multi-leader and multi-follower Stackelberg game	A two-stage P2P energy trading market is proposed. In the first stage, a multi-leader and multi-follower Stackelberg game establishes a comprehensive P2P market. In the second stage, network reconstruction is considered.	[53]
	A P2P energy trading market strategy is proposed based on the blockchain, and the Stackelberg game is used to establish leaders and followers. This strategy considers the market's supply and demand and provides accurate energy transfer signals, facilitating transactions between MG.	[54]
Cooperative game theory and blockchain	A P2P energy trading mechanism combining cooperative game theory and blockchain is proposed and verified by conducting comprehensive theoretical analysis and simulation experiments using a standard IEEE 14 bus system and real datasets.	[55]
Blockchain platform	A credit-based P2P electricity trading model in a blockchain environment is proposed to facilitate the local trading of RESs. The model is based on a blockchain platform where each participant is registered as a node and has a digital wallet. Each RES is equipped with a smart meter measuring production and consumption. In this model, participants can sell their excess electricity to other participating nodes or use it to offset their consumption.	[56]
Blockchain technology and smart contract	A P2P multi-energy market trading mechanism is proposed, which can simultaneously generate electricity and heat from RESs and enable users to trade their excess energy in a decentralized manner. The proposed mechanism utilizes blockchain technology to ensure secure and transparent transactions and smart contracts to automate the trading process. The trading mechanism is based on a bidding system, where users can place bids to buy or sell energy based on their energy needs or availability. Smart contracts automate the trading process, eliminating the need for manual intervention and ensuring that transactions are executed automatically when predefined conditions are met.	[57]
Distributed ledger technology	A P2P electricity trading method based on intelligent interconnection multi-regional regulatory distributed ledger technology is proposed, which promotes the reliability of P2P trading.	[58]
Decentralized market clearing	A new decentralized P2P energy trading platform is proposed. It mainly includes the market layer and the blockchain layer. The market layer is mainly used for transactions and clearing, while the blockchain layer is used for real-time settlement, and platform simulation experiments have been carried out.	[59]

Table 3. *Cont.*

Methods	Brief Description	Reference
Nash non-cooperative game	A Nash non-cooperative game P2P energy trading model based on residential and commercial prosumers is proposed to achieve price fairness in multi-energy transactions. This model considers the dynamics of the energy market, the energy consumption patterns of prosumers, and RES generation, which influences the energy supply. Prosumers with excess energy can sell it to other prosumers by using a blockchain-based decentralized platform, and buyers can purchase energy from the corresponding sellers by using the platform.	[60]
Stochastic decision-making framework and bilateral contracts	Using a stochastic decision-making framework and in the case of a P2P energy trading mechanism based on bilateral contracts, electricity is purchased from wind generators, and stored electricity is sold. Maximizing the profit of the upper wind power generators via a two-level stochastic model also minimizes the cost of the DR aggregator in the lower layer.	[61]
Double auction	A local electricity trading platform is proposed, which is based on the blockchain-distributed P2P double auction transaction mechanism, and the impact of trading mechanisms on distribution network control, operation, and planning is analyzed.	[62]
Cooperative game	An algorithm for local power exchange based on cooperative game theory has been proposed, which combines incentive mechanisms to facilitate M2M energy trading among local prosumers. The algorithm aims to maximize the community's social welfare by efficiently allocating energy trades and balancing energy demand and supply. The proposed algorithm provides a framework for efficient and fair M2M energy trading among local prosumers, leading to cost savings, reduced carbon footprint, and increased grid resilience.	[63]
Game-theoretic and a motivational psychology framework	A game-theoretic P2P energy trading scheme is developed, and a motivational psychology framework is introduced to analyze the psychological motivations of prosumers before they are persuaded to participate in energy trading.	[64]

By reviewing the research on P2P and M2M energy trading mechanisms or methods in power systems, we can observe the following: There is great interest in developing novel approaches to enable P2P and M2M energy trading methods in power systems. Blockchain technology is a promising solution for implementing P2P energy trading mechanisms due to its decentralized nature, security, and transparency. Various market mechanisms, such as auctions, negotiations, and smart contracts, have been proposed to implement P2P and M2M energy trading in power systems. Machine learning and artificial intelligence techniques have been explored to facilitate efficient trading mechanisms and enhance the reliability of energy trading systems. Integrating RESs and ESSs into energy trading frameworks can help improve efficiency and reduce energy costs. Standards and protocols for data exchange and communication are crucial for developing reliable and secure P2P and

M2M energy trading mechanisms in power systems. Overall, the research suggests that P2P and M2M energy trading mechanisms have the potential to revolutionize the energy market and provide significant benefits to consumers, producers, and the environment. However, further research and development are needed to overcome the technical, regulatory, and economic challenges associated with implementing these mechanisms.

4. Pricing in RESs, ESSs, and Controllable Loads Applications

4.1. Pricing Model in Today's Smart Grid

The pricing mechanism mainly considers the following principles: the price should be between on-grid and public electricity prices. Based on the basic principles of economics, price is inversely proportional to the ratio of supply and demand and must always maintain an economic balance [63]. Supply and demand are basic concepts in economics and play a crucial role in determining the price of goods and services. The law of supply states that the quantity of an interest supplied will generally increase as the price of the good increases, while the law of demand states that the quantity of a suitable demand will generally decrease as the price increases [64]. The supply and demand relationship reflects the pricing basis that all producers and consumers can jointly determine in a market. When the demand for electricity increases, the price tends to increase as generators struggle to keep up with the increased demand. Conversely, when there is less demand for electricity, the price tends to fall as generators try to attract more buyers. The relationship between supply and demand can significantly impact the economy. A surplus of energy can lead to lower prices and profits for producers, while a shortage can lead to higher prices and potentially higher profits. Therefore, producers must monitor the energy demand closely and adjust their production and pricing strategies accordingly. Similarly, consumers also need to assess the energy price and determine whether it is reasonable based on their perceived value and willingness to pay. Ultimately, an equilibrium is reached when the price of energy satisfies both the producers and consumers in a market.

The ratio of total electricity sales to total electricity purchases can express the supply and demand relationship [65]. The supply and demand ratio varies over time and is inversely proportional to price. When the supply–demand ratio is zero, there is no producer or consumer that sells electricity, and individuals can only purchase the required energy from the public grid. When the supply–demand ratio is equal to or greater than 1, this indicates an excess of electricity in the system, which needs to be sold back to the utility grid. The selling price between the prosumer and the public grid will dynamically change when the supply–demand ratio falls within the range of zero and one. If there is no excess electricity trading, it can only be purchased from the public grid. From the perspective of prosumers, if supply and demand are relatively small, buyers hope to reduce electricity consumption at a lower purchase price, and sellers also hope to reduce electricity consumption to sell more energy. Conversely, if the supply–demand ratio is large, sellers will increase power consumption for higher selling prices, and buyers will increase power consumption for more energy. Pricing in this model logically appeals to prosumers via DR.

In the case of electricity pricing, supply and demand factors are influenced by various factors such as weather conditions, time of day, season, and overall economic conditions [66]. When there is excess demand for electricity, the price of electricity tends to increase, while when there is excess supply, the price tends to decrease. In the case of RESs, the pricing mechanism should also consider the energy source's environmental benefits and long-term sustainability. Ultimately, the pricing mechanism must balance promoting efficient consumption and ensuring that electricity remains accessible and affordable for all.

4.2. Photovoltaic Prosumers

According to the feed-in tariffs of many countries, a typical photovoltaic prosumer metering configuration is two independent meters: one for photovoltaic power generation metering and one for a two-way measurement of the import and export of photovoltaic producers and consumers [67]. The photovoltaic power generation meter measures the amount of electricity generated by the photovoltaic system and is typically installed between photovoltaic panels and an inverter. This meter provides information on the amount of electricity generated from the photovoltaic system, which is used to calculate the feed-in tariff payment. The two-way meter measures the electricity consumed from the grid and the amount of electricity exported back to the grid by the photovoltaic prosumer. This meter is typically installed at the connection point between the prosumer's electrical installation and the grid. This meter provides information on the amount of electricity consumed from the grid and the amount of electricity exported to the grid, which are used to calculate the net electricity consumption and the feed-in tariff payment [68]. The photovoltaic prosumer pays for the electricity consumed from the grid and receives payment for the electricity exported back to the grid at a feed-in tariff rate. The feed-in tariff rate is typically higher than the retail electricity rate, incentivizing prosumers to generate more electricity from their photovoltaic system than they consume from the grid [69]. The prices designed for photovoltaic producers and consumers are power generation and import and export prices. The price of power generation is set according to the total power generation (self-consumption and public use) of the photovoltaic system, and the export price is usually lower than the import price. For this feed-in tariff, the self-use of photovoltaic energy is greatly encouraged.

Each photovoltaic prosumer consists of photovoltaic systems, energy management systems, controllable loads, and smart meters [70]. The photovoltaic system is a technology that converts sunlight into electricity. It consists of photovoltaic modules, inverters, and other components that ensure proper functioning. The energy management system is responsible for optimizing the energy generated by the photovoltaic system. It can monitor energy consumption, adjust loads based on demand, and store surplus energy. Controllable loads are appliances or devices that can be turned on or off based on energy availability. Smart meters measure energy consumption and production and transmit real-time data to the energy management system. It helps prosumers optimize their energy usage and enables energy providers to manage the grid efficiently. Photovoltaic prosumers can share excess energy between prosumers instead of trading directly with the public grid. The difference between PV prosumers comes from their different load curves and PV capacity so that they can be shared in the MG. In the energy trading system, there will be an intermediate person in charge of maintaining the balance of electricity and payment in the energy trading system. The person in charge buys electricity from the utility grid or photovoltaic prosumers with surplus power and then sells the electricity to other prosumers or the utility grid. Although feed-in tariffs can provide the basic impetus for energy transactions, an appropriate pricing mechanism is still required. Because electricity price is the basic requirement of prosumer billing, DR based on dynamic pricing is conducive to improving the utilization rate of photovoltaic energy.

Load forecasting technology is based on previous data, weather conditions, and RES availability in predicting the system's future energy demand. Load forecasting directed at the primary user aims to better forecast the future load of a given system over a specific time in order to ensure a balance between energy supply and demand. Demand forecasting is a key parameter for power system operation and planning. Larger errors in the forecasting model will increase operating costs, so an accurate load forecasting model becomes very important. The most used forecasting methods are multiple linear regression, stochastic time series, exponential smoothing, state space and Kalman filters, and artificial neural networks [71]. The battery controller only receives the electronic speed controller's control signal without considering the prosumer's net load [72]. For example, a battery may be discharged when the prosumer has excess photovoltaic generation. For instance, the battery may be discharged when the prosumer has excess photovoltaic generation. It can lead to inefficient battery use and potential loss of energy. Therefore, it is important to integrate a load controller that considers the prosumer's net load. This load controller can adjust the battery's charging and discharging rate based on the net load, optimizing the use of the battery, and ensuring maximum energy efficiency. Additionally, incorporating an energy management system can help monitor and regulate the overall energy consumption of the prosumer's home, helping them further reduce their energy expenses.

5. Pricing in Smart Grid with Demand-Side Ancillary Services Provision

Enabled by the growing installed capacities of DERs and energy storage systems, demand-side ancillary service provision from prosumers can help maintain the stability of the power system. This is achieved by actively controlling the energy assets at the demand side to follow the regulation curve sent by the grid operator. The regulation curve is essentially a frequency control signal, resulting from the area control error in a power system. Hence, following the regulation curve will help stabilize the grid frequency upon events such as sudden load changes and/or generator malfunctions. To motivate the prosumers to participate in the DSAS, the grid operator establishes incentive schemes that reward the entities that can provide DSAD. This involves an advanced pricing scheme for incentivization [73–75].

DR refers to the practice of managing electricity consumption in response to grid conditions, such as price, to balance demand and supply [76]. It is a flexible approach that allows consumers to reduce their energy consumption during peak hours voluntarily or when there is a supply shortage in the grid. It can help maintain grid stability, reduce the risk of blackouts, and avoid constructing additional power generators in order to meet peak demands. DR can also help reduce energy costs for consumers and improve energy efficiency by incentivizing them to use energy more wisely. Thus, it has become an important tool for grid operators and policymakers to manage the energy system.

D-FCAS, as a type of DR, comprises the active participation of electricity prosumers in providing ancillary services to help maintain the stability and reliability of the power grid. When there is a sudden increase or decrease in electricity demand or unexpected changes in the power output from DERs, D-FCAS can help mitigate the impact by rapidly adjusting electricity consumption to help balance supply and demand. This means that prosumers are actively involved in managing the power system. Demand-side management programs, including DR, time-of-use tariffs, and energy efficiency measures, incentivize prosumers to participate in providing ancillary services [77]. With the growing penetration of RESs, D-FCAS is becoming increasingly important with respect to maintaining the stability and reliability of the power grid and supporting the transition to a low-carbon energy system.

DSAS is important to both the supply and demand sides of the electricity market and is conducive to the interactions between the two parties. Therefore, it is necessary to formulate a pricing strategy for DSAS. This pricing strategy for DSAS should consider the cost of providing the service, the level of demand, the value of the service, competition, and regulatory requirements [78]. The cost of service provision includes equipment, maintenance, and labor. The price should be set to cover these costs and provide a reasonable profit margin for service providers. The level of demand for DSAS is an important factor in setting prices. If demand is high, the price may be higher, whereas if demand is low, the price may need to be lowered to encourage more participation. The value of DSAS to the electricity market should also be considered. For example, if the service helps to stabilize the grid and prevent blackouts, the price can be higher. Competition between service providers can also affect pricing. If many service providers offer the same service, prices may need to be lower to remain competitive. Regulatory requirements may also impact pricing. For example, if regulations require a certain level of participation in DSAS, prices may be lower to encourage more participation. By considering these factors, service providers can set fair and competitive prices while still covering costs and providing a reasonable profit margin.

5.1. Market Structure and Pricing

The market structure of DR is modelled based on the PJM electricity market [79] containing both day-ahead and hour-ahead decision-making processes. Prices are utilized to incentivize consumers to adjust their electrical loads and actively participate in the electricity market, facilitating DSAS and peak shaving. The PJM electricity market is an organized wholesale market that provides electricity to customers [80]. It operates as a competitive market where the price of electricity is determined by the supply and demand of electricity in real time. In this market structure, DR programs aim to manage peak demand by incentivizing customers to shift their electricity usage to times when electricity is cheaper, and demand is lower. By following this incentive, consumers can participate in the electricity market by reducing their demand during peak hours and increasing it during off-peak hours [81]. This creates a more efficient market by reducing the need for additional generation capacity during peak hours, and this reduces the likelihood of blackouts or power grid failures. DR programs enable consumers to benefit financially by reducing their electricity bills and participating in the electricity market. The programs also provide improved grid reliability, which helps utilities avoid investing in costly infrastructure upgrades. Encouraging participation in DR programs makes it possible to optimize the use of existing resources to meet the electricity demand and reduce the environmental impact of electricity use. The market structure of DR is based on the PJM electricity market, where prices are used to encourage consumers to participate in DR programs and manage peak demand. This enables DSAS and peak shaving, which creates a more efficient and less costly electricity market. By encouraging participation in DR programs, it is possible to optimize the use of existing resources, reduce the environmental impact of electricity use, and maintain grid reliability.

DR can be divided into economy-based and incentive-based DR. In the economy-based DR, the system operator will design different electricity price structures to motivate different types of users, such as residential and industrial [82]. Users can change their energy consumption patterns according to the pricing framework. End users are encouraged to participate in the DSAS scheme to reduce consumption when electricity prices increase and ensure safety and stability in the event of insufficient or faulty generation [83]. Figure 7 illustrates the implementation of DR among the main players in the power system. Generators and retail markets submit quotations to the wholesale market, which then provides market-clearing prices to both parties. These market-clearing prices reflect the prevailing supply and demand conditions of the wholesale market at a given time and are typically used as the basis for setting retail prices for electricity. In other words, generators and retailers submit their bids and offers to the wholesale market, and the market determines the price at which the energy will be sold. This price is based on some factors, including the cost of production, transmission and distribution, fuel prices, and market conditions. Retailers will then use this wholesale price as a starting point for setting their retail prices while considering their own costs and profit margins. Ultimately, the end user pays their energy bills based on this retail price, which is influenced by the wholesale market-clearing price. The user's participation in DR is the need to receive the information of the predicted price. Then, the user coordinates with the aggregator to achieve the global goal of reducing energy consumption. The user's participation is crucial to the success of DR programs, as it allows utilities to balance the supply and demand more effectively. This ultimately helps prevent power outages and reduces the cost of energy for everyone involved [67]. Additionally, users may be incentivized to participate in DR programs via other financial incentives. The DR of network integration is expected to minimize the cost of purchasing energy from the wholesale electricity market, network losses, and associated risk costs. It is achieved by optimizing the operation of assets and infrastructure, such as generators, storage systems, and transmission lines, and by leveraging advanced technology, data analytics, and modeling techniques to forecast demand and manage supply. Table 4 summarizes the research on pricing schemes related to DSAS within the past five years.

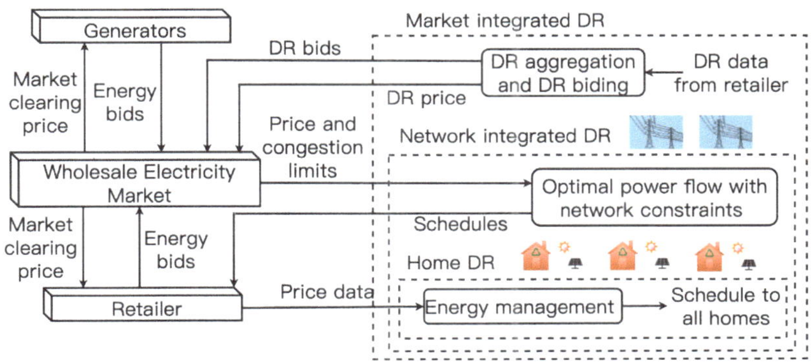

Figure 7. DR at all levels of the power system.

Table 4. Pricing strategies related to DSAS.

Pricing Strategies (Objectives)	Brief Description	Reference
Balance the supply and demand and minimize the cost of energy	A novel demand-side management framework is developed. The aggregator sets energy prices for flexible customers based on the predicted energy demand for the upcoming day. They then use a real-time market monitoring algorithm to adjust prices throughout the day in response to changes in market demand. The aggregator also buys energy from the day-ahead market, aiming to balance the supply and demand and minimize the cost of energy.	[84]
Increase user participation	User-centered design is an approach in which users' needs, preferences, and behaviors are considered in the design process. When designing a DSAS, it is important to consider the willingness of users to pay for the service and their time. Considering these factors, they can make the program more appealing and increase user participation, ultimately leading to a more effective DR program.	[85]
Manage flexible load requirements optimally	Combined with the time-of-use pricing strategy, a consumer perception pricing strategy (residential DR pricing strategy) is proposed. Each customer will receive electricity price signals based on their load requirements to manage flexible load requirements optimally.	[86]
Obtain optimal power supply and consumption strategies	A distributed real-time pricing strategy is proposed considering supply and demand interests. Real-time pricing is determined via information interaction between users and suppliers to obtain optimal power supply and consumption strategies.	[87]

Table 4. *Cont.*

Pricing Strategies (Objectives)	Brief Description	Reference
Reducing electricity usage during peak periods	A home energy management system introduces dynamic pricing to incentivize consumers to change how they operate on the user side, reducing electricity usage during peak periods.	[88]
Make electricity price forecasts more accurate	A dynamic pricing mechanism and a demand response program for industrial parks are proposed, and a dynamic price forecasting model uses long short-term memory technology to predict electricity prices with high accuracy. This technology can process large amounts of data and can detect patterns and trends that may otherwise be missed by traditional forecasting methods.	[89]
Benefit all stakeholders	A multi-objective optimization model for the dynamic price considering DR and multiple stakeholders' preferences. Optimizing supply and demand parties to jointly formulate dynamic prices and obtain dynamic price control strategies for different stakeholders in different situations.	[90]
Reducing peak demand	A model of residential DR based on real-time locational marginal price is proposed. This model utilizes real-time price signals to incentivize residential consumers to adjust their energy consumption patterns, reducing peak demand. The model considers location-specific prices, which indicate the cost of supplying energy to a particular location at a given time.	[91]
Improve market profitability	A probability distribution function of market price based on the demand–price allocation curve is proposed to estimate the optimal pricing method for retailers in the electricity market. Mainly considers the profitability of the retailer and the market-clearing price for bidding.	[92]
Reduce consumer costs	A reverse optimization pricing method is developed based on DSAS and day-ahead markets. Estimates the cost function of other consumers given historical market clearing prices and the capacity to generate electricity and then bids strategically based on these data.	[93]

5.2. Impact on Pricing Schemes

Table 5 shows the main positive impacts and limitations of DSAS on the electricity price scheme of the power system [10,94,95].

Table 5. DSAS impact on the pricing scheme in the power system.

Positive	Limitation
Cost reduction: DR programs encourage customers to reduce their electricity consumption during peak hours in exchange for incentives or lower rates.	The increased cost of electricity: DSAS can increase the cost of electricity by requiring additional resources to maintain power quality and reliability.
Improved grid stability: The grid's stability is improved by using DSAS, as they allow grid operators to manage electricity demand and supply better.	Distorted pricing signals: Ancillary services can distort the pricing signals in power markets, leading to market inefficiencies and resulting in sub-optimal allocation of resources.
Increased reliability: DSAS can increase the grid's reliability, as they can help prevent blackouts and other power disruptions by reducing demand during peak hours.	Uncertainty in market planning: This can result in increased investment costs in the short term to ensure adequate ancillary service capacity as well as the overbuilding of resources to ensure adequate capacity.
Encouraging RESs: DSAS can also encourage the use of RESs, and DASA refers to the ability to manage and adjust electricity consumption in real time to match the fluctuating electricity supply from variable RESs.	Reduced market competition: Ancillary services can disincentivize entry into the power market, as potential competitors may view the regulatory requirements surrounding ancillary services as too onerous or costly.
Enhanced energy efficiency: DSAS also supports energy efficiency by promoting energy conservation and reducing energy waste during peak hours.	Increased regulatory complexity: Ancillary services can also add to the regulatory complexity of power markets, requiring additional oversight and reporting requirements.

While DASA may introduce certain restrictions on power system pricing schemes, it is essential for the sustainability of the electricity market as they promote grid stability, energy efficiency, and RESs while reducing electricity price volatility and achieving environmental sustainability. DSAS allows for managing and controlling electricity supply and demand by integrating technologies such as energy storage, smart meters, and advanced control systems. With the growing adoption of renewable energy sources, DSAS is becoming increasingly important in ensuring the effective integration, management, and optimization of variable, intermittent sources into the grid.

6. Pricing Framework in Retail and Wholesale Markets in the Modern Smart Grid

The modern power trading market mainly comprises two types of markets: the wholesale market and the retail market. Wholesale market participants buy electricity from utility providers and sell it to retailers at wholesale prices that are generally not predictable due to fluctuations in supply and demand [96]. These participants include power generators, wholesale energy traders, and energy brokers. The wholesale electricity market plays a critical role in ensuring a reliable and affordable electricity supply for consumers. Providing a platform for energy trading, it allows for efficient price discovery and the allocation of resources. Participants in the wholesale market must navigate complex regulations and manage risk associated with market volatility. They also play a key role in facilitating RESs integration and DR programs. Wholesalers need to maintain relationships with their suppliers and retailers to ensure consistent supply and demand for their products. Whereas retailers profit by buying from wholesalers at variable prices and selling to end users at a fixed price in the retail market, wholesalers' profit by buying products in bulk at a lower price and selling them to retailers at a slightly higher price [97]. It allows retailers to purchase products at a price that allows them to profit while offering competitive prices to end users. It is necessary to establish various electricity price frameworks and evaluate the market to ensure good and potential energy transactions.

6.1. Pricing Framework in the Retail and Wholesale Markets

The setting of the pricing framework depends on the network's characteristics and the pricing strategy. Some current pricing models include unified pricing, block pricing, seasonal pricing, time-of-use pricing, peak hour pricing, key peak pricing, and real-time pricing. The setting of the pricing framework for power distribution is typically influenced by factors such as the type of customers being served (e.g., residential, commercial, or industrial), the level of demand for electricity in the area, the cost of delivering power to different areas, and regulatory requirements [98]. In general, the goal of setting a pricing framework for power distribution is to ensure that prices are set at a level that covers the cost of delivering electricity while encouraging the efficient use of the network and minimizing the impact on consumers.

Flat pricing means that even if demand changes, the price remains the same, and consumers will not pay high electricity bills for using electricity that was not planned or used during peak demand times [99]. Utility companies typically use this type of pricing for residential and small commercial customers. Flat pricing can give customers a sense of stability and predictability in their monthly bills, as they will always know what their electricity costs will be regardless of the time of day or year. However, it may not incentivize customers to reduce their energy usage during peak demand times, which can lead to increased strain on the electric grid during these times. This pricing strategy has a negative impact on utilities because their consumers pay the same rate for electricity no matter how much energy is consumed [100]. Block pricing helps segment consumers based on their electricity usage, and sloped rates increase prices as consumption increases while declining schemes have the opposite effects [101]. Utility companies can use this pricing strategy to encourage energy conservation among high-usage customers. For example, a utility company might have a block pricing system where the first 500 kilowatt hours (kWh) of electricity used monthly is charged at a lower rate: for example, USD 0.10 per kWh. Any usage above that amount is charged at a higher rate: for example, USD 0.20 per kWh. This encourages customers to use less electricity to stay within the lower-priced block. A declining scheme could also be used where the price per kWh decreases as usage increases. For instance, the first 200 kWh may be charged at USD 0.15 per kWh, the next 300 kWh may be charged at USD 0.12 per kWh, and anything above 500 kWh may be charged at USD 0.10 per kWh. This may encourage customers to use more electricity since the cost per kWh decreases as usage increases. Overall, block pricing helps segment consumers based on their electricity usage, encouraging those who use more energy to conserve and those who use less energy to consume more without incurring higher costs.

Seasonal pricing means that electricity prices are priced according to different seasons to meet different needs in different seasons [46]. This is usually carried out to reflect changes in demand for electricity caused by changes in weather patterns, economic activities, and other factors that affect consumers' energy consumption [85]. During peak seasons, especially in summer and winter, electricity demand is usually high due to the increased use of air conditioners, electric heaters, and other appliances. To balance the demand and supply of electricity, power companies usually charge higher prices during these periods to discourage excessive energy usage and help maintain a consistent electricity supply. On the other hand, during off-peak seasons such as spring and fall, when electricity demand is usually lower, power companies may offer lower prices to encourage energy consumption and utilization and avoid having an oversupply of electricity. Overall, seasonal pricing helps balance electricity demand, reduce energy wastage, and ensure that the electricity supply remains reliable and sustainable over time. Time-of-use pricing is a pre-declared price that varies depending on the time of day [102]. Electricity prices increase during peak load hours and decrease during off-peak load hours, encouraging customers to use electricity during non-peak hours when the demand for electricity is lower. This pricing strategy is designed to help utilities manage electricity demand and avoid overloading the power grid during peak demand periods [103]. This also helps create a fair pricing structure for customers who use electricity during off-peak hours. Time-of-use pricing can benefit customers who are able to shift their electricity usage to off-peak hours, reducing their overall electricity costs. However, it can be a challenge for customers who cannot easily shift their usage. Peak hour pricing electricity costs are much higher than at other times of the day. The difference from segment pricing is that the duration of peak hour pricing is shorter [104]. It typically applies only during certain times of the day when electricity demand is highest. This type of pricing is designed to encourage consumers to reduce their electricity usage during peak times, which helps reduce strain on the grid and prevent blackouts. Consumers can avoid high peak-hour prices by shifting their electricity usage to off-peak times. Many utilities also offer programs that allow customers to enroll in special rate plans that offer lower prices during off-peak hours [105]. Critical peak pricing is a dynamic pricing strategy in which utilities increase electricity prices in response to wholesale market price increases or emergencies, such as high demand or transmission constraints [84]. Critical peak pricing aims to reduce peak demand, avoid blackouts or brownouts, and minimize the need for expensive upgrades to the electricity grid. Utilities can reduce the strain on the grid by incentivizing consumers to lower their electricity usage during peak hours and maintain a more stable and efficient system. To participate in critical peak pricing programs, consumers typically need to enroll and agree to reduce their electricity usage during peak hours. Utilities may incentivize participants, such as bill credits or reduced rates during non-peak hours.

Real-time pricing is a pricing model that adjusts prices based on real-time market conditions [106]. In the case of energy usage, real-time pricing means that the cost of electricity changes regularly throughout the day as demand and supply change. For consumers, electricity may be cheaper during off-peak hours when demand is low and more expensive during peak hours when demand is high. Real-time pricing has been observed to encourage consumers to be more mindful of their energy use and to shift their consumption to times when it is less expensive, ultimately reducing energy demand during peak times. It also encourages energy providers to invest in RESs in order to reduce reliance on fossil fuels during peak hours, as prices for non-renewable energy sources can become prohibitively expensive. A brief scheme of real-time pricing is shown in Figure 8. The current research on real-time pricing mainly focuses on the following three aspects. The first is to focus on how users respond to real-time prices and realize demand with lower electricity prices. The second is to focus on real-time retail prices, regardless of users. The last aspect is to set the real-time retail electricity price based on users and retailers, matching supply and demand [106].

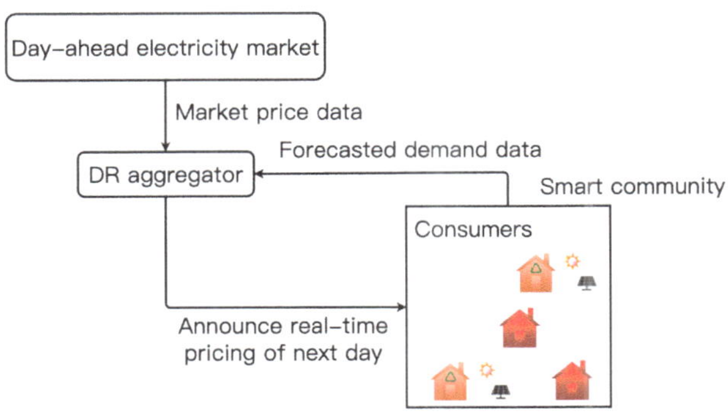

Figure 8. Graphical illustration of a real-time pricing strategy.

6.2. Limitation in Electricity Retail and Wholesale Markets

The limitations of the pricing framework for retail and wholesale markets in modern smart grids are as follows [107,108]:

- Lack of flexibility: The traditional pricing framework is rigid and lacks the flexibility to accommodate changes in customers' behavior and preferences. It cannot adapt to the dynamic energy demand and supply patterns in modern smart grids.
- Complexity: Pricing frameworks for smart grids are often complex and difficult to understand for the average consumer. This creates confusion and may discourage them from adopting energy-efficient practices.
- Inequity: Pricing frameworks may lead to inequity in terms of cost-sharing among customers, especially those who cannot afford to invest in energy-efficient appliances or heating systems.
- Data privacy concerns: Smart meters used to measure energy consumption and feed data into the pricing framework may raise privacy concerns among customers.
- Operational challenges: Implementing and maintaining pricing frameworks can be costly and time-consuming, especially for small utility companies.
- Resistance to change: Customers may resist changes in pricing structures due to familiarity and a lack of trust in new pricing mechanisms.
- Potential for market manipulation: Pricing frameworks have the potential to be manipulated by energy companies, leading to unfair pricing practices.

7. Future Research Directions

Researchers are actively studying pricing frameworks, methods, and strategies for power systems. However, despite ongoing efforts, this field still faces unresolved issues and challenges that require further investigation and exploration.

One of the main challenges is the pricing strategy after integrating variable RESs into the power system. RESs have intermittent generation patterns, making it difficult to forecast and plan for their output accurately.

Another challenge is determining the optimal pricing mechanism that promotes the efficient and cost-effective use of the power system while ensuring affordability for consumers. Some argue for a cost-based pricing approach, where prices reflect the true cost of providing electricity, while others advocate for market-based pricing approaches, where prices are determined by supply and demand.

Furthermore, there are concerns regarding power system pricing strategies' equity and distributional impacts. Certain pricing mechanisms, such as time-of-use pricing and peak/off-peak pricing, may disproportionately affect low-income households or those with limited ability to shift their energy usage during peak periods.

Moreover, the dynamic energy trading price models combined with grid pricing schemes incentivize end-user participation in community-based peer-to-peer energy trading and sharing.

In addition, new pricing frameworks need to be proposed for consumer-centric energy trading in order to provide win–win economic benefits via cooperation and stimulate multi-stakeholder activities and market dynamism. Such pricing frameworks should have higher flexibility, lower complexity, better equality, and higher data security.

It also includes improved energy efficiency, power dispatch, voltage support, and congestion management services for local grids; cross-regional energy sharing and trading; equipment depreciation costs; and transmission losses. The above problems and challenges can be used as directions for future research.

8. Conclusions

Over the past few decades, with the increase in RESs, the electricity market has transformed, and the production and consumption of energy have also undergone dramatic changes. Electricity price is a key factor that affects the interests of various stakeholders in the electricity market and has also attracted research attention. As the number of prosumers increases, providing a more decentralized and distributed grid becomes more important. The power system has also changed from a single vertical market and fixed electricity pricing model to a diversified market mainly composed of retail markets and wholesale markets. Current power systems have dynamic pricing schemes, and many pricing frameworks and strategies related to P2P and M2M energy trading and the provision of demand-side ancillary services have also been proposed. The framework and strategies for formulating electricity prices in various ways have brought advantages to the power system.

First, formulating electricity prices in different ways can encourage consumers to use electricity efficiently. Second, having new market components such as feed-in tariffs can increase the utilization of RESs. Third, dynamic pricing can help balance supply and demand and reduce the chance of blackouts and brownouts during peak demand periods. Fourth, innovative dynamic pricing models can incentivize utilities and prosumers to adopt new technologies and energy management systems. Fifth, appropriate pricing can reduce system costs and make the electricity market more competitive and cost-effective.

This paper reviews a large body of literature on pricing methods, strategies, and frameworks in the electricity market. The electricity market's reform and electricity price mechanism's development have been investigated from the 20th century to the present. The main components of electricity prices and the electricity market framework from the monopoly model to the single-buyer model, and to today's electricity market, were presented. Next, P2P and M2M energy trading and pricing strategies for DERs and load support were described. The pricing mechanisms of P2P and M2M energy trading and sharing and the current related research literature were reviewed. Recent technologies on energy trading platforms, blockchain, simulation experiments, game theory, optimization mechanisms, and algorithms were revisited. Then, pricing methodologies for RESs, batteries, and controllable load applications were elaborated. The market structure associated with DSAS and its advantages and disadvantages with respect to electricity system price schemes were detailed. We then presented the current pricing framework for retail and wholesale markets and the limitations that this framework imposes on the electricity system. Finally, future research directions for power system pricing methods, strategies, and frameworks were suggested.

From this review article, we know that government policies, regulations, technology, and market competition impact the power market framework and pricing strategies. Understanding the power market framework and pricing strategies can benefit policymakers, industries, and prosumers. Policymakers can use this knowledge to design effective policies and regulations that promote competition and innovation in the power market. Industry players can make informed decisions about investments, production, and pricing, while consumers can better manage their electricity consumption and expenses.

Author Contributions: Conceptualization, J.L. and S.S.Y.; methodology, J.L. and H.H.; formal analysis, J.L., H.H., S.S.Y. and H.T.; resources, S.S.Y. and H.T.; writing—original draft preparation, J.L.; writing—review and editing, H.H. and S.S.Y.; supervision, S.S.Y. and H.T. All authors have read and agreed to the published version of the manuscript.

Funding: This research received no external funding.

Data Availability Statement: This is a review paper. There was no original data generated.

Conflicts of Interest: The authors declare no conflict of interest.

References

1. Mahmood, A.; Javaid, N.; Razzaq, S. A review of wireless communications for smart grid. *Renew. Sustain. Energy Rev.* **2015**, *41*, 248–260. [CrossRef]
2. Li, Z.; Bahramirad, S.; Paaso, A.; Yan, M.; Shahidehpour, M. Blockchain for decentralized transactive energy management system in networked microgrids. *Electr. J.* **2019**, *32*, 58–72. [CrossRef]
3. Rasheed, M.B.; Qureshi, M.A.; Javaid, N.; Alquthami, T. Dynamic pricing mechanism with the integration of renewable energy source in smart grid. *IEEE Access* **2020**, *8*, 16876–16892. [CrossRef]
4. Rahimi, F.; Ipakchi, A. Demand response as a market resource under the smart grid paradigm. *IEEE Trans. Smart Grid* **2010**, *1*, 82–88. [CrossRef]

5. Lu, T.; Chen, X.; McElroy, M.B.; Nielsen, C.P.; Wu, Q.; Ai, Q. A reinforcement learning-based decision system for electricity pricing plan selection by smart grid end users. *IEEE Trans. Smart Grid* **2021**, *12*, 2176–2187. [CrossRef]
6. Wang, D.; Liu, B.; Jia, H.; Zhang, Z.; Chen, J.; Huang, D. Peer-to-peer Electricity Transaction Decisions of the User-side Smart Energy System Based on the SARSA Reinforcement Learning. *CSEE J. Power Energy Syst.* **2020**, *8*, 826–837.
7. Zhang, C.; Wu, J.; Zhou, Y.; Cheng, M.; Long, C. Peer-to-Peer energy trading in a Microgrid. *Appl. Energy* **2018**, *220*, 1–12. [CrossRef]
8. Bui, V.H.; Hussain, A.; Su, W. A Dynamic Internal Trading Price Strategy for Networked Microgrids: A Deep Reinforcement Learning-Based Game-Theoretic Approach. *IEEE Trans. Smart Grid* **2022**, *13*, 3408–3421. [CrossRef]
9. Al-Rubaye, S.; Choi, B.J. Energy load management for residential consumers in smart grid networks. In Proceedings of the 2016 IEEE International Conference on Consumer Electronics (ICCE), Las Vegas, NV, USA, 7–11 January 2016.
10. Yang, C.; Meng, C.; Zhou, K. Residential electricity pricing in China: The context of price-based demand response. *Renew. Sustain. Energy Rev.* **2018**, *81*, 2870–2878. [CrossRef]
11. Da Silva, P.P.; Cerqueira, P.A. Assessing the determinants of household electricity prices in the EU: A system-GMM panel data approach. *Renew. Sustain. Energy Rev.* **2017**, *73*, 1131–1137. [CrossRef]
12. Kolb, S.; Dillig, M.; Plankenbühler, T.; Karl, J. The impact of renewables on electricity prices in Germany-An update for the years 2014–2018. *Renew. Sustain. Energy Rev.* **2020**, *134*, 110307. [CrossRef]
13. Lago, J.; Marcjasz, G.; De Schutter, B.; Weron, R. Forecasting day-ahead electricity prices: A review of state-of-the-art algorithms, best practices and an open-access benchmark. *Appl. Energy* **2021**, *293*, 116983. [CrossRef]
14. Olivares, K.G.; Challu, C.; Marcjasz, G.; Weron, R.; Dubrawski, A. Neural basis expansion analysis with exogenous variables: Forecasting electricity prices with NBEATSx. *Int. J. Forecast.* **2022**, *39*, 884–900. [CrossRef]
15. Muqeet, H.A.; Munir, H.M.; Javed, H.; Shahzad, M.; Jamil, M.; Guerrero, J.M. An energy management system of campus microgrids: State-of-the-art and future challenges. *Energies* **2021**, *14*, 6525. [CrossRef]
16. Nowotarski, J.; Weron, R. Recent advances in electricity price forecasting: A review of probabilistic forecasting. *Renew. Sustain. Energy Rev.* **2018**, *81*, 1548–1568. [CrossRef]
17. Zhou, Y.; Lund, P.D. Peer-to-peer energy sharing and trading of renewable energy in smart communities—Trading pricing models, decision-making and agent-based collaboration. *Renew. Energy* **2023**, *207*, 177–193. [CrossRef]
18. Sampson, E.M.; Longe, O.M. Review of the Role of Energy Pricing Policies in Electricity Retail Market. In Proceedings of the 2021 IEEE PES/IAS PowerAfrica, Nairobi, Kenya, 23–27 August 2021.
19. Buckley, P. Prices, information and nudges for residential electricity conservation: A meta-analysis. *Ecol. Econ.* **2020**, *172*, 106635. [CrossRef]
20. Wang, Y.; He, Y.; Duan, Q.; Wang, H.; Bie, P. Distribution Locational Marginal Pricing for Active Network Management Considering High Penetration of Distributed Energy Resources. In Proceedings of the 2019 IEEE 3rd Conference on Energy Internet and Energy System Integration (EI2), Changsha, China, 8–10 November 2019.
21. Wang, D. Research on the fixed assets of the power grid company under the power transmission and distribution price reform. *Enterp. Reform Manag.* **2017**, *19*, 130–131.
22. Steiner, F. Regulation, industry structure, and performance in the electricity supply industry. Ph.D. Thesis, Stanford University, Stanford, CA, USA, August 2001.
23. Erdogdu, E. The impact of power market reforms on electricity price-cost margins and cross-subsidy levels: A cross country panel data analysis. *Energy Policy* **2011**, *39*, 1080–1092. [CrossRef]
24. Hattori, T.; Tsutsui, M. Economic impact of regulatory reforms in the electricity supply industry: A panel data analysis for OECD countries. *Energy Policy* **2004**, *32*, 823–832. [CrossRef]
25. Florio, M. Energy Reforms and Consumer Prices in the EU over twenty Years. *Econ. Energy Environ. Policy* **2014**, *3*, 37–51. [CrossRef]
26. Dillig, M.; Jung, M.; Karl, J. The impact of renewables on electricity prices in Germany—An estimation based on historic spot prices in the years 2011–2013. *Renew. Sustain. Energy Rev.* **2016**, *57*, 7–15. [CrossRef]
27. Winkler, J.; Gaio, A.; Pfluger, B.; Ragwitz, M. Impact of renewables on electricity markets—Do support schemes matter? *Energy Policy* **2016**, *93*, 157–167. [CrossRef]
28. Antweiler, W. A two-part feed-in-tariff for intermittent electricity generation. *Energy Econ.* **2017**, *65*, 458–470. [CrossRef]
29. Yang, C.; Ge, Z. Dynamic feed-in tariff pricing model of distributed photovoltaic generation in China. *Energy Procedia* **2018**, *152*, 27–32. [CrossRef]
30. Zhang, R.; Ni, M.; Shen, G.Q.; Wong, J.K. An analysis on the effectiveness and determinants of the wind power Feed-in-Tariff policy at China's national-level and regional-grid-level. *Sustain. Energy Technol. Assess.* **2019**, *34*, 87–96. [CrossRef]
31. Mohsenian-Rad, A.H.; Leon-Garcia, A. Optimal residential load control with price prediction in real-time electricity pricing environments. *IEEE Trans. Smart Grid* **2010**, *1*, 120–133. [CrossRef]
32. ActewAGL-What Makes Up Your Electricity Price? Available online: https://www.actewagl.com.au/beyond-energy/blog/what-makes-up-your-electricity-price (accessed on 8 May 2023).
33. Maiorano, A.; Song, Y.H.; Trovato, M. Imperfect competition: Modeling and analysis of oligopoly electricity markets. *IEEE Power Eng. Rev.* **1999**, *19*, 56–58. [CrossRef]

34. Sakya, I.M.R.; Simanjutak, U.R.; Schavemaker, P.H.; van der Sluts, L. Experience with internal competition using the single buyer model in Jawa Bali-Indonesia. In Proceedings of the 2006 Eleventh International Middle East Power Systems Conference, Minia, Egypt, 19–21 December 2006.
35. Bublitz, A.; Keles, D.; Zimmermann, F.; Fraunholz, C.; Fichtner, W. A survey on electricity market design: Insights from theory and real-world implementations of capacity remuneration mechanisms. *Energy Econ.* **2019**, *80*, 1059–1078. [CrossRef]
36. Hesamzadeh, M.R.; Hosseinzadeh, N.; Wolfs, P.J. A multi-criteria decision framework for optimal augmentation of transmission grid-Addressing a tool for sensitive zone detection in electricity market. *Int. J. Emerg. Electr. Power Syst.* **2008**, *9*, 1–16. [CrossRef]
37. Soto, E.A.; Bosman, L.B.; Wollega, E.; Leon-Salas, W.D. Peer-to-peer energy trading: A review of the literature. *Appl. Energy* **2021**, *283*, 116268. [CrossRef]
38. Zhou, Y.; Wu, J.; Long, C. Evaluation of peer-to-peer energy sharing mechanisms based on a multiagent simulation framework. *Appl. Energy* **2018**, *222*, 993–1022. [CrossRef]
39. Dehghanpour, K.; Nehrir, H. An agent-based hierarchical bargaining framework for power management of multiple cooperative microgrids. *IEEE Trans. Smart Grid* **2017**, *10*, 514–522. [CrossRef]
40. Ali, L.; Azim, M.I.; Peters, J.; Bhandari, V.; Menon, A.; Tiwari, V.; Green, J.; Muyeen, S.M. Application of a Community Battery-Integrated Microgrid in a Blockchain-Based Local Energy Market Accommodating P2P Trading. *IEEE Access* **2023**, *11*, 29635–29649. [CrossRef]
41. Lu, Z.; Bai, L.; Wang, J.; Wei, J.; Xiao, Y.; Chen, Y. Peer-to-Peer Joint Electricity and Carbon Trading Based on Carbon-Aware Distribution Locational Marginal Pricing. *IEEE Trans. Power Syst.* **2023**, *38*, 835–852. [CrossRef]
42. Xia, Y.; Xu, Q.; Huang, Y.; Du, P. Peer-to-peer energy trading market considering renewable energy uncertainty and participants' individual preferences. *Int. J. Electr. Power Energy Syst.* **2023**, *148*, 108931. [CrossRef]
43. May, R.; Huang, P. A multi-agent reinforcement learning approach for investigating and optimising peer-to-peer prosumer energy markets. *Appl. Energy* **2023**, *334*, 120705. [CrossRef]
44. Aznavi, S.; Fajri, P.; Shadmand, M.B.; Khoshkbar-Sadigh, A. Peer-to-peer operation strategy of PV equipped office buildings and charging stations considering electric vehicle energy pricing. *IEEE Trans. Ind. Appl.* **2020**, *56*, 5848–5857. [CrossRef]
45. Zhang, B.; Du, Y.; Lim, E.G.; Jiang, L.; Yan, K. Design and Simulation of Peer-to-Peer Energy Trading Framework with Dynamic Electricity Price. In Proceedings of the 2019 29th Australasian Universities Power Engineering Conference (AUPEC), Nadi, Fiji, 26–29 November 2019.
46. Tsao, Y.C.; Linh, V.T. A new three-part tariff pricing scheme for the electricity microgrid considering consumer regret. *Energy* **2022**, *254*, 124387. [CrossRef]
47. Saatloo, A.M.; Mirzaei, M.A.; Mohammadi-Ivatloo, B. A Robust Decentralized Peer-to-Peer Energy Trading in Community of Flexible Microgrids. *IEEE Syst. J.* **2023**, *17*, 640–651. [CrossRef]
48. Amin, W.; Ahmad, F.; Umer, K.; Khawaja, A.H.; Afzal, M.; Ahmad, S.A.; Chaitusaney, S. An Effective Pricing Mechanism for Electricity Trading Considering Customer Preference and Reserved Price in Direct P2P Electricity Market Under Uncertainty in Grid Supply. *IEEE Access* **2022**, *10*, 96197–96211. [CrossRef]
49. Melendez, K.A.; Das, T.K.; Kwon, C. A Nash-bargaining model for trading of electricity between aggregations of peers. *Int. J. Electr. Power Energy Syst.* **2020**, *123*, 106185. [CrossRef]
50. Reijnders, V.M.; Gerards, M.E.; Hurink, J.L. A hybrid electricity pricing mechanism for joint system optimization and social acceptance within energy communities. *Energy Rep.* **2022**, *8*, 13281–13292. [CrossRef]
51. Oh, E.; Son, S.Y. Peer-to-peer energy transaction mechanisms considering fairness in smart energy communities. *IEEE Access* **2020**, *8*, 216055–216068. [CrossRef]
52. Zhou, Y.; Cao, S. Coordinated multi-criteria framework for cycling aging-based battery storage management strategies for positive building–vehicle system with renewable depreciation: Life-cycle based techno-economic feasibility study. *Energy Convers. Manag.* **2020**, *226*, 113473. [CrossRef]
53. Li, J.; Ge, S.; Xu, Z.; Liu, H.; Li, J.; Wang, C.; Cheng, X. A network-secure peer-to-peer trading framework for electricity-carbon integrated market among local prosumers. *Appl. Energy* **2023**, *335*, 120420. [CrossRef]
54. Dong, J.; Song, C.; Liu, S.; Yin, H.; Zheng, H.; Li, Y. Decentralized peer-to-peer energy trading strategy in energy blockchain environment: A game-theoretic approach. *Appl. Energy* **2022**, *325*, 19852. [CrossRef]
55. Nguyen, D.H. Probability distribution of market clearing solution in peer-to-peer energy market. *Int. J. Electr. Power Energy Syst.* **2023**, *148*, 109014. [CrossRef]
56. Zhou, K.; Chong, J.; Lu, X.; Yang, S. Credit-Based Peer-to-Peer Electricity Trading in Energy Blockchain Environment. *IEEE Trans. Smart Grid* **2022**, *13*, 678–687. [CrossRef]
57. Wang, N.; Liu, Z.; Heijnen, P.; Warnier, M. A peer-to-peer market mechanism incorporating multi-energy coupling and cooperative behaviors. *Appl. Energy* **2022**, *311*, 118572. [CrossRef]
58. Ji, H.; Jian, J.; Yu, H.; Ji, J.; Wei, M.; Zhang, X.; Li, P.; Yan, J.; Wang, C. Peer-to-Peer Electricity Trading of Interconnected Flexible Distribution Networks Based on Distributed Ledger. *IEEE Trans. Ind. Inform.* **2021**, *18*, 5949–5960. [CrossRef]
59. Esmat, A.; de Vos, M.; Ghiassi-Farrokhfal, Y.; Palensky, P.; Epema, D. A novel decentralized platform for peer-to-peer energy trading market with blockchain technology. *Appl. Energy* **2021**, *282*, 116123. [CrossRef]
60. Jing, R.; Xie, M.N.; Wang, F.X.; Chen, L.X. Fair P2P energy trading between residential and commercial multi-energy systems enabling integrated demand-side management. *Appl. Energy* **2020**, *262*, 114551. [CrossRef]

61. Vahedipour-Dahraie, M.; Rashidizadeh-Kermani, H.; Shafie-Khah, M.; Siano, P. Peer-to-peer energy trading between wind power producer and demand response aggregators for scheduling joint energy and reserve. *IEEE Syst. J.* **2020**, *15*, 705–714. [CrossRef]
62. Hayes, B.P.; Thakur, S.; Breslin, J.G. Co-simulation of electricity distribution networks and peer to peer energy trading platforms. *Int. J. Electr. Power Energy Syst.* **2020**, *115*, 105419. [CrossRef]
63. McKenna, E.; Thomson, M. Photovoltaic metering configurations, feed-in tariffs and the variable effective electricity prices that result. *IET Renew. Power Gener.* **2013**, *7*, 235–245. [CrossRef]
64. Chiu, W.Y.; Sun, H.; Poor, H.V. Energy imbalance management using a robust pricing scheme. *IEEE Trans. Smart Grid* **2012**, *4*, 896–904. [CrossRef]
65. Jia, L.; Tong, L. Dynamic pricing and distributed energy management for demand response. *IEEE Trans. Smart Grid* **2016**, *7*, 1128–1136. [CrossRef]
66. Negash, A.I.; Haring, T.W.; Kirschen, D.S. Allocating the cost of demand response compensation in wholesale energy markets. *IEEE Trans. Power Syst.* **2014**, *30*, 1528–1535. [CrossRef]
67. Liu, N.; Yu, X.; Wang, C.; Li, C.; Ma, L.; Lei, J. Energy-sharing model with price-based demand response for microgrids of peer-to-peer prosumers. *IEEE Trans. Power Syst.* **2017**, *32*, 3569–3583. [CrossRef]
68. Huang, S.; Wu, Q. Dynamic tariff-subsidy method for PV and V2G congestion management in distribution networks. *IEEE Trans. Smart Grid* **2019**, *10*, 5851–5860. [CrossRef]
69. Hayat, M.A.; Shahnia, F.; Shafiullah, G.M. Replacing flat rate feed-in tariffs for rooftop photovoltaic systems with a dynamic one to consider technical, environmental, social, and geographical factors. *IEEE Trans. Ind. Inform.* **2018**, *15*, 3831–3844. [CrossRef]
70. Jiang, A.; Yuan, H.; Li, D. Energy management for a community-level integrated energy system with photovoltaic prosumers based on bargaining theory. *Energy* **2021**, *225*, 120272. [CrossRef]
71. Costanzo, G.T.; Zhu, G.; Anjos, M.F.; Savard, G. A system architecture for autonomous demand side load management in smart buildings. *IEEE Trans. Smart Grid* **2012**, *3*, 2157–2165. [CrossRef]
72. Palensky, P.; Dietrich, D. Demand side management: Demand response, intelligent energy systems, and smart loads. *IEEE Trans. Ind. Inform.* **2011**, *7*, 381–388. [CrossRef]
73. Liu, J.; Yu, S.S.; Hu, H.; Zhao, J.; Trinh, H. Demand-Side Regulation Provision of Virtual Power Plants Consisting of Interconnected Microgrids Through Double-Stage Double-Layer Optimization. *IEEE Trans. Smart Grid* **2023**, *14*, 1946–1957. [CrossRef]
74. Liu, J.; Hu, H.; Yu, S.S.; Trinh, H. Virtual Power Plant with Renewable Energy Sources and Energy Storage Systems for Sustainable Power Grid-Formation, Control Techniques and Demand Response. *Energies* **2023**, *16*, 3705. [CrossRef]
75. Chau, T.K.; Yu, S.S.; Fernando, T.; Iu, H.H.C. Demand-side regulation provision from industrial loads integrated with solar PV panels and energy storage system for ancillary services. *IEEE Trans. Ind. Inform.* **2017**, *14*, 5038–5049. [CrossRef]
76. Hwang, H.K.; Yoon, A.Y.; Kang, H.K.; Moon, S.I. Retail electricity pricing strategy via an artificial neural network-based demand response model of an energy storage system. *IEEE Access* **2020**, *9*, 13440–13450. [CrossRef]
77. Fraija, A.; Agbossou, K.; Henao, N.; Kelouwani, S.; Fournier, M.; Hosseini, S.S. A discount-based time-of-use electricity pricing strategy for demand response with minimum information using reinforcement learning. *IEEE Access* **2020**, *10*, 54018–54028. [CrossRef]
78. Sahraie, E.; Zakariazadeh, A.; Souchelmaie, M.B. Development of the Demand Response Programs in Smart Grids with Determination of the Ancillary Services Price for Demand-side. In Proceedings of the 2019 International Power System Conference (PSC), Tehran, Iran, 9–11 December 2019.
79. Ott, A.L. Experience with PJM market operation, system design, and implementation. *IEEE Trans. Power Syst.* **2003**, *18*, 528–534. [CrossRef]
80. PJM-Home. Available online: https://www.pjm.com/ (accessed on 13 May 2023).
81. Martinez-Pabon, M.; Eveleigh, T.; Tanju, B. Optimizing residential energy management using an autonomous scheduler system. *Expert Syst. Appl.* **2018**, *96*, 373–387. [CrossRef]
82. Muratori, M.; Rizzoni, G. Residential Demand Response: Dynamic Energy Management and Time-Varying Electricity Pricing. *IEEE Trans. Power Syst.* **2016**, *31*, 1108–1117. [CrossRef]
83. Vardakas, J.S.; Zorba, N.; Verikoukis, C.V. A survey on demand response programs in smart grids: Pricing methods and optimization algorithms. *IEEE Commun. Surv. Tutor.* **2014**, *17*, 152–178. [CrossRef]
84. Aguiar, N.; Dubey, N.; Gupta, V. Pricing Demand-Side Flexibility With Noisy Consumers: Mean-Variance Trade-Offs. *IEEE Trans. Power Syst.* **2023**, *38*, 1151–1161. [CrossRef]
85. Jiang, C.; Tseng, C.L.; Wang, Y.; Lan, Z.; Wen, F.; Chen, F.; Liang, L. Optimal Pricing Strategy for Data Center Considering Demand Response and Renewable Energy Source Accommodation. *J. Mod. Power Syst. Clean Energy* **2023**, *11*, 345–354. [CrossRef]
86. Rasheed, M.B.; R-Moreno, M.D. Minimizing pricing policies based on user load profiles and residential demand responses in smart grids. *Appl. Energy* **2022**, *310*, 118492. [CrossRef]
87. Zhang, L.; Gao, Y.; Zhu, H.; Tao, L. A distributed real-time pricing strategy based on reinforcement learning approach for smart grid. *Expert Syst. Appl.* **2022**, *191*, 116285. [CrossRef]
88. Munankarmi, P.; Wu, H.; Pratt, A.; Lunacek, M.; Balamurugan, S.P.; Spitsen, P. Home Energy Management System for Price-Responsive Operation of Consumer Technologies Under an Export Rate. *IEEE Access* **2022**, *10*, 50087–50099. [CrossRef]
89. Wu, J.; Wu, L.; Xu, Z.; Qiao, X.; Guan, X. Dynamic Pricing and Prices Spike Detection for Industrial Park With Coupled Electricity and Thermal Demand. *IEEE Trans. Autom. Sci. Eng.* **2022**, *19*, 1326–1337. [CrossRef]

90. Zhang, D.; Zhu, H.; Zhang, H.; Goh, H.H.; Liu, H.; Wu, T. Multi-objective optimization for smart integrated energy system considering demand responses and dynamic prices. *IEEE Trans. Smart Grid* **2021**, *13*, 1100–1112. [CrossRef]
91. Ding, T.; Qu, M.; Amjady, N.; Wang, F.; Bo, R.; Shahidehpour, M. Tracking equilibrium point under real-time price-based residential demand response. *IEEE Trans. Smart Grid* **2020**, *12*, 2736–2740. [CrossRef]
92. Wang, Y.; Wang, D.; Zhang, H.; Zhang, K. Optimal bidding of price-maker retailers with demand price quota curves under price uncertainty. *IEEE Access* **2020**, *8*, 120746–120756. [CrossRef]
93. Chen, R.; Paschalidis, I.C.; Caramanis, M.C.; Andrianesis, P. Learning from past bids to participate strategically in day-ahead electricity markets. *IEEE Trans. Smart Grid* **2019**, *10*, 5794–5806. [CrossRef]
94. Yan, X.; Ozturk, Y.; Hu, Z.; Song, Y. A review on price-driven residential demand response. *Renew. Sustain. Energy Rev.* **2018**, *96*, 411–419. [CrossRef]
95. da Silva, H.B.; Santiago, L.P. On the trade-off between real-time pricing and the social acceptability costs of demand response. *Renew. Sustain. Energy Rev.* **2018**, *81*, 1513–1521. [CrossRef]
96. Longe, O.M.; Ouahada, K.; Rimer, S.; Ferreira, H.C.; Han Vinck, A.J. Distributed optimisation algorithm for demand side management in a grid-connected smart microgrid. *Sustainability* **2017**, *9*, 1088. [CrossRef]
97. Golmohamadi, H.; Keypour, R. Retail energy management in electricity markets: Structure, challenges and economic aspects-a review. *Technol. Econ. Smart Grids Sustain. Energy* **2017**, *2*, 20. [CrossRef]
98. Ti, B.; Xiao, L.; Liu, J.; Yi, H.; Yang, B.; Li, G.; Liu, Z. Research on the Problems and Countermeasures on the Reform of Transmission and Distribution Price. In Proceedings of the 2018 2nd IEEE Conference on Energy Internet and Energy System Integration (EI2), Beijing, China, 20–22 October 2018.
99. Khalilisenobari, R.; Wu, M. Optimal participation of price-maker battery energy storage systems in energy and ancillary services markets considering degradation cost. *Int. J. Electr. Power Energy Syst.* **2022**, *138*, 107924. [CrossRef]
100. Ansarin, M.; Ghiassi-Farrokhfal, Y.; Ketter, W.; Collins, J. Cross-subsidies among residential electricity prosumers from tariff design and metering infrastructure. *Energy Policy* **2020**, *145*, 111736. [CrossRef]
101. Migden-Ostrander, J. Power sector transformation in state utility regulation: To boldly go where no regulator has gone before. *Electr. J.* **2019**, *32*, 106626. [CrossRef]
102. Hu, Y.; Li, Y.; Chen, L. Multi-objective optimization of time-of-use price for tertiary industry based on generalized seasonal multi-model structure. *IEEE Access* **2019**, *7*, 89234–89244. [CrossRef]
103. Chatterji, E.; Bazilian, M.D. Smart meter data to optimize combined roof-top solar and battery systems using a stochastic mixed integer programming model. *IEEE Access* **2020**, *8*, 133843–133853. [CrossRef]
104. Boiarkin, V.; Rajarajan, M.; Al-Zaili, J.; Asif, W. A novel dynamic pricing model for a microgrid of prosumers with photovoltaic systems. *Appl. Energy* **2023**, *342*, 121148. [CrossRef]
105. Long, C.; Wu, J.; Zhou, Y.; Jenkins, N. Peer-to-peer energy sharing through a two-stage aggregated battery control in a community Microgrid. *Appl. Energy* **2018**, *226*, 261–276. [CrossRef]
106. Abhyankar, N.; Lin, J.; Liu, X.; Sifuentes, F. Economic and environmental benefits of market-based power-system reform in China: A case study of the Southern grid system. *Resour. Conserv. Recycl.* **2020**, *153*, 104558. [CrossRef]
107. Hampton, H.; Foley, A.; Del Rio, D.F.; Smyth, B.; Laverty, D.; Caulfield, B. Customer engagement strategies in retail electricity markets: A comprehensive and comparative review. *Energy Res. Soc. Sci.* **2022**, *90*, 102611. [CrossRef]
108. Ribeiro, L.; Street, A.; Valladão, D.; Freire, A.C.; Barroso, L. Technical and economical aspects of wholesale electricity markets: An international comparison and main contributions for improvements in Brazil. *Electr. Power Syst. Res.* **2023**, *220*, 109364. [CrossRef]

designs

Article

Performance Explorations of a PMS Motor Drive Using an ANN-Based MPPT Controller for Solar-Battery Powered Electric Vehicles

Anjuru Viswa Teja [1], Wahab Razia Sultana [1] and Surender Reddy Salkuti [2,*]

1 School of Electrical Engineering, VIT University, Vellore 632014, Tamil Nadu, India; viswateja.anjuru2020@vitstudent.ac.in (A.V.T.); wraziasultana@vit.ac.in (W.R.S.)
2 Department of Railroad and Electrical Engineering, Woosong University, Jayang Dong, Dong-Gu, Daejeon 34606, Republic of Korea
* Correspondence: surender@wsu.ac.kr

Abstract: Solar energy can function as a supplementary power supply for other renewable energy sources. On average, Vellore region experiences approximately six hours of daily sunshine throughout the year. Solar photovoltaic (PV) modules are necessary to monitor and fulfill the energy requirements of a given day. An artificial neural network (ANN) based maximum power point tracking (MPPT) controller is utilised to regulate the solar photovoltaic (PV) array and enhance its output. The utilisation of this controller can enhance the efficiency of the module even in severe circumstances, where reduced current and torque ripples will be observed on the opposite end. The motorised vehicle has the capability to function at its highest torque level in different load scenarios as a result. The proposed method is expected to provide advantages in various electric vehicle (EV) applications that require consistent velocity and optimal torque to satisfy the load conditions. The study employs a solar battery that is linked to an SVPWM inverter and subsequently a DC-DC boost converter to supply power to the load. An Artificial Neural Network (ANN) based Maximum Power Point Tracking (MPPT) control system is proposed for a solar battery powered Electric Vehicle (EV) and the system's performance is evaluated by collecting and analysing data under adjustable load conditions to obtain constant parameters such as speed and torque. The MATLAB® Simulink® model was utilised for this purpose.

Keywords: permanent magnet synchronous motor; solar array; electric vehicles; DC-DC converter; maximum power point tracker; three phase inverter

Citation: Viswa Teja, A.; Razia Sultana, W.; Salkuti, S.R. Performance Explorations of a PMS Motor Drive Using an ANN-Based MPPT Controller for Solar-Battery Powered Electric Vehicles. *Designs* **2023**, *7*, 79. https://doi.org/10.3390/designs7030079

Academic Editor: Loránd Szabó

Received: 20 April 2023
Revised: 13 June 2023
Accepted: 14 June 2023
Published: 16 June 2023

1. Introduction

The power industry is exploring new forms of renewable energy to help meet rising electricity demands as natural ecological resources become exhausted. Using sources of renewable electricity is one way to reduce the amount of carbon dioxide in the atmosphere and combat global warming. Modern renewable energy sources are dominated by solar photovoltaic systems because of their simple construction. It has been discussed how solar panel systems may be positioned differently and function in a variety of designs [1]. It is possible to increase the effectiveness of solar systems using maximum power point (MPP) controllers and power electronics [2]. The most effective way to determine the maximum power output is to use of one of the many techniques that have been devised. Under rapidly changing climatic circumstances, the majority of MPP algorithms in use today are ineffective due to sluggish and imprecise tracking and oscillation. Their usage efficiency suffers as a result. To solve this problem, an ANN-based control mechanism is proposed. To lessen unequal irradiance on this PV module, two distinct MPPTs are utilized in this instance. A bypass diode is employed to boost energy harvesting and to remove hotspots, while the PV string is simply linked in series to the block diode to stop the reverse current from existing load [3]. The MPP's two-stage technique is used by the ANN-based MPPT to demonstrate its independence over time-dependent and transactional characteristics, since the MPP being monitored by PV led to circumstances without the need for time increments [4]. Differential evolution (DE), artificial neural networks (ANN), and conventional MPPTs can be used to address rapid changes in radiation and temperature, as well as the nonlinear properties of arrays [5]. When compared to the conventional incremental conductance method, our proposed MPPT approach significantly improves performance. This maximum power point tracking technique is based on ANN technology. When compared to Perturb and Obsevartion (P&O) controllers and incremental conductance, the abrupt shift in patterns occurs faster. A neural network can quickly improve efficiency, manage uncertainties, and solve issues without previous information. The learning algorithm changes layer weights in the artificial neural network to track MPPs more quickly [6]. NN controllers are efficient and responsive under both rapidly changing irradiance and partial shade.

Intelligent approaches outperform traditional ones, and artificial intelligence solves the disadvantages of Particle swarm Optimisation (PSO), Grey Wolf Optimisation (GWO), Genetic Algorithms (GA), Fuzzy Logic (FL), and others. Design and implementation complexity matter. In contrast to P and O's inability to match pace with the MPPs when solar irradiation is fluctuating rapidly, the ANN approach can swiftly do so. Furthermore, ANN has low-frequency oscillation near the MPP, while the P and O method exhibits strong oscillation. The introduction of a mean square error (MSE) increases the network's accuracy and increases the training network's efficacy [7]. The utilisation of the two-level genetic algorithm technique has been documented in the literature [8] for the purpose of assessing parameters that affect battery power, such as temperature, light intensity, and junction temperature. For the purpose of tracking and managing solar heat from such a hybrid system, Particle Swarm Optimization (PSO) is utilized [9]. For boost converters for isolated solar systems, a back- propagation-focused multilayer deep neural network with a 10° temperature change MPPT is employed to reduce long-term system losses and boost conversion efficiency [10]. Energy conversion efficiency ranges from 12 to 20 percent for photovoltaic modules. The associated load and the PV system both have an impact on energy conversion loss. An MPPT and DC-DC converter, which can only reach the load voltage at maximum power point levels, avoids this [11]. Our tropical country of India has not researched solar energy.

Utilising solar battery power technology [12], a Permanent Magnet Synchronous(PMS) motor drive is linked to an MPPT controller with an inverter to improve its electric driving system performance. A solar array's three-phase inverter compensates for output voltage and current distortion to increase electricity quality even in abnormal conditions [13]. When managing the PMS motor's torque to ensure optimal torque control, a genetic algorithm is used to generate better torque ripple performance [14]. The use of neural vectors for spatial vector modulation and a comparison of neural networks with traditional systems have both been presented [15]. The advantage of our ANN-created MPPT algorithm over other traditional algorithms is that it can control MPPs.

This study designs a MATLAB® Simulink® mathematical model of the proposed system. The mathematical model incorporates a permanent magnet synchronous motor, PV cell, battery, and parallel control method. This paper's main contributions:

- The study focuses on the topology of the proposed system, not its actual feasibility.
- The design attempts to accommodate realistic auxiliary losses to the extent that mathematical modelling permits. This is to guarantee a step toward resemblance and proximity to real-world outcomes to attempt to showcase the potential of the proposed system.
- The study contributes to proving the functioning of the suggested system in a test condition with diverse scenarios to display its robust architecture.

Previous methods for MPPT include P&O, Incremental Conductance (INC), CV, and CI (constant voltage, constant current). There are several issues with each of these methods, and they cannot function in partial shaded condition (PSC) environments. Observing the global MPP under PSCs is difficult for traditional algorithms; even if they do manage to achieve global maximum power point tracking (GMMPT), doing so takes an extensive amount of time and results in poor tracking accuracy and significant oscillation. This means that effective algorithms are crucial for GMPP extraction.

- ➢ Solar panels with MPPT have been utilised for commercial and residential purposes, but electric battery vehicles are the first mobile usage.
- ➢ The data that we used to train the ANN is from NASA's prediction of worldwide energy resources. https://power.larc.nasa.gov/data-access-viewer (accessed on 18 April 2023).
- ➢ Most literature utilises irradiance and panel temperature sensors which are costlier and may raise overall system costs. Such sensors are also inaccurate and fragile. The suggested ANNMPPT tracks maximum power points effectively and efficiently.

The MPPT technique's primary innovations:

- ANN-based MPPT algorithms do not require expensive irradiance and temperature sensors.
- The suggested technique increases tracking performance by readily integrating into the existing MPPT method.

This article combines optimisation with artificial intelligence to improve photovoltaic (PV) array tracking and power harvesting. Due to their exploration and optimisation skills, algorithms can be employed in artificial neural network training to find the most effective weights to minimise error.

Section 2 covers solar array mathematics. Section 3 presents the MPPT algorithm. Section 4 covers PMS motor drive mathematical modelling. Section 5 uses MPPTs and DC-DC converters to boost PV output and motor drive battery supply. Section 6 shows PMS-driven electric car ANN MPPT MATLAB Simulation results. Section 7 summarises the study as a whole.

2. Solar Array Mathematical Formulation

The photovoltaic cells are used in solar photovoltaic systems. Panels are formed by grouping cells, while arrays are created by grouping panels. Equations (1) and (2) specifically reference the ideal photovoltaic cells, which have been described by the fundamental mathematical Equation (2).

$$I_{mc} = I_{il} - [e\frac{\frac{qVmv}{kNTjt}}{Id} - 1] \quad (1)$$

where I_{mc} is the module current, I_{il} is the insolation light current, q is e-, k is Boltzmann's constant, N is the diode identity, and V_{mv} is photovoltaic cell voltage [16].

Since an actual PV module is made up of several PV cells, the fundamental equation for a PV system does not capture I-V characteristics [17]. Instead, extra parametric variables such as parallel and series resistances (R_p and R_s), which are shown in Figure 1, are needed. The solar cell's mathematical equation, given in Equation (2), provides the foundation for modeling PV modules. Figure 1 shows the equivalent circuit of PV module.

$$I_{mc} = I_{il} - I_{rs}\left[e^{q(\frac{V_{mv}+I_{mc}R_s}{N_d kNT_{jt}})} - 1\right] - \left[\frac{V_{mv} + I_{mc}R_s}{R_p}\right] \quad (2)$$

where I_{mc} is the module current in Amperes, R_s is series resistance in Ω, R_p is the shunt resistance in Ω, and V_{mv} is the module voltage. Current values at different irradiance and temperature is shown in Figure 2.

$$I_{il} = \frac{P_{irr}[I_s + T_{si}(T_{jt} - T_r)]}{P_r} \quad (3)$$

where P_{irr} is the applied irradiance in W/m^2, P_r is the reference irradiance, I_s is the short circuit current of the module, T_s is the temperature coefficient of I_s in A/°K, and T_{jt} and T_r are the applied and reference temperatures in kelvin.

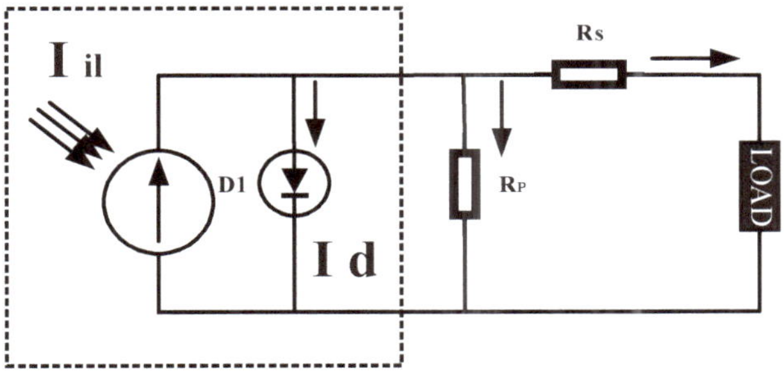

Figure 1. PV-equivalent circuit.

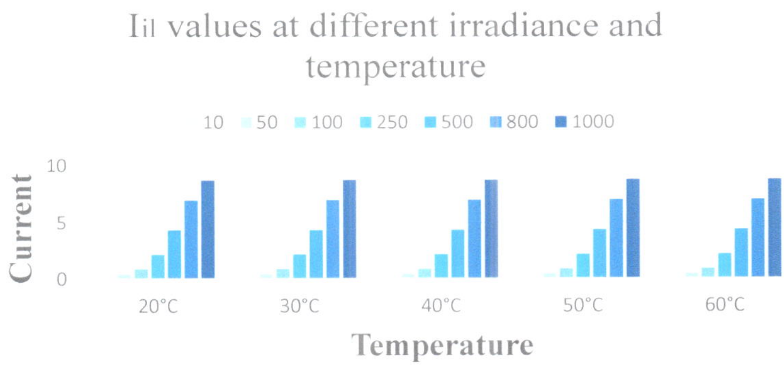

Figure 2. Current values with different irradiance as well as temperature.

Equation (4) gives the reverse saturation current of the module under nominal conditions as well as a reference temperature [18].

$$I_{rs} = \frac{I_s}{e^{\left(\frac{qV_0}{NkN_dT_{jt}}\right)} - 1} \tag{4}$$

where I_s is the saturation current and N is the total number of cells in a module. Even as the applied temperature rises, the voltage of the module drops, as shown by Equation (5).

$$V_{mv} = \frac{N_dkNT_{jt}}{q} \times ln\left[\frac{I_{il} - I_{rs}}{I_{rs}}\right] \tag{5}$$

Whereas the saturation current is as follows:

$$I_{st} = I_{rs}\left(\frac{T_a}{T_r}\right)^3 \times e^{\left\{\frac{qEgo}{Nk(\frac{1}{T_r} - \frac{1}{T_a})}\right\}} \tag{6}$$

$E_{go}$is the energy in the band gap in J/C and R_s is the shunt resistance, where small variations in the series resistance will have an impact on the output power of the PV, as the shunt resistance (R_s) is inversely related to the leakage current. Photovoltaic cells only generate about two watts when the circuit is open at between 0.5 and 0.7 volts [19]. To provide the necessary power, the cells must be linked in series. This single diode module array's basic current is determined by Equation (7).

$$I_A = N_p I_{il} - N_{pr} I_{rs} \left[e^{\frac{q}{N_d k N T_a} \left(\frac{V_{mv}}{N_{se}} + \frac{I_{mc} R_s}{N_{pr}} \right)} - 1 \right] - \left\{ \frac{1}{R_p} \left[\frac{N_{pr} V_{mv}}{N_{se}} + I_{mc} R_s \right] \right\} \quad (7)$$

N_{se} and N_{pr} refer to the number of series and parallel connected modules per string (two for series, six for parallel). Based on the 1soltech1STH-250 W polycrystalline solar module's data sheet specifications, at a maximum power of 305 W with 72 cells per module, having an open circuit voltage of 44.88 V with a short circuit current of 8.95 A and a voltage at a maximum point of 35.59 V with a current at a maximum power point of 8.57 A. The PV array was modeled for 1000 W/m^2 at 25 °C. The solar panel model in the simulation was constructed under typical test settings based on the aforementioned characteristics [20].

3. Stated MPPT Algorithm

The MPPT control technique optimises PV module performance under various environmental parameters like solar irradiance and temperature [21]. As is widely known, solar irradiation and temperature significantly affect the PV system's nonlinear output power characteristics as functions of irradiance and temperature curves. Daily sun irradiation fluctuates sharply. To maximise energy generation, the PV system's operating point must fluctuate when the PV array's MPP varies. Thus, MPPT maintains the PV array's MPP. MPPT automatically determines the current at maximum power point (I_{MPP}) or voltage at maximum power point (V_{MPP}) that a PV array should operate at to provide the maximum output power at maximum power point (P_{MPP}) at a certain temperature and irradiance level. Most MPPT methods respond to irradiance and temperature variations, while others operate more effectively in stable temperatures. Some MPPT methods are open-loop and need periodic fine-tuning, but most automatically adjust to array ageing. In our case, the module will be connected to a power converter that regulates the PV array-to-load current.

MPPT methods can precisely monitor the maximum power point (MPP); many techniques have been devised. The majority of MPPT algorithms currently in use suffer from the drawbacks of sluggish and incorrect tracking, as well as oscillation under quickly changing meteorological circumstances [22]. As a result, usage efficiency is decreased. This study offers an MPPT control technique based on an artificial neural network to address these drawbacks [23]. This technique surpasses all other conventional techniques in terms of system performance and effectiveness. A multilayer feedforward network is employed in this method [24]. To calculate the irradiance and temperature levels from the photovoltaic array current and voltage data, this study uses a deep neural network centered on MPPT incorporated with polyphase offline training. The ANN uses supervised learning to eliminate mistakes by calculating the necessary multiplication factor [25]. The methods described can deliver superior performance with fewer training sets under steady and transient conditions, even under rapidly varying environmental circumstances, since a power amplifier delivers the load with the highest possible output voltage [26]. In the simulation, a training network with a supervised learning feedforward algorithm is used to compensate for the solar array's nonlinearity. Figure 3 depicts the flow diagram for the proposed ANN-MPPT algorithm. According to photovoltaic cells' output characteristics, they have an ideal operating point at a specific cell temperature and level of illumination that is known as the most powerful point. Solar cells have variable maximum power points under various temperatures and light circumstances; even under identical conditions, the output power will fluctuate depending on the operating voltage used by the PV cell. It is therefore necessary to monitor the maximum power point of solar cells so that they can continuously operate at their maximum efficiency.

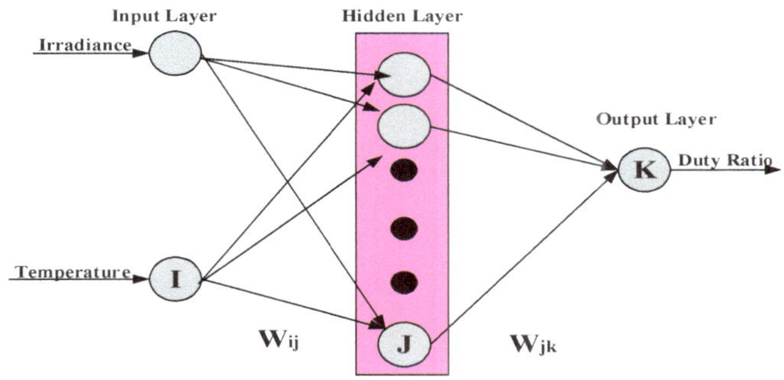

Figure 3. ANN structure in MPPT.

Currently, the perturbation and observation technique (P and O) and incremental conductivity approach (INC) are the most popular MPPT algorithms. P and O has a straightforward structure, fewer measurable parameters, and is simple to execute. However, selecting a step disruption that would satisfy the demands of tracking precision and speed is challenging. Furthermore, it frequently results in false positives. However, we have identified two circumstances in which the present change is minimal.

One instance is when a PV cell operates in the range of a continuous current source. The major control step can now be chosen using the aforementioned procedure. The second instance is that the MPP of the PV cell is circumvented. The control step currently needs to be brief, which contradicts the first instance. Consequently, this improved control algorithm is unable to produce more effective control results.

ANNs are AI technology, and AI outperforms conventional methods. Traditional approaches fail to monitor maximum power points and respond slowly to fast temperature and irradiance changes. Figure 3 shows an overview of MPPT ANN, Including solar temperature and irradiance inputs. The neural network targets the DC-DC converter duty ratio and will offer a duty ratio for each solar temperature and irradiance fluctuation to maximise each power point. Levenberg- Marquardt training algorithms create the network when the duty ratio and ANN are trained for different sun irradiation and temperature values. Adjusting layer weights during neural network training yields the desired results. Training weights are adjusted to follow objective values with minimal inaccuracy. ANN's performance is measured using mean squared error (MSE).

Training using the Levenberg-Marquardt method yields the network. In order to train the ANN, we compute the duty ratio for each possible set of solar irradiance levels as well as temperature values. During training, the neural network's weights in each layer are adjusted until the desired outputs are achieved.

Mean squared error (MSE) measures how well a regression line matches a set of data points. Its value matches the predicted squared error loss and risk function. Mean square error is calculated by averaging the squared errors from the data analysis in respect to a function. Throughout the training process, the weights are fine-tuned to obtain the most accurate tracking of the goal values as possible. Mean squared error is used as the ANN's performance metric. The mean squared error is calculated as follows:

$$F = \frac{1}{N}\sum_{i=1}^{N}[t_i - a_i]^2 \quad (8)$$

where t_i is the target and a_i is the present output.

Trial and error determine hidden layers. Output is boosted by the converter duty ratio. After training, any ANN algorithm can be applied. Learned incrementally, duty ratio data for the boost converter are recorded while changing irradiance and temperature inputs for the PV array to monitor solar panel power. To test the neural network, certain training points are kept.

3.1. *Effectiveness of a Conventional MPPT Controller with Distinct Irradiance (G) & Persistent Temperature (T)*

When the irradiance fluctuates between 100, 250, 500, 800, and 1000 W/m^2, the current, voltage, and irradiation levels of the PV are seen to immediately increase [27]. As a result, the PV array's overall net power increases. The conditions mentioned above are represented by the typical curves in Figure 4 below.

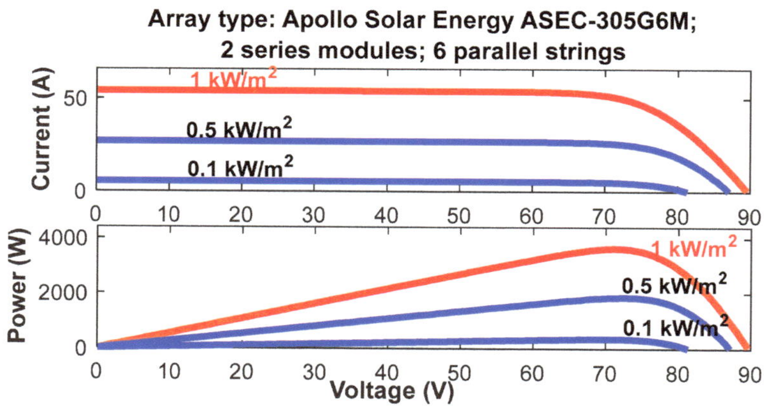

Figure 4. I-V/P-V properties under varying degrees of heat and light.

3.2. *Results of Standard MPPT Controller Testing with Variable T, Constant G*

The PV current increases as the photovoltaic array voltage quickly increases from 20 °C to 50 °C and eventually to 60 °C. This reduces PV array output. The curves shown in Figure 5 indicate these characteristics.

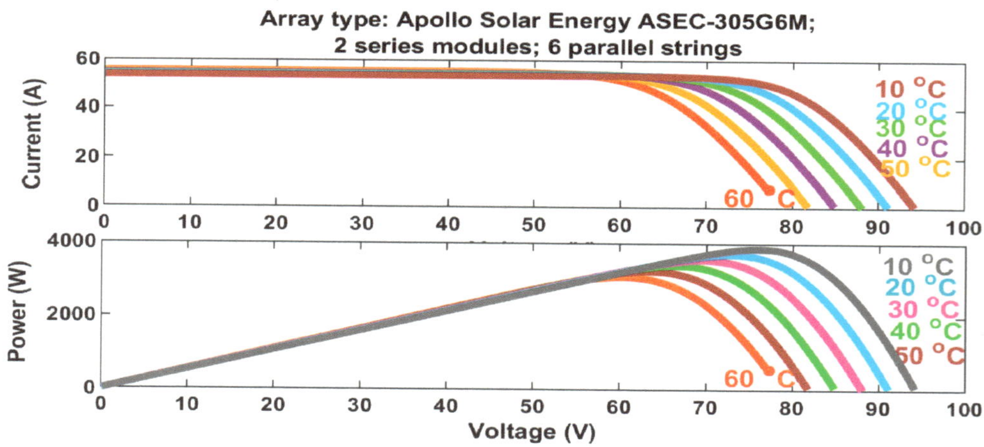

Figure 5. I-V/P-V characteristics with inconstant T and constant G.

3.3. *Performance of MPPT Algorithm Controller under Temperature and Irradiance Change*

When both temperature and irradiance levels are changing, the voltage decreases until the temperature rises, which causes the module current to increase. The opposite is also true. Additionally, there is not a significant increase in voltage until the irradiance increases, and vice versa.

4. Mathematical Modeling of PMSM

Permanent magnet synchronous motor (PMSM) drives are highly favoured by industry leaders in the automotive sector due to their potential for high power density, high speed, and excellent operating efficiency. PMSMs are categorised based on the location and style of the permanent magnets inside the rotors [28]. Three common types of PMSM are surface, embedded, and interior/buried. Both inset rotor and surface mounted PMSMs are referred to when the permanent magnets are exposed to an air gap. Due to the fact that the q-axis inductance may be substantially larger than the d-axis inductance [29], the magnets in the interior permanent magnet synchronous motor (IPMSM) are hidden within the rotor and have a greater capacity for flux-weakening. One of the major drawbacks of PMSM drives is their lack of excitation control. The internal Electro Magnetic Field (EMF) of a motor increases in tandem with its rotational velocity. As this style of operation often employs constant volts-per-hertz control, this type of behaviour is preferred in the so-called constant torque range. The related frequency converter's voltage limit, however, is reached as the speed rises. This is referred to as the motor's "flux-weakening operation," and it follows its "neutral" state. As the speed of the converter increases, it becomes essential to increase the internal voltage to match the external voltage [30]. Since the inverter has to commutate a larger current at higher speeds, the power factor of the motor changes to leading.

In Cartesian coordinates, Equation (9) is used to represent the voltage expression for the PMSM, ignoring magnetization circuit saturation, eddy losses, and hysteresis losses [31].

$$u_s = R_s i_s + \frac{d\lambda_s}{dt} \tag{9}$$

The equation for the motor's flux linkage is as follows:

$$\lambda_s = L_s i_s + \varphi f F_s(\theta) \tag{10}$$

The three windings' flux linkage is represented by s. The motor's voltages, resistances, and currents are represented by u_s, R_s, and i_s respectively [32]. The three-phase inductance is L_s, and the three-phase angle is F_s [33]. λ_s is considered the motor flux linkage.

$$i_s = \begin{bmatrix} i_A \\ i_B \\ i_C \end{bmatrix} \tag{11}$$

$$R_s = \begin{bmatrix} R & & \\ & R & \\ & & R \end{bmatrix} \tag{12}$$

$$\lambda_s = \begin{bmatrix} \lambda_A \\ \lambda_B \\ \lambda_C \end{bmatrix} \tag{13}$$

$$u_s = \begin{bmatrix} u_A \\ u_B \\ u_C \end{bmatrix} \tag{14}$$

$$F_s(\theta) = \begin{bmatrix} \sin\theta \\ \sin(\theta - \frac{2\pi}{3}) \\ \sin(\theta + \frac{2\pi}{3}) \end{bmatrix} \tag{15}$$

$$L_s = L_m \begin{bmatrix} 1 & \cos(\frac{2\pi}{3}) & \cos(\frac{4\pi}{3}) \\ \cos(\frac{4\pi}{3}) & 1 & \cos(\frac{2\pi}{3}) \\ \cos(\frac{4\pi}{3}) & \cos(\frac{2\pi}{3}) & 1 \end{bmatrix} + L_l \begin{bmatrix} 1 & & \\ & 1 & \\ & & 1 \end{bmatrix} \tag{16}$$

where L_m are mutual inductances, and L_l are inductance leakages. The motor's torque can be computed as:

$$T_e = \frac{1}{2} P_n \frac{\partial}{\partial \theta m} (i_s^T . \lambda_s) \quad (17)$$

The motor's pole pairs are denoted by P_n. The dynamic behaviour of the motor can be expressed by,

$$J \frac{d\omega_m}{dt} = T_e - T_L - B_{\omega m} \quad (18)$$

where J denotes the motor's spinning inertia, B denotes its damping coefficient, and T_L is its loading torque [34]. The motor's angular speed is denoted by ω_m. The motor's primary specifications are presented in Table 1.

Table 1. Motor Primary Specifications.

S. No.	Parameters	Quantities (SI)
1	Rated Capacity	500 W
2	Each coil's number of turns	60
3	Rated Current	8 A
4	The mover plate's width (w)	0.016 m
5	The coil area's width (c)	0.010 m
6	Number of poles (p)	8
7	The mover plate's height (h)	0.016 m
8	The stator's length (l)	0.350 m

Phasor diagrams analyse surface and interior-mounted permanent magnet motors, whose control methods have advantages and disadvantages. The unity power factor control technique optimises system volt-ampere (VA), whereas the constant mutual air gap flux connection controls restrict air gap flux linkage. Constant torque angle control reduces the power factor but relates the electromagnetic torque to the stator current magnitude; energy-saving maximum efficiency controls decrease motor net loss.

4.1. Drive Topology

Three-phase Voltage Source Inverter (VSI)s handle moderate-to-high power. These topologies provide three-phase output voltage with customizable phase, amplitude, and frequency [35]. Three-phase dc/ac voltage source inverters enable changeable frequency and ac voltage magnitudes in motor drives, active filters, power systems, and uninterruptible power supplies [36]. Figure 6 displays a six-modulation-switched three-phase inverter, featuring diode-bridge rectifiers, Inductor-Capacitor (LC) or Capacitor(C) filters, and utility power supplies providing input, represented by dc [37].

Figure 6 depicts the PMSM's three-phase inverter structure. S_1–S_6 represent the six Insulated Gate Bipolar Transistor (IGBT)s and their states [38]. S_1 and S_4 cannot switch on simultaneously for a single bridge in the case of a short circuit between V_{dc} and the ground. As a result, S_4 switches off as soon as S_1 does, and vice versa. As shown in Table 1, eight switching modes for three-phase alternative current motors are produced by six IGBTs. The eight vectors split the vector plane into six portions for the motor [39].

Two of the eight switching states represented in Table 2 result in output (ac) line voltages of zero. In this scenario, the higher or lower components are freely cycled by the ac line currents. The remaining states do not have zero ac line voltages. The inverter alternates between states to produce a certain voltage waveform. The subsequent ac line voltages are therefore made up of the discrete voltage values: V, 0, and V [40]. The modulating technique, which ensures the use of only the viable states, is used to select the states in order to produce the given waveform.

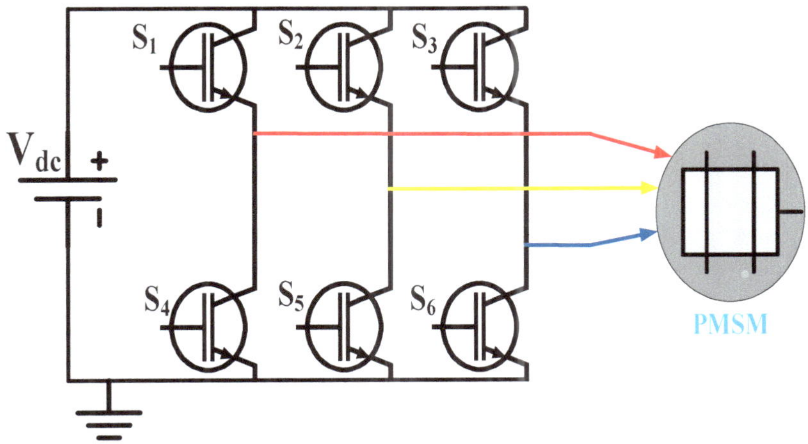

Figure 6. Inverter topology of PMSM.

$$\frac{V_{dc}}{2}(S_1 - S_2) = V_{an} + V_{no} \tag{19}$$

$$\frac{V_{dc}}{2}(S_2 - S_3) = V_{bn} + V_{no} \tag{20}$$

$$\frac{V_{dc}}{2}(S_3 - S_1) = V_{cn} + V_{no} \tag{21}$$

Table 2. Switching States of Inverter.

S_1	S_2	S_3	V_a	V_b	V_c
0	0	0	0	0	0
0	0	1	0	−V	V
0	1	0	−V	V	0
0	1	1	−V	0	V
1	0	0	V	0	−V
1	0	1	V	−V	0
1	1	0	0	V	−V
1	1	1	0	0	0

4.2. Three-Phase VSI Using Space Vector Pulse Width Modulation (SVPWM)

In three-phase inverters, space vector pulse width modulation (SVPWM) compares three sine waves phase-shifted by 120° to a very high-frequency carrier signal [41]. The signals are combined at a reference point where the comparator output is high when the sine wave, also known as the modulation signal, is greater than the saw tooth wave and low when the sine wave is smaller. Figure 7 shows this.

This section provides a sinusoidal pulse width modulation technique for a two-level converter that uses both solar panels and a battery pack as a Direct Current (DC) voltage source. To achieve a symmetric algorithm pulse with voltage balance, the fundamental concept and switching order of the SVPWM are provided. Through an ANN-based MPPT controller, this strategy is employed to regulate the inverter's output voltage [40]. The DC voltage is changed into a 3-phase instantaneous voltage reference using this SVPWM method, as shown in Figure 7.

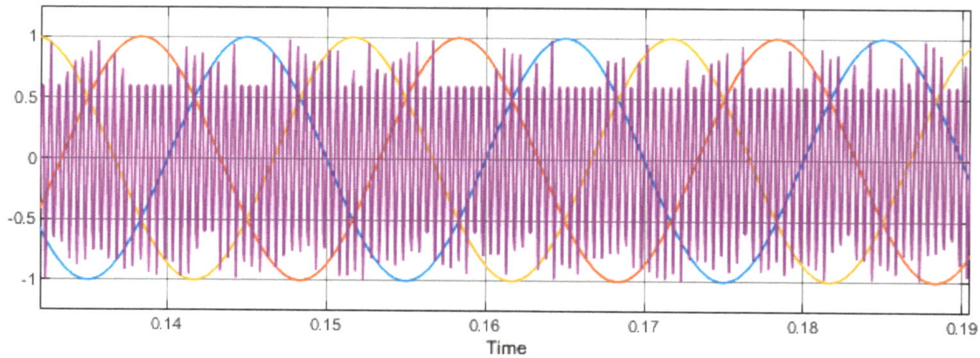

Figure 7. SVPWM used for three-phase VSI.

5. Proposed ANN MPPT Algorithm with PMSM as Load

The next design depicts the intended architecture of the system through a DC converter. The solar array has six 250-watt photovoltaic modules that are coupled in series as well as in parallel to generate greater output voltage and current [41]. Under different circumstances, the suggested MPPT algorithm, which is based on an ANN, obtains the most power possible from the photovoltaic system.

The power that reaches the solar array using the ANN MPPT algorithm is fed to the motor as well as a battery pack using a three-phase PWM-based voltage source inverter [42]. The energy stored in the battery is used when there is less power produced by the PV array due to dissimilar climatic conditions. As shown in Figure 8, a PMSM receives its three-phase power from VSI, and the drive axle of the vehicle is connected to the shaft of the motor along with the differential. The Matlab/Simulink model is shown in Figure 9.

The final configuration of the proposed ANN MPPT is shown in Figure 10, which consists of its inputs and outputs. The inner layers of the proposed ANN are shown in Figure 11.

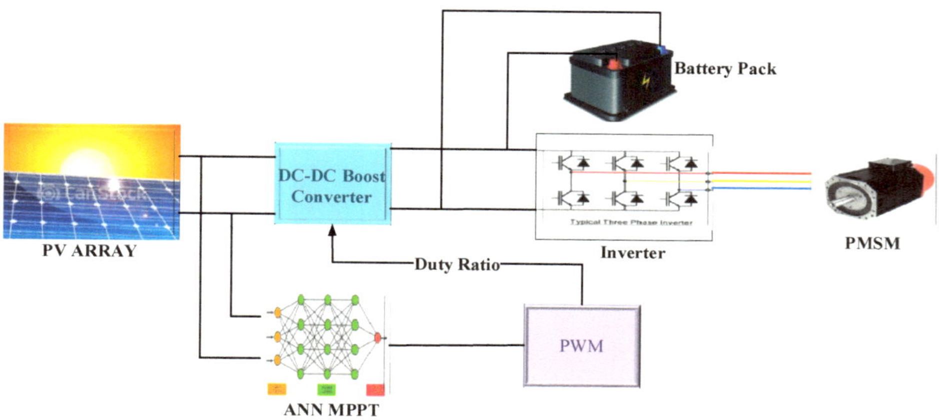

Figure 8. Diagram for proposed ANN MPPT.

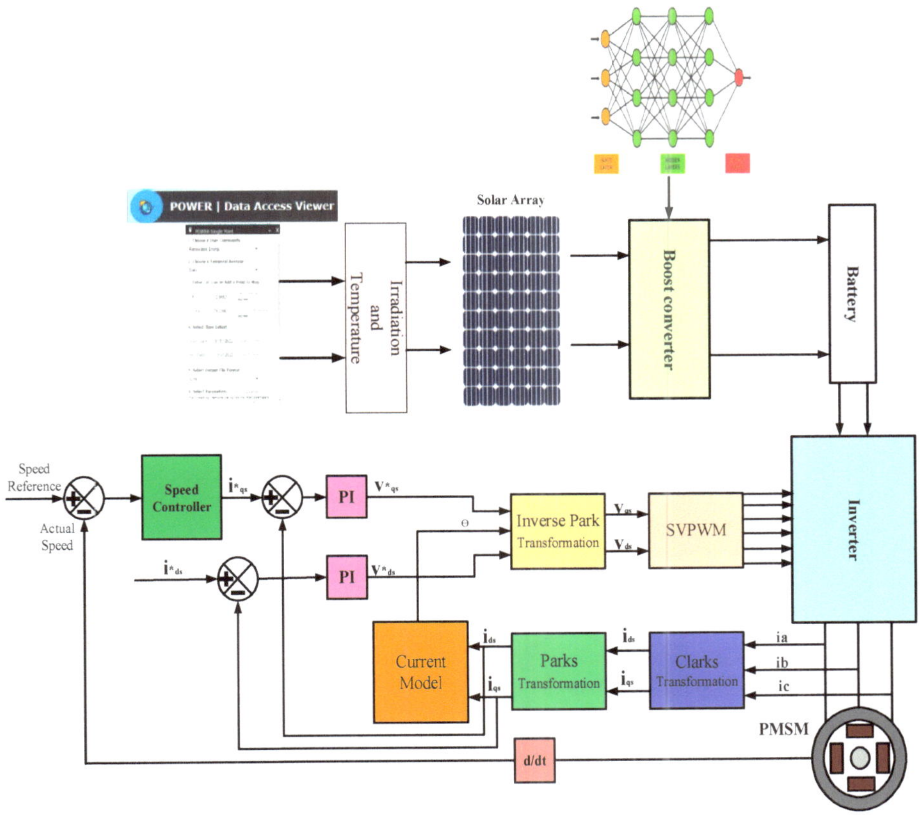

Figure 9. Solar-powered battery electric vehicle MATLAB/Simulink model.

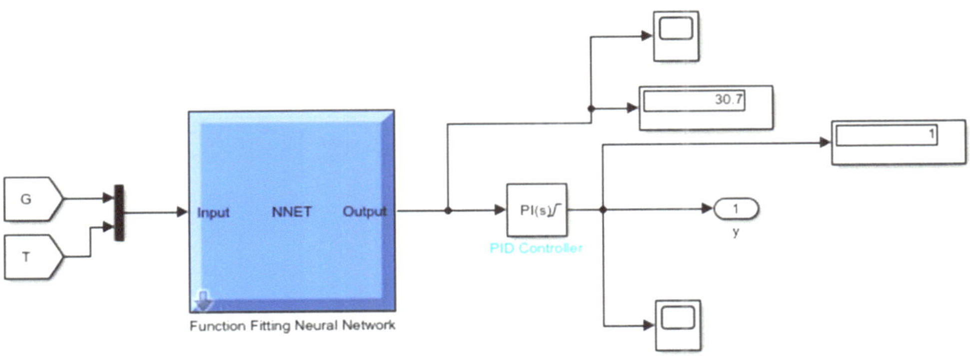

Figure 10. Final configuration of the proposed ANN MPPT.

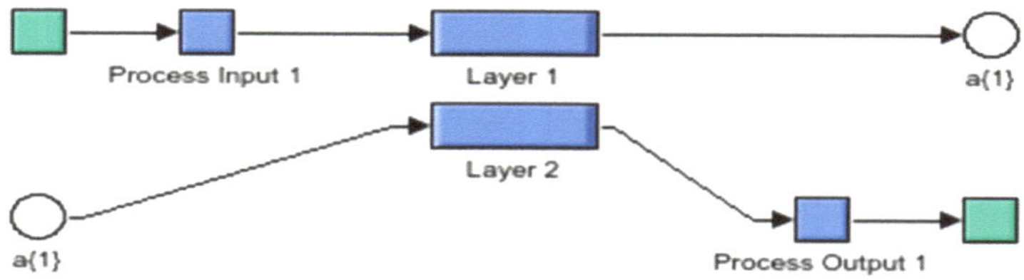

Figure 11. Layers inside ANN.

A MATLAB/Simulink model was developed for a proposed ANN MPPT-based solar-battery-powered electric vehicle driven by PMSM. Based on the irradiation (G)/temperature (T) the developed ANN MPPT extracts maximum power from the solar PV array, a DC-DC boost converter without a transformer boosts the power [42,43], which is further given as input to both batteries for storage and PMSM to drive the axle of the vehicle. The G and T values of an array are given as inputs to the neural network algorithm, and after several iterations, the output is given to DC Pulse Width Modulation (PWM) to generate gate pulses based on its duty cycle [44]. The DC link voltage is given to the battery pack at 80% State of Charge (SoC). A Space Vector Pulse Width Modulation (SVPWM) VSI gets its DC power from the battery and converts it into three-phase voltages given as input to the PMSM. The developed model gets maximum torque at different speeds using field-oriented control.

6. Results and Discussions

Simulink and MATLAB were used to create the suggested model. Solar irradiance (G) and temperature (T) are the module's inputs. In standard test conditions (STC), a module with 72 cells produced 250 watts of power. Two modules form a solar PV array. The simulation found that the MPP can produce 3100 watts with an open circuit voltage (VOC) of 44.88 volts and a short circuit current of 8.95 amps. Figure 12 shows these results.

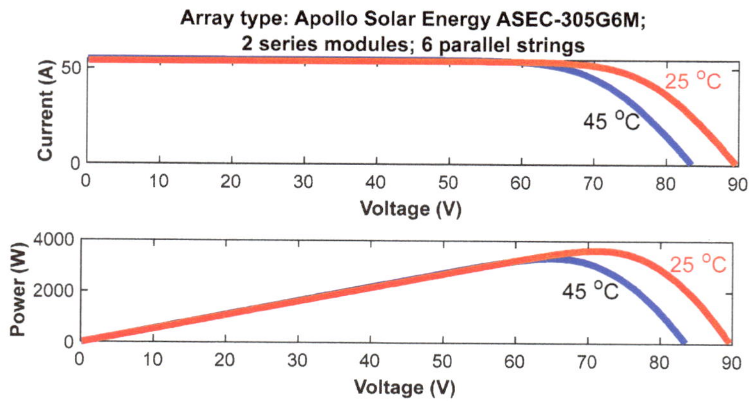

Figure 12. PV module I-V and P-V characteristics.

6.1. Results of a Simulation for an Asynchronous Motor Drive Using an Inverter

At 1400 rpm as a reference speed and 5 KHz as a switching frequency, the simulation goals were achieved. Here, a boost converter and an inverter are used by the motor driver to provide a 400 V supply. Figure 13 displays the inverter's output voltage.

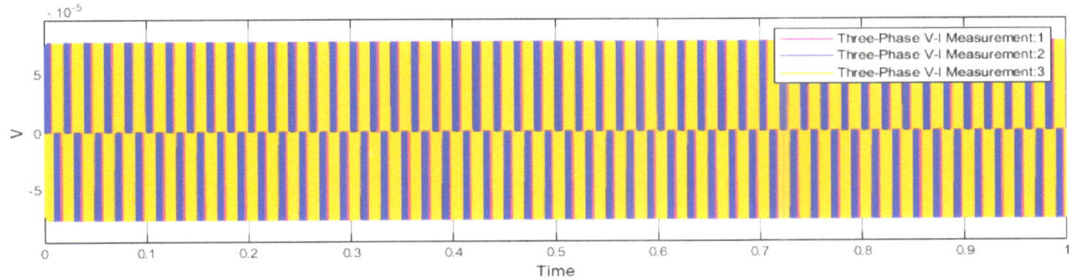

Figure 13. Three-Phase Inverter Output Voltage.

6.2. *Asynchronous Motor Drive Simulation Results in the on Position*

Starting ripple content and maximum torque currents are reduced to provide an early steady state for the PMS motor drive. The recommended MPPT produces 2.98 N-m, 7.596 A, and 1500 rpm of torque, a stator phase current, and speed. The torque's ripple content was 0.31, a significant improvement over previous methods. Figure 14 shows this.

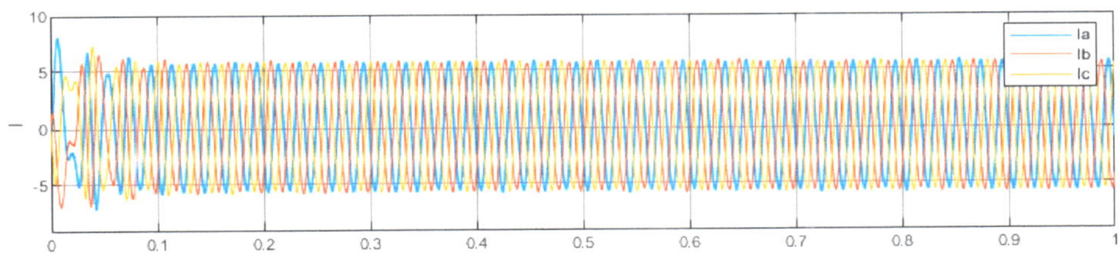

Figure 14. PMSM stator phase currents.

6.3. *Results of PMS Motor Drive Simulation*

Figure 15 shows the steady-state responses of the stator phase currents, torque, and speed of the proposed MPPT. In this figure, the proposed MPPT controller results in a more optimal response from the torque and speed.

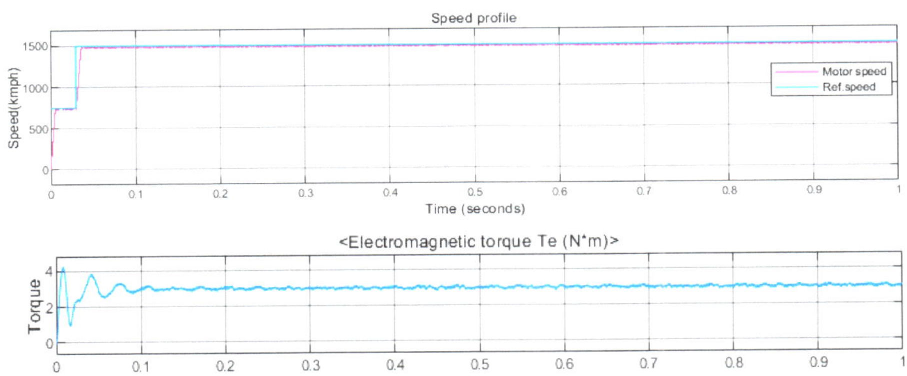

Figure 15. Speed and Torque characteristics of the proposed PMSM.

The ripple content of torque is reduced with the ANN block when compared to the torque without ANN; the comparison is shown in Figure 16.

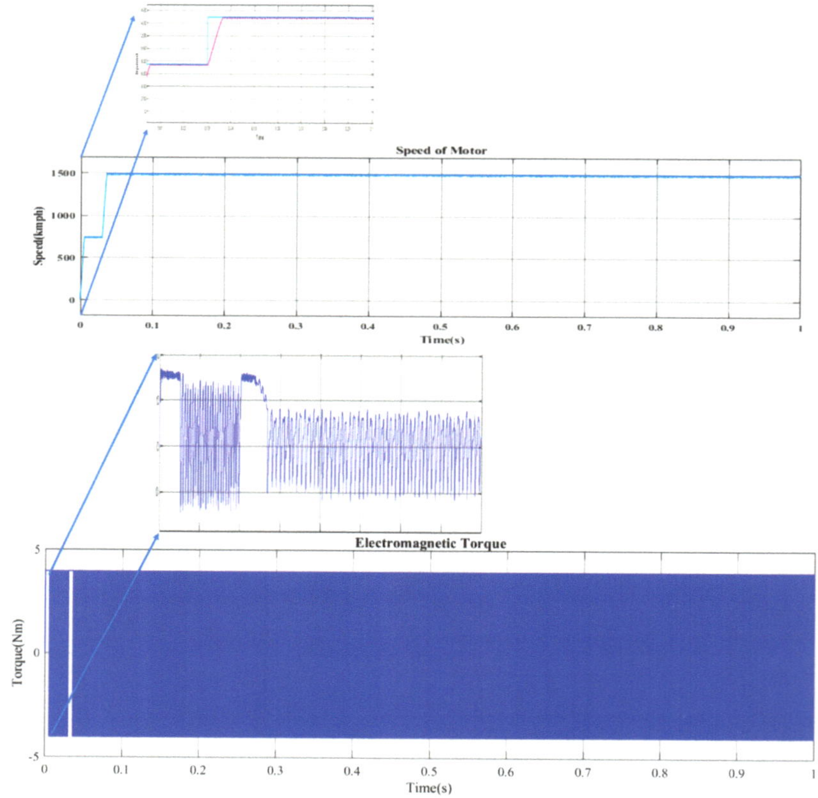

Figure 16. Speed and Torque characteristics of the PMSM without an ANN block.

6.4. Outcomes from the Results

- MPP is reached at minimum OCV (open circuit voltage) and Isc (short circuit current).
- While using the proposed MPPT, the rated torque and speed along with a stator phase current is achieved.
- The torque ripple content is reduced.
- Achieved sustainable improvement over other approaches currently in use.
- Attained better steady-state responses from torque and speed.

7. Conclusions

As the solution to energy demands and environmental problems, the usage of solar energy is indispensable. Using solar panels, the generation of power is encouraged by the MPPT algorithm. ANN-based MPPT controllers and PV array models were evaluated. The ANN-based MPPT controller reduced torque and stator phase current ripples. In this study, the proposed motor drive achieves both an early steady-state response and an improved speed response. As a consequence, the recommended MPPT with an ANN-based controller has considerably boosted the system's power usage and efficiency. This is crucial for an electric vehicle. The developed model will be scalable for different electric vehicle applications as well. A PV system with a DC-DC boost converter and a space vector modulated inverter may further improve power quality and performance.

Author Contributions: Conceptualization, A.V.T.; methodology, W.R.S. and S.R.S.; validation, A.V.T., W.R.S. and S.R.S.; formal analysis, A.V.T., S.R.S. and W.R.S.; investigation, A.V.T.; resources, W.R.S. and S.R.S.; data curation, S.R.S.; writing—original draft preparation, S.R.S.; A.V.T.; writing—review and editing, A.V.T., W.R.S. and S.R.S.; visualization, A.V.T.; supervision, W.R.S. and S.R.S. All authors have read and agreed to the published version of the manuscript.

Funding: WOOSONG UNIVERSITY's Academic Research Funding–2023.

Institutional Review Board Statement: Not applicable.

Informed Consent Statement: Not applicable.

Data Availability Statement: Not applicable.

Conflicts of Interest: The authors declare no conflict of interest.

Nomenclature

ANN	Artificial neural network
MPPT	Maximum power point tracking
SVPWM	Space vector pulse width modulation
I_{mc}	Module current
I_{is}	Insolation light current
q	Charge of electron
K	Boltzmann constant
N	Diode constant
V_{mv}	Voltage across PV cell
R_p	Parallel resistance
R_s	Series resistance
P_{irr}	Applied irradiance
T_s	Temperature coefficient
T_r	Reference Temperature
E_{go}	Energy in band gap
N_{se}	Series PV modules
N_{pr}	Parallel PV modules
I_{mpp}	Current at the maximum power point
V_{mpp}	Voltage at the maximum power point
P&O	Perturb and observation
INC	Incremental conductance
MSE	Mean squared error
G	Irradiance
T	Temperature
IPMSM	Interior permanent magnet synchronous motor
U_s	Motor voltage
R_S	Stator resistance
I_s	Stator currents
λ_s	Motor flux linkage
L_m	Mutual inductance
P_n	Motor pole pairs
J	Motor spinning inertia
B	Damping coefficient
T_L	Load torque
ω_m	Angular speed
VSI	Voltage source inverter
SOC	State of charge
VOC	Open circuit voltage

References

1. Pakkiraiah, B.; Sukumar, G.D. Research Survey on Various MPPT Performance Issues to Improve The Solar PV System Efficiency. *J. Solar Energy* **2016**, *2016*, 8012432. [CrossRef]
2. Mellit, A.; Kalogirou, S.A. Artificial intelligence techniques for photovoltaic applications: A review. *Prog. Energy Combust. Sci.* **2008**, *34*, 574–632. [CrossRef]
3. Saadi, A.; Moussi, A. Neural network use in the MPPT of photovoltaic pumping system. *J. Renew. Energy ICPWE* **2003**, *4*, 39–45.
4. Mei, Q.; Shan, M.; Liu, L.; Guerrero, J.M. A Novel Improved Variable Step-Size Incremental-Resistance MPPT Method for PV Systems. *IEEE Trans. Ind. Electron.* **2010**, *58*, 2427–2434. [CrossRef]
5. Tian, X.; Cai, Y.; Sun, X.; Zhu, Z.; Xu, Y. A Novel Energy Management Strategy for Plug-in Hybrid Electric Buses Based on Model Predictive Control and Estimation of Distribution Algorithm. *IEEE/ASME Trans. Mechatron.* **2022**, *27*, 4350–4361. [CrossRef]
6. Tian, X.; Cai, Y.; Sun, X.; Zhu, Z.; Wang, Y.; Xu, Y. Incorporating Driving Style Recognition into MPC for Energy Management of Plug-In Hybrid Electric Buses. *IEEE Trans. Transp. Electr.* **2022**, *9*, 169–181. [CrossRef]
7. Faraji, R.; Rouholamini, A.; Naji, H.R.; Fadaeinedjad, R.; Chavoshian, M.R. FPGA-based real time incremental conductance maximum power point tracking controller for photovoltaic systems. *IET Power Electron.* **2014**, *7*, 1294–1304. [CrossRef]
8. Montecino, D.A.; Perez, C.A.; Bowyer, K.W. Two-Level Genetic Algorithm for Evolving Convolutional Neural Networks for Pattern Recognition. *IEEE Access* **2021**, *9*, 126856–126872. [CrossRef]
9. Bendib, B.; Krim, F.; Belmili, H.; Almi, M.F.; Bolouma, S. An intelligent MPPT approach based on neural-network voltage estimator and fuzzy controller, applied to a stand-alone PV system. In Proceedings of the IEEE 23rd International Symposium on Industrial Electronics (ISIE), Istanbul, Turkey, 1–4 June 2014; pp. 404–409.
10. Al-Amoudi, A.; Zhang, L. Application of radial basis function networks for solar-array modelling and maximum power-point prediction. *IEE Proc.-Gener. Transm. Distrib.* **2000**, *147*, 310–316. [CrossRef]
11. Bahgat, A.; Helwa, N.; Ahmad, G.; El-Shenawy, E. Maximum power point traking controller for PV systems using neural networks. *Renew. Energy* **2005**, *30*, 1257–1268. [CrossRef]
12. Attia, H. High performance PV system based on artificial neural network MPPT with PI controller for direct current water pump applications. *Int. J. Power Electr. Drive Syst.* **2019**, *10*, 1329–1338. [CrossRef]
13. Algazar, M.M.; Abd El-Halim, H.; Salem, M.E.E.K. Maximum power point tracking using fuzzy logic control. *Int. J. Electr. Power Energy Syst.* **2012**, *39*, 21–28. [CrossRef]
14. Kassem, A.M. MPPT control design and performance improvements of a PV generator powered DC motor-pump system based on artificial neural networks. *Int. J. Electr. Power Energy Syst.* **2012**, *43*, 90–98. [CrossRef]
15. Jain, S.; Agarwal, V. New current control based MPPT technique for single stage grid connected PV systems. *Energy Convers. Manag.* **2007**, *48*, 625–644. [CrossRef]
16. Villalva, M.G.; Filho, E.R. Dynamic analysis of the input-controlled buck converter fed by a photovoltaic array. *Sba Controle Automação Soc. Bras. Autom.* **2008**, *19*, 463–474. [CrossRef]
17. Won, C.Y.; Kim, D.H.; Kim, S.C.; Kim, W.S.; Kim, H.S. A new maximum power point tracker of photovoltaic arrays using fuzzy controller. In Proceedings of the 1994 Power Electronics Specialist Conference-PESC'94, Taipei, Taiwan, 24 June 1994; Volume 1, pp. 396–403.
18. Beale, M.H.; Hagan, M.T.; Demuth, H.B. *Neural Network Toolbox User's Guide*; The MathWorks Inc: Natick, MA, USA, 1992.
19. Rajalakshmi, M.; Razia Sultana, W. Intelligent hybrid battery management system for electric vehicle. *Artif. Intell. Tech. Electr. Hybrid Electr. Veh.* **2020**, *1*, 179–206. [CrossRef]
20. Sultana, W.R.; Sahoo, S.K.; Saikiran, K.S.; Reddy, G.R.; Reddy, P.H. A computationally efficient finite state model predictive control for cascaded multilevel inverter. *Ain Shams Eng. J.* **2016**, *7*, 567–578. [CrossRef]
21. Razia Sultana, W.; Sahoo, S.K.; Prabhakar Karthikeyan, S.; Jacob Raglend, I.; Harsha Vardhan Reddy, P.; Rajasekhar Reddy, G.T. Elimination of harmonics in seven-level cascaded multilevel inverter using particle swarm optimization technique. In *Artificial Intelligence and Evolutionary Algorithms in Engineering Systems*; Springer: New Delhi, India, 2015; pp. 265–274.
22. Murali, A.; Wahab, R.S.; Gade, C.S.R.; Annamalai, C.; Subramaniam, U. Assessing Finite Control Set Model Predictive Speed Controlled PMSM Performance for Deployment in Electric Vehicles. *World Electr. Veh. J.* **2021**, *12*, 41. [CrossRef]
23. Shiau, J.-K.; Wei, Y.-C.; Chen, B.-C. A Study on the Fuzzy-Logic-Based Solar Power MPPT Algorithms Using Different Fuzzy Input Variables. *Algorithms* **2015**, *8*, 100–127. [CrossRef]
24. Rauh, A.; Frenkel, W.; Kersten, J. Kalman Filter-Based Online Identification of the Electric Power Characteristic of Solid Oxide Fuel Cells Aiming at Maximum Power Point Tracking. *Algorithms* **2020**, *13*, 58. [CrossRef]
25. Remoaldo, D.; Jesus, I.S. Analysis of a Traditional and a Fuzzy Logic Enhanced Perturb and Observe Algorithm for the MPPT of a Photovoltaic System. *Algorithms* **2021**, *14*, 24. [CrossRef]
26. Bhukya, L.; Kedika, N.R.; Salkuti, S.R. Enhanced Maximum Power Point Techniques for Solar Photovoltaic System under Uniform Insolation and Partial Shading Conditions: A Review. *Algorithms* **2022**, *15*, 365. [CrossRef]
27. Shahparasti, M.; Savaghebi, M.; Adabi, M.E.; Ebel, T. Dual-Input Photovoltaic System Based on Parallel Z-Source Inverters. *Designs* **2020**, *4*, 51. [CrossRef]
28. Pillay, P.; Krishnan, R. Modeling of permanent magnet motor drives. *IEEE Trans. Ind. Electron.* **1988**, *35*, 537–541. [CrossRef]
29. Boby, K.; Kottalil, A.M.; Ananthamoorthy, N.P. Mathematical modelling of pmsm vector control system based on SVPWM with pi controller using Matlab. *Int. J. Adv. Res. Electr. Electron. Instrum. Eng.* **2013**, *2*, 689–695.

30. Karabacak, M.; Eskikurt, H.I. Design, modelling and simulation of a new nonlinear and full adaptive backstepping speed tracking controller for uncertain PMSM. *Appl. Math. Model.* **2012**, *36*, 5199–5213. [CrossRef]
31. Bouksaim, M.; Mekhfioui, M.; Srifi, M.N. Design and Implementation of Modified INC, Conventional INC, and Fuzzy Logic Controllers Applied to a PV System under Variable Weather Conditions. *Designs* **2021**, *5*, 71. [CrossRef]
32. Figueiroa, V.; Torres, J.P.N. Simulation of a Small Smart Greenhouse. *Designs* **2022**, *6*, 106. [CrossRef]
33. Priyadarshi, N.; Padmanaban, S.; Mihet-Popa, L.; Blaabjerg, F.; Azam, F. Maximum Power Point Tracking for Brushless DC Motor-Driven Photovoltaic Pumping Systems Using a Hybrid ANFIS-FLOWER Pollination Optimization Algorithm. *Energies* **2018**, *11*, 1067. [CrossRef]
34. Cortés, B.; Tapia, R.; Flores, J.J. System-Independent Irradiance Sensorless ANN-Based MPPT for Photovoltaic Systems in Electric Vehicles. *Energies* **2021**, *14*, 4820. [CrossRef]
35. Guerra, M.I.S.; de Araújo, F.M.U.; Dhimish, M.; Vieira, R.G. Assessing Maximum Power Point Tracking Intelligent Techniques on a PV System with a Buck–Boost Converter. *Energies* **2021**, *14*, 7453. [CrossRef]
36. Derbeli, M.; Napole, C.; Barambones, O.; Sanchez, J.; Calvo, I.; Fernández-Bustamante, P. Maximum Power Point Tracking Techniques for Photovoltaic Panel: A Review and Experimental Applications. *Energies* **2021**, *14*, 7806. [CrossRef]
37. Khan, M.J.; Kumar, D.; Narayan, Y.; Malik, H.; Márquez, F.P.G.; Muñoz, C.Q.G. A Novel Artificial Intelligence Maximum Power Point Tracking Technique for Integrated PV-WT-FC Frameworks. *Energies* **2022**, *15*, 3352. [CrossRef]
38. Subramani, P.; Mani, S.; Lai, W.-C.; Ramamurthy, D. Sustainable Energy Management and Control for Variable Load Conditions Using Improved Mayfly Optimization. *Sustainability* **2022**, *14*, 6478. [CrossRef]
39. Malkawi, A.M.A.; Odat, A.; Bashaireh, A. A Novel PV Maximum Power Point Tracking Based on Solar Irradiance and Circuit Parameters Estimation. *Sustainability* **2022**, *14*, 7699. [CrossRef]
40. Sibtain, D.; Gulzar, M.M.; Shahid, K.; Javed, I.; Murawwat, S.; Hussain, M.M. Stability Analysis and Design of Variable Step-Size P&O Algorithm Based on Fuzzy Robust Tracking of MPPT for Standalone/Grid Connected Power System. *Sustainability* **2022**, *14*, 8986. [CrossRef]
41. Ali, A.I.M.; Alaas, Z.M.; Sayed, M.A.; Almalaq, A.; Farah, A.; Mohamed, M.A. An Efficient MPPT Technique-Based Single-Stage Incremental Conductance for Integrated PV Systems Considering Flyback Central-Type PV Inverter. *Sustainability* **2022**, *14*, 12105. [CrossRef]
42. Aljanad, A.; Mohamed, A.; Khatib, T.; Ayob, A.; Shareef, H. A Novel Charging and Discharging Algorithm of Plug-in Hybrid Electric Vehicles Considering Vehicle-to-Grid and Photovoltaic Generation. *World Electr. Veh. J.* **2019**, *10*, 61. [CrossRef]
43. Chandran, V.; Patil, C.K.; Karthick, A.; Ganeshaperumal, D.; Rahim, R.; Ghosh, A. State of Charge Estimation of Lithium-Ion Battery for Electric Vehicles Using Machine Learning Algorithms. *World Electr. Veh. J.* **2021**, *12*, 38. [CrossRef]
44. Banda, G.; Kolli, S.G. An Intelligent Adaptive Neural Network Controller for a Direct Torque Controlled eCAR Propulsion System. *World Electr. Veh. J.* **2021**, *12*, 44. [CrossRef]

designs

MDPI

Review

The Status of On-Board Hydrogen Storage in Fuel Cell Electric Vehicles

Julián A. Gómez and Diogo M. F. Santos *

Center of Physics and Engineering of Advanced Materials, Laboratory for Physics of Materials and Emerging Technologies, Chemical Engineering Department, Instituto Superior Técnico, Universidade de Lisboa, 1049-001 Lisbon, Portugal; julian.gomez.mayorga@tecnico.ulisboa.pt
* Correspondence: diogosantos@tecnico.ulisboa.pt

Abstract: Hydrogen as an energy carrier could help decarbonize industrial, building, and transportation sectors, and be used in fuel cells to generate electricity, power, or heat. One of the numerous ways to solve the climate crisis is to make the vehicles on our roads as clean as possible. Fuel cell electric vehicles (FCEVs) have demonstrated a high potential in storing and converting chemical energy into electricity with zero carbon dioxide emissions. This review paper comprehensively assesses hydrogen's potential as an innovative alternative for reducing greenhouse gas (GHG) emissions in transportation, particularly for on-board applications. To evaluate the industry's current status and future challenges, the work analyses the technology behind FCEVs and hydrogen storage approaches for on-board applications, followed by a market review. It has been found that, to achieve long-range autonomy (over 500 km), FCEVs must be capable of storing 5–10 kg of hydrogen in compressed vessels at 700 bar, with Type IV vessels being the primary option in use. Carbon fiber is the most expensive component in vessel manufacturing, contributing to over 50% of the total cost. However, the cost of FCEV storage systems has considerably decreased, with current estimates around 15.7 $/kWh, and is predicted to drop to 8 $/kWh by 2030. In 2021, Toyota, Hyundai, Mercedes-Benz, and Honda were the major car brands offering FCEV technology globally. Although physical and chemical storage technologies are expected to be valuable to the hydrogen economy, compressed hydrogen storage remains the most advanced technology for on-board applications.

Keywords: hydrogen storage; fuel cell electric vehicles; hydrogen economy; compressed hydrogen

Citation: Gómez, J.A.; Santos, D.M.F. The Status of On-Board Hydrogen Storage in Fuel Cell Electric Vehicles. *Designs* **2023**, *7*, 97. https://doi.org/10.3390/designs7040097

Academic Editors: Surender Reddy Salkuti and Dibin Zhu

Received: 18 June 2023
Revised: 19 July 2023
Accepted: 27 July 2023
Published: 2 August 2023

1. Introduction

The transport sector has become the second largest contributor to the energy crisis, with a share of 29% of the total final energy consumption and 20.3% of greenhouse gas (GHG) global emissions [1,2]. It has the highest dependency index on fossil fuels compared to other industries, with 37% of global carbon dioxide (CO_2) emissions originating from transportation end-use. In the past decade, CO_2 emissions from the transportation sector have exhibited the fastest growth due to the increasing demand and the limited low-carbon emission technology options available for the industry [3].

The four main contributors to global CO_2 emissions from transport are road, aviation, shipping, and rail transportation. As illustrated in Figure 1, nearly 75% of these emissions stem from road vehicles, most of which come from cars and buses (45% of emissions), and the remaining 30% originate from trucks. Aviation, shipping, and rail transportation contribute approximately 12%, 11%, and 2% of the total CO_2 emissions, respectively [4]. The growth of transport demand is expected to continue globally, driven by increasing populations and the economic development of non-developed countries. According to estimates from the U.S. Energy Information Administration in 2016, the shares of world transportation energy use were nearly equal between OECD and non-OECD countries in 2020.

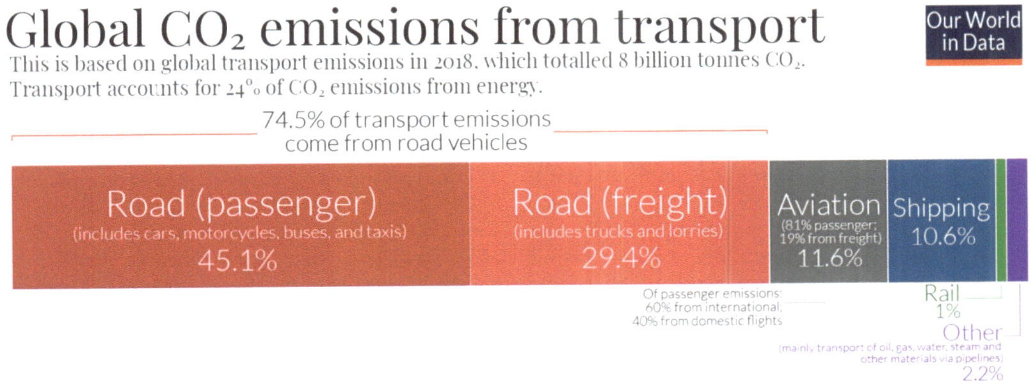

Figure 1. Global CO_2 emissions from transport by mode of transportation [4].

The latter is projected to experience a sustained increase in demand for transportation fuels. This trend is expected to result in non-OECD countries accounting for as much as 61% of global transport energy consumption by 2040 [5]. To address the increasing need for energy consumption and reduce GHG emissions in the transport sector, electric mobility has emerged as a strong option for clean transportation. This includes revolutionary automotive technologies such as battery electric vehicles (BEVs), hybrid electric vehicles (HEVs), and fuel cell electric vehicles (FCEVs). Among these, hydrogen-based fuel cell technology for EVs has demonstrated high potential in storing and converting chemical energy into electricity, offering advantages such as high energy conversion rates, efficient drivetrain, and zero CO_2 emissions compared to conventional gasoline vehicles [6].

Recent research has demonstrated that FCEVs have achieved comparable performance to internal combustion engine vehicles (ICEVs) and are anticipated to experience increased adoption in the market due to their numerous attractive features, including superior energy efficiency, reduced maintenance requirements, and emission-free operation. However, hydrogen on-board processing and storage still represent a significant barrier to the widespread commercialization of hydrogen fuel cell vehicles. To operate an FCEV, a proton-exchange membrane fuel cell (PEMFC) is supplied with hydrogen, which is then utilized to produce electricity that powers an electric motor.

One of the principal challenges facing the transportation industry is the safe and effective storage of hydrogen. Compressed hydrogen storage technology has emerged as the most promising on-board storage method due to its high performance and practicality. Nonetheless, other storage technologies, such as liquid and cryo-compressed hydrogen storage, are still in the early stages of development. To enable the practical utilization of hydrogen, it is necessary to enhance the energy density of hydrogen storage systems, reduce costs, and improve interoperability among vehicle systems [7].

The usage of hydrogen for on-board applications is expected to rise over the next decade, leading to the greater affordability of FCEVs and related technologies. It is anticipated that various physical or chemical storage technologies will contribute significantly to the hydrogen economy, with improvements made to hydrogen storage and FCEVs performance by leveraging the unique characteristics of each technology. Government policies in different countries are also expected to significantly impact the growth of the hydrogen economy, particularly for the on-board applications of hydrogen. This paper aims to provide a comprehensive review of on-board hydrogen storage technologies in FCEVs, covering their current developmental status as well as the potential prospects, advantages, and drawbacks associated with implementing hydrogen storage. Moreover, this report presents data analyses regarding safety, cost-effectiveness, CO_2 emissions, and feasibility of hydrogen storage in the market.

2. Fuel Cell Electric Vehicles

The automotive industry has developed electric-assist models for vehicle electrification to meet future goals and improved pollution regulations. These models include start/stop technology, HEVs, plug-in HEVs (PHEVs), BEVs, and FCEVs, representing the final stage of automotive electrification adaptation [8].

Since its inception, the FCEV has undergone extensive research and development, leading to significant changes and improvements. The initial prototype of the FCEV was pioneered by General Electric in 1966, utilizing an alkaline fuel cell as its power source and two cryogenic tank vessels for liquid hydrogen and liquid oxygen [9,10]. FCEVs share several key components with BEVs, including power controllers, inverters, and electric motors. However, the primary point of differentiation between the two lies in their energy source. While BEVs rely on battery-stored energy, FCEVs utilize fuel cells, which offer several advantages over batteries. Notably, fuel cells are both lighter and smaller, and can generate electricity for as long as the fuel supply lasts. Given these similarities and differences, both battery and fuel cell technologies are expected to coexist in the future. While BEVs are well suited for short-range and small vehicles, FCEVs are particularly well suited to medium-large and long-range vehicles [10,11].

2.1. Fuel Cell Types for On-Board Applications

Over time, various hydrogen fuel cell types have been developed, aiming to achieve greater efficiency and find applications that best align with their technical specifications. When it comes to on-board vehicle applications, selecting the optimal fuel cell depends on various factors, including efficiency, durability, cost, and fuel compatibility. PEMFC using hydrogen as the fuel is the most employed technology for vehicle applications due to its exceptional efficiency and low operating temperatures, typically around 80 °C. However, a significant drawback of PEMFCs is their sensitivity to fuel impurities. The performance of PEMFCs can be significantly compromised by the presence of impurities in the fuel stream, thereby requiring a high purity level for the hydrogen gas used [12].

Several fuel cell types have recently been tested for on-board applications with promising outcomes. These include solid-oxide fuel cells (SOFC), direct methanol fuel cells (DMFC), alkaline fuel cells (AFC), molten carbonate fuel cells (MCFC), and phosphoric acid fuel cells (PAFC). Table 1 compares the technical specifications, benefits and drawbacks, operational parameters, and fuel cell efficiency of the six main types of fuel cells. All reported fuel cells have been tested in FCEVs for commercial or research purposes [13].

Due to the challenges related to hydrogen storage, handling, and transportation, much research is ongoing on the use of liquid fuels that can replace hydrogen to power the so-called direct liquid fuel cells (DLFCs). From these, the DMFC is the most well-known and developed. The DMFC has been explored as an alternative to traditional internal combustion engines in on-board passenger vehicle applications. However, there are several reasons why methanol fuel cells are not commonly used in this context. These include the fact that methanol has lower energy density than gasoline or hydrogen, methanol fuel cells have lower overall energy efficiency than other fuel cell technologies, and safety concerns. The latter is related to the high flammability and toxicity of methanol. Therefore, although modern DMFCs are designed with safety features, there are still concerns about the handling, storage, and transportation of methanol.

In contrast, the PEMFC offers notable benefits in terms of high energy density, rapid start-up, and a lightweight, cost-effective, and compact design. The core structural components of a PEMFC are the anode electrode, cathode electrode, catalyst layers, proton-conducting polymer electrolyte membrane, gas diffusion layers, and two bipolar plates (Figure 2). These elements constitute the fundamental operating mechanism of this type of fuel cell.

Table 1. Comparison of fuel cells (FCs) used in fuel cell electric vehicles (FCEVs) [13]. Reprinted with permission from Elsevier.

Type	PEMFC	PAFC	AFC	SOFC	MCFC	DMFC
Advantages	-Low operating temperature -Fast charging, less corrosion, simplified electrolyte management	-Higher efficiency with CHP -Increased tolerance to fuel impurities	-Higher performance -Faster cathode reaction and start -Lower material cost -Low operating temperature	-Suitable for CHP -Higher efficiency -Fuel flexibility -Hybrid/gas turbine cycle	-Suitable for CHP -Higher efficiency -Fuel flexibility	-Low cost due to lack of fuel reformer
Disadvantages	-Expensive catalyst cost -Sensitive to impurities	-Expensive catalyst cost -Long start time	-Sensitive to CO_2 levels in O_2 and H_2	-High operation temperature, corrosion, and breakdown -Long start time	-High operation temperature -Long start -Power density	-Intermediates poisoning catalyst surface
Efficiency (%)	40–60	40	60	60	50	40
Fuel	Hydrogen	Hydrogen	Hydrogen	Hydrogen	Hydrogen	Methanol
Operating Temp. (°C)	<100	150–200	90–100	500–1000	600–700	60–200
Cell Voltage (V)	1.1	1.1	1	0.8–1	0.7–1	0.2–0.4
Stack Power (kW)	1–250	50–100	1–100	1–3000	300–3000	0.001–100

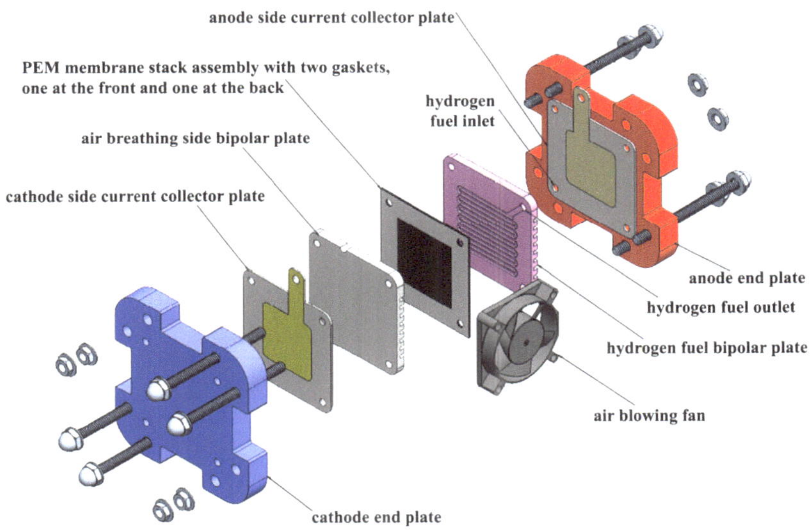

Figure 2. Schematic representation of the PEMFC architecture [9]. Reprinted with permission from Elsevier.

The process in a fuel cell can be divided into two electrochemical half-reactions. These half-reactions are spatially separated, forcing electrons transferred from the fuel to flow through an external circuit to perform useful work before completing the chemical reaction. An electrolyte membrane is used to achieve this spatial separation of the half-reactions between the two electrodes. The membrane is composed of a solid polymer film that has been acidified, typically a perfluorinated compound known as Nafion (thickness around 50–175 μm). The anode is where the hydrogen oxidation reaction (HOR) occurs (Equation (1)), with hydrogen releasing two electrons and producing H^+ ions, known as protons. These electrons flow through the bipolar plates interconnected with the external circuit. Simultaneously, the protons undergo migration through the polymer electrolyte membrane from the anode to the cathode, the electrode where the oxygen reduction reaction (ORR) occurs (Equation (2)). The ORR involves recombining the electrons with the protons, producing water and generating heat [9,14]. The sum of the two half-reactions leads to the fuel cell's overall reaction (Equation (3)).

$$\text{Anode reaction:} \quad H_{2(g)} \rightarrow 2H^{+} + 2e^{-} \tag{1}$$

$$\text{Cathode reaction:} \quad 1/2O_2 + 2H^{+} + 2e^{-} \rightarrow H_2O \tag{2}$$

$$\text{Overall reaction:} \quad H_{2(g)} + 1/2O_{2(g)} \rightarrow H_2O \tag{3}$$

PEMFCs exhibit some drawbacks regarding their cost, performance, and materials. Ongoing research efforts are focused on addressing the technological challenges involved in reducing the cost of fuel cell components and improving their durability. One of the present challenges in fuel cell technology is the slow electrochemical reactions that occur at low temperatures. To increase the kinetics of the reaction, a catalyst is required, typically platinum, which is an expensive material. To overcome this, multiple attempts have been made to minimize the use of platinum by optimizing the particle size of platinum, formulating mixtures of platinum and other metals, or identifying substitute metals [15].

2.2. Topological Classification of FCEVs

FCEVs are divided into two main categories depending on the system's configuration: fully FCEVs and fuel cell hybrid electric vehicles (FCHEVs). As mentioned before, in FCEVs, energy generation is produced by the fuel cell to supply the electric motor and other components directly. On the other hand, FCHEVs have different energy generation/storage units to support the fuel cell stack [13,16]. Batteries, supercapacitors (SC), superconducting magnetic energy storage systems (SMES), photovoltaic (PV) panels, and flywheels are some of the auxiliary energy units utilized in FCHEVs to support the FC stacks in hybrid systems [16]. Figure 3 presents different system configurations for FCEVs, taking into account the hybridization components for energy generation and storage.

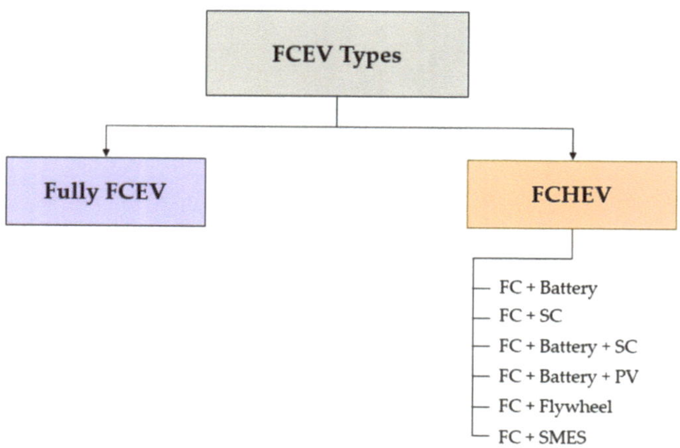

Figure 3. Topological classification of FCEVs according to power supplies.

The aim of hybridizing vehicles is to address certain drawbacks of fully electric vehicles. Battery integration seeks to provide an initial high current to start the electric motor, thus avoiding the low efficiency zone of the fuel cell (FC). On the other hand, SC hybridization aims to support the FC's transient power demand during vehicle operation. The combination of these two supplementary units offers the aforementioned benefits, ensuring a continuous energy supply and a dynamic response of the FC during transient events.

Another promising option is to incorporate alternative power sources alongside the FCs as additional energy generators. Photovoltaic (PV) panels generate DC current that can directly power the electric motor or charge the battery. Furthermore, flywheels can be used as an alternative energy storage system to harness rotational speed and convert mechanical energy into electricity, thereby supporting power supply during transient events. Lastly, the last configuration aims to capitalize on the energy storage capabilities of SMES. This is achieved through the creation of a magnetic field using a DC current flow and a superconducting coil.

2.3. FCEV Components

Although an FCEV's most important component is the fuel cell stack, since it converts the fuel into electricity, it is not the only component of the system. The key components of a hydrogen FCEV include the auxiliary and main batteries, a DC/DC converter, an electric motor, a fuel tank, a fuel filler, a power electronics controller, a thermal system, and a transmission [17]. These various components perform specific functions to ensure the vehicle's smooth operation. All these components operate simultaneously, requiring precise design and control. The fuel cell stack is the primary component of a fuel cell electric vehicle, as it converts the main fuel source, hydrogen, into electricity that powers the electric motor. The PEMFC is the most commonly used fuel cell, as it operates at low temperatures (60–80 °C) and has a high power density.

Usually, there are two batteries in the car, the first one, the auxiliary battery, is a low-voltage battery used to jump-start the vehicle until the main battery is available and to power some of the car's components. The main battery is a high-voltage battery that stores the energy provided by the regenerative braking and serves as a backup option for the electric motor. The electric motor drives the car's wheels and is powered by the fuel cell and the main battery. The DC-DC converter converts high-voltage DC to low-voltage DC from the main battery to recharge the auxiliary battery and power components that run on low voltage [17].

The fuel tank serves as the storage for the hydrogen in the fuel cell electric vehicle, enabling it to be utilized later by the fuel cell for power generation. The fuel filler, typically in the form of a nozzle, facilitates the filling of the tank with hydrogen or alternative fuels, such as methanol, for specific fuel cell configurations. On the other hand, the transmission functions as a gearbox to transfer mechanical power from the motor to drive the vehicle's wheels. The power electronics controller plays a crucial role in managing the energy flow from the fuel cell and battery to the electric motor. It regulates the amount of electrical power supplied to the motor and controls the flow of the electric current, while also managing the motor's speed to produce varying torque levels, enabling the car to operate at different speeds. Finally, the thermal system regulates and maintains the optimal temperature of the fuel cell, batteries, motor, electronic components, and other car parts [17]. Figure 4 shows the location of these various components.

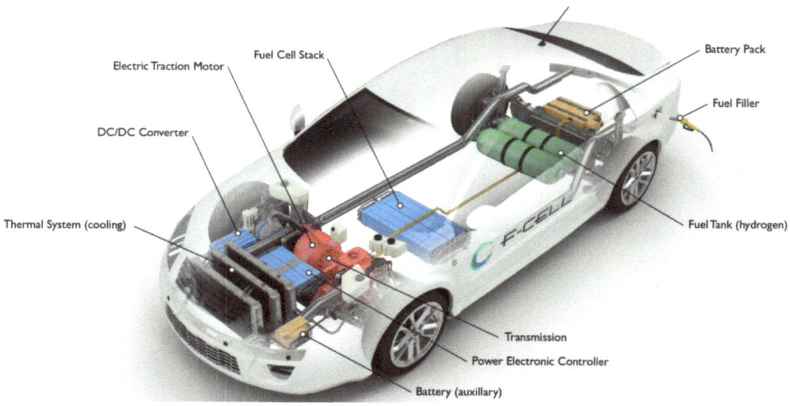

Figure 4. Location of the components included in a hydrogen FCEV [17].

3. Hydrogen Storage Techniques for On-Board Applications

In a context strongly influenced by sustainable development and resource optimization, green energy as a way of powering vehicles has become one of the main priorities of the automotive industry. Hydrogen has emerged as one of the most promising energy sources. However, the limited range has posed a significant challenge to developing hydrogen engines in the automotive sector. Fortunately, innovations in the design and construction of hydrogen storage tanks are being developed with great potential, making it possible to travel long distances with a hydrogen vehicle soon.

As with other traditional fuels, safe and efficient hydrogen storage is essential for its utilization as an energy carrier. The U.S. Department of Energy (DOE) has identified the primary technical difficulty for hydrogen storage in transportation as the capacity to store sufficient hydrogen to meet the driving range requirement (>500 km) while meeting the vehicle's weight, volume, efficiency, safety, and cost limitations [18]. To utilize hydrogen as an energy carrier, it must first be transported and stored. Various methods are available to store hydrogen, each with its own advantages and limitations, making them suitable for different applications. Hydrogen storage for vehicle applications can be classified into two main categories: physical storage and chemical storage, as illustrated in Figure 5. This section comprehensively analyzes the four primary hydrogen storage technologies for on-board applications.

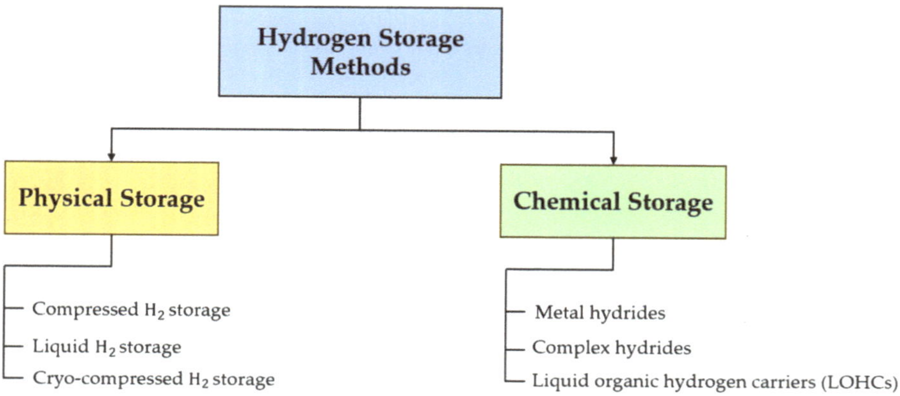

Figure 5. Hydrogen storage methods classification for on-board vehicle applications.

3.1. Physical Hydrogen Storage

Concerning physical hydrogen storage, hydrogen can be stored as a high-pressure gas or a low-temperature cryogenic liquid. The storage system can be classified into liquid, cryo-compressed, and compressed categories based on the storage pressure and temperature. These methods aim to enhance the gravimetric and volumetric storage density of hydrogen. To achieve a driving range greater than 500 km in FCEV, a hydrogen storage capacity of about 5–6 kg is essential [6]. H_2 can be stored as a liquid at −253 °C or as a high-pressure gas at 700 bar; the main differences for each type of storage technique are shown in Figure 6.

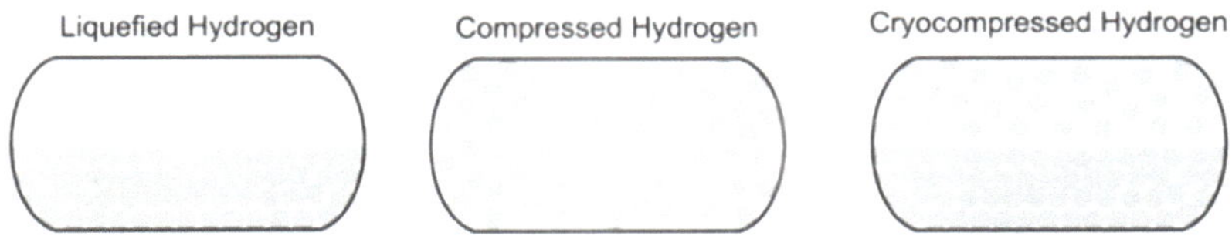

Figure 6. Simplified concept of typical physical hydrogen storage methods [19]. Reprinted with permission from Elsevier.

3.1.1. Compressed Hydrogen Storage

Under normal temperature and pressure, the density of the H_2 gas is very low, near about 0.08238 kg/m^3, e.g., for storing 5 kg of hydrogen, which implies a volume of around 60 m^3 and energy content of 600 MJ (166.65 kWh). For the same weight and energy content, the required gasoline volume is 0.019 m^3. Thus, it is clear that for efficient storage, hydrogen density should be increased by reducing the volume under normal temperature and pressure conditions. Compressed storage is the most established storage technology for hydrogen; it involves the physical storage of hydrogen gas in high-pressure vessels that are commonly classified into four standard types: Type I, II, III, and IV, depending on the lightweight and low-cost vessel material that can withstand the high-pressure requirements, the resilience of the material to resist hydrogen diffusion, and the likely damage caused by the stored hydrogen [19,20]. Figure 7 shows the different types of pressurized vessels that are currently commercially available.

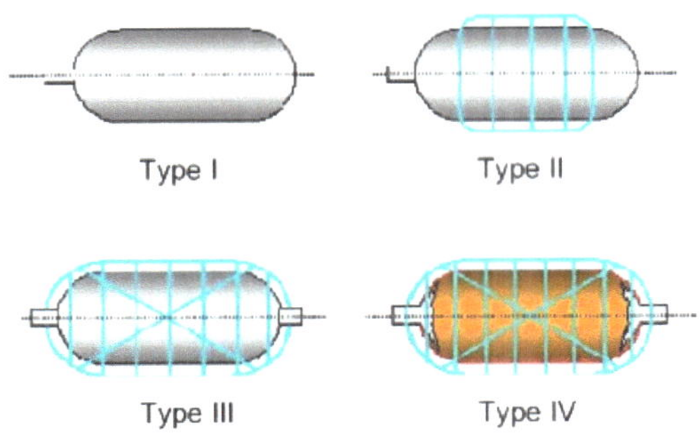

Figure 7. Schematic representation of pressure vessel types used for compressed hydrogen storage [21]. Reprinted with permission from Elsevier.

When considering a storage technique for hydrogen, there are several important factors to keep in mind. First, the material composition of the tank should be lightweight, inexpensive, and strong enough to meet the required stress, strain, and safety specifications. In addition to the material composition, the tank's geometry is also a crucial consideration. Hydrogen should ideally be stored in cylindrical vessels, as spherical vessels can be difficult to fit on-board. Finally, the material's thermal conductivity must be high enough to manage exothermic heat during the tank filling (compressed storage offers high rates of hydrogen filling and release). Concerning the standard vessels, they can be classified according to the following types [6,20]:

Type I: The cheapest option, these vessels are made of metallic materials capable of withstanding pressure up to 200–300 bar [21]. The metal wall must be relatively thick for high hydrogen pressures or densities. This increases the vessel's weight and substantially decreases the net hydrogen gravimetric energy density. Type I offers an extremely low gravimetric energy density of about 1 wt.%. Steel or aluminum alloy are the most common materials used for these tanks [19].

Type II: In these vessels, the metallic cylindrical section is wrapped with a fiber resin composite. Type II vessels weigh 30–40% less than Type I vessels but are 50% more expensive [19,21]. Due to their low hydrogen storage density, Type I and Type II vessels are unsuitable for on-board applications.

Type III: Carbon fiber composite tanks, also known as Type III vessels, consist of a carbon-fiber-reinforced plastic (CFRP) shell with a metallic liner, typically made of aluminum. Type III vessels have a hydrogen storage capacity that is 25% and 75% greater than that of Type I and Type II vessels [21]. These tanks are highly durable and lightweight but have low thermal conductivity, which can present challenges during hydrogen compression and release due to the low heat release rate. Type III tanks are generally suitable for hydrogen storage at up to 450 bar. However, it can also be used for pressures up to 700 bar [19].

Type IV: Type IV high-pressure vessels are composed entirely of composite materials, similar to Type III. However, the main difference lies in the liner material used in these tanks. In Type III vessels, the liner is mostly metal, contributing to at least 5% of the mechanical strength. In contrast, Type IV vessels predominantly use polymeric liners, such as high-density polyethylene (HDPE), with little to no metal content. These tanks are also suitable for storing hydrogen at 700 bar pressure [19].

An advanced type of hydrogen storage vessel, known as Type V, has been proposed, although it is not yet commercially available. This design is an enhancement of Type IV, incorporating reinforcing space-filling skeletons to achieve even higher hydrogen volumetric and gravimetric densities [21].

Hydrogen storage at 700 bars in Type III or Type IV vessels offers a practical solution with a refueling time of less than 3 min and a driving range of 500 km. Several vehicles with such tanks, including the Honda FCX Clarity, Toyota Mirai, Hyundai Tucson, and Hyundai NEXO, are already available for sale. Table 2 summarizes the materials, normal operating pressure, cost, and gravimetric density for each type of pressure vessel.

Table 2. Pressure vessel materials, operating pressure, cost, and gravimetric density according to type [22].

Type	Materials	Typical Pressure (bar)	Cost ($/kg)	Gravimetric Density (wt.%)
I	All-metal construction	300	83	1.7
II	Mostly metal, composite wrapped in the hoop direction	200	86	2.1
III	Metal liner, full composite overwrap	700	700	4.2
IV	All composite construction	700	633	5.7 (Toyota Mirai)

However, the public acceptance of on-board pressurized vessels is limited due to the potential risk of explosion resulting from sudden shocks. Nevertheless, a study revealed that a gasoline leak in a vehicle on fire might be more catastrophic than a hydrogen leak in an FCEV. In the event of a compressed hydrogen tank leak in the open air, the hydrogen, being lighter than air, will quickly dissipate harmlessly into the atmosphere without having the chance to combust. Of course, this situation may be more problematic in enclosed spaces, such as tunnels or parking spaces, where hydrogen can accumulate on the ceiling. The hydrogen fuel infrastructure, including transportation and dispensation (filling stations), is also growing [19]. In the commercial market, FCEVs utilizing compressed hydrogen are increasingly prevalent and demonstrate operational viability. Notably, the Toyota Mirai and Hyundai Nexo vehicles employ compressed H_2 at 700 bar pressure, featuring three Type IV cylinders, and can travel over 600 km on a single charge [20].

The Type IV storage vessel has become the preferred option for FCEV manufacturers. As mentioned, it primarily comprises three layers of materials designed to withstand severe internal and external loading conditions. The inner layers consist of a high-density polyethylene (HDPE) liner that serves as a hydrogen barrier and provides the shell for overwrapping the outer layer. HDPE's mechanical and thermal properties make it an ideal material for manufacturing Type IV tanks, as it offers thermal stability up to 120 °C, excellent chemical resistance, and toughness to the tank structure. Therefore, it is capable of maintaining its properties during a large number of charging and discharging cycles [6]. Figure 8 shows a schematic representation of a Type IV vessel for the automotive industry.

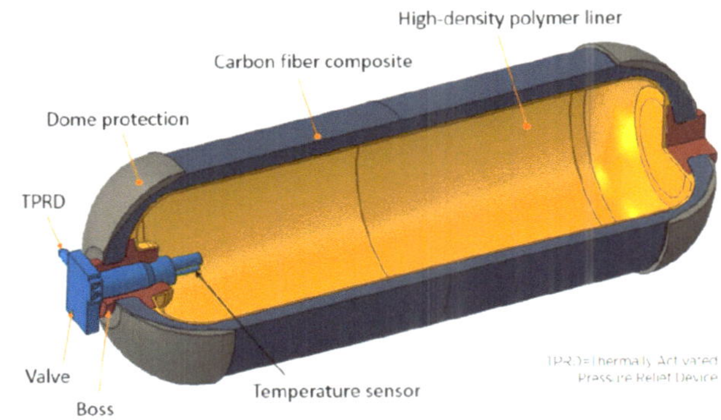

Figure 8. A schematic of a Type IV composite overwrapped pressure vessel designed for compressed hydrogen storage on board FCEVs [22].

The liner of a composite pressure vessel is typically fabricated using various methods such as injection, blow, compressing, and rotational molding processes. These techniques are commonly employed to manufacture hard and high-strength plastic products for commercial applications. Injection molding, in which melted plastic is injected into a mold cavity, is generally preferred for mass-producing small products. Blow molding is similar to injection molding, but the melted plastic is extruded vertically into a molten tube and cooled to form a hollow part. Rotational molding, on the other hand, involves rotating the plastic in hollow molds and cooling it with water to harden and form a hollow part [6]. This technique entails the bi-axial rotation of the metallic mold in a heated oven to produce stress-free parts.

Rotational molding is a cost-effective and efficient process that produces less waste, making it an ideal choice for low-to-high-volume production. This method is particularly suitable for manufacturing hollow shapes of varying sizes, resulting in improved part-wall-thickness distribution compared to other molding processes, such as injection and extrusion molding. The outer shell of Type IV tanks is typically made of fiber-reinforced polymer with epoxy resin [6]. Various composite materials are commonly used to fabricate Type IV tanks, and their mechanical properties are shown in Table 3.

Carbon fiber with epoxy resin is widely preferred for Type IV tanks due to its strength, flexibility, and high translation efficiency. However, the high manufacturing cost associated with carbon fiber, which contributes to approximately 62% of the total weight of the tank and has a higher per kg cost, has led to the adoption of the filament winding technique for wrapping the fiber [6]. This technique is a popular method for producing symmetric composite tanks, tubes, cylinders, and domes. It involves depositing fibers in a specifically oriented pattern that matches the direction of stresses and loads in the structure. As shown in Figure 9, the fibers are placed on polymer shells in layers called "ply" at defined angles, known as ply angles, in helical and hoop directions.

Table 3. Mechanical properties of materials for outer shell [6]. Reprinted with permission from CRC Press.

Fiber Type	Tensile Strength (MPa)	Tensile Modulus (GPa)	Density (g cm^{-3})	Specific Heat Capacity (J kg^{-1} K^{-1})	Thermal Conductivity (W m^{-1} K^{-1})	Composite Strength (MPa)
T300 carbon	3530	230	1.76	777	10.5	-
T700S carbon	4900	230	1.80	752	9.6	2830
T700G carbon	4900	240	1.80	752	9.6	-
T800	5880	294	1.80	752	11.3	-
Flax	8.25	23	1.45	-	-	345
Basalt	50	80	1.80	-	-	1600

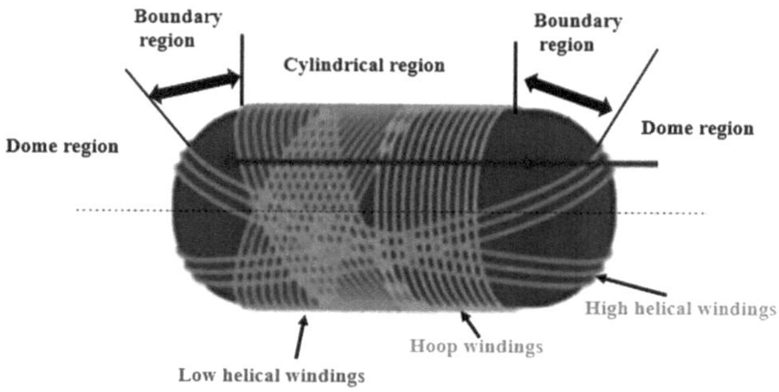

Figure 9. Winding pattern on cylindrical and dome sections of the tank [6]. Reprinted with permission from CRC Press.

3.1.2. Liquid Hydrogen Storage

Currently, compressed hydrogen is the most widely adopted hydrogen storage technology. However, its low energy density and safety concerns have motivated researchers to investigate alternative methods like liquid hydrogen. The aerospace and nuclear industries were the early adopters of liquid hydrogen due to its high energy density and purity. The development of liquid hydrogen production and storage systems was accelerated by the research and production of rocket propulsion based on liquid oxygen and liquid hydrogen during the 1930s and 1950s. As a result, industrial-scale systems were constructed to meet the increasing demand for liquid hydrogen in these sectors [23].

As previously mentioned, hydrogen exhibits lower energy density per unit volume than many other fuels, translating into a need for larger storage tanks to store the same amount of energy. To address this issue, the liquefaction of hydrogen is a potential solution. Compressed hydrogen at 350 and 700 bar and 288 K has a density of 24.5 and 41.5 g/L, respectively, and heating values of 2.94 and 4.97 MJ/L. In contrast, hydrogen in liquid form at 1 and 3.5 bar and at its normal boiling point of 20 K (−253 °C) has a heating value of 8.50 and 7.68 MJ/L, respectively, and a density of 70.9 and 64.0 kg/m^3, respectively [19]. Remarkably, liquid hydrogen is approximately 1.8 times denser than high-pressure hydrogen at 700 bar and 288 K, highlighting its potential advantages in energy storage and transportation [24].

Aspects such as increased density over high-pressure gas storage, reduced weight and pressure of tanks, and enhanced safety considerations, have made liquid hydrogen an attractive alternative to compressed hydrogen. Although liquid hydrogen can only exist at low temperatures and pressures, the corresponding tanks can be lighter as they operate at lower pressures. Moreover, liquid hydrogen storage presents a more compact and cost-effective storage and transportation solution than compressed hydrogen [21,25]. As a result of its higher gravimetric and volumetric density, liquid hydrogen is now widely regarded as the optimal choice for transportation and distribution, offering significant advantages in terms of economics, technical feasibility, and energy content [23]. Liquid hydrogen also has several drawbacks that need to be addressed. One of the most significant issues with liquid hydrogen is the energy-intensive liquefaction process, which can consume up to 30% of the energy content of the stored hydrogen. Furthermore, to minimize boil-off, it is necessary to maintain constant pressure in the storage tank and ensure that it is well insulated. A cooling and venting system should also be in place to achieve this.

Liquid hydrogen storage has several disadvantages, one of which is the low efficiency of the liquefaction process. Around 30 to 35% of the energy value of hydrogen is consumed during the liquefaction process, which is approximately three times more than the energy required to compress hydrogen. Therefore, advancements in hydrogen liquefaction are necessary to make liquid hydrogen a viable option for hydrogen storage. Additionally, it is worth noting that liquid hydrogen storage is still an emerging technology that requires further development [6,21,23,26].

To store hydrogen in liquid form, it is essential to maintain the hydrogen temperature below its boiling point of 20 K to avoid overpressure in the storage container. This requires precise temperature control and the implementation of cooling systems. Efficient insulation of storage tanks is also crucial to minimize the loss of hydrogen through evaporation. However, it is important to note that insulation is never perfect, and heat transfer from the environment to the tank is inevitable, increasing pressure inside the container. Given that liquid hydrogen is stored at low pressures, the pressure must be regulated to prevent overpressure, which can be achieved by venting hydrogen from the tank through a valve, a process known as "boil-off" [22]. Thus, a cryogenic storage vessel with effective insulation is necessary to prevent the loss of hydrogen through venting and extend the storage time without any loss. Despite the high level of insulation in cryogenic storage vessels, "boil-off" is an inevitable phenomenon, resulting in 2 to 3% loss of hydrogen per day. This presents a significant challenge for liquid hydrogen storage, as "boil-off" cannot be entirely prevented but only minimized. The loss of hydrogen through "boil-off" impacts energy efficiency and cost and poses a safety concern, particularly in confined spaces where a hydrogen leakage may occur [6,22,23,26].

Liquid hydrogen is stored in cryogenic tanks that are cylindrical or spherical double-walled vessels, where the inner and outer walls are separated by multiple layers of vacuum to provide thermal insulation, as shown in Figure 10.

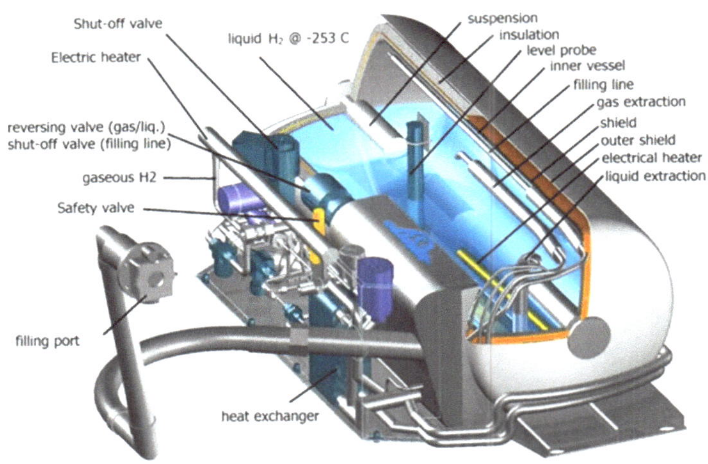

Figure 10. Scheme of the main components of a typical tank for liquid hydrogen storage [27].

The construction of these vessels involves using specific materials to minimize conductive and radiant energy flows from the environment into the tank, along with vacuum-insulated jackets to further reduce thermal losses. These materials must possess the ability to withstand the low temperatures of liquid hydrogen, prevent hydrogen gas permeation, and resist hydrogen embrittlement. By implementing these design features, modern large-scale cryogenic tanks can effectively mitigate the "boil-off" effect, reducing it to as low as 0.1% per day of the total hydrogen stored [21,23,26,27]. The "boil-off" rate of liquid hydrogen is influenced not only by the thermal insulation of the tank but also by the size and shape of the vessel. Larger tanks are generally preferred, requiring a smaller proportion of insulation mass and volume than smaller tanks. Tank shape is also an essential factor, as a reduced surface area of the liquid can minimize the heat flow from the surroundings into the liquid. Cylindrical or spherical shapes are typically preferred due to their lower surface-area-to-volume ratios [21,22].

Cryogenic storage operates at low working pressures, which presents opportunities to utilize novel materials that can reduce the tank's weight and improve other performance characteristics. This could result in tanks with similar specific energy storage quantities as conventional fuel tanks. In addition to these performance benefits, using new materials and design features can also improve environmental conditions and safety, while minimizing costs by selecting cost-effective materials and processes [28]. Regarding insulation, two main types are commonly employed. The first one is the vacuum-jacketed system. This method involves using multiple layers of a low-emissivity and high-reflectivity material separated by thin fiberglass sheets. The outer container is designed to maintain a low pressure, which results in excellent insulation. However, a main drawback of this method is that the system can fail if the vacuum is lost, resulting in a significant waste of liquid hydrogen through "boil-off". The second option is to apply rigid closed-cell foam insulation to the tank's exterior. While this method may exhibit a higher density and thermal conductivity, it can still be a viable choice depending on the specific application [29].

Liquid hydrogen storage systems are particularly appealing for flight and space applications and large-scale hydrogen transportation and delivery. This is due to their high volumetric and gravimetric energy storage densities, making them attractive to these sectors. The short-term utilization of hydrogen and the fact that high-power consumption or cost are not major concerns further increase the appeal of liquid hydrogen storage systems in mobility. Trailers and ships have already been employed for liquid hydrogen distribution, with ongoing efforts to enhance transport capacity.

Achieving successful liquid hydrogen storage relies on several factors, including improving insulation to minimize "boil-off," designing efficient cryogenic tanks, and optimizing the liquefaction process. However, a significant challenge today and in the near future is the lack of infrastructure for producing and storing liquid hydrogen. Significant developments in hydrogen infrastructure are necessary to realize a hydrogen-based mobility system. Additionally, managing cryogenic liquids requires implementing safety measures and technologies not widely used today [23].

3.1.3. Cryo-Compressed Hydrogen Storage

Cryo-compressed hydrogen storage has become an exciting option to take advantage of the main characteristics of compressed and liquid hydrogen storage principles. This method allows hydrogen to be stored at elevated pressures above ambient (1 bar) and at temperatures equivalent to or lower than its boiling point (−233 °C). Compressing hydrogen to 350 bar at −233 °C can increase its gravimetric and volumetric density from 70 g/L at 1 bar to 90 g/L, resulting in higher storage efficiency [6,26]. This approach results in a higher energy storage capacity per unit volume, mitigating issues such as the demand for high pressures and volumes in compressed hydrogen storage, as well as boil-off losses associated with liquid hydrogen storage.

As previously discussed, the on-board application of cryo-compressed hydrogen can be accomplished using Type III or Type IV vessels. However, due to the lower operating pressures of cryo-compressed hydrogen storage (typically 300 bar) compared to compressed hydrogen storage (700 bar), the requirement for more costly carbon fiber composites may be reduced, making Type III vessels a preferred option for this storage approach [21]. The vessel used for cryogenic hydrogen storage must be designed to withstand extreme conditions of low temperature and high pressure, ensuring effective containment of the cryogenic fluid.

This allows the storage system to be fueled with various hydrogen storage methods, including liquid hydrogen (LH_2), cold-compressed hydrogen, ambient-temperature compressed hydrogen, or cryo-compressed hydrogen. This versatility gives the cryo-compressed storage system a significant advantage over other storage techniques [22,23]. The Lawrence Livermore National Laboratory researchers have developed an illustrative model of a cryo-compressed storage tank, as depicted in Figure 11.

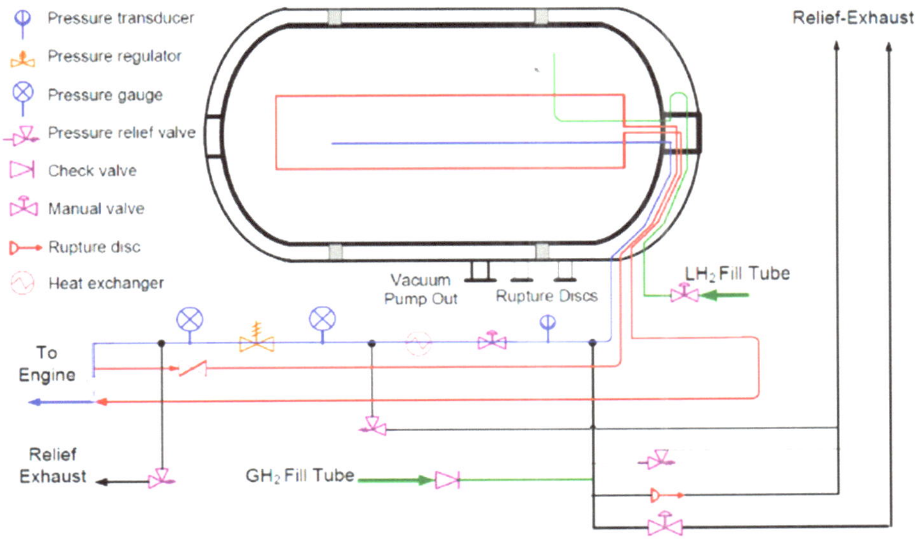

Figure 11. Type III cryo-compressed H_2 storage tank schematic representation [22].

The tank comprises an inner high-pressure vessel with a metallic liner serving as a barrier between the composite overwrap and the hydrogen, thereby preventing gas leakages. The metallic liner is then wrapped in a carbon-fiber-coated metal, creating a robust outer layer. Additionally, the tank comprises a vacuum space lined with multiple layers of highly reflective metalized plastic, which functions as high-performance thermal insulation. Finally, a metallic outer jacket is included to limit heat transfer between the stored hydrogen and the surrounding environment [20,30].

Ensuring safety during the storage and consumption of hydrogen is of paramount importance, particularly for on-board applications. While cryo-compressed hydrogen storage offers several safety advantages over liquid and compressed hydrogen storage techniques, the physical damage to the cryo-pressured vessel due to a vehicular accident remains a potential risk. Such an event could lead to the sudden expansion of cryo-compressed hydrogen to atmospheric pressure. The potential mechanical energy that can be released by a sudden expansion of high-pressure hydrogen gas has been analyzed and presented in Figure 12. The diagram illustrates the expansion energies of normal H_2 at 300 K, as well as the expansion energies of para-H_2 at temperatures of 60 K and 150 K.

Hydrogen can exist in two different forms based on the orientation of its nuclear spin: para-H_2 and ortho-H_2. At very low temperatures, around 20 K, hydrogen is primarily in the para-H_2 state, which is the low-energy form and stable under these conditions. However, as the temperature increases from 20 K to room temperature, the equilibrium shifts towards the ortho-H_2 state [31].

The results indicate that the maximum mechanical energy release occurs at almost ambient temperature (300 K) with a value of 0.6 kWh/kgH_2 at 300 bar, and this maximum release remains relatively constant with increasing pressure. In contrast, temperature strongly influences the theoretical maximum mechanical energy. Specifically, at a pressure of 300 bar, the maximum mechanical energy release is significantly reduced by almost 92% (0.05 kWh/kgH_2) at a temperature of 60 K. These findings suggest that cryo-compressed hydrogen storage may be a safer option, as the potential damage of a sudden rupture is reduced at lower temperatures [31].

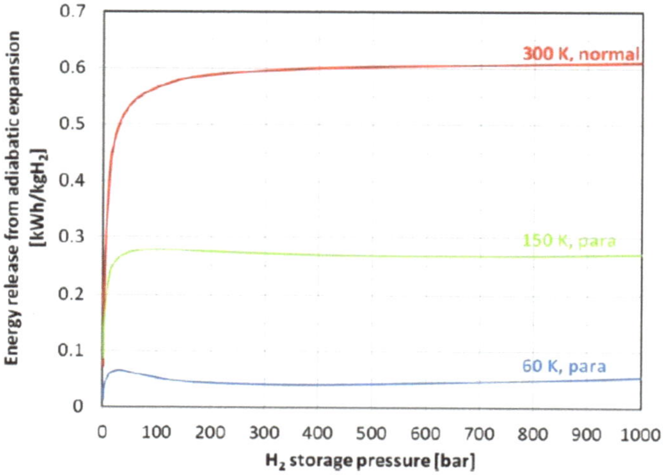

Figure 12. Maximum mechanical energy released upon instantaneous expansion of H_2 gas. [31]. Reprinted with permission from Elsevier.

The combination of the low burst energy and high hydrogen storage density at cryogenic temperatures presents a suitable solution for developing smaller tanks that can withstand automotive collisions. Furthermore, as previously noted, the cryo-compressed vessel is designed with various layers that offer an expansion volume that permits sudden release at a significantly reduced pressure. The low energy content of cold hydrogen, along with the effective dissipation power mitigation, can produce highly secure systems that cause minimal damage, even in the event of abrupt failure.

Cryo-compressed hydrogen storage technology can fill the storage tank with compressed, cryo-compressed, or liquid hydrogen, presenting several advantages over other hydrogen storage methods, including a greater storage capacity and enhanced safety indicators. Despite these benefits, cryo-compressed tanks are not yet commercially feasible due to the availability of infrastructure and the cost associated with this storage technique.

Table 4 comprehensively summarizes the various types of physical hydrogen storage methods for on-board vehicle applications. It presents key information such as the type of vessel used, operating pressure and temperature, application, as well as their respective advantages and disadvantages. This table offers a convenient overview of the different storage options, aiding in understanding and comparing the characteristics of compressed H_2 storage, liquid H_2 storage, and cryo-compressed H_2 storage for on-board vehicle applications.

3.1.4. Safety Standards for On-Board Applications

Compressed hydrogen storage for on-board vehicle applications combines robustness and safety advantages. Hydrogen tanks are engineered to withstand high pressures, undergo rigorous testing, and adhere to stringent safety standards, ensuring the system's integrity and durability. To make sure that FCEVs are safe, there are international standards that define certain requirements. These standards have specific criteria for hydrogen storage systems.

Table 4. Comparison of physical hydrogen storage methods for on-board vehicle applications.

Storage Method	Type of Vessel	Operating Pressure (bar)	Operating Temperature (K)	Application	Advantages	Disadvantages
Compressed Hydrogen Storage	Type III Type IV	700	<350	Small-scale storage	-Relatively low-cost and mature technology -Lightweight and compact	-Requires high-pressure tanks, leading to increased weight -Lower energy density compared to liquid hydrogen storage.
Liquid Hydrogen Storage	Type III	1–3.5	20–40	Large-scale storage	-High energy density -Can be stored at lower pressures compared to compressed hydrogen	-Cryogenic temperature requires specialized insulation -Requires significant space due to the need for insulation and larger storage tanks
Cryo-Compressed Hydrogen Storage	Type III Type IV	350	20–40	Large-scale storage	-Higher energy density compared to compressed hydrogen storage -Enables efficient storage with reduced losses	-Requires both cryogenic temperature and high-pressure vessels, increasing complexity and cost -Not commercially available

One of these is the GTR 13 standard. It establishes that the hydrogen permeation rate of the compressed hydrogen storage system should be less than 46 N cm^3 h^{-1} L^{-1} under specific operating conditions: the system should be tested at 1.15 times the nominal working pressure (NWP) and a temperature of 55 °C. Another safety standard that applies to the hydrogen storage tanks in the system is ISO 19881: 2018; Gaseous hydrogen—Land vehicle fuel containers; ISO®: Geneva, Switzerland. Available online: https://www.iso.org/standard/65029.html (accessed on 28 June 2023). According to ISO 19881, the steady-state permeability of Type IV hydrogen storage tanks should be less than 6 N cm^3 h^{-1} L^{-1} at the nominal working pressure (NWP) and the ambient temperature (293 K) [32]. Moreover, the lighter-than-air nature of hydrogen minimizes the risk of fire or explosion in the event of a leak, as it quickly disperses instead of pooling on the ground like gasoline. Advanced safety features like pressure relief devices and rupture discs further prevent over-pressurization and enable controlled hydrogen release during emergencies, enhancing overall safety.

For that reason, equipping FCEV tanks with safety valves and pressure relief devices is essential to prevent overpressurization. These devices release excess pressure in the event of an anomaly, preventing potential tank ruptures. Regular maintenance and inspection of these valves are crucial to ensure safety standards.

3.1.5. Underground Hydrogen Storage

Natural underground formations, including aquifers, depleted natural gas fields, and artificial caverns, such as salt caverns, present a potential solution for hydrogen storage. Aquifers are particularly attractive due to their water-bearing permeable rock or sand layers that can trap hydrogen injected at high pressure. In addition to aquifers, hydrogen can also be stored in the porous rock found in natural gas caverns [33]. These hydrocarbon reservoirs, located deep beneath the subsurface, are known for their porous and permeable nature because most recoverable products have already been extracted. Depleted hydrocarbon reservoirs have a history of success as gas storage options, as they are known for storing hydrocarbons, such as natural gas, and have well-established geological structures. Despite this, there is a potential risk to the purity of injected gas if the remaining gas in the reservoir is not properly managed, affecting the integrity of the stored hydrogen [34,35].

On the other hand, salt caverns offer secure and stable underground storage facilities for various materials, including oil, natural gas, and hydrogen. Formed by dissolving salt formations by injecting fresh water under high pressure, salt caverns are typically found in underground salt domes. The pressure inside a salt cavern is critical and needs to be carefully monitored as it is affected by the amount of gas stored within it. With appropriate management, salt caverns provide a reliable and safe way to store hydrogen underground over extended periods [35]. A typical salt cavern can be up to 2000 m deep, 1,000,000 m^3 in volume, 300 to 500 m in height, and 50 to 100 m in diameter, allowing for vast hydrogen storage [34].

These underground formations could provide a secure and cost-effective option for large-scale hydrogen storage, enabling its integration with intermittent renewable energy sources and decarbonizing various sectors, including transportation, heating, and power generation.

3.2. Chemical Storage

Currently, major automotive manufacturers prioritize on-board compressed hydrogen gas storage (at 700 bar) due to its fast-refueling capability (within <3 min) and its established technological maturity. However, to overcome the limitations of the physical storage methods, solid-state hydrogen storage could play an important role due to its potential advantage of offering higher volumetric densities. To achieve higher hydrogen storage density with compactness, researchers are currently exploring the potential of solid materials that can physically absorb or chemically react with the gas. Another relevant way to store hydrogen is using absorption or adsorption processes via two different mechanisms known as chemisorption and physisorption, which allow H_2 to bind to the surface of a specific material. Physisorption involves weak molecular hydrogen bonding to the surface material with weak Van der Waal forces. In contrast, chemisorption involves dissociating H_2 molecules into H atoms and their subsequent migration to the material to occupy the interstitial site to form new and strong chemical bonds [36].

Chemical hydrogen storage has been found to have some advantages over other methods. One of the most notable advantages is its better volumetric densities, which is crucial for on-board applications. However, chemical hydrogen storage poses challenges and drawbacks, such as slow kinetics, low gravimetric densities, low reversibility, and high dehydrogenation temperatures or pressures. Despite these limitations, studies and efforts are being made to improve different chemical hydrogen storage solutions for specific applications. One way to address reaction constraints is through the use of catalysts, which can enhance kinetics and thermodynamic stability for a metal hydride. It is important to note that, while catalysts can improve storage efficiency, they do not absorb hydrogen and can potentially reduce storage capacity by occupying space previously intended for hydrogen [37]. For material-based technologies to be suitable for on-board hydrogen storage, they must possess certain key characteristics. These may include high hydrogen storage capacity, rapid charging and discharging capabilities, good reversibility, stability over multiple hydrogen uptake and release cycles, and rapid kinetic properties.

Chemical hydrogen storage relies on a storage carrier that, under certain conditions of temperature and pressure, can absorb or react with hydrogen (hydrogenation) to form a stable compound that is stable under atmospheric conditions. Upon demand, the hydrogen can be released (dehydrogenation) by altering the pressure and temperature conditions. Various chemical methods are available for hydrogen storage, including metal hydrides such as intermetallic compounds, complex hydrides, chemical hydrides, liquid organic hydrogen carriers, and nanostructures. However, these storage techniques are currently in their early research stages and are not technically and economically feasible for automotive applications [37].

3.2.1. Metal Hydrides

Metal hydrides are being explored as a potential solution for reversible on-board hydrogen storage, as they can release hydrogen at the relatively low temperatures and pressures required for fuel cells. Hydrides are chemical compounds in which hydrogen combines with another element. Several metals can form metal hydrides through a reversible reaction with hydrogen or electrochemical reactions. The formation of metal hydrides upon reaction with H_2 can be expressed by Equation (4),

$$M + x/2\,H_2 \leftrightarrow MH_x + Q \tag{4}$$

where M represents the metal or an intermetallic compound, MH_x is the respective hydride formed in the reaction, Q the heat of the reaction, and x represents the ratio of hydrogen to metal, i.e., x = cH [H/M]. As a consequence of the reduction in entropy of the hydride relative to both the metal and gaseous hydrogen phase, the formation of hydrides is typically accompanied by the release of heat at ambient and elevated temperatures, while the reverse reaction involving the release of hydrogen is an endothermic process. Therefore, the release or desorption of hydrogen necessitates the provision of heat. Metals can be hydrogenated through either molecular hydrogen gas or hydrogen atoms obtained from an electrolyte. In the case of gas-phase loading, the process involves multiple stages of hydrogen reacting with the metal to form the hydride, which must be carefully considered [38]. Magnesium hydride (MgH_2) has been widely investigated as a potential hydrogen storage material owing to its high gravimetric storage capacity of 10.1 wt.%. However, its limited reversibility within a practical hydrogen pressure range makes it unsuitable for on-board hydrogen storage [6].

3.2.2. Complex Hydrides

The formation of complex hydrides occurs through the reaction of atomic hydrogen containing complex anions such as amides, alanates, and borohydrides with the metal cations such as group 1 or 2 elements. The release of hydrogen from these compounds occurs by decomposing the "host" into two or more components. Alanates, amides, and borohydrides are among the complex hydrides currently investigated as promising hydrogen storage materials.

Due to the utilization of lighter materials, complex hydrides exhibit higher gravimetric densities than intermetallic hydrides. However, these materials face drawbacks as they require high hydrogenation and dehydrogenation temperature (500 K–700 K) and are susceptible to decomposition [6]. Table 5 presents some of the parameters required for the hydrogenation and dehydrogenation process for some of the most common complex hydrides used in this storage technique.

Table 5. H_2 absorption/desorption properties of various complex hydrides [39]. Reprinted with permission from Elsevier.

Complex Hydride	Temperature (K)	Pressure (MPa)	Hydrogen Storage Capacity (wt.%)
$LiBH_4$ nanocomposite	T_{des}: 573	-	6.5
$LiBH_4 + LiBH_{3.75}F_{0.25}$	T_{des}: 373	0.1	9.6
$LiBH_4 + SiO_2$	T_{des}: 373	5.0	13.5
$NaAlH_4$ + 2.0 mol.% $Ti(OBu)_4$	T_{abs}: 423 T_{des}: 433	11.4	4.0
$NaAlH_4$ + 1.0 mol.% $TiCl_3$	T_{abs}: 323–383	-	5.6
$NaAlH_4$ + 1.0 mol.% Ti	T_{abs}: 443 T_{des}: 423	15.4	5.6
$NaAlH_4$ + 4.0 mol.% Ti	T_{des}: 373	-	4.8
$NaAlH_4$ $Ti(OBu^n)_4$ and $Fe(OEt)_2$	T_{des}: 374	8.8	4.0
$NaAlH_4$ + 4.0 mol.% Ti	T_{des}: 450	2.5	1.7
Na_3AlH_6 + 2.0 mol.% $TiCl_3$	T_{abs}: 473 T_{des}: 543	6.0	2.1
Na_3AlH_6 + 2.0 mol.% $Ti(OBu)_4$	T_{abs}: 443 T_{des}: 503	3.0	2.3
$NaAlH_4$ + porous carbon	T_{des}: 673	10.0	7.0
$NaAlH_4$ + non-porous carbon	T_{des}: 673	10.0	6.3

A common challenge with complex hydrides is the occurrence of multi-step dehydrogenation reactions that require different temperature and pressure conditions, adding complexity to on-board applications. Complex hydrides can achieve high energy densities with the appropriate doping, such as tetrahydroborates with SiO_2 reaching capacities up to 13.5 wt.%. However, the technology suffers from slow kinetics and high thermodynamic stability, limiting its practicality for most on-board hydrogen storage applications.

3.2.3. Liquid Organic Hydrogen Carriers (LOHCs)

LOHCs are utilized for hydrogen storage through a chemical reaction. These carriers are unsaturated organic compounds containing double or triple carbon bonds that undergo hydrogenation by binding with hydrogen. This exothermic reaction occurs at elevated pressure and temperature (30–50 bar and 150–200 °C, respectively), facilitated by a catalyst. Hydrogen storage is achieved through reversible hydrogenation and dehydrogenation of the carbon double bonds. Hydrogenation is an exothermic process that occurs at high pressure and temperature, whereas dehydrogenation is an endothermic process that takes place at atmospheric pressure. Both processes are catalyst-dependent, with catalysts playing a significant role in facilitating the reaction [6].

The hydrogen storage capacity of unsaturated organic compounds is typically around 6 wt.%. The molecular bonding of hydrogen in LOHC materials significantly increases their volumetric density, making their transportation and distribution similar to the actual crude oil. This allows for the utilization of existing fuel infrastructure without the need for significant modifications to refueling, transport, and storage facilities, which gives an advantage to the widespread adoption of this technology. Higher volumetric density LOHC materials are typically cost-effective and easily accessible. They generally possess non-toxic properties and low dehydrogenation temperatures; however, the hydrogenation and dehydrogenation of LOHCs require intensive energy, converting this option into non-suitable for on-board applications with the current technology.

4. Perspectives of Hydrogen Storage Technologies in EV Market

The preceding sections presented an overview of current hydrogen storage technologies. Based on the analysis, it was established that compressed hydrogen storage stands as the primary technology for commercial vehicles. Nevertheless, it is noteworthy to emphasize the significance of the progress in alternative hydrogen storage technologies, such as liquid or cryo-compressed hydrogen storage and material-based hydrogen storage, which are expected to play a crucial role in facilitating the transition towards a hydrogen economy.

For a more comprehensive evaluation of the maturity levels of the discussed technologies, the Energy Technology Perspectives (ETP) for Clean Energy Technology guide may serve as a valuable resource. This guide offers information on around 500 individual designs and components of diverse energy systems, providing insights into the technology readiness level (TRL), development and deployment plans, cost and performance improvement targets, and leading stakeholders for each technology. Table 5 presents the data obtained from this tool regarding hydrogen storage, displaying the TRL on the left-hand side through a scale ranging from 1 (representing an initial idea) to 11 (representing a fully developed technology) [40].

Table 6 illustrates that compressed hydrogen storage in pressure vessels has reached a high level of technological maturity (TRL 11). In contrast, liquid hydrogen storage is currently in the final stages of commercialization (TRL 8), indicating some opportunities for further development. Furthermore, hydrogen storage in metal hydrides is still in the prototyping phase (TRL 4), suggesting considerable potential for advancement in this area.

Table 6. H_2 storage technology readiness levels/ETP clean energy technology guide (adapted from [40]).

Technology Readiness Level (TRL)	Storage Technology	Importance for Net-Zero Emissions
11	above-ground physical storage/ammonia storage	High
11	above-ground physical storage/pressure vessel	High
9–10	underground geological storage/salt caverns	High
7–9	above-ground physical storage/liquid hydrogen storage tank	High
5	underground geological storage/lined hard rock cavern	Moderate
5	underground geological storage/fast-cycling salt cavern	High
4	underground geological storage/depleted gas fields	High
3–5	materials-based storage/metal hydrides	Moderate
3	underground geological storage/aquifer	High
2–3	materials-based storage/adsorbents	Moderate

Although compressed hydrogen storage is considered a fully developed technology, technical objectives still encourage further development for this storage technique in on-board applications. These targets were established by the U.S. DRIVE Partnership, a collaborative initiative comprising the DOE, the U.S. Council for Automotive Research (USCAR), energy and utility companies, and various organizations. Table 7 summarizes the achieved technical parameters for compressed hydrogen storage compared to the technical performance targets established by the DOE for hydrogen storage systems in light-duty vehicles.

Table 7. Comparison between the achieved technical parameters for compressed hydrogen storage (Type III and Type IV vessels) and the performance targets set by the U.S. DOE for hydrogen storage systems in light-duty vehicles [41].

Performance and Cost Metric	Units	Type III Vessel	Type IV Vessel	DOE Targets		
		350 bar	700 bar	2020 Target	2025 Target	Ultimate Target
System gravimetric capacity	wt.%	5.5	5.2	4.5	5.5	6.5
System volumetric capacity	g-H_2/L	17.6	26.3	30	40	50
Storage system cost	$/kWh	15.4	18.7	10	9	8
WTT efficiency (LHV)	%	56.5	54.2	60	60	60

To achieve DOE Technical Targets for hydrogen storage in light-duty vehicles, several key aspects need to be addressed. Firstly, advanced materials like carbon fiber composites or carbon nanotubes should be explored, aiming for high hydrogen storage capacities, fast kinetics, and long-term stability. Secondly, innovative tank designs should be developed, considering factors like shape, internal structure, and materials used, while leveraging advanced manufacturing techniques for enhanced strength and efficiency. Thirdly, optimizing system integration with other vehicle components ensures efficient hydrogen delivery, minimal energy losses, and the effective utilization of stored hydrogen. Additionally, improving hydrogen delivery efficiency and reducing system costs through cost-effective manufacturing methods and materials ensures durability and reliability.

The conventional method of hydrogen storage presents a significant obstacle to advancing hydrogen-based transportation systems. The primary challenge arises from the requirement for large-volume compressed gas storage tanks, which are impractical for systems with weight and space limitations. Using higher-volume tanks would increase costs and potentially lead to safety concerns. Thus, developing alternative storage methods that allow for high-density hydrogen storage while addressing these concerns is crucial for the widespread adoption of hydrogen as a transportation fuel. Research efforts focus on exploring new materials and technologies that enable more efficient and cost-effective hydrogen storage systems. To achieve a minimum range of 500 km with readily accessible refueling capacity, it would be beneficial for fuel cell vehicles to overcome these challenges. Although certain light-duty fuel cell electric vehicles with compressed gas on-board storage have been introduced in the market, developing such systems for light-duty vehicles remains a formidable task [41].

5. Data Analysis

5.1. Type IV Vessel Manufacturing Costs

Type IV pressure vessels for hydrogen storage in FCEVs have a plastic liner overwrapped with a high-strength carbon fiber composite material. The use of carbon fiber composites provides Type IV pressure vessels with significantly lower weight than metal pressure vessels would have, making them a desirable option for FCEVs manufacturers. However, the high cost of the carbon fiber composite material is a significant drawback. The cost of high-strength carbon fiber comes primarily from the cost of the precursor fiber and the conversion of this fiber to carbon fiber. To reduce the cost of high-strength carbon fiber, researchers focus on developing lower-cost precursor fibers and considering cost-reduction strategies for conversion processes. Additionally, efforts are ongoing to develop alternative lower-cost and high-performance fiber and resin materials, composite additives for improved performance, and alternative pressure vessel manufacturing processes. Identifying alternative materials for balance-of-plant components is another important aspect of reducing the cost of compressed hydrogen storage systems. Overall, the continued research and development of cost-effective and high-performance Type IV pressure vessels and related materials are necessary to widely adopt FCEVs as a viable transportation alternative [18].

Design for Manufacture and Assembly (DFMA) is a commonly used methodology for cost analysis in compressed hydrogen storage systems. The DFMA process involves modeling specific manufacturing steps to project the material and manufacturing costs of the complete system. The first step in performing cost analysis is defining the system design and the specific manufacturing and assembly steps required to achieve the specified design. Once these steps are specified, the material inputs, capital equipment costs, and operating expenses are computed, and the estimated cost is the sum of the material, manufacturing, and assembly costs. This approach allows for a comprehensive understanding of the costs associated with the manufacturing and assembly of compressed hydrogen storage systems, and it can inform decisions related to optimizing the design and manufacturing processes for cost efficiency. This methodology is not only based on the raw material and manufacturing costs of the system; it also includes income tax rates, average labor rates, and electricity utility costs that can influence the final cost of the storage systems [42]. This section presents and discusses a comprehensive analysis of a Type IV 700 bar hydrogen storage vessel costs made by Strategic Analysis Inc.

Between 2013 and 2015, several improvements in the design of Type IV pressure vessels for hydrogen storage were implemented to reduce manufacturing costs. These changes led to a 12% reduction in the overall system cost. One key area of focus was the materials used to manufacture the vessels. Researchers replaced the higher-density epoxy resin with a lower-cost and lower-density resin. Additionally, they used a lower-cost carbon fiber made from polyacrylonitrile with a methyl acrylate co-monomer precursor based on a high-volume textile fiber process. Another important design change during this period was the integration of balance-of-plant (BOP) components. This integration enabled more efficient hydrogen storage and delivery, reducing the overall system cost and improving system performance. In 2015, two additional improvements were identified that significantly reduced the projected cost of compressed hydrogen storage systems. Firstly, a switch to aluminum for selected BOP components was proposed. Aluminum is a lightweight, high-strength material that is readily available and can be cost-effective to manufacture. Secondly, incorporating Toyota's design and manufacturing improvements was crucial in reducing the cost of compressed hydrogen storage systems. Toyota proposed an alternative liner geometry that eliminates high-angle helical winding and an alternate winding scheme. Additionally, replacing T-700 carbon fiber with higher modulus T-720 carbon fiber further improved the design. These improvements resulted in a 3.2% reduction in the overall system cost [42,43].

According to the DOE, there is a promising outlook for cost reductions in hydrogen storage systems. The DOE predicts that the cost of hydrogen storage systems will decrease from 22.94 \$/kWh for 10,000 systems per year to 14.07 \$/kWh for 500,000 systems per year. This represents a significant cost reduction and reflects the ongoing efforts to identify and implement design and material improvements for compressed hydrogen storage systems. As the production volume of these systems increases, economies of scale will likely drive down manufacturing costs, making hydrogen a more attractive alternative fuel for transportation. Furthermore, the DOE's predictions highlight the potential for continued technological advancements, which will further drive down costs and improve overall system efficiency. Figure 13 illustrates the cost distribution among the various components involved in the manufacturing process of a Type IV vessel tank. To ensure an accurate comparison, all costs were adjusted to reflect the 2019 value while considering the cost variations for crucial components, such as carbon fiber [43].

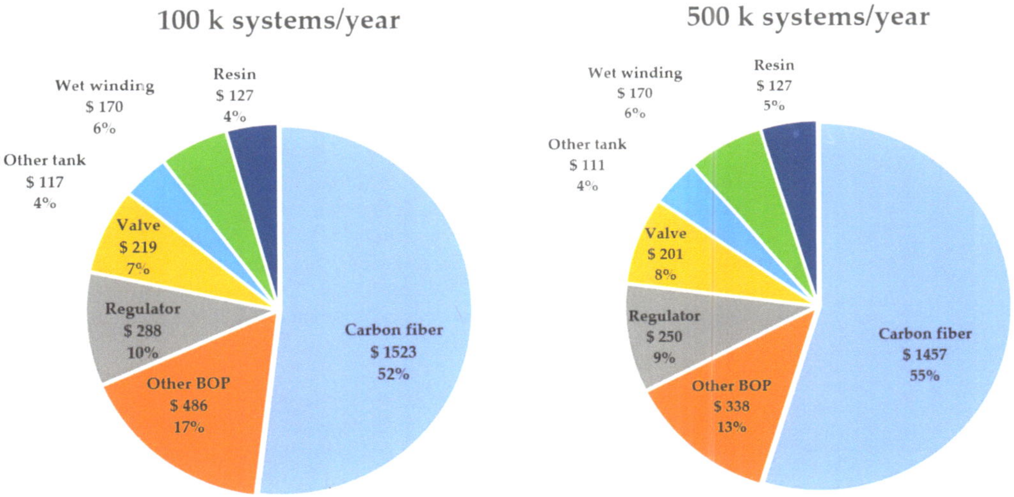

Figure 13. Hydrogen storage system costs by percentage of the total cost (annual production of 100 k units shown on the left and 500 k units shown on the right) (adapted from [43]).

Figure 13 highlights that in both investigated scenarios, the BOP expenses constitute approximately 35% of the total system cost, while the cost of carbon fiber dominates the budget, accounting for nearly 53%. The remaining percentage is distributed among the resin component and the wet winding process. These findings underscore the crucial role of carbon fiber in the cost structure of the studied system, suggesting the importance of its optimization and cost reduction in enhancing the economic feasibility of the overall system. Moreover, the identified cost distribution patterns can serve as a valuable guide for future cost analysis and optimization efforts.

Current Situation and Future Goals

Referring to the most recent report (2019) released by the DOE's Hydrogen and Fuel Cells Program, the current price of the system is estimated to be 15.7 \$/kWh, with the presented value denoted in 2016 dollars to facilitate meaningful comparisons across time frames, while factoring out the effects of inflation. The program aims to further reduce the cost while increasing the system's capacity. The ultimate target set by the program is to achieve a price of 8 \$/kWh, representing a significant reduction of less than 50% of the 2019 cost. These objectives highlight the critical role of continuous research and development efforts to enhance the economic feasibility of fuel cell technologies and lead the way for their widespread adoption in various sectors [43].

5.2. *Available Brands for FCEVs*

The FCEV market has recently witnessed significant growth as companies continue to produce and test different fuel cell systems. This section lists the most promising vehicles and their specifications, which are crucial to evaluating their potential. Some of these models are still in the prototype stage and have not been commercialized yet. The Environmental Protection Agency (EPA) is the organization that is responsible for determining the range of an electric car, truck, or SUV through a series of tests on a dynamometer, which are designed to simulate real-world driving conditions (EPA-estimated range).

5.2.1. Toyota Mirai

The 2021 Toyota Mirai (SEDAN) features an estimated range of 647 km, certified by the EPA. The vehicle boasts 182 horsepower (hp) and is powered by a 114 kW fuel cell stack. The hydrogen fuel is stored in three high-pressure carbon-fiber-reinforced polymer tanks at 700 bar, with a combined capacity of approximately 5.6 kg. The Mirai also has a fast charging time of 5 min. The vehicle uses a solid-polymer electrolyte fuel cell with a maximum speed of 300 km h^{-1}, accelerating from 0 to 100 in 9.2 s. The starting price for the vehicle is $50,000 in the United States [44,45].

5.2.2. Hyundai Nexo

The Hyundai Nexo 2022 (SUV) by Hyundai has an EPA-estimated range of 667 km and is equipped with a 161 hp motor. It is powered by a 95 kW fuel cell stack and uses compressed hydrogen gas at 700 bar, stored in 3 tanks that hold 2.1 kg of H_2 each. The Hyundai NEXO also has a fast refueling time of about 5 min. The starting price for the vehicle is $60,135. The Hyundai Nexo employs a PEMFC similar to the Toyota Mirai. The vehicle's maximum speed is 178 km/h, and it can accelerate from 0 to 100 in 8.4 s [46].

5.2.3. Honda Clarity

The Honda Clarity 2022 (SEDAN) by Honda is equipped with a PEMFC that uses compressed hydrogen gas at 700 bar as fuel stored in 2 tanks that can hold approximately 5.4 kg of hydrogen. It has an EPA-estimated range of up to 580 km and generates 174 hp with a 103kW fuel cell stack. The vehicle also boasts a quick charging time of 3 min. The Honda Clarity is priced at $59,000 and has a maximum speed of 165 km/h, with an acceleration from 0 to 100 in 8.1 s [47–49].

5.2.4. Gumpert Nathalie

The Gumpert Nathalie 2021 is an FCEV developed by Gumpert AIWAYS, powered by methanol rather than the conventional compressed hydrogen fuel. This FCEV stands out from other cars of its kind with its impressive 540 hp generated by its four electric motors, one for each wheel, making it a high-end performance vehicle, or supercar. The EPA estimates Nathalie's range to be up to 1200 km in Eco mode, with a 15 kW fuel cell and 3–5 min charging time. Methanol is stored in a 65L tank, and while energy is produced through a chemical reaction, the end products are water and CO_2, which can be considered a drawback of this fuel type. The Nathalie is priced at $460,000, with only 500 units slated for production. The vehicle is equipped with a direct methanol fuel cell and can reach a maximum speed of 300 km h^{-1}, with acceleration from 0 to 100 taking only 2.5 s [50,51].

5.2.5. Changan DEEPAL SL03

The Changan DEEPAL SL03 is a FCEV with an EPA-estimated range of 730 km. It is equipped with a 218 hp motor and a 160 kW fuel cell stack. The vehicle is capable of fast charging in as little as 3 min and is powered by compressed hydrogen stored in tanks capable of holding up to 5 kg of gas. The starting price for this vehicle is $100,000. However, insufficient details are available regarding the quantity and pressure of the hydrogen fuel, the number of tanks, the speed capabilities, and the specific fuel cell technology used [52,53].

5.2.6. BMW iX5 Hydrogen

The BMW iX5 Hydrogen 2022 (SUV) by BMW is estimated by the EPA to have a range of at least 482 km, totaling 374 hp. The fuel cell stack has a capacity of 125 kW, and the vehicle can be charged in 3–4 min. Compressed hydrogen gas at 700 bar is used as fuel and stored in 2 tanks with a total capacity of 6 kg of hydrogen. The starting price is $70,000, and only 100 vehicles were produced. The maximum speed is 210 km h^{-1}, and the acceleration from 0 to 100 takes 5.3 s. Although the type of fuel cell is not specified, it is expected to use a PEMFC due to the use of compressed hydrogen gas at 700 bar [54,55].

5.2.7. Nissan e-NV200 SOFC e-BIO

The Nissan e-NV200 SOFC e-BIO is a bioethanol-powered fuel cell electric vehicle developed by Nissan. It employs a solid oxide fuel cell that oxidizes bioethanol to generate electricity. However, due to the high operating temperatures of the solid oxide fuel cell, a sophisticated cooling system is required (details of which are currently unavailable). The vehicle has an estimated range of at least 600 km and is equipped with a 30 L ethanol tank. As the technology is still in the testing phase, no further information regarding the FCEV is available [56,57].

Table 8 provides a comprehensive overview of the relevant details of the FCEVs currently available in the market. While some of these models are not yet available for purchase by the general public, an evaluation of their specifications can still be conducted. In accordance with the findings of the previous analysis, compressed hydrogen remains the predominant fuel option due to the inherent difficulties associated with other fuel alternatives. The PEMFC is the preferred choice for powering FCEVs. However, alternative fuels such as methanol and bioethanol are currently undergoing testing. FCEVs using these fuels employ different fuel cell technologies, namely direct methanol fuel cell and solid oxide fuel cell, respectively.

Table 8. Specifications of commercially available FCEV models.

Model	Technology/Fuel	Autonomy (km)	Power (HP)	Fuel Tank Capacity (kg)	Max Speed (km/h)	Price ($)	Country of Origin	Fuel Cell Type	Fuel Cell Power (kW)	Source
Toyota Mirai	Compressed/ Hydrogen	647	182	5.6	300	50,000	Japan	PEMFC	114	[44,45]
Hyundai NEXO	Compressed/ Hydrogen	667	161	6.3	178	60,135	South Korea	PEMFC	95	[46]
Honda Clarity	Compressed/ Hydrogen	580	174	5.4	165	59,000	Japan	PEMFC	100	[47–49]
Gumpert Nathalie	Methanol	600–1200	540	65 L	300	460,000	Germany, China	DMFC	130	[50,51]
Changan DEEPAL SLo3	Compressed/ Hydrogen	730	218	5	N/S	100,000	China	N/A	N/A	[52,53]
BMW iX5 Hydrogen	Compressed/ Hydrogen	482	374	6	210	70,000	Germany	PEMFC	N/A	[54,55]
Nissan e-NV200 SOFC e-BIO	Bio-Ethanol	600	N/S	30 L	N/A	N/A	U.S., Brazil	SOFC	N/A	[56,57]

The charging time, although critical, is not included in Table 7 as it is generally consistent for fuel cell vehicles, typically requiring approximately 3 to 5 min. The range of fuel cell cars is typically around 600 km, with 2 to 3 tanks storing 5 to 6 kg of hydrogen, depending on the model. As such, the range of fuel cell cars is generally considered acceptable, and not a major concern. Methanol and bio-ethanol fuel cell vehicles are still undergoing testing, and their impact on the fuel cell electric vehicle market will be of interest as these technologies evolve.

5.3. Hydrogen Storage Market for On-Board Applications

Hydrogen storage has emerged as a prominent solution for powering fuel cells in various transportation applications. All commercially available hydrogen vehicles are equipped with a 700 bar storage tank that facilitates hydrogen supply into the fuel cell stack, powering the vehicle. To fulfill the minimum driving range requirements, it is necessary to have an on-board hydrogen storage capacity of 5–13 kg of hydrogen. Automotive manufacturers typically incorporate two or three hydrogen storage tanks into their fuel cell vehicles, which are situated between the front and rear suspension. These tanks must meet stringent safety standards as they are pressurized up to 875 bar. In 2014, an estimated 4000 hydrogen-powered fuel cell vehicles were in operation, projected to increase to 2,000,000 annually by 2023 [58,59].

The hydrogen storage tanks market has attracted significant investment from key players focused on developing composite pressure vessels for hydrogen-powered vehicles. Among the major manufacturers of storage tanks for transportation are Iljin Hy Solutions (Jeonju-si, Republic of Korea), Hexagon Composites (Ålesund, Norway), Quantum Fuel Systems (Ventura, CA, USA), BMW (Munich, Germany), Luxfer Group (Manchester, UK), NPROXX (Heerlen, Netherlands), Worthington Industries Inc (Columbus, OH, USA), and Faurecia (Nanterre, France) [58,59]. Table 9 presents technical sizing and operational data obtained from prominent vessel manufacturers. Many of these storage tank manufacturers participate in strategic development programs in collaboration with national governments and principal fuel cell producers to expedite the implementation of hydrogen technology in the automotive industry.

Table 9. Technical information of manufacturers of Type IV tanks for on-board applications.

	Iljin Hy Solutions [60]	Hexagon Purus [61]	Worthington Industries Inc. [62]	Quantum Fuel Systems [63]	Faurecia [64]
Country of origin	South Korea	Norway	U.S.	U.S.	France
H_2 capacity (kg)	2.1	3.1	200–411 L	3.1	2.75
Max length (mm)	900	1050	N/A	721	1110
Max diameter (mm)	350	440	N/A	504	350
Service pressure (bar)	700	700	700	700	700
Container temperature (°C)	N/A	N/A	80	−40 to +82	80
Tank weight (kg)	N/A	59	N/A	51	37.5
Further information	Hyundai NEXO tank supplier. More than 70,000 tanks have been sold by the end of 2021	BMW, Robert Bosch GmbH, TesTneT Engineering, and Hexagon Purus partnership to develop innovative hydrogen storage systems for future FCEVs	Involved in the innovation of the latest technology integration within the hydrogen storage tank requirements in the market	Development programs with GM, Honda, Toyota, Suzuki, BMW, and DOE	Selected to partner on the hydrogen fuel cell to equip a Bavarian fleet of heavy-duty trucks provided by MAN

5.4. ICEV, BEV, and FCEV Comparison

Internal combustion engines are complex systems comprising several interrelated components, including cylinders, pistons, fuel injectors, and spark plugs. During combustion, fuel (typically gasoline or diesel) is mixed with air and ignited by the spark plug. This reaction generates thermal energy, creating mechanical energy to drive the engine's pistons. As the pistons move, they turn a crankshaft, which powers the car's wheels. The exhaust gases produced during this process are released through the cylinders and expelled through the exhaust system. The efficiency of this process is affected by several factors, including the fuel type and quality, combustion chamber design, and engine operating conditions.

Unlike those with internal combustion engines, fully electric vehicles rely solely on the energy stored in the battery to propel the wheels. They utilize a single-speed transmission system to optimize the electric motor's output power. In contrast to conventional vehicles that require refueling at a gas station once the fuel supply is depleted, electric car drivers can recharge their battery by connecting to an external electricity source, such as a charging station or a household outlet. This reliance on electricity as the sole energy source offers the potential to reduce carbon emissions and dependence on fossil fuels in the transportation sector.

FCEVs and BEVs have demonstrated their competitive potential with ICEVs and are expected to become increasingly prevalent in the market. This is largely due to their desirable characteristics, including higher energy efficiency, reduced maintenance requirements, faster acceleration, and emission-free and noiseless operation. Various studies have demonstrated the superior energy efficiency of electric vehicles over ICEVs, with FCEVs and BEVs showing significant advantages. Additionally, the reduced need for regular maintenance in electric vehicles, particularly in the case of FCEVs, can result in lower operating costs for users. Furthermore, the quiet and emission-free operation of electric vehicles offers significant environmental benefits, particularly in urban areas where air pollution is a pressing issue.

Comparative performance and cost data for BEVs, ICEVs, and FCEVs technologies are available and can be found in Table 10. These metrics are essential for making informed decisions regarding vehicle purchases and can aid in adopting electric vehicles as a viable alternative to traditional ICEVs.

Table 10. Information overview of internal combustion engine vehicles (ICEVs), battery electric vehicles (BEVs), and fuel cell electric vehicles (FCEVs).

	Internal Combustion Engine Vehicle	Battery Electric Vehicle	Fuel Cell Electric Vehicle	Source
Overall efficiency (%)	19	75	30	[65]
Technology cost ($/kW)	23.1	130	53	[65]
Refueling cost ($/kWh)	1.24	0.14–0.30	0.24	[65]
Fuel consumption (at 100 km)	6.85 L	17.3 kWh	1.05 kg	[66]
Range Autonomy (km)	650	200–400	up to 600	[65]
Operating hours (h)	10,000	7900	3000	[67]
Costs per hour of Operation ($/h)	0.58	3.91	2.62	[67,68]
TTW CO_2 Emissions	203 (kg CO_2/5 kg H_2)	101 (kg CO_2/5 kg H_2)	0 (kg CO_2/5 kg H_2)	

Some of the data compiled in Table 10 have been reported by Dash et al. [65] and Yang et al. [66]. Although remarkably important, these data lack specific details on the vehicle model and the methodology used to derive the efficiency values. While Dash et al. [65] provide an overall efficiency for each vehicle type, Yang et al. [66] do not mention the source or method of obtaining their values. Additionally, the vehicle models are not specified in these works. However, the fuel consumption data is based on a simulation using the New European Driving Cycle (NEDC), accounting for the energy consumption of auxiliaries and a 10% penalty for aggressive driving patterns [62]. It is worth noting that the fuel consumption data for battery electric vehicles (BEVs) does not include the 10% losses during charging. Concerning the fuel consumption for the gasoline-powered ICEVs, it refers to the gasoline consumption of an average of the Mercedes A-170 and the Mercedes CDI A-160.

Furthermore, the operating hours in Table 10 are not accompanied by details regarding the source, methodology, or vehicle models used [67]. However, referring to the sources cited in the work by Handwerker et al. [67], it can be found that the operating hours of internal combustion engine vehicles (ICEVs) were attributed to a BMW, while the BEV operating hours were related to a Tesla (without specifying a particular model). As for the fuel cell electric vehicle (FCEV), no specific vehicle reference was provided, only the utilization of a PEMFC. Finally, regarding the TTW (tank-to-wheel) CO_2 emissions [68], the ICEV pertains to a gasoline car (model unspecified), the FCEV refers to the Toyota Mirai, and the BEV refers to the Tesla Model 3.

As shown in Table 9, both BEVs and FCEVs demonstrate markedly higher overall efficiency when compared to ICEVs. However, despite the superior performance of the former, ICEVs remain the most widely used technology due to their relatively lower cost. Nonetheless, the overall cost of BEVs and FCEVs technologies is expected to decrease in the coming years, as these vehicles become increasingly attractive to consumers and investors alike. This trend is expected to be driven by various factors, including advancements in battery and fuel cell technologies, improvements in manufacturing processes, and economies of scale resulting from increased adoption. Furthermore, adopting electric vehicles will likely have significant environmental benefits, including reduced GHG emissions and improved air quality. Developing and promoting BEVs and FCEVs technologies are crucial in achieving sustainable transportation systems and mitigating the negative impacts of fossil fuel use.

While BEVs may initially appear to have lower CO_2 emissions than their gasoline-powered counterparts, their overall lifecycle emissions can be significantly influenced by various factors, including manufacturing, mining, driving, disposal, and recycling. To accurately evaluate the true environmental impact of BEVs, it is, therefore, necessary to consider their complete lifecycle [69]. Research conducted by Volkswagen comparing the lifetime CO_2 emissions of battery-powered and diesel-powered vehicles found that the battery-powered car only becomes more environmentally friendly than its diesel counterpart after driving 100,000 km. Thus, it is crucial to consider the entire lifecycle of electric vehicles when assessing their environmental impact [67].

On the other hand, FCEVs have shown significant potential for decarbonizing the transportation sector, as the only byproduct of their operation is water. However, the widespread adoption of FCEVs depends on the availability of green hydrogen, produced by water electrolysis using electricity from renewable sources. Most (95%) hydrogen production relies on steam reforming, producing carbon emissions. Therefore, renewable electricity sources must be used for hydrogen production to achieve a 100% green hydrogen industry for the transportation sector. This highlights the critical role of renewable energy in decarbonizing the transportation sector and the importance of developing sustainable energy systems to support the widespread adoption of low-emission vehicles.

6. Conclusions

The transport sector is responsible for a significant portion of the total final energy consumption. Specifically, road transport for passengers accounts for the largest share, at 45.1% of the sector's energy consumption. Using hydrogen as a fuel in the transport sector can offer a promising solution for reducing GHG emissions. Hydrogen has several advantages, such as abundance and high energy density, making it a promising alternative to power vehicles rather than conventional fossil fuels. There are different types of fuel cells, but PEMFCs are the most commonly used for on-board applications due to their high efficiency and ease of operation under typical vehicle operating conditions.

Hydrogen has a low volumetric density at ambient conditions, requiring storage systems to increase its capacity. Physical and chemical storage are two approaches used for hydrogen storage. Physical storage options include compressed, liquid, and cryogenic-compressed hydrogen. Compressed hydrogen is the most commonly used for on-board applications, owing to its better conversion efficiency and practicality. On the other hand, chemical storage has limited applicability for on-board hydrogen storage due to practical limitations such as low density, high cost, or difficulties in the reversibility of the chemical reactions. Efficient and cost-effective hydrogen storage remains a critical challenge that requires further research and development to make the use of hydrogen as a fuel in various applications more viable and widespread.

Numerous car manufacturers are strongly interested in compressed hydrogen storage systems for FCEVs. The safety of such systems is an essential consideration, and current technologies have achieved satisfactory standards. Going forward, the main challenge will be to enhance their cost-effectiveness and practicality. The cost of FCEVs storage systems has declined significantly, with current estimates at around 15.7 $/kWh. However, with sustained progress and the expansion of system production, this cost could potentially decrease to 8 $/kWh. Notably, FCEVs vehicles are now less expensive to produce and operate than BEVs, boasting greater range autonomy, albeit with lower overall efficiency.

A critical obstacle to achieving a rapid transition towards hydrogen mobility is reducing the costs associated with fuel cells and storage systems. Significant research and development efforts are required to optimize the performance and efficiency of these components, as well as to identify new, more cost-effective materials and manufacturing processes. Improving the durability and reliability of fuel cells and storage systems will also be key to reducing costs, as this will help to extend their service life and reduce maintenance and replacement expenses.

The transition to a hydrogen fuel cell vehicle industry carries wide-ranging implications beyond the mere feasibility and cost of the vehicles. It is essential to conduct a comprehensive evaluation of the economic and environmental impacts of this shift. This includes assessing the potential benefits of reduced GHG emissions, improved air quality, and reduced dependence on fossil fuels. At the same time, it is also important to consider the potential challenges associated with developing and deploying the supporting infrastructure required to enable the widespread adoption of hydrogen fuel cell vehicles, such as refueling stations and storage facilities.

Overall, this study has effectively presented an exhaustive overview of hydrogen storage technologies, highlighting their current state of the art. A thorough analysis of vessel manufacturing costs and market alternatives has also been conducted. With the help of this study, it is evident that addressing potential issues with hydrogen storage technology will be critical for facilitating the integration of the hydrogen economy into the transportation sector. These challenges include increasing the hydrogen storage capacity, upgrading safety protocols, and developing cost-effective solutions. By overcoming these obstacles, a reliable and efficient hydrogen-powered transportation infrastructure can be established.

7. Future Outlook

Hydrogen fuel is a promising pathway to address the energy and environmental challenges facing the world today. As such, hydrogen storage has become increasingly critical, not only for developing cleaner and more sustainable technologies, but also for fostering stronger markets, creating new job opportunities, and driving the growth and development of communities worldwide. While compressed hydrogen storage remains the primary method of storing and transporting hydrogen, recent research has identified metal hydrides as a promising alternative that could offer several advantages over compressed storage. Metal hydrides have the potential to address issues such as low energy density and safety concerns, making them an exciting prospect for the future of hydrogen storage. However, it should be noted that metal hydrides are still in their early stages of development, and further research is needed to optimize their performance and commercial viability. With continued investment and research, metal hydrides could play a key role in realizing the full potential of hydrogen fuel and contributing to a more sustainable future.

As the world faces significant challenges related to climate change and the need for sustainable energy sources, the discussion around hydrogen technologies has gained considerable attention. The decision-making process regarding the future of hydrogen technologies carries substantial weight, as it will impact the planet and future generations. While some may view the current discussions as overly built-up, they might represent a hopeful path toward a sustainable future. It is important to recognize the potential of hydrogen technologies and consider all aspects of their implementation, from technological advancements to economic and societal impacts. Therefore, it is critical for governments and decision makers to carefully evaluate and consider the findings and recommendations of scientific research, such as the work we have presented in this study, to make informed decisions about the future of hydrogen technologies. Students, researchers, engineers, and scientists must continue working on sustainability and clean energy topics to implement new solutions and alleviate our global dependency on fossil fuels.

Author Contributions: Conceptualization, J.A.G.; methodology, J.A.G. and D.M.F.S.; investigation, J.A.G.; writing—original draft preparation, J.A.G.; writing—review and editing, D.M.F.S.; supervision, D.M.F.S. All authors have read and agreed to the published version of the manuscript.

Funding: Fundação para a Ciência e a Tecnologia (FCT, Portugal) is acknowledged for funding a research contract in the scope of programmatic funding UIDP/04540/2020 (D.M.F. Santos).

Data Availability Statement: The data presented in this study are available on request from the corresponding author.

Conflicts of Interest: The authors declare no conflict of interest.

References

1. IEA World Energy Balances 2020: Overview. Available online: https://iea.blob.core.windows.net/assets/23f096ab-5872-4eb0-91c4-418625c2c9d7/World_Energy_Balances_Overview_2020_edition.pdf (accessed on 30 April 2023).
2. Global Carbon Dioxide Emissions from 1970 to 2021, by Sector. Available online: https://www.statista.com/statistics/276480/world-carbon-dioxide-emissions-by-sector/ (accessed on 30 April 2023).
3. IEA World Energy Outlook 2021. Available online: https://iea.blob.core.windows.net/assets/4ed140c1-c3f3-4fd9-acae-789a4e14a23c/WorldEnergyOutlook2021.pdf (accessed on 28 April 2023).
4. Hannah Ritchie Cars, Planes, Trains: Where Do CO2 Emissions from Transport Come From? Available online: https://ourworldindata.org/co2-emissions-from-transport (accessed on 30 April 2023).
5. EIA International Energy Outlook 2016. Available online: https://www.eia.gov/outlooks/ieo/pdf/0484(2016).pdf (accessed on 30 April 2023).
6. Sapre, S.; Pareek, K.; Rohan, R. *Compressed Hydrogen in Fuel Cell Vehicles: On-Board Storage and Refueling Analysis*, 1st ed.; CRC Press: Boca Raton, FL, USA, 2022; ISBN 978-1-00-324431-8.
7. Whiston, M.M.; Lima Azevedo, I.M.; Litster, S.; Samaras, C.; Whitefoot, K.S.; Whitacre, J.F. Hydrogen Storage for Fuel Cell Electric Vehicles: Expert Elicitation and a Levelized Cost of Driving Model. *Environ. Sci. Technol.* **2021**, *55*, 553–562. [CrossRef] [PubMed]
8. Tanç, B.; Arat, H.T.; Baltacıoğlu, E.; Aydın, K. Overview of the next Quarter Century Vision of Hydrogen Fuel Cell Electric Vehicles. *Int. J. Hydrogen Energy* **2019**, *44*, 10120–10128. [CrossRef]
9. Wilberforce, T.; El-Hassan, Z.; Khatib, F.N.; Al Makky, A.; Baroutaji, A.; Carton, J.G.; Olabi, A.G. Developments of Electric Cars and Fuel Cell Hydrogen Electric Cars. *Int. J. Hydrogen Energy* **2017**, *42*, 25695–25734. [CrossRef]
10. U.S. Department of Energy Alternative Fuels Data Center: Fuel Cell Electric Vehicles. Available online: https://afdc.energy.gov/vehicles/fuel_cell.html (accessed on 30 April 2023).
11. Pollet, B.G.; Staffell, I.; Shang, J.L.; Molkov, V. 22- Fuel-Cell (Hydrogen) Electric Hybrid Vehicles. In *Alternative Fuels and Advanced Vehicle Technologies for Improved Environmental Performance*; Elsevier: Amsterdam, The Netherlands, 2014; pp. 685–735; ISBN 978-0-85709-522-0.
12. Bernay, C.; Marchand, M.; Cassir, M. Prospects of Different Fuel Cell Technologies for Vehicle Applications. *J. Power Sources* **2002**, *108*, 139–152. [CrossRef]
13. İnci, M.; Büyük, M.; Demir, M.H.; İlbey, G. A Review and Research on Fuel Cell Electric Vehicles: Topologies, Power Electronic Converters, Energy Management Methods, Technical Challenges, Marketing and Future Aspects. *Renew. Sustain. Energy Rev.* **2021**, *137*, 110648. [CrossRef]
14. O'hayre, R.; Cha, S.W.; Colella, W.; Prinz, F.B. Chapter 1: Introduction. In *Fuel Cell Fundamentals*; John Wiley & Sons, Inc.: Hoboken, NJ, USA, 2016; pp. 1–24. ISBN 978-1-119-19176-6.
15. Weber, A.Z.; Lipman, T.E. Fuel Cells and Hydrogen Production: Introduction. In *Fuel Cells and Hydrogen Production: A Volume in the Encyclopedia of Sustainability Science and Technology*, 2nd ed.; Lipman, T.E., Weber, A.Z., Eds.; Encyclopedia of Sustainability Science and Technology Series; Springer: New York, NY, USA, 2019; pp. 1–8. ISBN 978-1-4939-7789-5.
16. Luo, Y.; Wu, Y.; Li, B.; Qu, J.; Feng, S.; Chu, P.K. Optimization and Cutting-edge Design of Fuel-cell Hybrid Electric Vehicles. *Int. J. Energy Res.* **2021**, *45*, 18392–18423. [CrossRef]
17. U.S. Department of Energy Alternative Fuels Data Center: How Do Fuel Cell Electric Vehicles Work Using Hydrogen? Available online: https://afdc.energy.gov/vehicles/how-do-fuel-cell-electric-cars-work (accessed on 30 April 2023).
18. Hydrogen and Fuel Cell Technologies Office Hydrogen Storage. Available online: https://www.energy.gov/eere/fuelcells/hydrogen-storage (accessed on 30 April 2023).
19. Usman, M.R. Hydrogen Storage Methods: Review and Current Status. *Renew. Sustain. Energy Rev.* **2022**, *167*, 112743. [CrossRef]
20. Michler, T.; Lindner, M.; Eberle, U.; Meusinger, J. 3—Assessing Hydrogen Embrittlement in Automotive Hydrogen Tanks. In *Gaseous Hydrogen Embrittlement of Materials in Energy Technologies*; Elsevier: Amsterdam, The Netherlands, 2012; pp. 94–125; ISBN 978-1-84569-677-1.

21. Langmi, H.W.; Engelbrecht, N.; Modisha, P.M.; Bessarabov, D. Chapter 13-Hydrogen Storage. In *Electrochemical Power Sources: Fundamentals, Systems, and Applications*; Elsevier: Amsterdam, The Netherlands, 2022; pp. 455–486. ISBN 978-0-12-819424-9.
22. Rivard, E.; Trudeau, M.; Zaghib, K. Hydrogen Storage for Mobility: A Review. *Materials* **2019**, *12*, 1973. [CrossRef]
23. Stetson, N.T.; McWhorter, S.; Ahn, C.C. 1-Introduction to Hydrogen Storage. In *Compendium of Hydrogen Energy*; Elsevier: Amsterdam, The Netherlands, 2016; pp. 3–25. ISBN 978-1-78242-362-1.
24. Abe, J.O.; Popoola, A.P.I.; Ajenifuja, E.; Popoola, O.M. Hydrogen Energy, Economy and Storage: Review and Recommendation. *Int. J. Hydrogen Energy* **2019**, *44*, 15072–15086. [CrossRef]
25. Yang, M.; Hu, S.; Yang, F.; Xu, L.; Bu, Y.; Yuan, D. On-Board Liquid Hydrogen Cold Energy Utilization System for a Heavy-Duty Fuel Cell Hybrid Truck. *World Electr. Veh. J.* **2021**, *12*, 136. [CrossRef]
26. Hwang, H.T.; Varma, A. Hydrogen Storage for Fuel Cell Vehicles. *Curr. Opin. Chem. Eng.* **2014**, *5*, 42–48. [CrossRef]
27. Rao, A.G.; Yin, F.; Werij, H. Energy Transition in Aviation: The Role of Cryogenic Fuels. *Aerospace* **2020**, *7*, 181. [CrossRef]
28. Tzimas, E.; Filiou, C.; Peteves, S.D.; Veyret, J.-B. Hydrogen Storage: State-of-the-Art and Future Perspective. Available online: https://publications.jrc.ec.europa.eu/repository/handle/JRC26493 (accessed on 30 April 2023).
29. Arat, H.T.; Sürer, M.G. State of Art of Hydrogen Usage as a Fuel on Aviation. *Eur. Mech. Sci.* **2017**, *2*, 20–30. [CrossRef]
30. Ahluwalia, R.K.; Peng, J.K.; Roh, H.S.; Hua, T.Q.; Houchins, C.; James, B.D. Supercritical Cryo-Compressed Hydrogen Storage for Fuel Cell Electric Buses. *Int. J. Hydrogen Energy* **2018**, *43*, 10215–10231. [CrossRef]
31. Aceves, S.M.; Petitpas, G.; Espinosa-Loza, F.; Matthews, M.J.; Ledesma-Orozco, E. Safe, Long Range, Inexpensive and Rapidly Refuelable Hydrogen Vehicles with Cryogenic Pressure Vessels. *Int. J. Hydrogen Energy* **2013**, *38*, 2480–2489. [CrossRef]
32. Su, Y.; Lv, H.; Zhou, W.; Zhang, C. Review of the Hydrogen Permeability of the Liner Material of Type IV On-Board Hydrogen Storage Tank. *World Electr. Veh. J.* **2021**, *12*, 130. [CrossRef]
33. OECD. *The Future of Hydrogen: Seizing Today's Opportunities*; Organisation for Economic Co-operation and Development: Paris, France, 2019.
34. A Review on Underground Hydrogen Storage: Insight into Geological Sites, Influencing Factors and Future Outlook—ScienceDirect. Available online: https://www.sciencedirect.com/science/article/pii/S2352484721014414 (accessed on 18 July 2023).
35. Folkson, R. 6—Hydrogen as an Energy Vector for Transportation Vehicles. In *Alternative Fuels and Advanced Vehicle Technologies for Improved Environmental Performance*, 2nd ed.; Folkson, R., Sapsford, S., Eds.; Woodhead Publishing Series in Energy; Woodhead Publishing: Sawston, UK, 2022; pp. 151–171. ISBN 978-0-323-90979-2.
36. Ma, L.-J.; Han, T.; Jia, J.; Wu, H.-S. Cooperative Physisorption and Chemisorption of Hydrogen on Vanadium-Decorated Benzene. *RSC Adv.* **2020**, *10*, 37770–37778. [CrossRef]
37. Ley, M.B.; Jepsen, L.H.; Lee, Y.-S.; Cho, Y.W.; Bellosta Von Colbe, J.M.; Dornheim, M.; Rokni, M.; Jensen, J.O.; Sloth, M.; Filinchuk, Y.; et al. Complex Hydrides for Hydrogen Storage—New Perspectives. *Mater. Today* **2014**, *17*, 122–128. [CrossRef]
38. Dornheim, M. Thermodynamics of Metal Hydrides: Tailoring Reaction Enthalpies of Hydrogen Storage Materials. In *Thermodynamics—Interaction Studies—Solids, Liquids and Gases*; Moreno Pirajn, J.C., Ed.; InTechOpen: London, UK, 2011; ISBN 978-953-307-563-1.
39. Rusman, N.A.A.; Dahari, M. A Review on the Current Progress of Metal Hydrides Material for Solid-State Hydrogen Storage Applications. *Int. J. Hydrogen Energy* **2016**, *41*, 12108–12126. [CrossRef]
40. IEA ETP Clean Energy Technology Guide—Data Tools. Available online: https://www.iea.org/data-and-statistics/data-tools/etp-clean-energy-technology-guide (accessed on 30 April 2023).
41. Xu, Z.; Zhao, N.; Hillmansen, S.; Roberts, C.; Yan, Y. Techno-Economic Analysis of Hydrogen Storage Technologies for Railway Engineering: A Review. *Energies* **2022**, *15*, 6467. [CrossRef]
42. James, B.D.; Houchins, C.; Huya-Kouadio, J.M.; DeSantis, D.A. *Final Report: Hydrogen Storage System Cost Analysis*; Strategic Analysis Inc.: Arlington VA, USA, 2016; pp. DOE-SA--0005253, 1343975.
43. Jesse, A.; Houchins, C. Rajesh Ahluwalia Onboard Type IV Compressed Hydrogen Storage System—Cost and Performance Status. Available online: https://www.hydrogen.energy.gov/pdfs/19008_onboard_storage_cost_performance_status.pdf (accessed on 30 April 2023).
44. 2023 Toyota Mirai Features and Specs | Toyota.Com. Available online: https://www.toyota.com/mirai/2023/features/ (accessed on 30 April 2023).
45. 2023 Toyota Mirai Review, Pricing, and Specs. Available online: https://www.caranddriver.com/toyota/mirai (accessed on 30 April 2023).
46. 2023 Hyundai NEXO Fuel Cell | Specs. Available online: https://www.hyundaiusa.com/us/en/vehicles/nexo (accessed on 30 April 2023).
47. Honda Global | Clarity Fuel Cell—Picture Book. Available online: https://global.honda/innovation/FuelCell/Clarity-Fuel-Cell-picturebook.html (accessed on 30 April 2023).
48. Specifications—2021 Honda Clarity Fuel Cell. Available online: https://www.hondainfocenter.com/2021/Clarity-Fuel-Cell/Feature-Guide/Specifications/ (accessed on 30 April 2023).
49. 2021 Honda Clarity Review, Pricing, and Specs. Available online: https://www.caranddriver.com/honda/clarity (accessed on 30 April 2023).
50. GUMPERT Nathalie First Edition—Roland Gumpert. Available online: https://www.rolandgumpert.com/en/ (accessed on 30 April 2023).

51. Deirdre Meet Nathalie: A New Supercar That Runs on a Methanol Fuel Cell. Available online: https://www.evolving-science.com/environment/supercar-runs-methanol-fuel-cell-00945 (accessed on 30 April 2023).
52. Arnes, B. Changan Dreams of Being the Tesla of Hydrogen—Green Hydrogen News 2022. Available online: https://energynews.biz/changan-dreams-of-being-the-tesla-of-hydrogen/ (accessed on 15 June 2023).
53. Suzan, China's First Mass-Produced Hydrogen Fuel Cell Car: Changan DEEPAL SL3 Listed | China Car News, Reviews and More 2023. Available online: https://www.chinapev.com/changan/chinas-first-mass-produced-hydrogen-fuel-cell-car-changan-deepal-sl3-listed/ (accessed on 30 June 2023).
54. Rubio, O. BMW i Hydrogen NEXT está en Fase de Pruebas en Carreteras de Europa. Available online: https://greenracingnews.com/es/bmw-i-hydrogen-next-esta-en-fase-de-pruebas-en-carreteras-de-europa/ (accessed on 30 April 2023).
55. 2023 BMW IX5 Hydrogen First Drive Review: More Than Just a Science Project. Available online: https://www.cnet.com/roadshow/news/2023-bmw-ix5-hydrogen-first-drive-review/ (accessed on 30 April 2023).
56. 2016—NISSAN e-NV200 SOFC e-BIO | Fabricante NISSAN | PlanetCarsZ. Available online: https://www.planetcarsz.com/carros/nissan-e-nv200-sofc-e-bio-2016 (accessed on 30 April 2023).
57. Nissan E-Bio Fuel-Cell: First Solid Oxide Fuel-Cell van Debuts. Available online: https://www.drive.com.au/news/nissan-e-bio-fuel-cell-first-solid-oxide-fuel-cell-van-debuts/ (accessed on 30 April 2023).
58. Markets and Markets Hydrogen Storage Market by Storage Form, Type of Storage, Application, Region. Available online: https://www.marketsandmarkets.com/Market-Reports/hydrogen-storage-market-15698551.html (accessed on 30 April 2023).
59. Global Automotive Hydrogen Storage Tank Market 2022–2027 | April 2023 Updated. Available online: https://mobilityforesights.com/product/automotive-hydrogen-storage-tank-market/ (accessed on 30 April 2023).
60. Hydrogen Type 4 Cylinder 700 Bar 51L | Hyfindr.Com. Available online: https://hyfindr.com/marketplace/components/hydrogen-tanks/hydrogen-type-4-cylinder-700-bar-51l/ (accessed on 30 June 2023).
61. Hexagon Purus Hexagon Purus Downloads. Available online: https://hexagonpurus.com/downloads (accessed on 30 April 2023).
62. Worthington Industries Hydrogen and CNG Gas Cylinders Manufacturer—Type 4. Available online: https://www.worthingtonindustries.eu/gas-cylinders/high-pressure/type-4/ (accessed on 30 April 2023).
63. Quantum Fuel Systems Hydrogen Cylinder General Specifications. Available online: https://www.qtww.com/wp-content/uploads/2019/01/H2-Tank-Specifications-Sept-2021-All-Tanks.pdf (accessed on 20 April 2023).
64. Furecia H2 Mobility Faurecia Hydrogen Strategy. Available online: https://www.faurecia.com/sites/groupe/files/investisseurs/2019%2006%2014%20-%20Kepler%20Cheuvreux%20Event%20-%20Hydrogen%20-%20pour%20envoi.pdf (accessed on 20 April 2023).
65. Dash, S.K.; Chakraborty, S.; Roccotelli, M.; Sahu, U.K. Hydrogen Fuel for Future Mobility: Challenges and Future Aspects. *Sustainability* **2022**, *14*, 8285. [CrossRef]
66. Yang, Z.; Wang, B.; Jiao, K. Life Cycle Assessment of Fuel Cell, Electric and Internal Combustion Engine Vehicles under Different Fuel Scenarios and Driving Mileages in China. *Energy* **2020**, *198*, 117365. [CrossRef]
67. Handwerker, M.; Wellnitz, J.; Marzbani, H. Comparison of Hydrogen Powertrains with the Battery Powered Electric Vehicle and Investigation of Small-Scale Local Hydrogen Production Using Renewable Energy. *Hydrogen* **2021**, *2*, 76–100. [CrossRef]
68. Wong, E.Y.C.; Ho, D.C.K.; So, S.; Tsang, C.-W.; Chan, E.M.H. Life Cycle Assessment of Electric Vehicles and Hydrogen Fuel Cell Vehicles Using the GREET Model—A Comparative Study. *Sustainability* **2021**, *13*, 4872. [CrossRef]
69. Lieutenant, K.; Borissova, A.V.; Mustafa, M.; McCarthy, N.; Iordache, I. Comparison of "Zero Emission" Vehicles with Petrol and Hybrid Cars in Terms of Total CO_2 Release—A Case Study for Romania, Poland, Norway and Germany. *Energies* **2022**, *15*, 7988. [CrossRef]

designs

Article

Field Experiment for a Prequalification Scheme for a Distribution System Operator on Distributed Energy Resource Aggregations

Jung-Sung Park [1] and Bal-Ho Kim [2,*]

[1] Research Institute, Korea Electric Power Corporation, Daejeon 305-760, Republic of Korea; jindulfa@kepco.co.kr
[2] Electronic and Electrical Engineering, Hongik University, 94, Wausan-ro, Mapo-gu, Seoul 04066, Republic of Korea
* Correspondence: bhkim@hongik.ac.kr; Tel.: +82-2-320-1462

Abstract: The purpose of this paper is to summarize and share the field experiment results of KEPCO's project consortium to create a TSO-DSO-DERA interaction scheme. The field experiment was conducted based on the prequalification algorithm proposed in previous research from the same consortium, and was designed to verify the validity of the algorithm under realistic grid conditions. In addition, during the course of the field experiment, it was found that points that were missed or not given much importance in the existing prequalification algorithm could affect the completeness of the overall system, and then practical improvements were made to improve this. The demonstration results confirm that the proposed algorithm is effective in real-world grid environments and can help DSOs to ensure the reliability of the distribution system while supporting DERA's participation in the wholesale market using the proposed prequalification scheme.

Keywords: prequalification; distribution system operator; distributed energy resource aggregator; distribution constraints; wholesale market participation

Citation: Park, J.-S.; Kim, B.-H. Field Experiment for a Prequalification Scheme for a Distribution System Operator on Distributed Energy Resource Aggregations. *Designs* **2023**, *7*, 134. https://doi.org/10.3390/designs7060134

Academic Editors: Dibin Zhu and Surender Reddy Salkuti

Received: 6 November 2023
Revised: 13 November 2023
Accepted: 15 November 2023
Published: 17 November 2023

1. Introduction

In order to drive cost-effective investment in distribution grids while increasing shares of renewable energy, authorities and utilities are exploring ways to increase hosting capacities using connect-and-manage connection arrangements or non-wire alternatives with battery storage [1]. In the past, traditional connect-and-forget connection arrangements have been based on a worst-case scenario, with limited hosting capacities that would not cause a problem without any operational action. However, recent changes in grid connection policy have led to a situation in which distribution system operators (DSOs) are forced to actively take operational measures in the event of distribution constraints, even though this allows for more renewable energy to be accommodated in the same line capacity.

Meanwhile, regulators around the world are considering ways to increase the visibility of distributed energy resources (DERs) and integrate them into the governance of wholesale markets by providing higher levels of economic compensation to DERs that contribute to the grid and improve system reliability [2–4]. This work contains ideas for institutionalizing participation in wholesale electricity markets by aggregating many small DERs that physically connect the distribution grid through new market participants such as distributed energy resource aggregators (DERAs) or virtual power plants (VPPs). This research was motivated by the fact that the space in which DERAs are financially re-warded is the wholesale market at the transmission level, but the resources they manage are widely physically distributed throughout the distribution grid. If enough DERs are connected to the grid to potentially cause distribution constraints, and there is interest in participating in the wholesale market through a DERA, the DSO needs to prequalify the DERA's bidding actions. The prequalification process allows for DSOs to receive DERA's day-ahead and real-time market bids in advance, review them for potential grid constraint events, and implement a system where only bids that pass the process are forwarded to the wholesale market. In the absence of such a system, DERA may act only to maximize revenue in the wholesale market, without considering distribution constraints, causing a situation where the reliability of the distribution grid is significantly threatened.

Previous studies have implemented and studied various forms of such schemes. In one study, the DSO calculated power flows based on the VPP's bids and responded by performing grid reconfiguration if it anticipated a constraint problem. If there was still a problem, it calculated the sensitivity of each VPP's resources and the scale of the violation and delivered it to the VPP. The VPP would then adjust their scheduling and submit it to the DSO. This case is characterized by the utilization of a periodic rolling horizon update method to respond to future uncertainties [5]. Another study proposed a scheme in which the DSO and transmission system operator (TSO) each operated their own balancing market. DER bids went directly to the TSO market bids if they did not cause problems on the distribution grid, and if they did, they were cleared within the DSO's balancing market first and the adjusted bids went to the TSO market [6]. On the other hand, refs. [7,8] pro-posed a system for DSOs to verify the bids of DERAs based on robust optimization to handle uncertainties in generation and load. These two papers are unique in that they pro-pose an interaction scheme with a design based on the legal authority of each entity in the real world, and they present the idea of converting robust optimization based on chance-constrained optimization into a cooperative solution based on linearized information exchange between entities. In the most recent work, the DSO calculates and communicates the expected distribution system constraints according to the VPP's bidding plan, and the VPP also estimates its own distribution system constraints based on multiple scenario forecasts, and utilizes them in actual operations. In this paper, the process of the DSO communicating distribution system constraints to the VPP occurs only once and does not include an iteration process [9].

In addition to the above theoretical research, field experiments are also being conducted in various countries. The EU's SmartNet project sought to design an optimized interaction scheme between TSO-DSOs and proposed five representative TSO-DSO coordination schemes: a centralized AS market model, local AS market model, shared balancing responsibility model, common TSO-DSO AS market model, and integrated flexibility market model [10]. OMI Polo Español S.A., an electricity market operator in Iberia, con-ducted the IREMEL project to define the operating model for the regional electricity market and to design and demonstrate mechanisms to address distribution system constraints that arise from the participation of distributed resources in the wholesale market [11]. The COOIDINET project in Spain, Sweden, and Greece is a demonstration of DSOs and TSOs working together to implement a cooperative system in a reliable and efficient manner. In the project, a TSO-DSO cooperation system was designed in which congestion management in the distribution system was first addressed at the local DSO level, with priority access to local flexibility resources, and only unused bids were forwarded to the higher-level system operator [12]. These demonstration projects took into account the market rules and power industry structure in each jurisdiction in which the demonstration was conducted.

In Korea, a demonstration study of the TSO-DSO-DERA coordination scheme was conducted from August 2019 to August 2023, led by KEPCO, considering the current Korean power industry structure and market rules. In particular, this project focused on the interaction between DSO and DERAs who want to participate in the TSO market, and demonstrated a prequalification scheme for two distribution lines of KEPCO. The references [7,8] are the deliverables from this demonstration project. In addition to the day-ahead prequalification, this project also included a demonstration of the direct control scheme of the DSO in a real-time situation, which is beyond the scope of this paper and is not described. This paper aims to summarize and share practical algorithm improvements and the lessons learned based on the algorithm in the previous study in [7], as well as validation through real-world demonstration sites.

In Section 2, the roles and responsibilities of relevant actors in the proposed scheme are presented. Sections 3 and 4 describe the overall scheme of the proposed prequalification process and details of the internal prequalification algorithm of DSO, respectively. Section 5 provides the results of the field experiment of the prequalification scheme on the real demonstration sites in Korea. Section 6 presents the concluding remarks of this study.

2. Roles and Responsibilities of Relevant Actors

In this section, we describe the roles and responsibilities of each actor involved in the DSO prequalification scheme assumed in this study. This can be depicted as shown in Figure 1.

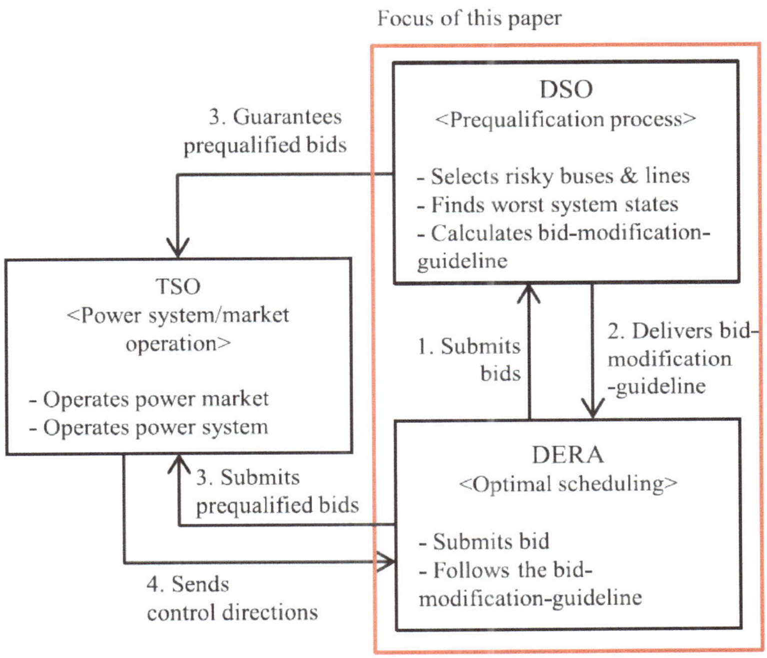

Figure 1. Roles and responsibilities of each relevant actor.

The governance and actor-specific roles and responsibilities assumed in this study are not based on an exploration of all the theoretically possible cases, as in other previous studies, but rather on the premise of minimal effort to remain as much as possible within the current structure and the recently revised laws and regulations.

In Korea, the DSO, the entity that actually implements and operates the prequalification scheme, is the state-owned power company, KEPCO. It directly operates transmission, distribution, and sales businesses and has six power generation companies as subsidiaries with a portfolio of large traditional power generation sources [13].

Individual arrangements for output control between the DERs or DERAs representing them and DSOs are needed to resolve the potential distribution constraints caused by DERs.

Because DERA is a private entity that recruits DERs to directly participate in the wholesale market and earn revenue, it is not responsible for the reliability of the distribution or transmission grid; however, through the national grid code and the network access agreement with KEPCO, it can be disconnected due to the decision of a certain TSO or DSO if it threatens the reliability of the network. From DERA's perspective, it has a motivation to coordinate with DSOs to trade-off the risk of total disconnection for the bounded risk of partial curtailment.

In South Korea, the Korea Power Exchange (KPX) operates the wholesale market and transmission system, similar to independent system operators in the United States.

They are not directly responsible for coordinating and managing the distribution system constraints caused by DERs unless the constraints extend to transmission system issues, as the Electricity Business Act delegates operational responsibility for power systems below 154 kV to KEPCO. This situation in Korea can be seen in Table 1.

Table 1. Operator in power system by voltage levels in Korea.

Voltage Level	Operator
≥154 kV	ISO (KPX)
70 kV and 22.9 kV Dedicated transmission line	TO (KEPCO)
≤22.9 kV distribution line	DSO (KEPCO)

3. The DSO's Overall Prequalification Scheme for the Market Participation of DERA

The DSO prequalification module designed in the demonstration project interacts with DERA's bidding module, as shown in Figure 2. Before the DERA submits a bid to the wholesale market, it requests a prequalification procedure for the bid to the DSO, and the DSO's prequalification module uses an internal algorithm to review the potential constraints of the distribution system based on the expected generation of the DER under the DERA's bid.

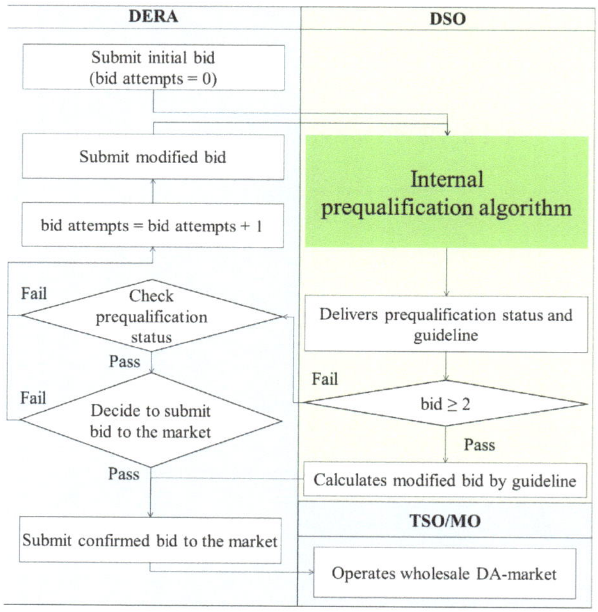

Figure 2. Overall interaction process of DSO and DEAR in the proposed prequalification scheme.

Here is a breakdown of the entire process, as shown in the figure above.

(1) Submission of initial bids to DSO by DERA

The DERA submits bids to the DSO prior to its participation in the wholesale market between 00:00 and 09:00 on D-1. The DERA determines and submits initial energy and reserve bids based on the generation forecast of the DERs managed by itself and its bidding strategy. It is allowed to continuously modify the bids during these hours.

(2) Execution of the internal prequalification algorithm of DSO

The proposed day-ahead prequalification algorithm is executed as a module of the DSO's internal distribution operation system. The algorithm is completed within 15 min, between 09:00 and 09:15 on D-1. The algorithm performs a robust optimization by considering the initial bids submitted by DERAs, the load of each distribution node predicted by the DSO itself, and the forecast error range of each node. The DSO performs the optimization with the goal of maximizing the sum of the DERAs' 24 h generation in order to minimize the output curtailment of the DERAs. The result of the algorithm is the maxi-mum biddable capacity of the DERs at each node and the upper and lower bounds of the biddable output for storage resources. The resulting bid modification guideline values are passed to the DERA. The DERA is obliged to comply with the modification guidelines for bids that are determined to threaten the reliability of the distribution system according to the grid access agreement with the DSO.

(3) Iteration of DERA's bid modification and DSO's re-prequalification under the bid modification guidelines

The DERA will modify the output setpoints of each resource in light of the bid modification guidelines received from the DSO and resubmit them to the DSO. The first modified bid submission will take place during D-1 09:15–09:30, and the corresponding prequalification algorithm will take place between 09:30 and 09:45. The second modified bid submission will take place between D-1 09:45 and 10:00, and the corresponding prequalification algorithm will take place between D-1 10:00 and 10:15. Since the second modified bid submission is the third bid submitted by the DERA, if it does not pass prequalification by this time, the DSO will be forced to apply the last provided bid modification guidelines to end the process. This is because, due to time constraints in real-world systems' operation, it is not possible to provide an infinite number of chances, so in this demonstration study, we set a total of three chances, including the initial bidding after trial and error. This can be customized according to the market operation timeline of each jurisdiction applying prequalification.

(4) Qualified DERA bids enter the existing transmission-level wholesale market

Qualified bids will be delivered to the transmission-level wholesale market as the one of the bids from conventional market participants.

4. Efficient DSO Prequalification Algorithm Based on Robust Optimization

In this section, we detail the formulation of the prequalification and the algorithm to solve it efficiently, based on a robust optimization approach, so that the DSO can support the market participation of DERAs while maintaining the reliability of the distribution system. For this purpose, we use the following nomenclature in Table 2.

Table 2. Sets, indices, parameters and variables.

Sets and Indices	
b, i, j, k	Subscripts for buses
l	Subscripts for lines
b_{rsk}, l_{rsk}	Subscripts for risky buses and lines
t, a	Subscripts for DER and DERA
s, s'	Subscripts for iterations
$\mathcal{N}, \mathcal{L}$	Set of buses and lines
$\mathcal{N}_{rsk\_ovv}$, $\mathcal{N}_{rsk\_udv}$, $\mathcal{L}_{rsk\_uof}$, $\mathcal{L}_{rsk\_dof}$	Set of overvoltage, undervoltage risky buses and upper overflow, down overflow risky lines
$\mathcal{G}_i^a \supset \mathcal{S}_i^a, \mathcal{R}_i^a$	Set of DER, storage, and DER with reserve of DERA a on bus i
$\mathcal{A}$	Set of DERA
$\mathcal{X}_{PF}^s$, $\mathcal{X}_{V\_OV}^s$, $\mathcal{X}_{V\_UV}^s$, $\mathcal{X}_{F\_UF}^s$, $\mathcal{X}_{F\_DF}^s$, $\mathcal{X}_{bid}^s$	Set of result of power flow, worst overvoltage state, worst undervoltage state, worst upper overflow state, worst down overflow state and revised bidding in iteration s

Table 2. *Cont.*

Parameters	
G_{ij}, B_{ij}	Conductance and susceptance of the line between bus i and j [p.u.]
$\overline{V}$, $\underline{V}$	Upper and lower limit of the voltage magnitude [p.u.]
$\overline{S}_{f,l}$	Upper rate capacity of line l [p.u.]
$\overline{P}_{g,t}$	Rate capacity of DER t [p.u.]
$\overline{\Delta P}_{g,t}, \underline{\Delta P}_{g,t}$	Increase and decrease controllable amount of DRES t [p.u.]
$p^{ini}_{g,t}, q^{ini}_{g,t}$	Active and reactive power of DRES t in initial bidding of CVPP [p.u.]
$r^{ini}_{u,t}, r^{ini}_{d,t}$	Ramp up and down reserve power of DRES t in initial bidding of CVPP [p.u.]
$\overline{p}^{ini}_{g,t}, \underline{p}^{ini}_{g,t}$	Upper and lower active power of DRES t in initial bidding of CVPP [p.u.]
$\overline{P}_{d,i}, \overline{Q}_{d,i}$	Demand forecast of bus i [p.u.]
$\underline{\mathrm{pf}}$	A lower limit of the power factor
$\sigma^{high}_{g,t}$, $\sigma^{low}_{g,t}$	Upper and lower uncertainty range of output of DER t [p.u.]
$\sigma^{high}_{d,t}$, $\sigma^{low}_{d,i}$	Upper and lower uncertainty range of demand forecast of bus i [p.u.]
$OV^s_{i,a}$, $UV^s_{i,a}$	Overvoltage and Undervoltage violation on bus i of DERA a in iteration s [p.u.]
$UF^s_{l,a}$, $DF^s_{l,a}$	Upper overflow and down overflow violation on line l of DERA a in iteration s [p.u.]
$Sn^s_{ovp,i,j}, Sn^s_{ovq,i,j}$	Overvoltage magnitude sensitivity of active and reactive power on bus j to bus i to in iteration s at overvoltage worst case
$Sn^s_{uvp,i,j}, Sn^s_{uvq,i,j}$	Undervoltage magnitude sensitivity of active and reactive power on bus j to bus i to in iteration s at undervoltage worst case
$Sn^s_{usp,l,k}, Sn^s_{usq,l,k}$	Squared complex flow sensitivity of active and reactive power on bus k to line l to in iteration s at upper overflow worst case
$Sn^s_{dsp,l,k}, Sn^s_{dsq,l,k}$	Squared complex flow sensitivity of active and reactive power on bus k to line l to in iteration s at down overflow worst case
$Allc^s_{ov,i,a}, Allc^s_{uv,i,a}$	Allocation factor of overvoltage and undervoltage violation on bus i of DERA a in iteration s
$Allc^s_{uf,l,a}, Allc^s_{df,l,a}$	Allocation factor of upper and down overflow violation on line l of DERA a in iteration s
ϵ_V, ϵ_F, ϵ_{Bid}	Criteria of overvoltage security, overflow security and convergence of bid revision
Decision Variables	
$p^s_{g,t}$, $q^s_{g,t}$	Active and reactive power of DER t in iteration s [p.u.]
$r^s_{u,t}$, $r^s_{d,t}$	Ramp up and ramp down reserve power of DER t in iteration s [p.u.]
$\overline{p}^s_{g,t}, \underline{p}^s_{g,t}$	Upper and lower active power of DER t in iteration s [p.u.]
$\Delta p^s_{g,t}$, $\Delta q^s_{g,t}$	Adjustment of active and reactive power of DER t in iteration s [p.u.]
$\overline{\Delta p}^s_{g,t}$, $\underline{\Delta p}^s_{g,t}$	Adjustment of Upper and lower active power of DER t in iteration s [p.u.]
p^s_i, q^s_i	Active and reactive injected power on bus i in iteration s [p.u.]
$\lvert v^s_i \rvert$, θ^s_i	Magnitude and angle of voltage on bus i in iteration s [p.u.]
$\zeta^s_{g,i}$, $\zeta^s_{d,i}$	Forecast error rate of the DER output and demand on bus i in iteration s
$s^s_{f,l}$, $p^s_{f,l}$, $q^s_{f,l}$	Complex, active and reactive power flow on line l in iteration s [p.u.]
$\theta^s_{ij}, \theta^s_l$	Angle of voltage between bus i and bus j or on line l in iteration s [p.u.]
$u^s_{g,t}$	Dropout binary variable of DER t in iteration s
Auxiliary Variables	
$p^{s,ab}_{g,t}$, $q^{s,ab}_{g,t}$	Absolute value of active and reactive power difference of DER t between initial bid and iteration s [p.u.]
$r^{s,ab}_{u,t}$, $r^{s,ab}_{d,t}$	Absolute value of Ramp up and ramp down reserve power difference of DER t between initial bid and iteration s [p.u.]
$p^{s,dch}_{st,t}$, $q^{s,dch}_{st,t}$	Active and reactive discharging power of storage DER t in iteration s [p.u.]
$p^{s,ch}_{st,t}$, $q^{s,ch}_{st,t}$	Active and reactive charging power of storage DER t in iteration s [p.u.]
$v^s_{st,t}$	Charging state binary variable of storage DER t in iteration s
Binary Variables	
$\zeta^s_{g,i}$, $\zeta^s_{d,i}$	Forecast error rate of the DER output and demand on bus i in iteration s

4.1. Original Formulation for Prequalification

We assumed that the distribution grid had the following three properties:

- DSOs are regulated entities and strive to be neutral, and as public entities, they strive to maximize social welfare;
- The load is inflexible and is supplied with electricity at a fixed tariff. It is important to note that this is not very different from the actual empirical environment in Korea;
- Only DERs based on renewable energy sources with zero-marginal-cost characteristics are connected to the distribution grid, and they are also price-takers.

According to the first property, the basic objective function of DSO can be formulated as social welfare maximization. If we restrict a region of the distribution system under the DSO's jurisdiction to one ecosystem, the social welfare of this ecosystem can be expressed as the sum of the surplus of customers and producers. In this environment, consumer surplus is invariant with respect to the amount of electricity generation because the second property assumes that customers are price-takers and not price-responsive. Producers' surplus is constrained by the physical environment of the distribution grid, and their sur-plus can be maximized by generating as much as possible under these constraints. Considering the uncertainty inherent in the output of DERs, the social welfare maximization problem in this environment can be formulated as a chance-constrained problem with the following equations, as has already been presented in [7]:

$$\max_{X} \mathbb{E}\left[\sum_{i \in \mathcal{G}} \widetilde{p}_{g,i}\right] \tag{1a}$$

subject to

$$\widetilde{p}_{g,i} = p_{g,i} \cdot \xi_{g,i},\ \forall i \in \mathcal{G} \tag{1b}$$

$$\widetilde{q}_{g,i} = q_{g,i} \cdot \xi_{g,i},\ \forall i \in \mathcal{G} \tag{1c}$$

$$\widetilde{P}_{d,i} = \overline{P}_{d,i} \cdot \xi_{d,i},\ \forall i \in \mathcal{N} \tag{1d}$$

$$\widetilde{Q}_{d,i} = \overline{Q}_{d,i} \cdot \xi_{d,i},\ \forall i \in \mathcal{N} \tag{1e}$$

$$\Pr\{|v_i(\xi)| \le \overline{V}\} \ge 1 - \epsilon,\ \ \forall i \in \mathcal{N} \backslash \{1\} \tag{1f}$$

$$\Pr\left\{\left|s_{f,l}(\xi)\right|^2 \le \overline{S}^2{}_{f,l}\right\} \ge 1 - \epsilon,\ \ \forall i \in \mathcal{L} \tag{1g}$$

$$G\left(\boldsymbol{p}_{g}, \boldsymbol{q}_{g}, |v_i(\xi)|, |s_{f,l}(\xi)|, \boldsymbol{x}(\xi), \xi\right) \le 0 \tag{1h}$$

$$H\left(\boldsymbol{p}_{g}, \boldsymbol{q}_{g}, |v_i(\xi)|, |s_{f,l}(\xi)|, \boldsymbol{x}(\xi), \xi\right) = 0 \tag{1i}$$

Assume that $\boldsymbol{p}^*{}_g$ and $\boldsymbol{q}^*{}_g$ are the optimal solution of Equation (1); then, we can obtain $\xi^{*,i}$ and $\xi^{*,l}$, which can maximize $|v_i(\xi)|$ and $|s_{f,l}(\xi)|$, respectively. $\boldsymbol{p}^*{}_g$ and $\boldsymbol{q}^*{}_g$ will satisfy the chance-constraints Equation (1f,g). Then, in this optimal case, the chance constraints can be rewritten into a deterministic form, which can be presented in Equation (2):

$$|v_i(\xi)| \le \overline{V},\ \forall i \in \mathcal{N} \backslash \{1\} \tag{2a}$$

$$\left|s_{f,l}(\xi)\right|^2 \le \overline{S}^2{}_{f,l},\ \forall i \in \mathcal{L} \tag{2b}$$

The essence of the algorithm proposed in reference [7] is that a simple examination of the overvoltage and overflow cases is sufficient to find a solution that is close to the optimal solution of Equation (1). The bus voltages in the grid are positively correlated with each other, and so are the line flows on each line. This has been proven in section V-D of [7]. Applying this tendency, we can find the worst overvoltage case $\xi^{v,\,rsk}$ by finding the scenario that maximizes the sum of the voltages of each risky bus for $\mathcal{N}_{rsk}$, the set of buses that are likely to exceed the constraint, and the worst overflow case $\xi^{f,\,rsk}$ by finding the scenario that maximizes the sum of the usage rates for $\mathcal{L}_{rsk}$. The same principle can be applied to the undervoltage. Once the power generation vectors of each node corresponding to $\xi^{v,\,rsk}$ and $\xi^{f,\,rsk}$ have been determined, the effect of the change in the active and reactive power of each node on the voltage of each node, the change in the line flow of each line around these values can be formulated as a linear approximation using the sensitivity matrix from the power flow equation, and this approximation can be substituted into Equation (2) to find the bid-modification guideline, which is the active and reactive power setpoint for each node that does not exceed the constraints of the distribution system. This idea is the main finding of reference [7]. Using this idea, Ref. [7] tackles a complex full robotization problem like Equation (1), but instead of solving a complex problem, they propose an efficient algorithm that consists of a series of simple computations to obtain the desired solution.

4.2. Efficient Prequalification Algorithm Implemented in the Field Experiment

In this paper, we adapted the algorithm from reference [7] for the field experiment and summarized it using step-by-step formulas. The specific differences from the previous paper are summarized in the Section 4.3.

- **Step 1: Selection of risky nodes and lines**

The DSO conducts a power flow calculation of the distribution system under its jurisdiction, based on the bidding information submitted by DERA and the own-load forecast of the DSO. From the results of the power flow calculation, the risky set is determined by determining which lines exceed the overvoltage/undervoltage risk threshold or the reverse/forward overflow risk threshold.

$$\theta_1^s = 0,\ |v_1^s| = 1 \tag{3a}$$

$$p_i^s = |v_i^s| \sum\nolimits_{j\in\mathcal{N}} \left|v_j^s\right| \left(G_{ij}\cos\theta_{ij}^s + B_{ij}\sin\theta_{ij}^s\right),\ \forall i \in \mathcal{N} \tag{3b}$$

$$q_i^s = |v_i^s| \sum\nolimits_{j\in\mathcal{N}} \left|v_j^s\right| \left(G_{ij}\sin\theta_{ij}^s - B_{ij}\cos\theta_{ij}^s\right),\ \forall i \in \mathcal{N} \tag{3c}$$

$$p_{f,l}^s = |v_i^s| \left|v_j^s\right| \left(G_{ij}\cos\theta_{ij}^s + B_{ij}\sin\theta_{ij}^s\right) - G_{ij}|v_i^s|^2,\ \forall(i,j) = \forall l \in \mathcal{L} \tag{3d}$$

$$q_{f,l}^s = |v_i^s| \left|v_j^s\right| \left(G_{ij}\sin\theta_{ij}^s - B_{ij}\cos\theta_{ij}^s\right) + B_{ij}|v_i^s|^2,\ \forall(i,j) = \forall l \in \mathcal{L} \tag{3e}$$

$$\left|s_{f,l}^s\right|^2 = \left(p_{f,l}^s\right)^2 + \left(q_{f,l}^s\right)^2,\ \forall l \in \mathcal{L} \tag{3f}$$

$$p_i^s = \sum\nolimits_{t\in\mathcal{G}_i} p_{g,t}^s - \overline{P}_{d,i},\ \forall i \in \mathcal{N} \tag{3g}$$

$$q_i^s = \sum\nolimits_{t\in\mathcal{G}_i} q_{g,t}^s - \overline{Q}_{d,i},\ \forall i \in \mathcal{N} \tag{3h}$$

$$\mathcal{N}_{rsk\_ovv}^s = \left\{ b_{rsk}^{s'} \middle| b_{rsk}^{s'} = \underset{b\in\mathcal{N}}{\operatorname{argmax}} \left(\left|v_b^{s'}\right| - \overline{v} \right),\ \left|v_b^{s'}\right| \in \left|v^{s'}\right| \in \mathcal{X}_{PF}^{s'},\ \forall s' \le s \right\} \tag{4a}$$

$$\left(\mathcal{N}^{s}_{rsk\_udv} = \left\{ b^{s'}_{rsk} \middle| b^{s'}_{rsk} = \underset{b \in \mathcal{N}}{\operatorname{argmin}} \left(\left| v^{s'}_{b} \right| - \overline{v} \right), \ \left| v^{s'}_{b} \right| \in \left| \boldsymbol{v}^{s'} \right| \in \mathcal{X}^{s'}_{PF}, \forall s' \leq s \right\} \right. \tag{4b}$$

$$\mathcal{L}^{s}_{rsk\_uof} = \left\{ l^{s'}_{rsk} \middle| l^{s'}_{rsk} = \underset{l \in \mathcal{L}}{\operatorname{argmax}} \left(\frac{\left| s^{s'}_{f,l} \right| - \overline{S}_{f,l}}{\overline{S}_{f,l}} \right), \ \left| s^{s'}_{f,l} \right| \in \left| \boldsymbol{s}^{s'} \right| \in \mathcal{X}^{s'}_{PF}, \forall s' \leq s, p^{s'}_{f,l} \in R^{-} \right\} \tag{4c}$$

$$\mathcal{L}^{s}_{rsk\_dof} = \left\{ l^{s'}_{rsk} \middle| l^{s'}_{rsk} = \underset{l \in \mathcal{L}}{\operatorname{argmax}} \left(\frac{\left| s^{s'}_{f,l} \right| - \overline{S}_{f,l}}{\overline{S}_{f,l}} \right), \ \left| s^{s'}_{f,l} \right| \in \left| \boldsymbol{s}^{s'} \right| \in \mathcal{X}^{s'}_{PF}, \forall s' \leq s, p^{s'}_{f,l} \in R^{+} \right\} \tag{4d}$$

- **Step 2: Exploration of the worst-case scenario considering uncertainty ranges**

The DSO considers the forecast error of DERs and loads, and the amount each DER responds to the frequency control signal from the TSO. For each of the four cases, overvoltage, undervoltage, reverse overflow, and forward overflow, the worst-case scenario is explored, in which each value of the risky nodes and lines within the considered uncertainty ranges causes the greatest violation of grid constraints.

$$\max_{\mathcal{X}^{s}_{V\_OV}} \sum\nolimits_{i \in \mathcal{N}^{s}_{rsk\_ovv}} \left| v^{s}_{i} \right| \tag{5a}$$

subject to

(3a-3c)

$$1 - \sigma^{low}_{d,i} \leq \xi^{s}_{d,i} \leq 1 + \sigma^{high}_{d,i}, \ \forall i \in \mathcal{N} \tag{5b}$$

$$1 - \frac{\sum_{t \in \mathcal{G}_i - \mathcal{R}_i} p^{s}_{g,t} \sigma^{low}_{g,i} + \sum_{t \in \mathcal{R}_i} r^{s}_{d,t}}{\sum_{t \in \mathcal{G}_i} p^{s}_{g,t}} \leq \xi^{s}_{g,i} \leq 1 + \frac{\sum_{t \in \mathcal{G}_i - \mathcal{R}_i} p^{s}_{g,t} \sigma^{high}_{g,i} + \sum_{t \in \mathcal{R}_i} r^{s}_{u,t}}{\sum_{t \in \mathcal{G}_i} p^{s}_{g,t}}, \ \forall i \in \mathcal{N} \tag{5c}$$

$$p^{s}_{i} = \xi^{s}_{g,i} \sum\nolimits_{t \in \mathcal{G}_i} p^{s}_{g,t} - \overline{P}_{d,i} \xi^{s}_{d,i}, \ \forall i \in \mathcal{N} \tag{5d}$$

$$q^{s}_{i} = \xi^{s}_{g,i} \sum\nolimits_{t \in \mathcal{G}_i} q^{s}_{g,t} - \overline{Q}_{d,i} \xi^{s}_{d,i}, \ \forall i \in \mathcal{N} \tag{5e}$$

$$\min_{\mathcal{X}^{s}_{V\_UV}} \sum\nolimits_{i \in \mathcal{N}^{s}_{rsk\_udv}} \left| v^{s}_{i} \right| \tag{6}$$

subject to

(3a-3c, 5b-5e)

$$\max_{\mathcal{X}^{s}_{F\_UF}} \sum\nolimits_{l \in \mathcal{L}^{s}_{rsk\_uof}} \left| s^{s}_{f,l} \right|^{2} / \left| \overline{S}_{f,l} \right|^{2} \tag{7}$$

subject to

(3a-3f, 5b-5e)

$$\max_{\mathcal{X}^{s}_{F\_DF}} \sum\nolimits_{l \in \mathcal{L}^{s}_{rsk\_dof}} \left| s^{s}_{f,l} \right|^{2} / \left| \overline{S}_{f,l} \right|^{2} \tag{8}$$

subject to

(3a-3f, 5b-5e)

- **Step 3: Check of grid constraint violation**

The DSO verifies that no grid constraints are violated in all worst-case scenarios found in Step 2. If any violations have occurred, go to Step 4; if none exist, go directly to Step 6.

$$OV_i^s = |v_i^s| - \overline{v},\ \ \forall i \in \mathcal{N}, |v_i^s| \in |\boldsymbol{v}^s| \in \mathcal{X}_{V_{OV}}^s \tag{9a}$$

$$UV_i^s = \underline{v} - |v_i^s|,\ \ \forall i \in \mathcal{N}, |v_i^s| \in |\boldsymbol{v}^s| \in \mathcal{X}_{V\_UV}^s \tag{9b}$$

$$UF_l^s = \left|s_{f,l}^s\right|^s - \left|\overline{S}_{f,l}\right|^2,\ \ \forall l \in \mathcal{L}, \left|s_{f,l}^s\right| \in \left|\boldsymbol{s}_f^s\right| \in \mathcal{X}_{F\_UF}^s \tag{9c}$$

$$DF_l^s = \left|s_{f,l}^s\right|^s - \left|\overline{S}_{f,l}\right|^2,\ \ \forall l \in \mathcal{L}, \left|s_{f,l}^s\right| \in \left|\boldsymbol{s}_f^s\right| \in \mathcal{X}_{F\_DF}^s \tag{9d}$$

$$max\boldsymbol{OV}^s \leq \epsilon_V \tag{10a}$$

$$max\boldsymbol{UV}^s \leq \epsilon_V \tag{10b}$$

$$max\boldsymbol{UF}^s \leq \epsilon_F \tag{10c}$$

$$max\boldsymbol{DF}^S \leq \epsilon_F \tag{10d}$$

- **Step 4: Calculation of the sensitivity matrix of each DERA and allocation of violation information**

The DSO calculates the sensitivity matrix regarding the impacts of DERs on each node and line for each worst-case scenario. Based on this information, the DSO allocates the violation information to each DERA.

$$\begin{bmatrix} \Delta\boldsymbol{\theta}_{-1}^s \\ \Delta|\boldsymbol{v}_{-1}|^s \end{bmatrix} = \begin{bmatrix} \boldsymbol{J}_{p\theta}^s(\boldsymbol{x}) & \boldsymbol{J}_{p|v|}^s(\boldsymbol{x}) \\ \boldsymbol{J}_{q\theta}^s(\boldsymbol{x}) & \boldsymbol{J}_{q|v|}^s(\boldsymbol{x}) \end{bmatrix}^{-1} \Bigg|_{\boldsymbol{x}=\mathcal{X}_{V\_OV}^s} \begin{bmatrix} \Delta\boldsymbol{p}_{-1}^s \\ \Delta\boldsymbol{q}_{-1}^s \end{bmatrix} = \begin{bmatrix} \boldsymbol{Sn}_{o\theta p}^s(\boldsymbol{x}) & \boldsymbol{Sn}_{ovp}^s(\boldsymbol{x}) \\ \boldsymbol{Sn}_{o\theta q}^s(\boldsymbol{x}) & \boldsymbol{Sn}_{ovq}^s(\boldsymbol{x}) \end{bmatrix} \begin{bmatrix} \Delta\boldsymbol{p}_{-1}^s \\ \Delta\boldsymbol{q}_{-1}^s \end{bmatrix} \tag{11a}$$

$$\begin{bmatrix} \Delta\boldsymbol{\theta}_{-1}^s \\ \Delta|\boldsymbol{v}_{-1}|^s \end{bmatrix} = \begin{bmatrix} \boldsymbol{J}_{p\theta}^s(\boldsymbol{x}) & \boldsymbol{J}_{p|v|}^s(\boldsymbol{x}) \\ \boldsymbol{J}_{q\theta}^s(\boldsymbol{x}) & \boldsymbol{J}_{q|v|}^s(\boldsymbol{x}) \end{bmatrix}^{-1} \Bigg|_{\boldsymbol{x}=\mathcal{X}_{V\_UV}^s} \begin{bmatrix} \Delta\boldsymbol{p}_{-1}^s \\ \Delta\boldsymbol{q}_{-1}^s \end{bmatrix} = \begin{bmatrix} \boldsymbol{Sn}_{o\theta p}^s(\boldsymbol{x}) & \boldsymbol{Sn}_{ovp}^s(\boldsymbol{x}) \\ \boldsymbol{Sn}_{o\theta q}^s(\boldsymbol{x}) & \boldsymbol{Sn}_{ovq}^s(\boldsymbol{x}) \end{bmatrix} \begin{bmatrix} \Delta\boldsymbol{p}_{-1}^s \\ \Delta\boldsymbol{q}_{-1}^s \end{bmatrix} \tag{11b}$$

$$Sn_{usp,l,k}^s = \frac{\partial\left|s_{f,l}\right|^2}{\partial p_k} = 2\left(p_{f,l}^s\frac{\partial p_{f,l}}{\partial p_k} + q_{f,l}^s\frac{\partial q_{f,l}}{\partial p_k}\right)\Bigg|_{\boldsymbol{x}=\mathcal{X}_{F\_UF}^s},\ \ \forall l \in \mathcal{L} \tag{12a}$$

$$Sn_{usq,l,k}^s = \frac{\partial\left|s_{f,l}\right|^2}{\partial q_k} = 2\left(p_{f,l}^s\frac{\partial p_{f,l}}{\partial q_k} + q_{f,l}^s\frac{\partial q_{f,l}}{\partial q_k}\right)\Bigg|_{\boldsymbol{x}=\mathcal{X}_{F\_UF}^s},\ \ \forall l \in \mathcal{L} \tag{12b}$$

$$Sn_{dsp,l,k}^s = \frac{\partial\left|s_{f,l}\right|^2}{\partial p_k} = 2\left(p_{f,l}^s\frac{\partial p_{f,l}}{\partial p_k} + q_{f,l}^s\frac{\partial q_{f,l}}{\partial p_k}\right)\Bigg|_{\boldsymbol{x}=\mathcal{X}_{F\_DF}^s},\ \ \forall l \in \mathcal{L} \tag{12c}$$

$$Sn_{dsq,l,k}^s = \frac{\partial\left|s_{f,l}\right|^2}{\partial q_k} = 2\left(p_{f,l}^s\frac{\partial p_{f,l}}{\partial q_k} + q_{f,l}^s\frac{\partial q_{f,l}}{\partial q_k}\right)\Bigg|_{\boldsymbol{x}=\mathcal{X}_{F\_DF}^s},\ \ \forall l \in \mathcal{L} \tag{12d}$$

$$OV_i^s = \sum\nolimits_{a\in\mathcal{A}} OV_{i,a}^s,\ \ \forall i \in \mathcal{N} \tag{13a}$$

$$UV_i^s = \sum\nolimits_{a\in\mathcal{A}} UV_{i,a}^s,\ \ \forall i \in \mathcal{N} \tag{13b}$$

$$UF_l^s = \sum\nolimits_{a\in\mathcal{A}} UF_{l,a}^s,\ \ \forall l \in \mathcal{L} \tag{13c}$$

$$DF^s_l = \sum_{a\in\mathcal{A}} DF^s_{l,a}, \ \forall l \in \mathcal{L} \tag{13d}$$

$$OV^s_{i,a} = \frac{OV^s_i * Allc^s_{ov,i,a}}{\sum_{a\in\mathcal{A}} Allc^s_{ov,i,a}}, \ \forall i \in \mathcal{N} \tag{14a}$$

$$UV^s_{i,a} = \frac{UV^s_i * Allc^s_{uv,i,a}}{\sum_{a\in\mathcal{A}} Allc^s_{uv,i,a}}, \ \forall i \in \mathcal{N} \tag{14b}$$

$$UF^s_{l,a} = \frac{UF^s_l * Allc^s_{uf,l,a}}{\sum_{a\in\mathcal{A}} Allc^s_{uf,l,a}}, \ \forall l \in \mathcal{L} \tag{14c}$$

$$DF^s_{l,a} = \frac{DF^s_l * Allc^s_{df,l,a}}{\sum_{a\in\mathcal{A}} Allc^s_{df,l,a}}, \ \forall l \in \mathcal{L} \tag{14d}$$

$$Allc^s_{ov,i,a} = \sum_{j\in\mathcal{N}} \left(Sn^s_{ovp,i,j} \sum_{t\in\mathcal{G}^a_j} p^s_{g,t} \right), \ \forall i \in \mathcal{N}, \forall a \in \mathcal{A} \tag{15a}$$

$$Allc^s_{uv,i,a} = \sum_{j\in\mathcal{N}} \left(Sn^s_{uvp,i,j} \sum_{t\in\mathcal{G}^a_j} p^s_{g,t} \right), \ \forall i \in \mathcal{N}, \forall a \in \mathcal{A} \tag{15b}$$

$$Allc^s_{uf,l,a} = \sum_{j\in\mathcal{N}} \left(Sn^s_{usp,l,k} \sum_{t\in\mathcal{G}^a_k} p^s_{g,t} \right), \ \forall l \in \mathcal{L}, \forall a \in \mathcal{A} \tag{15c}$$

$$Allc^s_{df,l,a} = \sum_{k\in\mathcal{N}} \left(Sn^s_{dsp,l,k} \sum_{t\in\mathcal{G}^a_k} p^s_{g,t} \right), \ \forall l \in \mathcal{L}, \forall a \in \mathcal{A} \tag{15d}$$

- **Step 5: Calculation of maximum allowable bids based on sensitivity matrix and constraints violations**

This step calculates the maximum allowable bids for each DERA, taking into account the sensitivity of the DERs calculated in Step 4 and the scale of the constraint violations distributed among the DERAs. The objective function of the DERA is set to minimize the difference between the initial bid and the revised bid based on the iterative process.

$$-OV^s_{i,a} \geq \sum_{j\in\mathcal{N}} \left(Sn^s_{ovp,i,j} \sum_{t\in\mathcal{G}^a_j} \overline{\Delta p^s_{g,t}} + Sn^s_{ovq,i,j} \sum_{t\in\mathcal{G}^a_j} \Delta q^s_{g,t} \right), \ \forall i \in \mathcal{N} \tag{16a}$$

$$-OV^s_{i,a} \geq \sum_{j\in\mathcal{N}} \left(Sn^s_{ovp,i,j} \sum_{t\in\mathcal{G}^a_j} \underline{\Delta p^s_{g,t}} + Sn^s_{ovq,i,j} \sum_{t\in\mathcal{G}^a_j} \Delta q^s_{g,t} \right), \ \forall i \in \mathcal{N} \tag{16b}$$

$$-UV^s_{i,a} \geq \sum_{j\in\mathcal{N}} \left(Sn^s_{uvp,i,j} \sum_{t\in\mathcal{G}^a_j} \overline{\Delta p^s_{g,t}} + Sn^s_{uvq,i,j} \sum_{t\in\mathcal{G}^a_j} \Delta q^s_{g,t} \right), \ \forall i \in \mathcal{N} \tag{16c}$$

$$-UV^s_{i,a} \geq \sum_{j\in\mathcal{N}} \left(Sn^s_{uvp,i,j} \sum_{t\in\mathcal{G}^a_j} \underline{\Delta p^s_{g,t}} + Sn^s_{uvq,i,j} \sum_{t\in\mathcal{G}^a_j} \Delta q^s_{g,t} \right), \ \forall i \in \mathcal{N} \tag{16d}$$

$$-UF^s_{l,a} \geq \sum_{k\in\mathcal{N}} \left(Sn^s_{usp,l,k} \sum_{t\in\mathcal{G}^a_k} \overline{\Delta p^s_{g,t}} + Sn^s_{usq,l,k} \sum_{t\in\mathcal{G}^a_k} \Delta q^s_{g,t} \right), \ \forall l \in \mathcal{L} \tag{16e}$$

$$-UF^s_{l,a} \geq \sum_{k\in\mathcal{N}} \left(Sn^s_{usp,l,k} \sum_{t\in\mathcal{G}^a_k} \underline{\Delta p^s_{g,t}} + Sn^s_{usq,l,k} \sum_{t\in\mathcal{G}^a_k} \Delta q^s_{g,t} \right), \ \forall l \in \mathcal{L} \tag{16f}$$

$$-DF^s_{l,a} \geq \sum_{k\in\mathcal{N}} \left(Sn^s_{dsp,l,k} \sum_{t\in\mathcal{G}^a_k} \overline{\Delta p^s_{g,t}} + Sn^s_{dsq,l,k} \sum_{t\in\mathcal{G}^a_k} \Delta q^s_{g,t} \right), \ \forall l \in \mathcal{L} \tag{16g}$$

$$-DF^s_{l,a} \geq \sum_{k\in\mathcal{N}} \left(Sn^s_{dsp,l,k} \sum_{t\in\mathcal{G}^a_k} \underline{\Delta p^s_{g,t}} + Sn^s_{dsq,l,k} \sum_{t\in\mathcal{G}^a_k} \Delta q^s_{g,t} \right), \ \forall l \in \mathcal{L} \tag{16h}$$

$$\min_{\mathcal{X}_{bm}^{s}} \sum_{t\in\mathcal{G}^a-\mathcal{S}^a}\left(\left|p_{g,t}^{ini}-p_{g,t}^{s}\right|+\left|q_{g,t}^{ini}-q_{g,t}^{s}\right|\right)+m\sum_{t\in\mathcal{S}^a}\left(\left|p_{g,t}^{ini}-p_{g,t}^{s}\right|+\left|q_{g,t}^{ini}-q_{g,t}^{s}\right|\right)+\sum_{t\in\mathcal{R}^a}\left(\left|r_{u,t}^{ini}-r_{u,t}^{s}\right|+\left|r_{d,t}^{ini}-r_{d,t}^{s}\right|\right) \tag{17a}$$

subject to

(16)

$$\overline{p_{g,t}^{s}}=\begin{cases}p_{g,t}^{s}+r_{u,t}^{s}, & \forall t\in\mathcal{R}^a\\ p_{g,t}^{s}, & \forall t\in\mathcal{G}^a-\mathcal{R}^a\end{cases} \tag{17b}$$

$$\underline{p_{g,t}^{s}}=\begin{cases}p_{g,t}^{s}-r_{u,t}^{s}, & \forall t\in\mathcal{R}^a\\ p_{g,t}^{s}, & \forall t\in\mathcal{G}^a-\mathcal{R}^a\end{cases} \tag{17c}$$

$$\overline{p_{g,t}^{s}}=\overline{p_{g,t}^{s-1}}+\overline{\Delta p_{g,t}^{s}},\ \forall t\in\mathcal{G}^a \tag{17d}$$

$$\underline{p_{g,t}^{s}}=\underline{p_{g,t}^{s-1}}+\underline{\Delta p_{g,t}^{s}},\ \forall t\in\mathcal{G}^a \tag{17e}$$

$$\left(\overline{p_{g,t}^{ini}}-\underline{\Delta P}_{g,t}\right)u_{g,t}^{s}\le\overline{p_{g,t}^{s}}\le\left(\overline{p_{g,t}^{ini}}+\overline{\Delta P}_{g,t}\right)u_{g,t}^{s},\ \forall t\in\mathcal{G}^a \tag{17f}$$

$$\left(\overline{p_{g,t}^{ini}}-\underline{\Delta P}_{g,t}\right)u_{g,t}^{s}\le\underline{p_{g,t}^{t}}\le\left(\overline{p_{g,t}^{ini}}+\overline{\Delta P}_{g,t}\right)u_{g,t}^{s},\ \forall t\in\mathcal{G}^a \tag{17g}$$

$$\left(q_{g,t}^{ini}-\underline{\Delta Q}_{g,t}\right)u_{g,t}^{s}\le q_{g,t}^{s}\le\left(q_{g,t}^{ini}+\overline{\Delta Q}_{g,t}\right)u_{g,t}^{s},\ \forall t\in\mathcal{G}^a \tag{17h}$$

$$max\left(\left|\overline{p_{g,t}^{s}}\right|,\left|\underline{p_{g,t}^{s}}\right|\right)^2+\left(q_{g,t}^{s}\right)^2\le\overline{P}_{g,t}^2,\ \forall t\in\mathcal{G}^a \tag{17i}$$

$$\left|q_{g,t}^{s}\right|\le\left|p_{g,t}^{s}tan\left(cos^{-1}\left(\underline{pf}\right)\right)\right|,\ \forall t\in\mathcal{G}^a \tag{17j}$$

$$p_{g,t}^{s,ab}\ge p_{g,t}^{ini}-p_{g,t}^{s},\ \forall t\in\mathcal{G}^a \tag{18a}$$

$$p_{g,t}^{s,ab}\ge -p_{g,t}^{ini}+p_{g,t}^{s},\ \forall t\in\mathcal{G}^a \tag{18b}$$

$$q_{g,t}^{s,ab}\ge q_{g,t}^{ini}-q_{g,t}^{s},\ \forall t\in\mathcal{G}^a \tag{18c}$$

$$q_{g,t}^{s,ab}\ge -q_{g,t}^{ini}+q_{g,t}^{s},\ \forall t\in\mathcal{G}^a \tag{18d}$$

$$r_{u,t}^{s,ab}\ge r_{u,t}^{ini}-r_{u,t}^{s},\ \forall t\in\mathcal{R}^a \tag{18e}$$

$$r_{u,t}^{s,ab}\ge -r_{u,t}^{ini}+r_{u,t}^{s},\ \forall t\in\mathcal{R}^a \tag{18f}$$

$$r_{d,t}^{s,ab}\ge r_{d,t}^{ini}-r_{d,t}^{s},\ \forall t\in\mathcal{R}^a \tag{18g}$$

$$r_{d,t}^{s,ab}\ge -r_{d,t}^{ini}+r_{d,t}^{s},\ \forall t\in\mathcal{R}^a \tag{18h}$$

$$q_{g,t}^{s}\ge\frac{tan\left(cos^{-1}\left(\underline{pf}\right)\right)}{\sqrt{tan\left(cos^{-1}\left(\underline{pf}\right)\right)^2+1}-1}\left(\overline{p_{g,t}^{s}}-\overline{P}_{g,t}\right),\ \forall t\in\mathcal{G}^a \tag{19a}$$

$$q_{g,t}^{s} \leq -\frac{tan\left(cos^{-1}\left(\underline{pf}\right)\right)}{\sqrt{tan\left(cos^{-1}\left(\underline{pf}\right)\right)^{2}+1-1}}\left(\overline{p_{g,t}^{s}}-\overline{P}_{g,t}\right),\ \forall t\in\mathcal{G}^{a} \tag{19b}$$

$$q_{g,t}^{s} \geq -\frac{tan\left(cos^{-1}\left(\underline{pf}\right)\right)}{\sqrt{tan\left(cos^{-1}\left(\underline{pf}\right)\right)^{2}+1-1}}\left(\underline{p_{g,i}^{s}}+\overline{P}_{g,t}\right),\ \forall t\in\mathcal{G}^{a} \tag{19c}$$

$$q_{g,t}^{s} \leq \frac{tan\left(cos^{-1}\left(\underline{pf}\right)\right)}{\sqrt{tan\left(cos^{-1}\left(\underline{pf}\right)\right)^{2}+1-1}}\left(\underline{p_{g,i}^{s}}+\overline{P}_{g,t}\right),\ \forall t\in\mathcal{G}^{a} \tag{19d}$$

$$q_{g,t}^{s} \geq \begin{cases} -p_{g,t}^{s}tan\left(cos^{-1}\left(\underline{pf}\right)\right), & if\ p_{g,t}^{s}\geq 0 \\ p_{g,t}^{s}tan\left(cos^{-1}\left(\underline{pf}\right)\right), & otherwise \end{cases},\ \forall t\in\mathcal{G}^{a} \tag{20a}$$

$$q_{g,t}^{s} \leq \begin{cases} p_{g,t}^{s}tan\left(cos^{-1}\left(\underline{pf}\right)\right), & if\ p_{g,t}^{s}\geq 0 \\ -p_{g,t}^{s}tan\left(cos^{-1}\left(\underline{pf}\right)\right), & otherwise \end{cases},\ \forall t\in\mathcal{G}^{a} \tag{20b}$$

$$q_{g,t}^{s} \geq -p_{g,t}^{s}tan\left(cos^{-1}\left(\underline{pf}\right)\right),\ \forall t\in\mathcal{G}^{a}-\mathcal{S}^{a} \tag{21a}$$

$$q_{g,t}^{s} \leq p_{g,t}^{s}tan\left(cos^{-1}\left(\underline{pf}\right)\right),\ \forall t\in\mathcal{G}^{a}-\mathcal{S}^{a} \tag{21b}$$

$$p_{g,t}^{s} = p_{st,t}^{s,dch}-p_{st,t}^{s,ch},\ \forall t\in\mathcal{S}^{a} \tag{22a}$$

$$q_{g,t}^{s} = q_{st,t}^{s,dch}-q_{st,t}^{s,ch},\ \forall t\in\mathcal{S}^{a} \tag{22b}$$

$$p_{st,t}^{s,dch} \leq \overline{P}_{g,t}v_{st,t}^{s},\ \forall t\in\mathcal{S}^{a} \tag{22c}$$

$$p_{st,t}^{s,ch} \leq \overline{P}_{g,t}(1-v_{st,t}^{s}),\ \forall t\in\mathcal{S}^{a} \tag{22d}$$

$$q_{st,t}^{s,dch} \geq -p_{st,t}^{s,dch}tan\left(cos^{-1}\left(\underline{pf}\right)\right),\ \forall t\in\mathcal{S}^{a} \tag{23a}$$

$$q_{st,t}^{s,dch} \leq p_{st,t}^{s,dch}tan\left(cos^{-1}\left(\underline{pf}\right)\right),\ \forall t\in\mathcal{S}^{a} \tag{23b}$$

$$q_{st,t}^{s,ch} \geq -p_{st,t}^{s,ch}tan\left(cos^{-1}\left(\underline{pf}\right)\right),\ \forall t\in\mathcal{S}^{a} \tag{23c}$$

$$q_{st,t}^{s,ch} \leq p_{st,t}^{s,ch}tan\left(cos^{-1}\left(\underline{pf}\right)\right),\ \forall t\in\mathcal{S}^{a} \tag{23d}$$

$$\min_{\mathcal{X}_{bid}^{s}} \sum\nolimits_{t\in\mathcal{G}^{a}-\mathcal{S}^{a}}\left(p_{g,t}^{s,ab}+q_{g,t}^{s,ab}\right)+m\sum\nolimits_{t\in\mathcal{S}^{a}}\left(p_{g,t}^{s,ab}+q_{g,t}^{s,ab}\right)+\sum\nolimits_{t\in\mathcal{G}^{a}}\left(r_{u,t}^{s,ab}+r_{d,t}^{s,ab}\right) \tag{24}$$

subject to

(16, 17b-17h, 18, 19, 21-23)

- **Step 6: Check the convergence of modified bids**

This step checks that the bid value is not too different from the bid value of the previous iteration if it passed the criteria in Step 3. Since the power flow equation is nonlinear and we consider it by linearizing it in the form of sensitivity, there may be an error from the actual optimal value. To reduce this, the following iteration step is performed:

$$max\left\{\left|\overline{p_{g,t}^{s}}-\overline{p_{g,t}^{s-1}}\right|\middle|t\in\mathcal{G}^{a}\right\}\leq\epsilon_{Bid} \tag{25a}$$

$$max\left\{\left|\underline{p_{g,t}^{s}}-\underline{p_{g,t}^{s-1}}\right|\middle|t\in\mathcal{G}^{a}\right\}\leq\epsilon_{Bid} \tag{25b}$$

$$max\left\{\left|q_{g,t}^{s}-q_{g,t}^{s-1}\right|\middle|t\in\mathcal{G}^{a}\right\}\leq\epsilon_{Bid} \tag{25c}$$

- **Step 7: Additional calculation of upper and lower bounds for outputs of storage resources**

In this step, the DSO calculates and communicates to DERAs the upper and lower bounds of storage output that will not affect the reliability of the grid.

$$\max_{\substack{\boldsymbol{p}_{st},\ \boldsymbol{q}_{st}\\ \Delta\boldsymbol{p}_{st},\ \Delta\boldsymbol{q}_{st}}} \sum\nolimits_{t\in\mathcal{S}^{a}} p_{st,t} \tag{26a}$$

subject to

$$p_{st,t}=p_{g,t}^{end}+\Delta p_{st,t},\ \forall t\in\mathcal{S}^{a} \tag{26b}$$

$$q_{st,t}=q_{g,t}^{end}+\Delta q_{st,t},\ \forall t\in\mathcal{S}^{a} \tag{26c}$$

$$\overline{p_{g,t}^{ini}}-\underline{\Delta P}_{g,t}+r_{d,t}^{end}\leq p_{st,t}\leq\overline{p_{g,t}^{ini}}+\overline{\Delta P}_{g,t}-r_{u,t}^{end},\ \forall t\in\mathcal{S}^{a} \tag{26d}$$

$$q_{g,t}^{ini}-\underline{\Delta Q}_{g,t}\leq q_{st,t}\leq q_{g,t}^{ini}+\overline{\Delta Q}_{g,t},\ \forall t\in\mathcal{S}^{a} \tag{26e}$$

$$-OV_{i,a}^{end}\geq\sum\nolimits_{j\in\mathcal{N}}\left(Sn_{ovp,i,j}^{end}\sum\nolimits_{t\in\mathcal{S}_{j}^{a}}\Delta p_{st,t}+Sn_{ovq,i,j}^{end}\sum\nolimits_{t\in\mathcal{S}_{j}^{a}}\Delta q_{st,t}\right),\ \forall i\in\mathcal{N} \tag{26f}$$

$$-UV_{i,a}^{end}\geq\sum\nolimits_{j\in\mathcal{N}}\left(Sn_{uvp,i,j}^{end}\sum\nolimits_{t\in\mathcal{S}_{j}^{a}}\Delta p_{st,t}+Sn_{uvq,i,j}^{end}\sum\nolimits_{t\in\mathcal{S}_{j}^{a}}\Delta q_{st,t}\right),\ \forall i\in\mathcal{N} \tag{26g}$$

$$-UF_{l,a}^{end}\geq\sum\nolimits_{k\in\mathcal{N}}\left(Sn_{usp,l,k}^{end}\sum\nolimits_{t\in\mathcal{S}_{k}^{a}}\Delta p_{st,t}+Sn_{usq,l,k}^{end}\sum\nolimits_{t\in\mathcal{S}_{k}^{a}}\Delta q_{st,t}\right),\ \forall l\in\mathcal{L} \tag{26h}$$

$$-DF_{l,a}^{end}\geq\sum\nolimits_{k\in\mathcal{N}}\left(Sn_{dsp,l,k}^{end}\sum\nolimits_{t\in\mathcal{S}_{k}^{a}}\Delta p_{st,t}+Sn_{dsq,l,k}^{end}\sum\nolimits_{t\in\mathcal{S}_{k}^{a}}\Delta q_{st,t}\right),\ \forall l\in\mathcal{L} \tag{26i}$$

$$max\left(\left(p_{st,t}+r_{u,t}^{end}\right)^{2},\left(p_{st,t}-r_{d,t}^{end}\right)^{2}\right)+\left(q_{st,t}\right)^{2}\leq\overline{P}_{g,t}^{2},\ \forall t\in\mathcal{S}^{a} \tag{26j}$$

$$\left|q_{st,t}\right|\leq\left|p_{st,t}tan\left(cos^{-1}\left(\underline{pf}\right)\right)\right|,\ \forall t\in\mathcal{S}^{a} \tag{26k}$$

$$q_{st,t}\geq\frac{tan\left(cos^{-1}\left(\underline{pf}\right)\right)}{\sqrt{tan\left(cos^{-1}\left(\underline{pf}\right)\right)^{2}+1}-1}\left(p_{st,t}+r_{u,t}^{end}-\overline{P}_{g,t}\right),\ \forall t\in\mathcal{S}^{a} \tag{27a}$$

$$q_{st,t}\leq-\frac{tan\left(cos^{-1}\left(\underline{pf}\right)\right)}{\sqrt{tan\left(cos^{-1}\left(\underline{pf}\right)\right)^{2}+1}-1}\left(p_{st,t}+r_{u,t}^{end}-\overline{P}_{g,t}\right),\ \forall t\in\mathcal{S}^{a} \tag{27b}$$

$$q_{st,t} \geq -\frac{tan\left(cos^{-1}\left(\underline{pf}\right)\right)}{\sqrt{tan\left(cos^{-1}\left(\underline{pf}\right)\right)^2+1-1}}\left(p_{st,t} - r_{d,t}^{end} + \overline{P}_{g,t}\right),\ \forall t \in \mathcal{S}^a \tag{27c}$$

$$q_{st,t} \leq \frac{tan\left(cos^{-1}\left(\underline{pf}\right)\right)}{\sqrt{tan\left(cos^{-1}\left(\underline{pf}\right)\right)^2+1-1}}\left(p_{st,t} - r_{d,t}^{end} + \overline{P}_{g,t}\right),\ \forall t \in \mathcal{S}^a \tag{27d}$$

$$p_{st,t} = p_{st,t}^{dch} - p_{st,t}^{ch},\ \forall t \in \mathcal{S}^a \tag{28a}$$

$$q_{st,t} = q_{st,t}^{dch} - q_{st,t}^{ch},\ \forall t \in \mathcal{S}^a \tag{28b}$$

$$p_{st,t}^{dch} \leq \overline{P}_{g,t} v_{st,t},\ \forall t \in \mathcal{S}^a \tag{28c}$$

$$p_{st,t}^{ch} \leq \overline{P}_{g,t}(1 - v_{st,t}),\ \forall t \in \mathcal{S}^a \tag{28d}$$

$$q_{st,t}^{dch} \geq -p_{st,t}^{dch} tan\left(cos^{-1}\left(\underline{pf}\right)\right),\ \forall t \in \mathcal{S}^a \tag{29a}$$

$$q_{st,t}^{dch} \leq p_{st,t}^{dch} tan\left(cos^{-1}\left(\underline{pf}\right)\right),\ \forall t \in \mathcal{S}^a \tag{29b}$$

$$q_{st,t}^{ch} \geq -p_{st,t}^{ch} tan\left(cos^{-1}\left(\underline{pf}\right)\right),\ \forall t \in \mathcal{S}^a \tag{29c}$$

$$q_{st,t}^{ch} \leq p_{st,t}^{ch} tan\left(cos^{-1}\left(\underline{pf}\right)\right),\ \forall t \in \mathcal{S}^a \tag{29d}$$

A flow chart representation of the entire algorithm from step 1 to step 6 is shown in Figure 3.

4.3. Improvements in the Implemented Algorithm Compared to Previous Work

This demonstration study tried to improve some settings and shorten the execution time of the proposed algorithm in the previous theoretical studies through trial and error in the practical implementation. In this section, we will express and explain the meaning of the additions for each group of formulas in the previous section compared to that of the algorithm in [7].

- **Equations (4b,d), (6), (8), (9b,d) and (10b,d)**

When selecting cases that exceed the risk criteria, overvoltage, undervoltage, reverse overcurrent, and forward overcurrent are all checked by adding Equation (4b,d).

- **Equation (5c)**

While the previous algorithm in [7] treated the participation of DERs in ancillary services as an uncertainty for the DSO, this study changes the formula to consider the reserve bidding as well as the uncertainty of DERs.

- **Equations (13a–d), (14a–d), (15a–d) and (16a–h)**

During the course of the demonstration project, the DERAs participating in the demonstration claimed that if too much discretion is given to them, this will entail high costs and project management uncertainty in the project's early stages, and that they preferred a bid-modification guideline in the form of a range of absolute output values provided by the DSO. This allowed for the iteration process to proceed but limited the information exchanged by modifying the initial algorithm to a range of absolute output values in a matrix. In addition, while the original study only considered a single DERA, we added formulas to distribute the information in the case of multiple DERAs by calculating their contribution to the number of violations.

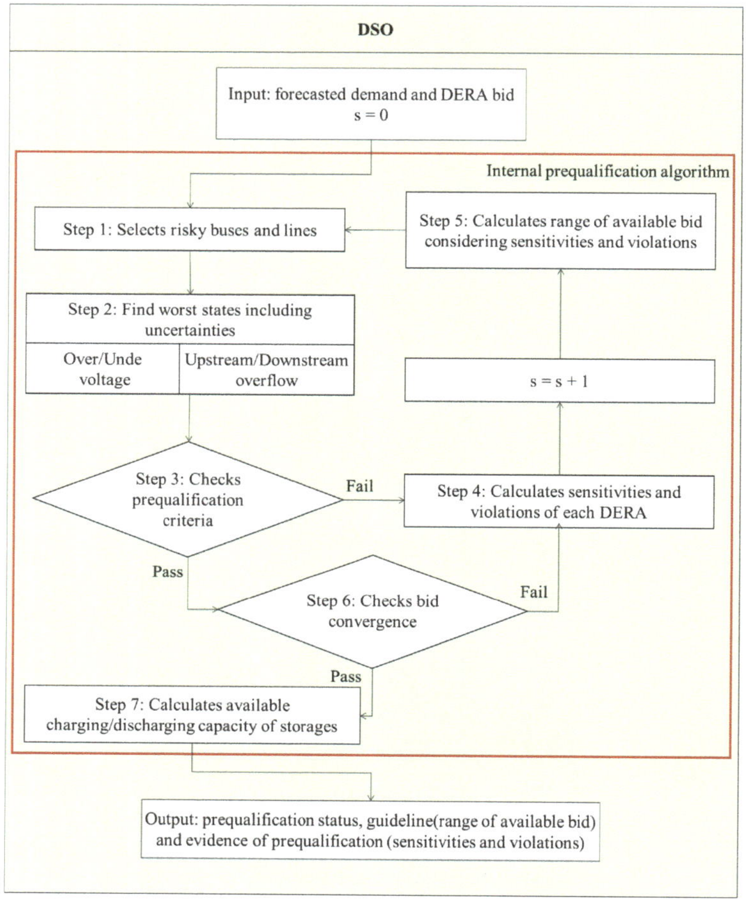

Figure 3. Flow chart of internal prequalification algorithm.

- **Equation (17a–j)**

When the DERA has storage and strategically intends to reduce its output at a certain time and increase its sales at another time, the DSO makes a decision regardless of the DERA's intention. This was a violation of the DERA's free-will and discretion, so we added the above equations to modify the prequalification to best honor the DERA's strategy in the original bid.

- **Equations (18a–h), (19a–d), (20a,b), (21a,b), (22a–d), (23a–d) and (24)**

In this paper, auxiliary variables were used to linearize the objective function and conditional constraints in absolute value form in the DSO control procedure, and the power factor operating ranges of the DERs were linearized to avoid considering regions that cannot be solved in the first place. These measures were taken to improve the speed of the algorithm during the validation process.

- **Equation (25a–c)**

In addition, the real-world demonstration results showed that the process of calculating corrective bids in the algorithm did not converge and oscillated, with large deviations in value, so we added a function that finds a corrective bid that does not cause a violation and terminates itself when this happens.

- **Equations (26a–k), (27a–d), (28a–d) and (29a–d)**

We devised a method that allows for the DERA to schedule storage resources for its own benefit but allows for DSOs to calculate and communicate the upper and lower biddable output limits in a contextualized manner that does not threaten the grid.

5. Results of Field Experiment

We conducted a demonstration using a real distribution line as the target site to verify the validity of the proposed prequalification scheme. In this section, we describe the configuration of the demonstration site and show that the proposed scheme works well in the real world through the demonstration results.

5.1. System Configurations

The demonstration was conducted on 22.9 kV lines named Oh-ryong D/L and Cheong-sa D/L in the city of Gwangju, Korea. The grid diagrams of these D/L are shown in Figures 4 and 5. There are 7.858 MW of photovoltaic system (PV) connected to Oh-ryong D/L, of which 4.8 MW are already under firm-connection arrangements with KEPCO and cannot be contractually controlled. These resources are shown in the grid diagram with a dotted line. In addition, a total of 17.3 MW/69.2 MWh of energy storage system (ESS) is connected, of which 6 MW/24 MWh are not participating in the demonstration, and have existing contractual charge and discharge scheduling arrangements with the utility.

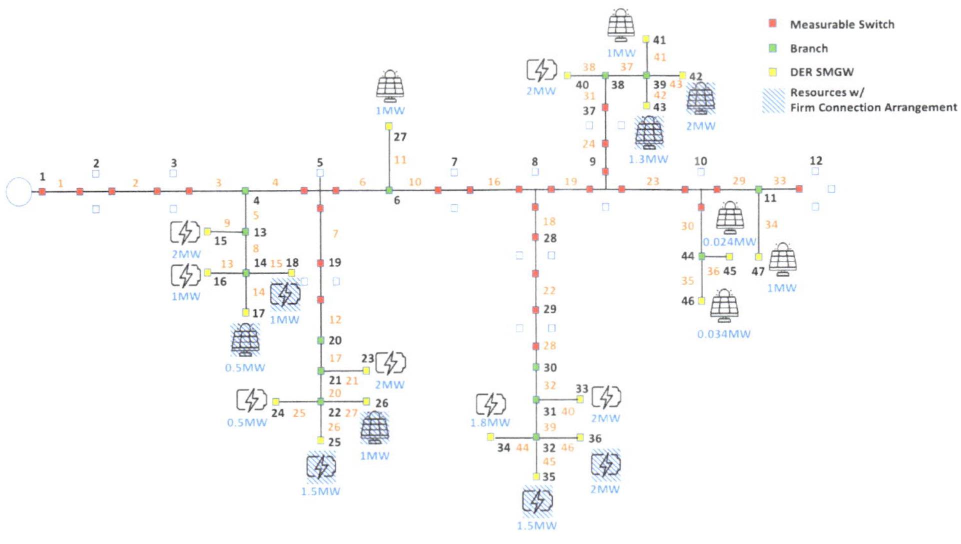

Figure 4. Network configuration of demonstration site: Oh–ryong D/L.

There is 2.117 MW of the photovoltaic system (PV) connected to Cheong–sa D/L, of which 2.5 MW is already under firm-connection arrangements with KEPCO and cannot be contractually controlled. These resources are shown in the grid diagram with a dotted line. In addition, a total of 18.2 MW/73 MWh of energy storage system (ESS) is connected, of which 2.5 MW/10 MWh does not participate in the demonstration and has existing contractual charge and discharge scheduling arrangements with the utility.

These resources are also shown on the grid map with oblique lines on the item. If all the connected PV and ESS were to fully output, the total generation of D/L would exceed the capacity of 10 MW, which could potentially result in a grid constraint violation.

This situation is not allowed under KEPCO's existing connection regulations, but we temporarily set it up like this for the demonstration purposes. As there was no device to measure the output of each DER, the voltage of the point, and the phase angle, or to deliver the information to the operating system, this project installed smart meter gateway devices in the yellow-colored part of the grid map and collected the information.

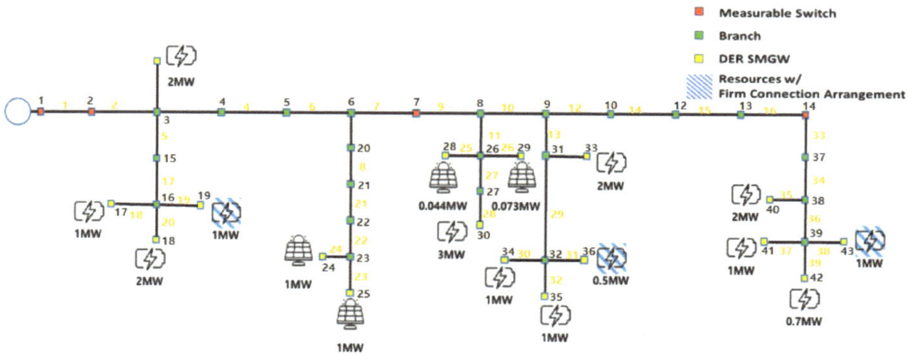

Figure 5. Network configuration of demonstration site: Cheong–sa D/L.

5.2. Scenario Configuration

First, it was assumed that only one DERA was participating in the prequalification that took place the day before. This was necessary because only one entity was involved in the project as the DERA. This DERA wanted to bid on the wholesale market using its contracted DERs, and before doing so, it submitted its bid information to the DSO for prequalification to ensure that this did not cause problems with the reliability of the distribution system. The missed resources, which are marked in the oblique lines in Figures 4 and 5, and are not included in this scheme, are recognized as negative loads.

In this study, we show the operation results for 24 h on a sunny day in the winter season during the demonstration period. On that day, the net-load of the entire D/L had the following pattern when measured at node 1 in the network shown in the following Figure 6, and the solar power generation had the pattern shown in the following Figure 7.

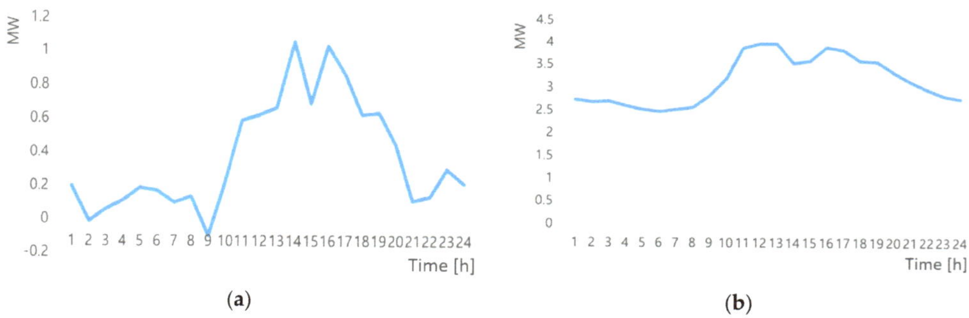

Figure 6. (**a**) Net–load pattern of the Oh–ryong and (**b**) net–load pattern of the Cheong–sa D/L.

In this situation, the DERA, which wanted to participate in the wholesale market by aggregating the resources existing in this system, acted according to the procedure described in Section 3 of this paper.

The scheme proposed in this study takes a robust approach to manage the uncertainty of load and generation, with 5% uncertainty from the day-ahead forecast values, and the robustness of the scheme is verified by the actual load and generation realized in the real-time operation stage. As a constraint of the distribution system, we set criteria to keep the voltage between 0.95 p.u. and 1.05 p.u. Additionally, in step 1 of Section 4, which selects the risky bus and risky line, the voltage was less than or equal to 0.96; more than 1.04 was needed to select the risky bus, and the voltage was more than 60% of the line capacity before selecting the risky line. These appropriate criteria were established after a significant amount of trial and error.

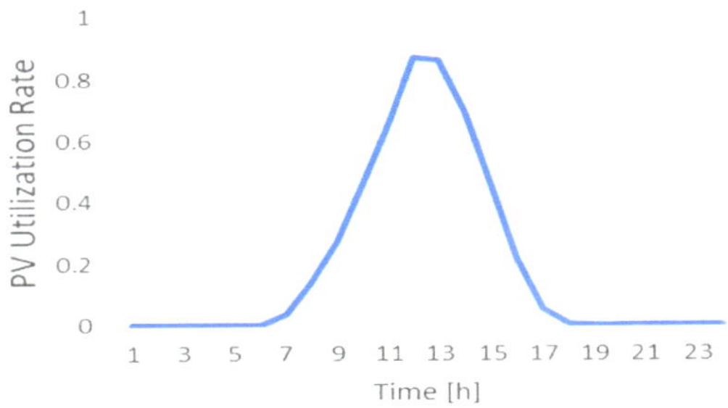

Figure 7. PV utilization rate pattern on a sunny day in the winter season.

5.3. Verification of the Proposed Scheme

For the Oh–ryong D/L, out of the total 24 h of the prequalification scheme operation, we show the results for 3 a.m., 10 a.m., 11 a.m. and 5 p.m., when constraint violations occurred due to DERA's initial bids, for voltages or line flows, respectively.

It can be seen from Figure 8 that the site did not experience any explicit voltage violations from the initial bids for these four time slots, guidelines were communicated to resolve possible constraints on line flows, and the DERA responded with modified bids, resulting in a margin on the voltage side. The red line in Figure 8 is the voltage limit range, and the blue line is the voltage of each node.

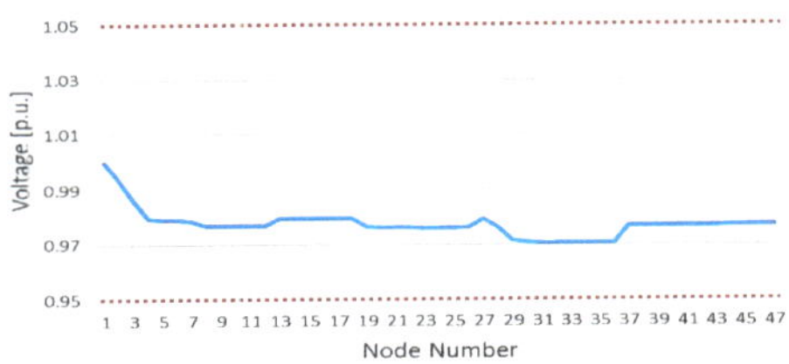

(**a-1**) Node voltages with the initial bids at 3 a.m.

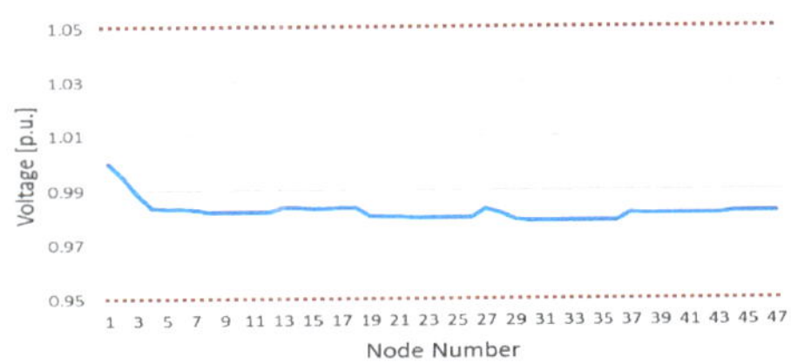

(**a-2**) Node voltages with the 1st modified bids at 3 a.m.

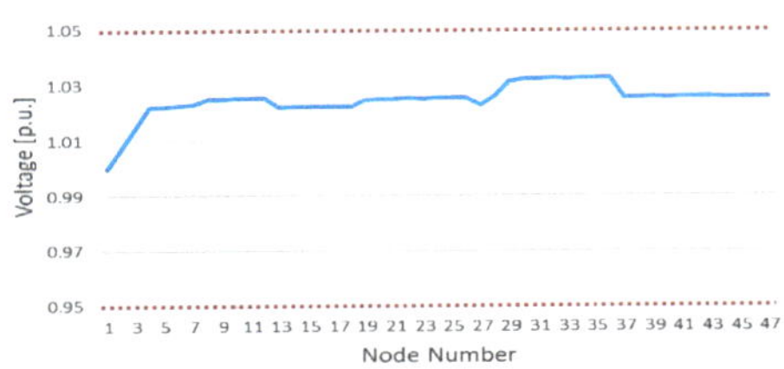

(**b-1**) Node voltages with the initial bids at 10 a.m.

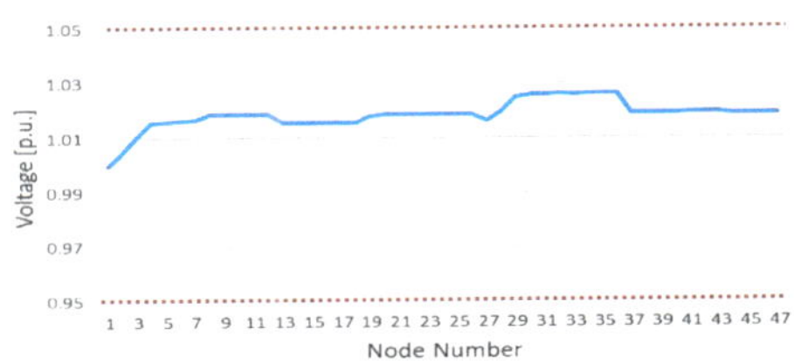

(**b-2**) Node voltages with the 1st modified bids at 10 a.m.

Figure 8. *Cont.*

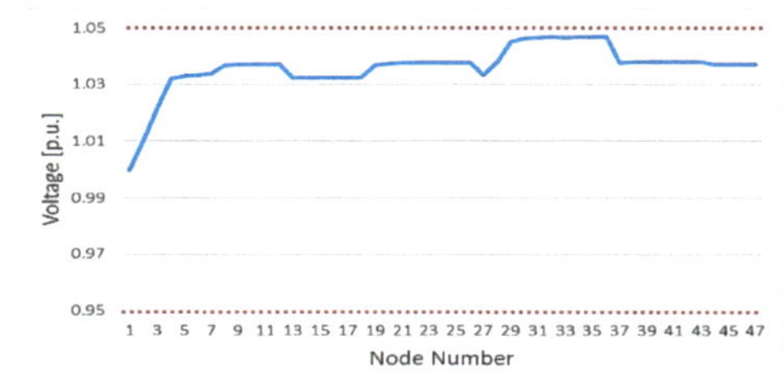

(**c-1**) Node voltages with the initial bids at 11 a.m.

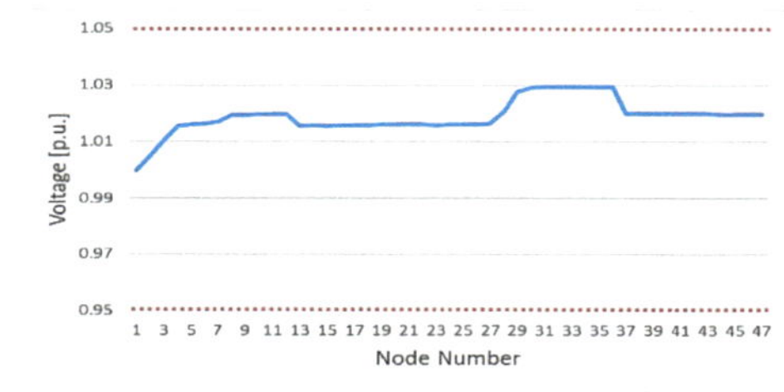

(**c-2**) Node voltages with the 1st modified bids at 11 a.m.

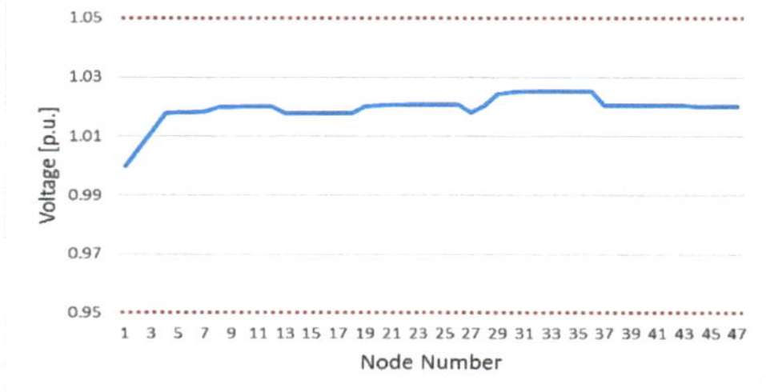

(**d-1**) Node voltages with the initial bids at 5 p.m.

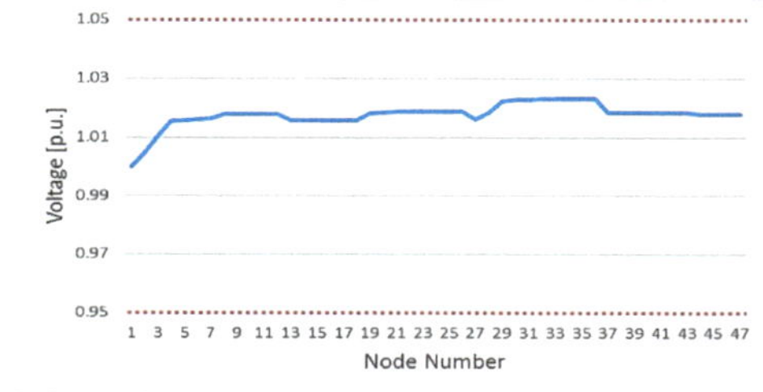

(**d-2**) Node voltages with the 1st modified bids at 5 p.m.

Figure 8. Node voltages with the initial bids (-1) and the 1st modified bids (-2) (Oh–ryong D/L).

However, as shown in Figure 9, the initial bids of the DERA were analyzed to cause line flow violations in all four time slots. The result at 3 a.m. was a forward overflow and all other time slots showed reverse overflows. The DSO predicted this situation in advance through the prequalification algorithm proposed in this paper and provided bid modification guidelines to DERA. This information was provided to the DERA in the form of the guidelines shown in Table 3, and the DERA modified its bid within the ranges presented in the guidelines and resubmitted it to the DSO. The result was that all constraints were resolved in the next iteration, as shown in Figure 9. The red line in Figure 9 is the distribution line flow limit range, and the blue line is the line flow of each node.

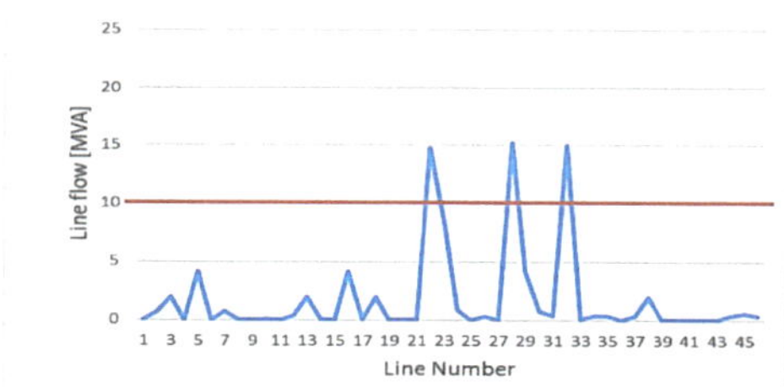

(**a-1**) Line flows with the initial bids at 3 a.m.

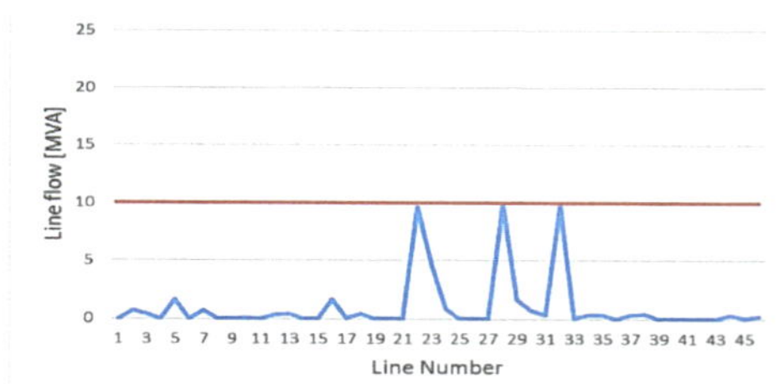

(**a-2**) Line flows with the 1st modified bids at 3 a.m.

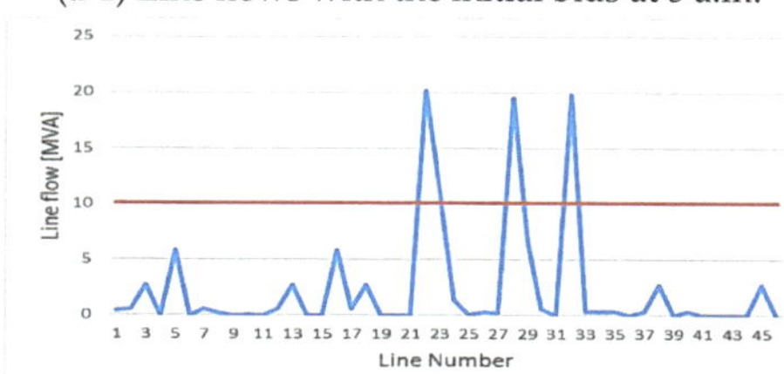

(**b-1**) Line flows with the initial bids at 10 a.m.

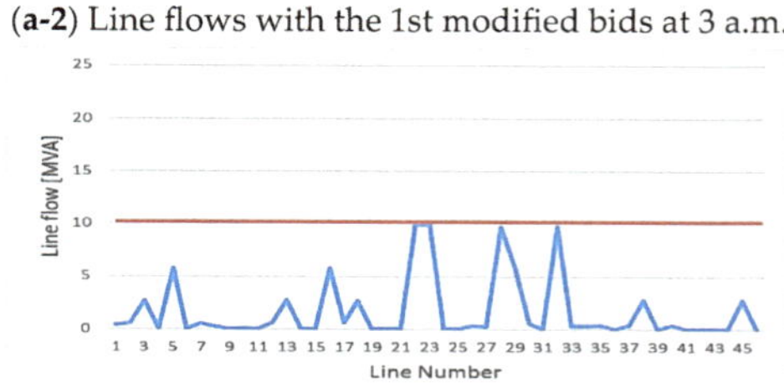

(**b-2**) Line flows with the 1st modified bids at 10 a.m.

Figure 9. *Cont.*

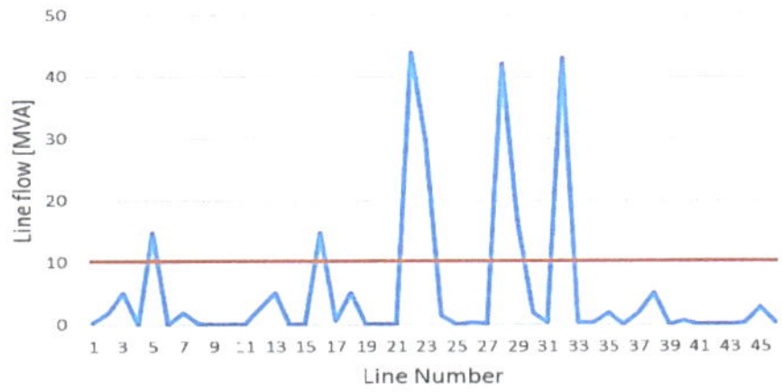

(**c-1**) Line flows with the initial bids at 11 a.m.

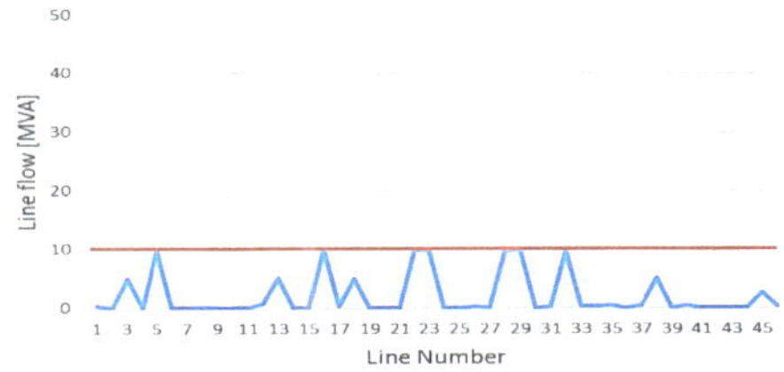

(**c-2**) Line flows with the 1st modified bids at 11 a.m.

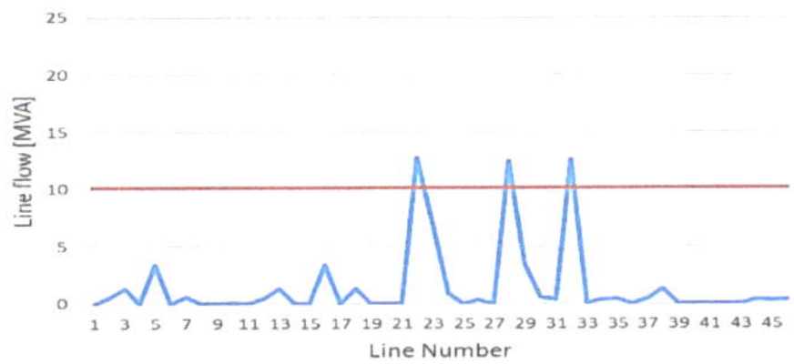

(**d-1**) Line flows with the initial bids at 5 p.m.

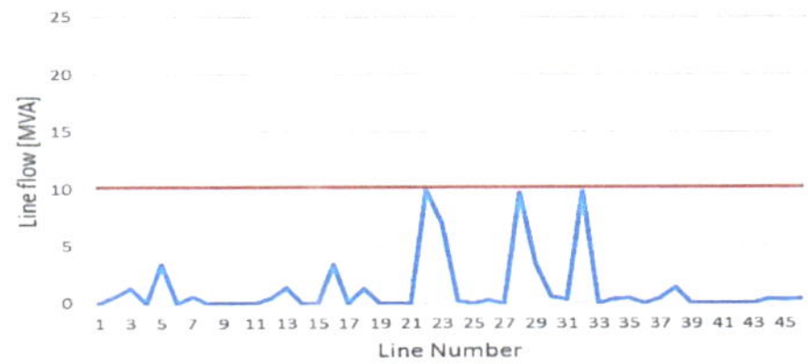

(**d-2**) Line flows with the 1st modified bids at 5 p.m.

Figure 9. Line flows with the initial bids (-1) and the 1st modified bids (-2) (Oh–ryong D/L).

Table 3. Example of bid modification guidelines for DERA from the prequalification algorithm of DSO at 11 a.m.

DER Index	DER Type	Hour Index	Max. Allowable Generation Power [kW]	Max. Allowable Discharging Power [kW]	Max. Allowable Charging Power [kW]
15	ESS	11	-	−1909.22	−1940.00
16	ESS	11	-	970.00	−970.00
23	ESS	11	-	−1940.00	−1940.00
24	ESS	11	-	−43.71	−485.00
27	PV	11	8.77	-	0
33	ESS	11	-	1940.00	−1940.00
34	ESS	11	-	1746.00	−1746.00
40	ESS	11	-	0	−1940.00
41	PV	11	746.10	0	0
45	PV	11	17.91	0	0
46	PV	11	25.37	0	0
47	PV	11	454.26	0	0

In the Cheong–sa D/L, out of the total 24 h of the prequalification scheme operation, we will show the results for 3 a.m., 10 a.m., 11 a.m., 1 p.m., 2 p.m., 5 p.m. and 6 p.m., when constraint violations occurred due to DERA's initial bids, for voltages or line flows, respectively.

It can be seen from Figure 10 that the site did not experience any explicit voltage violations from the initial bids for these four time slots, that guidelines were communicated to resolve possible constraints on line flows, and the DERA responded with modified bids, resulting in a margin on the voltage side. The red line in Figure 10 is the voltage limit range, and the blue line is the voltage of each node.

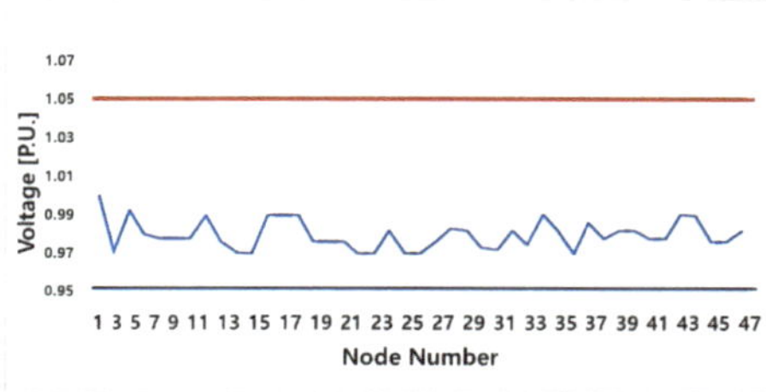

(**a-1**) Node voltages with the initial bids at 3 a.m.

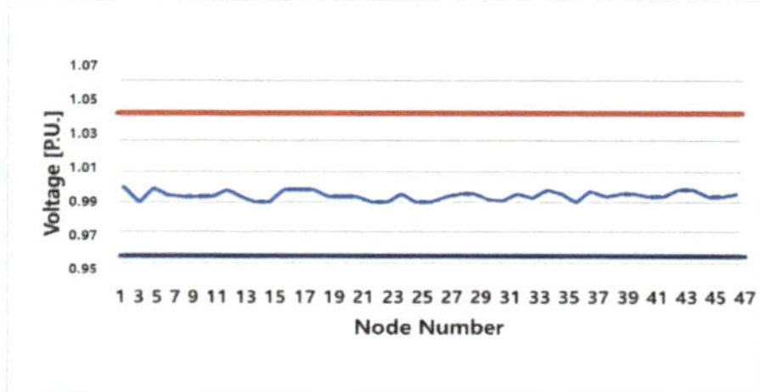

(**a-2**) Node voltages with the 1st modified bids at 3 a.m.

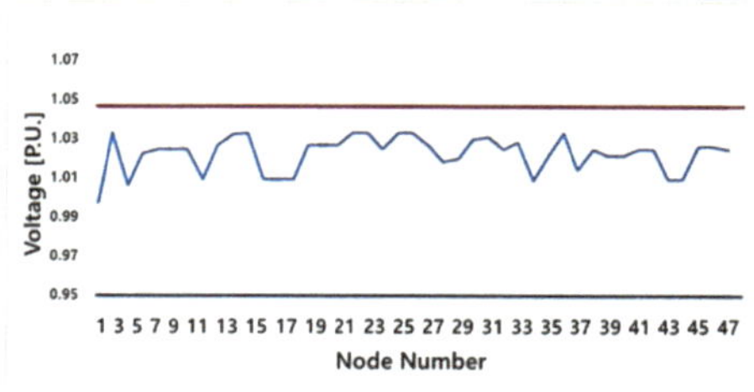

(**b-1**) Node voltages with the initial bids at 10 a.m.

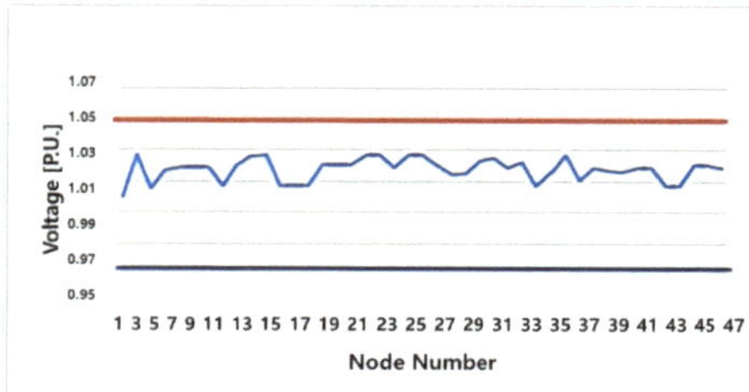

(**b-2**) Node voltages with the 1st modified bids at 10 a.m.

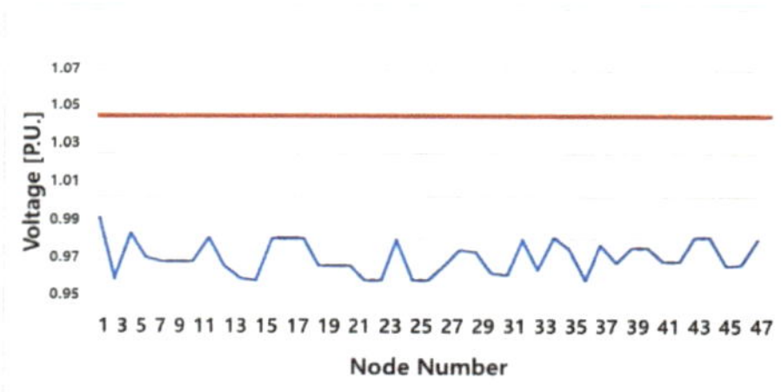

(**c-1**) Node voltages with the initial bids at 11 a.m.

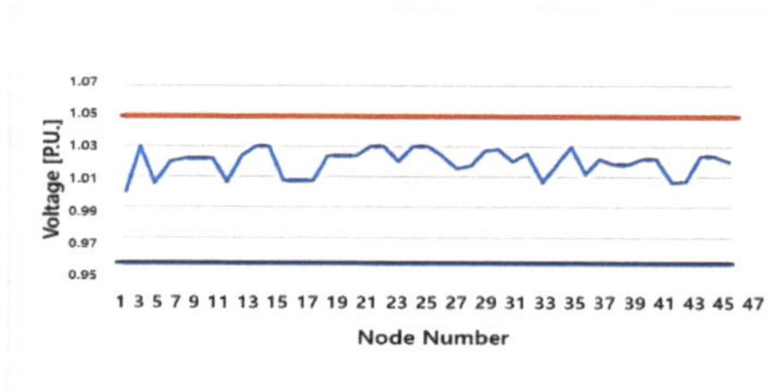

(**c-2**) Node voltages with the 1st modified bids at 11 a.m.

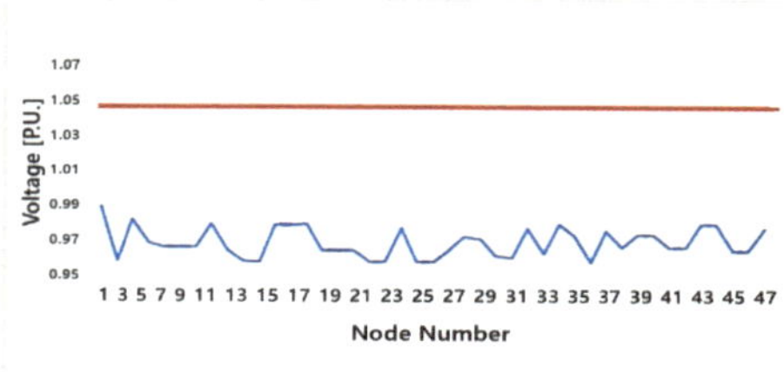

(**d-1**) Node voltages with the initial bids at 1 p.m.

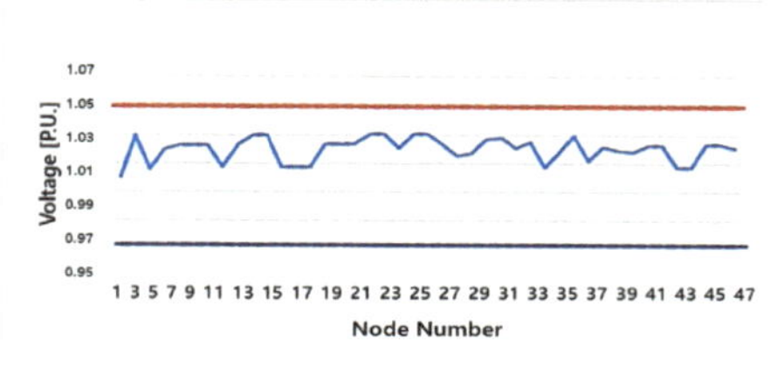

(**d-2**) Node voltages with the 1st modified bids at 1 p.m.

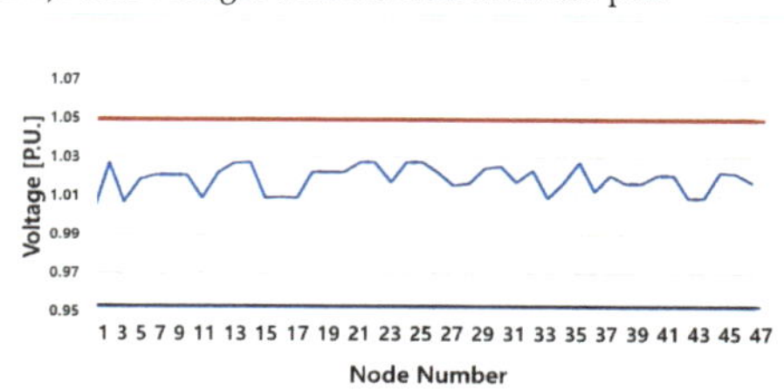

(**e-1**) Node voltages with the initial bids at 2 p.m.

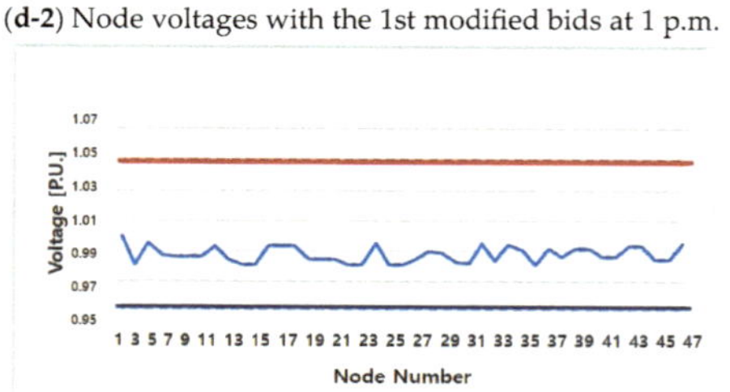

(**e-2**) Node voltages with the 1st modified bids at 2 p.m.

Figure 10. *Cont.*

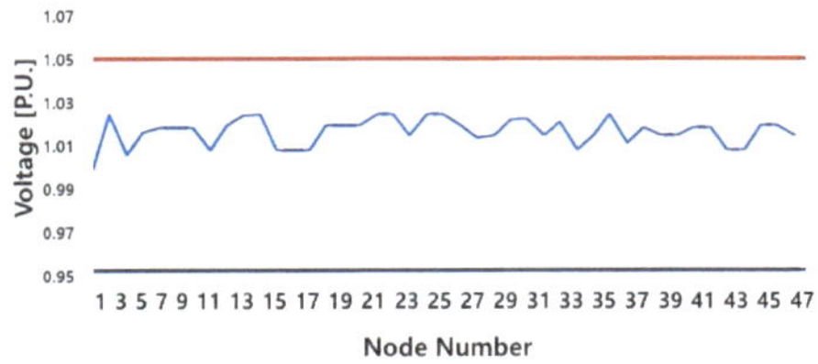

(**f-1**) Node voltages with the initial bids at 5 p.m.

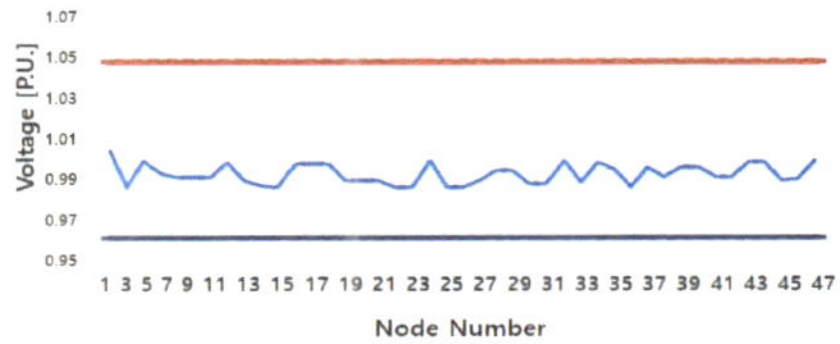

(**f-2**) Node voltages with the 1st modified bids at 5 p.m.

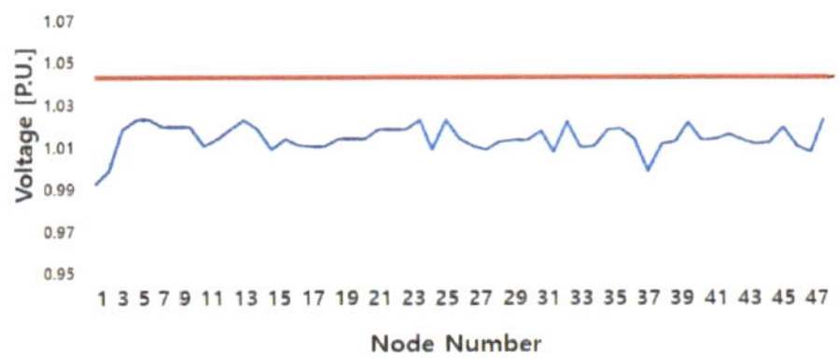

(**g-1**) Node voltages with the initial bids at 6 p.m.

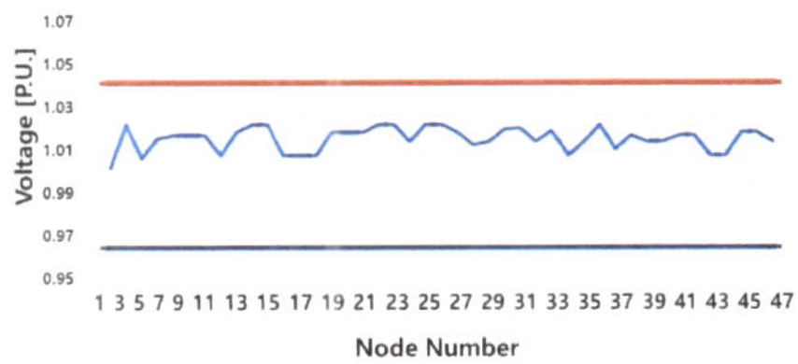

(**g-2**) Node voltages with the 1st modified bids at 6 p.m.

Figure 10. Node voltage with the initial bids (-1) and the 1st modified bids (-2) (Cheong–sa D/L).

However, as shown in Figure 11, the initial bids of the DERA were analyzed to cause line flow violations in all seven time slots. The result was that all constraints were resolved in the next iteration, as shown in Figure 11. The red line in Figure 11 is the distribution line flow limit range, and the blue line is the line flow of each node.

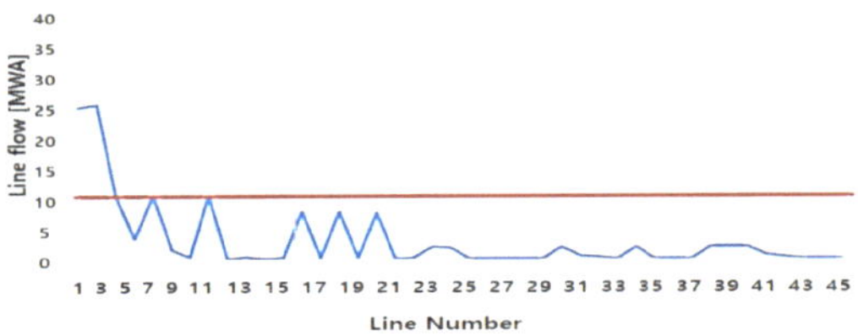

(**a-1**) Line flows with the initial bids at 3 a.m.

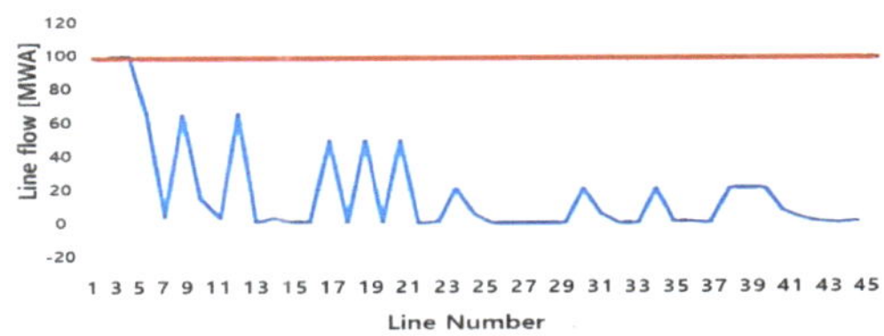

(**a-2**) Line flows with the 1st modified bids at 3 a.m.

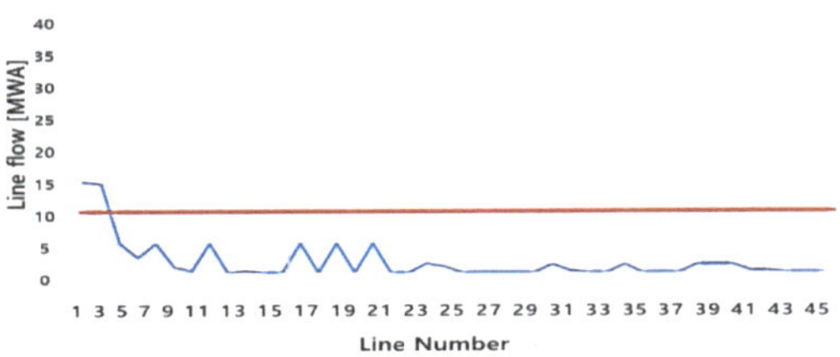

(**b-1**) Line flows with the initial bids at 10 a.m.

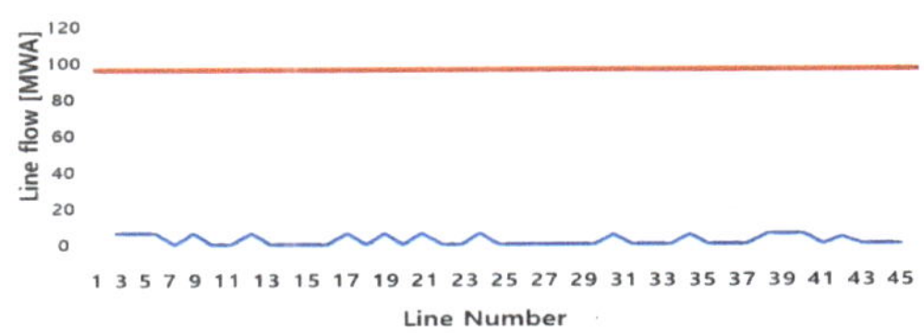

(**b-2**) Line flows with the 1st modified bids at 10 a.m.

Figure 11. *Cont.*

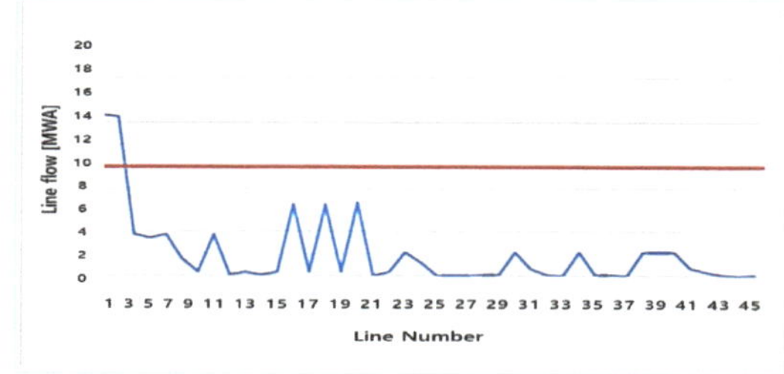

(**c-1**) Line flows with the initial bids at 11 a.m.

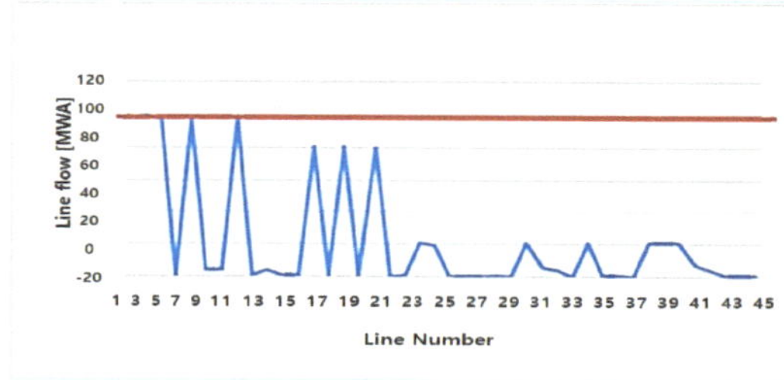

(**c-2**) Line flows with the 1st modified bids at 11 a.m.

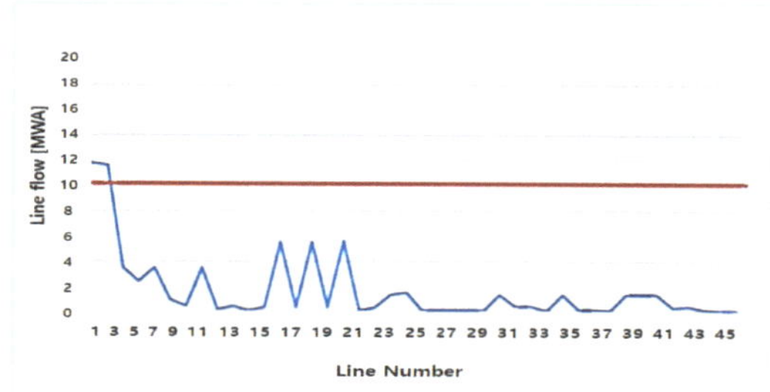

(**d-1**) Line flows with the initial bids at 1 p.m.

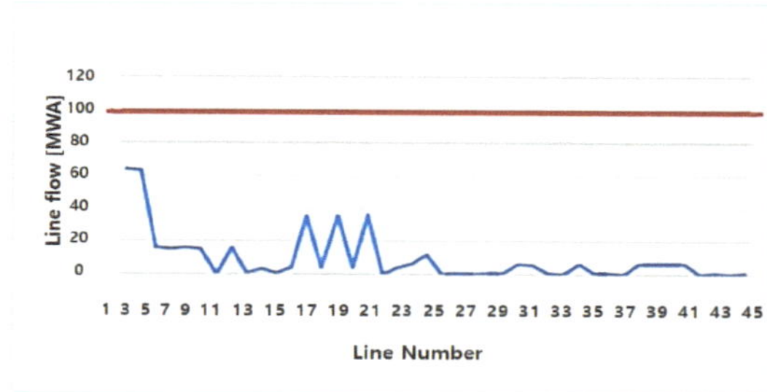

(**d-2**) Line flows with the 1st modified bids at 1 p.m.

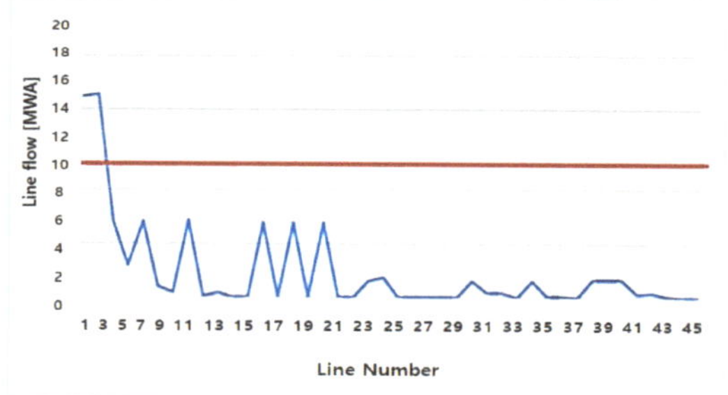

(**e-1**) Line flows with the initial bids at 2 p.m.

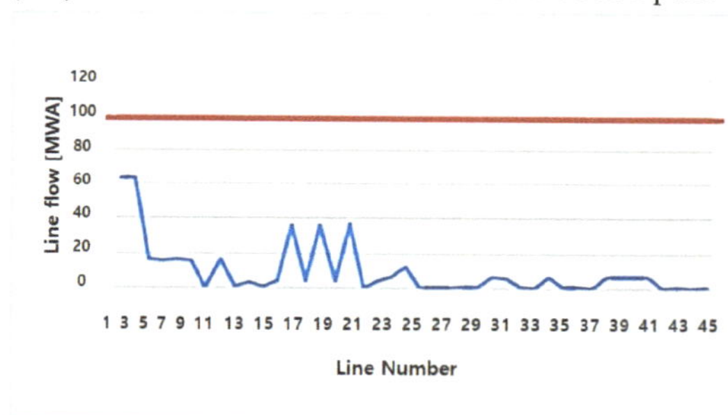

(**e-2**) Line flows with the 1st modified bids at 2 p.m.

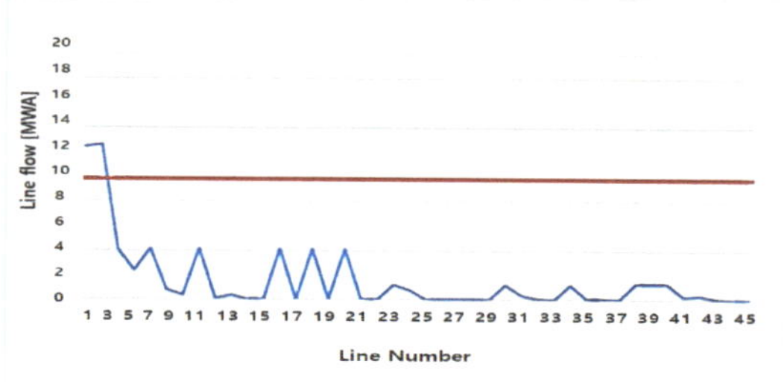

(**f-1**) Line flows with the initial bids at 5 p.m.

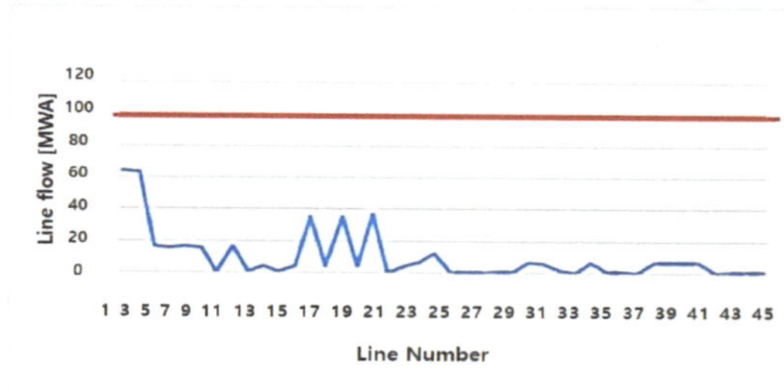

(**f-2**) Line flows with the 1st modified bids at 5 p.m.

Figure 11. *Cont.*

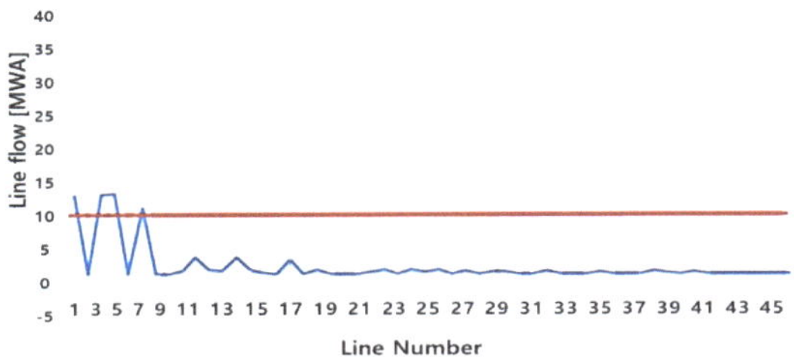

(g-1) Line flows with the initial bids at 6 p.m.

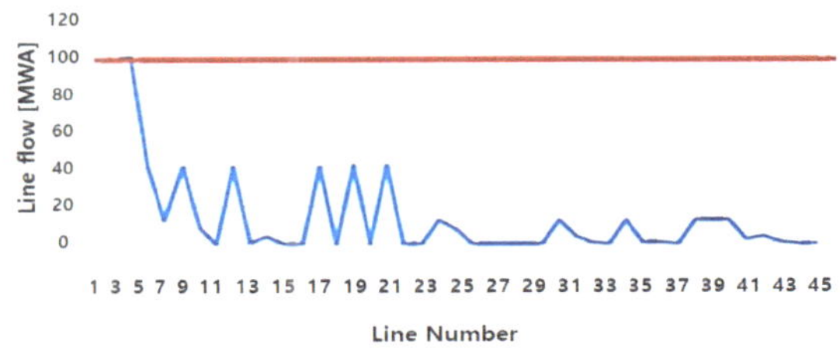

(g-2) Line flows with the 1st modified bids at 6 p.m.

Figure 11. Line flows with the initial bids (-1) and the 1st modified bids (-2) (Cheong–sa D/L).

Through this process, the DERA was validated by the DSO so that its bid did not threaten the reliability of the distribution grid. What we wanted to verify through this demonstration was whether the overall algorithm results could ensure the reliability of the distribution system in a real system, and whether the interaction process between DSO and DERA could be completed within a limited time. It can be concluded that satisfactory results were obtained for the above situations.

6. Conclusions

This paper aims to summarize and share the demonstration result of KEPCO's project consortium to create a TSO-DSO-DERA interaction system. It describes the process of improving the theoretical algorithm proposed in papers [7,8] to apply to the actual demonstration site and verification through actual field experiments.

In the process of the demonstration project, the roles and responsibilities of each assumed actor, in theory, did not change, but unlike the premise of the existing algorithm, which emphasizes DERA's discretion, in reality, the DERA, which participated in the demonstration project, did not welcome too much discretion and information-sharing, which is an impressive lesson learned. Of course, this tendency is obviously due to the fact that such a system does not exist at present, and the relevant knowledge is in its infancy. In addition, it was not easy to achieve the same algorithm performance as in the lab test environment in the field environment, as expected, and to solve this problem, mathematically relaxing the non-linear components as much as possible and applying computational techniques for parallel processing according to time index played an important role.

Based on the results of the mentioned demonstration project, a new government-funded project has been launched to expand the application of D/L across Jeju-island in Korea and will utilize the lessons learned and research results derived from this study.

Author Contributions: J.-S.P. and B.-H.K. conceived and designed the experiments; J.-S.P. and B.-H.K. performed the simulation and the experiments; J.-S.P. analyzed the data; J.-S.P. and B.-H.K. wrote the paper. All authors have read and agreed to the published version of the manuscript.

Funding: This research was funded by Energy Technology Evaluation and Planning (KETEP) (grant number: RS-2023-00237679) in 2023.

Data Availability Statement: The authors hereby declares that the data presented in this study will be presented upon request from the corresponding author.

Acknowledgments: This work was supported by a Korea Institute of Energy Technology Evaluation and Planning (KETEP) grant funded by the Korean government (MOTIE) (No. RS-2023-00237679, Development of Demonstration Zone for New Electricity Service Model).

Conflicts of Interest: One of the authors comes from Research Institute, Korea Electric Power Corporation.

References

1. Ismael, S.M.; Abdel Aleem, S.H.E.; Abdelaziz, A.Y.; Zobaa, A.F. State-of-the-ort of hosting capacity in modern power systems with distributed generation. *Renew. Energy* **2019**, *130*, 1002–1020. [CrossRef]
2. US Federal Energy Regulatory Commission. *FERC Order No. 2222: Participation of Distributed Energy Resource Aggregations in Markets Operated by Regional Transmission Organizations and Independent System Operators*; US Federal Energy Regulatory Commission: Washington, DC, USA, 2020.
3. The European Parliament and The Council of The European Union. Regulation (EU) 2019/943 of the European Parliament and of the Council of 5 June 2019 on the Internal Market for Electricity. *Off. J. Eur. Union* **2019**, *158*, 54–124.
4. The European Parliament and The Council of The European Union. Directive (EU) 2019/944 of the European Parliament and of the Council of 5 June 2019 on common rules for the internal market for electricity and amending Directive 2012/27/EU. *Off. J. Eur. Union* **2019**, *158*, 125–199.
5. Koraki, D.; Strunz, K. Wind and solar power integration in electricity markets and distribution networks through service-centric virtual power plants. *IEEE Trans. Power Syst.* **2018**, *33*, 473–485. [CrossRef]
6. Edmunds, C.; Galloway, S.; Elders, I.; Bukhsh, W.; Telford, R. Design of a DSO-TSO balancing market coordination scheme for decentralised energy. *IET Gener. Transm. Distrib.* **2020**, *14*, 707–718. [CrossRef]
7. Moon, H.S.; Jin, Y.G.; Yoon, Y.T.; Kim, S.W. Prequalification Scheme of a Distribution System Operator for Supporting Wholesale Market Participation of a Distributed Energy Resource Aggregator. *IEEE Access* **2021**, *9*, 80434–80450. [CrossRef]
8. Jeong, C.M.; Kim, S.W.; Park, J.S.; Yoon, Y.T.; Moon, H.S. Robust prequalification process of a distributed system operator considering N-1 contingency. *IET Renew. Power Gener.* **2022**, *16*, 2694–2710. [CrossRef]
9. Park, S.-Y.; Park, S.W.; Son, S.-Y. Optimal VPP operation considering network constraints uncertainty of DSO. *IEEE Access* **2023**, *11*, 8523–8530. [CrossRef]
10. Gerard, H.; Rivero, E.; Six, D. Basic Schemes for TSO-DSO Coordination and Ancillary Services Provision—D 1.3. (Version 1.1). Available online: https://smartnet-project.eu/publications/index.html#tab-id-2 (accessed on 19 January 2017).
11. Omie. Modelo de Funcionamiento de los Mercados Locales de Electricidad: Proyecto IREMEL. 2019. Available online: https://www.omie.es/sites/default/files/2019-12/modelo_de_funcionamiento_mercados_locales_electricidad_v2_0.pdf (accessed on 25 June 2019).
12. Stevens, N.; Merckx, K.; Crucifix, P.; Gómez, I.; Santos-Mugica, M.; Díaz, Á.; Sanjab, A.; Kessels, K.; Rivero, E.; Mou, Y.; et al. D2.1—Markets for DSO and TSO Procurement of Innovative Grid Services: Specification of the Architecture, Operation and Clearing Algorithms. (Version 1.0). 2021. Available online: https://private.coordinet-project.eu/files/documentos/6033b5fe475cdCoordiNet_WP2_D2.1_Markets%20for%20DSO%20and%20TSO%20procurement%20of%20innovative%20grid%20services_V1.0_20.02.2021.pdf (accessed on 20 February 2021).
13. Küfeoğlu, S.; Kim, S.W.; Jin, Y.G. History of electric power sector restructuring in South Korea and Turkey. *Electr. J.* **2019**, *32*, 106666. [CrossRef]

designs

MDPI

Review

A Review of Uncertainties in Power Systems—Modeling, Impact, and Mitigation

Hongji Hu, Samson S. Yu * and Hieu Trinh

School of Engineering, Deakin University, 75 Pigdons Rd, Waurn Ponds, Geelong, VIC 3216, Australia; hj.hu@ieee.org (H.H.); hieu.trinh@deakin.edu.au (H.T.)
* Correspondence: samson.yu@deakin.edu.au

Abstract: A comprehensive review of uncertainties in power systems, covering modeling, impact, and mitigation, is essential to understand and manage the challenges faced by the electric grid. Uncertainties in power systems can arise from various sources and can have significant implications for grid reliability, stability, and economic efficiency. Australia, susceptible to extreme weather such as wildfires and heavy rainfall, faces vulnerabilities in its power network assets. The decentralized distribution of population centers poses economic challenges in supplying power to remote areas, which is a crucial consideration for the emerging technologies emphasized in this paper. In addition, the evolution of modern power grids, facilitated by deploying the advanced metering infrastructure (AMI), has also brought new challenges to the system due to the risk of cyber-attacks via communication links. However, the existing literature lacks a comprehensive review and analysis of uncertainties in modern power systems, encompassing uncertainties related to weather events, cyber-attacks, and asset management, as well as the advantages and limitations of various mitigation approaches. To fill this void, this review covers a broad spectrum of uncertainties considering their impacts on the power system and explores conventional robust control as well as modern probabilistic and data-driven approaches for modeling and correlating the uncertainty events to the state of the grid for optimal decision making. This article also investigates the development of robust and scenario-based operations, control technologies for microgrids (MGs) and energy storage systems (ESSs), and demand-side frequency control ancillary service (D-FCAS) and reserve provision for frequency regulation to ensure a design of uncertainty-tolerance power system. This review delves into the trade-offs linked with the implementation of mitigation strategies, such as reliability, computational speed, and economic efficiency. It also explores how these strategies may influence the planning and operation of future power grids.

Keywords: renewable energy; modeling uncertainties; mitigation approaches; uncertainty-tolerance; reliability; economic efficiency

Citation: Hu, H.; Yu, S.S.; Trinh, H. A Review of Uncertainties in Power Systems—Modeling, Impact, and Mitigation. *Designs* **2024**, *8*, 10. https://doi.org/10.3390/designs8010010

Academic Editors: Surender Reddy Salkuti and Dibin Zhu

Received: 20 December 2023
Revised: 3 January 2024
Accepted: 16 January 2024
Published: 18 January 2024

1. Introduction

The development and evolution of identified uncertainties in power systems have continued to progress over the years. Understanding this evolution is crucial for comprehending the challenges and developments in the field of electrical power generation, transmission, and distribution. The introduction provides a broad summary of the progression of uncertainty. From the late 19th century to the mid-20th century, power systems were marked by relative simplicity, featuring centralized power generation in local power plants [1]. Uncertainty is mainly caused by demand and occasional technical failures. As the power systems grew, the unpredictability of load and outages required manual adjustments. The mid-20th century marked the growth of interconnected power grids to facilitate electricity transfer across regions [2]. As grids expanded, the uncertainty associated with power transfer, transmission losses, and voltage regulation became prominent concerns. The late 20th century saw the integration of RESs, such as wind and solar, into the grid [3].

This introduced a new dimension of uncertainty related to RES generation patterns. The late 20th century and early 21st century witnessed the liberalization of energy markets. Market dynamics, with price fluctuations and market manipulation, added an economic layer of uncertainty to the power system [4].

Advancements in technology introduced both opportunities and challenges in the 21st century. On one hand, grid automation, advanced sensors, and smart grids improved system control and efficiency. On the other hand, these technologies introduced concerns about cybersecurity and the need for increased resilience against cyber threats. In many regions, aging infrastructure has become a significant concern, leading to uncertainty regarding the reliability and lifespan of power system components. Uncertainties raised from the integration of RESs and AMIs have been a serious threat for different Australian regions with dramatic impacts in terms of power system outages [5]. Independent system operators, transmission and distribution network service providers, regulators, and policymakers are collectively working to enhance system resilience through various approaches. AMI networks are being deployed worldwide with penetration rates in Victoria, Australia, reaching as high as 30% for solar customers in 2021 [6]. While existing security designs offer assurance against successful cyber-attacks, the potential consequences of a breach remain high. An analysis of a load drop attack on an AMI system using a cyber–physical model is given in [7].

A crucial distinction in uncertainty sources lies in their nature as continuous or discrete. Variables like renewable generation and prices, allowing any value in a range, are modeled as continuous random variables [8]. They exhibit infinite support, as seen in normally distributed random variables. Conversely, discrete events such as component outages are represented by discrete random variables [9]. The choice of probability distribution depends on the time scale; short-term variations may use a Gaussian distribution, while long-term scenarios follow a Weibull distribution [10]. Uncertainty characterization also varies with decision-makers; one company perceives bids as uncertain, while the system operator views them as deterministic. Figure 1 illustrates the spatial and temporal scale of planning activities in electricity markets, providing a visual representation [11]. Nodal LMP represents the market price at a particular node or location within the power grid, indicating the cost of delivering one additional unit of electricity at a specific point in the power system based on the supply and demand conditions at that location. Uncertainty significantly grows with the expansion of the system size and longer estimated time scales. In this case, the decision variables for long-term transmission level planning and short-term local network operation are different. In long-term planning, the focus may shift toward assessing system adequacy through extreme scenarios rather than considering the entire distribution.

In the future, the primary focus could be on climate change and environmental concerns [12]. As climate change emerges as a pressing issue, there is an increasing emphasis on reducing carbon emissions. With a high penetration of RESs in the power system, weather-related uncertainties have a great impact on the reliability, stability and efficiency of the power system from a wide range of aspects including unpredicted short-term events and long-term climate changes. Moreover, the consequences of the uncertainties extend far beyond mere inconveniences. They can result in extensive blackouts, leading to cascading power system failures, which can, in turn, force the utility company into bankruptcy [13]. Climate Central found that severe weather was responsible for 80% of large-scale power outages. Among these instances, 59% were attributed to heavy rainfall and thunderstorms [14]. Weather-related uncertainties have a significant impact on the entire energy supply chain, particularly power generation, transmission, and distribution [15]. Understanding these uncertainties should not only be limited to technical experts but also include policymakers and the general public to ensure informed decision making in the energy sector [16]. In this paper, we categorize key weather-related uncertainties, examine their effects on power systems, and explore suitable methods for estimating and forecasting these uncertainties as well as strategies for compensation and mitigation. The mitigation approaches increase the balance requirements with associated costs. Reconfiguration, adaptations in infrastructure, grid management, and energy storage operation are key components of a strategy to mitigate the effects of weather-related uncertainties on power systems [17]. Unfortunately, there is no unified control strategy to tackle all possible uncertainties. In addition to integrating uncertainty into public energy research and development decisions, the decision made by the policymakers cannot be only made for the near future but must also consider the possible uncertainties. The prospects of the power system mainly involve digitalization and AI [18]. The power system is increasingly adopting data-driven technologies and artificial intelligence AI to improve forecasting, grid management, and demand-side response, offering potential solutions to long-standing challenges.

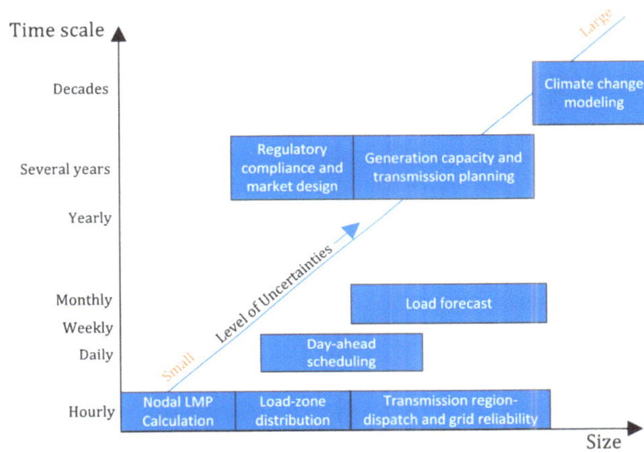

Figure 1. Raised level of uncertainties with the horizon: The X-axis corresponds to increasing geographic size and the Y-axis represents the increasing time scale.

The power system in Australia boasts a vast geographic reach accommodating diverse energy sources and decentralized population centers. As Australia strives to address its unique challenges, the power system is at a juncture, requiring strategies for sustainable and resilient future development. The existing review articles for power system uncertainties lack an event-specific analysis of impacts, comprehensive modeling and mitigation approaches, and potential limitations and future directions for the newly raised uncertainties in modern power networks. Moreover, in this paper, the mitigation methods focused on emerging technologies, particularly in terms of the coordinated operation of DERs and data-driven methods such as predictive maintenance. This review illustrates the evolving nature of uncertainty in the power system, which underscores the importance of adaptability and innovation to meet the changing demands of modern electricity networks.

The rest of this paper is organized as follows. Section 2 offers a description of the impact of uncertainties including weather-related events, cyber-attacks, and asset management. Section 3 provides a comprehensive review of the existing modeling approaches for the aforementioned uncertainties. Section 4 reviews the mitigation approaches, especially the emerging technologies which may ultimately result in options that reduce the reliance on extensive powerlines for those uncertainties. Section 5 outlines the limitations and future directions of mitigation strategies, weighing the trade-offs among reliability, economic efficiency, and computational speed. Finally, conclusions are drawn in Section 6.

2. Impact of the Uncertainties

The electricity grid faces vulnerabilities stemming from outdated system flexibility, aging assets, unpredictable weather patterns, and cyber–physical security threats. In the United States, weather disasters have been a primary cause of power outages, with a total of 178 incidents from the 1980s to 2014, including 8 in 2014 alone. The cumulative damages from these events surpassed the USD 1 trillion [19]. Numerous industries came to a standstill for hours, and in some cases, days. Several individuals requiring specialized healthcare lost their lives due to prolonged power outages and the inability to promptly restore electricity [19]. Anticipated climate change is poised to escalate the frequency, intensity, and duration of extreme weather events [20]. While rain and floods do not directly endanger overhead transmission lines, they pose a threat to substation equipment like switchgear and control cubicles. The combination of rain with strong winds or lightning, however, can emerge as a significant threat to overhead lines. In addition, uncertainties in the evolving nature of cyber threats can reduce the reliability of critical infrastructures. These disruptive events can be classified into three categories: (1) the uncertainties that will mainly affect the power generation; (2) factors like rainfall, bushfires, and aging infrastructure that affect both power generation and one or more electrical infrastructure components; and (3) cyber-attacks which affect the power system via communication links.

2.1. Uncertainties Affecting the Generation

Incorporating solar and wind energy sources into power systems has gained growing significance in the effort to reduce carbon emissions to reach a sustainable energy future. However, the intermittency and generation uncertainties associated with these RESs could pose challenges to the operation of power systems. Since wind turbines and solar power systems were first connected to the power grid in 1941 and 1954, respectively [21,22], they have experienced rapid and continuous growth. Numerous countries have introduced incentive and tax credit programs to facilitate their achievement of approved national and local RPSs. Consequently, the strong growth of wind and solar power has driven advancements in wind turbine and PV technologies, culminating in reduced costs. Figure 2 illustrates the evolution of renewable energy generation from various sources between 1965 and 2022 with solar and wind generation emerging as the primary contributors to renewable energy in 2022 [23]. Solar power has gradually surpassed wind generation due to the promotion and adoption of rooftop solar panels. According to the AEMO's estimated step change scenario in their Integrated System Plan 2022, renewables are expected to make up 98% of the total annual generation by 2050 with both distributed (rooftop) and utility-scale solar and wind energy taking dominance in the generation mix [24]. Therefore, it is essential to accurately estimate the impact of high penetration of wind and solar power on the system operation and implement effective mitigation measures [25].

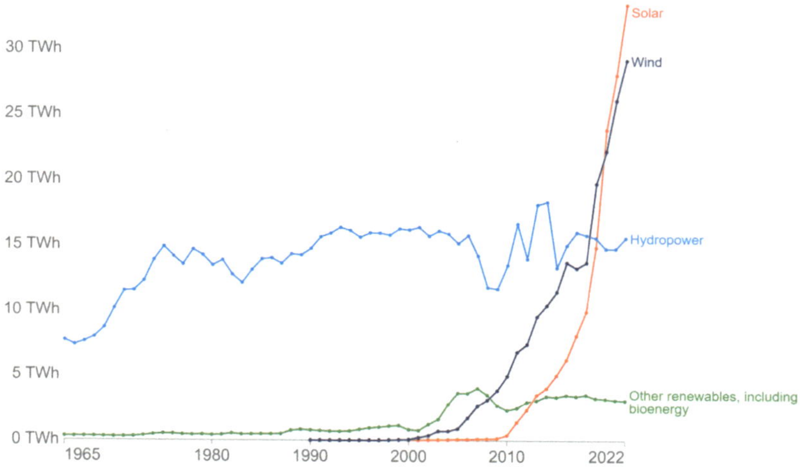

Figure 2. Australian renewable energy generation by sources.

The impact of integrating solar and wind power into a power system is contingent on two primary factors: penetration level and system flexibility. A higher penetration level indicates that a larger portion of energy generation comes from RESs like solar and wind. This can lead to reduced reliance on fossil fuels and decreased carbon emissions, leading to a more sustainable future. However, the increased penetration level may also strain grid stability due to fluctuations in supply and demand [26]. In terms of system flexibility, it is defined as the ability to balance supply and demand in real-time operation, which is crucial when integrating solar and wind power into the grid [27]. Enhancing system flexibility can be achieved through the incorporation of ESSs, D-FCAS programs, and MG technologies, as elaborated in Section 4. In general, the impacts of wind and solar power integration can be categorized as follows:

- The intermittent performance of wind and solar generation may lead to the sub-optimal operation of conventional generation units during the unit commitment process [28]. The cost of high-RES penetration on thermal units has been further investigated in [29,30].
- The utilization of induction generators in wind turbines inherently leads to the absorption of reactive power from the grid. For fixed-speed turbines, fluctuations in reactive power output can result in voltage fluctuations. Nevertheless, modern variable-speed turbines, e.g., DFIGs, can offer reactive power support through suitable interfacing methods [31].
- Fast ramping and frequent start-up generation units are required to provide reserves for wind and solar power fluctuations, which could also increase the operational cost of power systems [32,33]. Moreover, the lack of system flexibility with integrated wind and solar power has the potential to trigger blackouts in the power system. In February 2021, the cold period in Texas led to the icing of the turbine blades, which prevented the turbines from operating, causing a total power outage [34]. Researchers in [35] found that over 60% of solar penetration will cause blackout due to load imbalance according to the simulation of the Kythnos power system. Consequently, protective measures are essential to maintain system frequency and prevent cascading failure events.

Short-term impacts that are associated with operational time scale are listed above, whereas long-term impacts that involve planning for peak load periods are listed below:

- Transmission congestions can occur when the wind and solar generation is away from the load center [36]. Moreover, the high-RES penetration level will lead to higher transmission capacity, thereby increasing distribution losses.
- Due to the intermittency and variability of solar and wind generation, it is challenging to rely solely on these resources to meet load demand during peak hours, especially during cloudy or windless periods. In this case, the paper emphasizes the importance of demand-side ancillary services in Section 4, which can be achieved by either encouraging customers to adjust energy usage during peak hours or establishing VPP to optimize load curtailment.

2.2. Uncertainties Affecting the Network Assets

Both wildfires and rainfall have a strong impact on the overall reliability of the power system, and the frequency and intensity of these events can affect the long-term planning of power systems. Wildfire risks to the grid have emerged as a global concern in recent years. During early 2020, numerous wildfires erupted in all states in Australia, destroying more than 10 million hectares of land. Strong wind associated with wildfires can cause transmission lines to sag or break, which interrupts the electricity supply over long distances. At the peak of wildfires, approximately 20,000 households experienced power supply disconnections in New South Wales [37]. Moreover, the Brazilian National Institute for Space Research detected 103,000 wildfires in the Brazilian Amazon, reflecting an annual Increase of 16% [38]. Similarly, the increased frequency and duration of excessive rainfall can damage transmission lines, towers, and substations, leading to power outages. The heavy rainfall could also influence the hydropower generation by increasing water flow in rivers. This paper specifically delves into the analysis of potential modeling and mitigation strategies concerning rainfall and wildfires with a primary emphasis on infrastructure damage that encompasses the destruction of the power transmission network. It is important to note that the examination of the impact and corresponding mitigation strategies for natural disasters, i.e., hurricanes, is beyond the scope of this study.

In addition to external factors such as wildfires and heavy rainfall causing failures, the health condition of aging assets must also be taken into account, as it is a significant contributor to common asset failures. The impact of electric asset failure can range from minor to sever, which is contingent upon specific circumstances. A minor failure might result in a temporary and localized power outage causing inconvenience for the customers, which has limited widespread. In contrast, a severe failure could lead to a widespread and prolonged blackout, affecting a large number of populations and causing great economic losses. Moreover, severe failures could have cascading effects, impacting various sectors including healthcare, communication, transportation, and manufacturing. Understanding the spectrum of potential consequences is crucial for developing uncertainty mitigation strategies to ensure the resilience of the power system. It should be emphasized that the consequences aspect of risk is highly case-specific for asset failure [39].

Apart from the direct consequences of asset failures, they can also precipitate social unrest and political change. For instance, the poorly implemented electric deregulation in California contributed to Governor Gray Davis losing a recall election, which was marked by a surge in consumer prices and rolling blackouts statewide in 2003 [40]. Following the events at Fukushima Daiichi in 2011, the political landscape in Japan remains uncertain despite tentative steps to restart some of the nation's reactors [41].

2.3. *Uncertainties Affecting the Communication Link*

According to [42], one of the unifying concepts for cyber-attack is a cause of system failure including intentional, malicious, and human-induced faults in both software and hardware. If cyber security is not managed and controlled effectively, the cyber-attack can reduce the stability of the operation network as evidenced by the U.K. power outage in 2019 [43] and the ERCOT incident in 2009 [44]. In March 2019, a cyber-attack targeted a wind plant in Salt Lake City, Utah, U.S., resulting in the loss of operator control over wind turbines [45]. Subsequently, The U.S. Department of Energy emphasized the critical necessity for dedicated efforts to identify vulnerabilities, raise awareness, and develop strategies to protect wind energy infrastructure against cyber-attacks [46].

A typical multi-layer architecture of a smart grid is demonstrated in Figure 3 including application, communication, physical, and data acquisition layers. The objective of the application layer is to participate in the energy and reserve market and control the actuators based on their status. Typically, the most significant consequences of an attack occur when the attacker gains access to SCADA systems and initiates control actions [47]. Attackers can manipulate raw data measurements, introducing unnoticed errors into estimates of state variables such as bus voltage magnitude, phases, and line power. This risk arises when attackers exploit the tolerance for small errors within state estimation approaches, posing a significant threat to the security of the power system [48]. Moreover, the false estimation can lead to incorrect decisions making the real-time prices of the electricity market only profitable to the attackers.

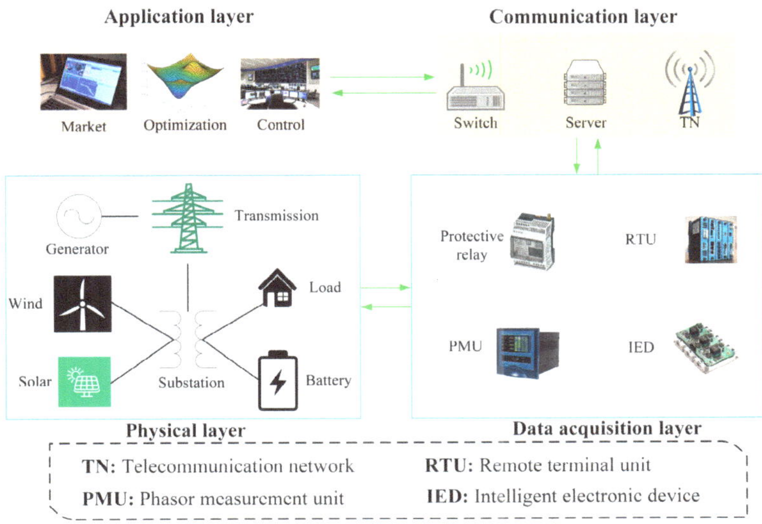

Figure 3. Physical and communication structure of the smart grid [46].

3. Uncertainty Modeling

System operators, generation companies, and consumers depend on various input data to establish parameters in a mathematical optimization model, enabling them to make informed decisions. However, a large number of these parameters are uncertain. For instance, wind and solar energy generation is affected by weather forecast uncertainty, and the electricity prices are influenced by the RESs generation and other participants in the market. Researchers in [31] highlight the challenges in integrating large-scale renewable energy generation into the grid due to its variability and the need for accurate forecasting models to reduce uncertainty. This paper focuses on the review of current forecasting and estimation models for weather conditions, which could serve as inputs for both uncertainty mitigation strategies and modeling processes for wildfire and aging assets.

Forecasting methods for wind and solar energy may vary depending on many factors, including the available weather information, prediction horizon and resolutions, and historical data. According to [49–52], the existing forecast methods can be categorized based on forecasting models, forecasting horizon, and performance metrics, as depicted in Figure 4. The physical models are deterministic and depend on meteorological parameters, e.g., pressure and temperature. This model uses complex mathematical equations to simulate the atmosphere's behavior, which is complex and time consuming [49]. Statistical models are the combinations of AR and MA models, e.g., ARMA, SARIMA, and ARIMA [53]. These models utilize the historical data of the time series to estimate the future value of solar and wind generation. The key benefit of statistical models is their ability to provide highly accurate short-term forecasts, as indicated in reference [54]. Following [54], the formation of ARMA(p, q), where p stands for the number of flags in the AR model and q stands for the number of flags in the MA model, is illustrated as follows:

$$\hat{P}_t = \sum_{k=1}^{p} \alpha_k P_{t-k} + \sum_{k=1}^{q} \theta_k \epsilon_{t-k} + \epsilon_t \tag{1}$$

P_t and ϵ_t are the power generation and the corresponding error at time t, respectively, and α_k and θ_k are the coefficients that can be trained by the historical data. Compared with ARMA, ARIMA(p, d, q) added another variable d, which converts non-stationary data into stationary, making the model more versatile for a wider range of time-series data and suitable for medium-term predictions. Moreover, SARIMA is an extension of ARIMA by capturing the seasonal variations for long-term wind and solar power prediction [55].

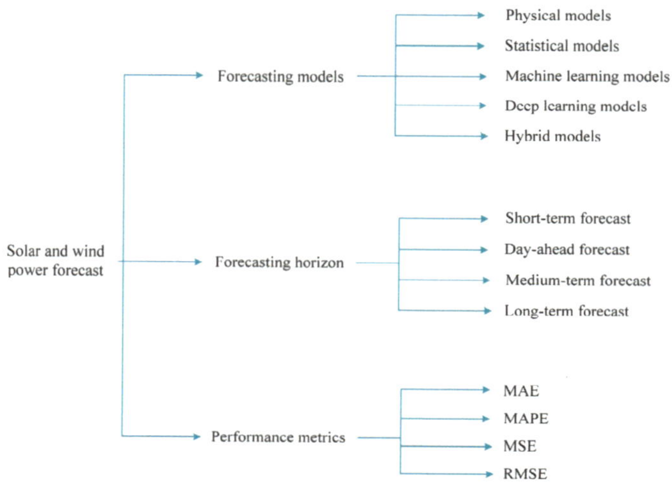

Figure 4. Solar and wind power forecast classification.

Machine learning models mainly include logistic regression, LR, decision trees, k-nearest neighbors, SVM, and ANNs. These models also belong to the realm of supervised machine learning where the output is modeled as a function of inputs. Normally, ANNs require a big data set with huge training time, and the results can vary a lot depending on the input data. Compared with other ANNs, BPNs offer greater efficiency in learning and are relatively simple to be implemented. The BPN model is not only capable of making smooth predictions but also of identifying short-term patterns within time-series data [49]. Researchers in [56] proved that the SVM-based forecasting model is more accurate than the conventional forecasting methods for solar radiation prediction. Another solution for solar and wind energy forecast is using deep learning techniques which are capable of obtaining high accuracy in various applications [57–60]. Deep learning models commonly involve using large amounts of data and complex patterns, e.g., CNNs, RNNs, and LSTM networks. The RNN-LSTM model for solar power forecast is applied in [61,62], where the model consists of two LSTM layers, two hidden layers and one output layer. The Adam optimizer is selected to optimize the weight parameters in each layer of the RNN-LSTM model due to its adaptive learning rate capability, and the activation function for the LSTM layers are hyperbolic tangent. The inputs for the solar forecast model include time, wind speed, air pressure, humidity, temperature, wind direction, and pyranometer. Multiple forecasting models can be combined to generate a hybrid model, e.g., the combination of ARIMA and ANN, to give a better prediction.

At a given location, the normalized standard deviation of forecasting errors of a single solar or wind farm tends to rise as the forecast horizon increases, and the forecasting methods that are suitable for short-term forecasts may not be applicable for long-term prediction. Therefore, the forecasting methods can be classified based on the prediction horizon, including (1) short-term forecast: from a few minutes to a couple of hours ahead; (2) day-ahead forecast: 24 h to 48 h ahead; (3) medium-term forecast: one week to two weeks ahead; and (4) long-term forecast: a month to several months ahead [63]. As the uncertainties for solar and wind power are not very large throughout the lifetime of a wind turbine, i.e., within a 10% deviation, seasonal variations are generally more predictable than annual ones [49,52]. In addition, the solar and wind energy associated with weather systems are challenging to predict beyond a few days, whereas the diurnal variations that occur within the day are more predictable.

The performance metrics used in the current literature are MSE, RMSE, MAE, and MAPE. An overview of current solar and wind energy forecasting methods is given in Table 1.

Table 1. Solar and wind forecasting methods.

Reference	Brief Description	Input	Error Metrics
Brahma and Wadhvani, 2020 [64]	The LSTM-based deep learning model is utilized for solar irradiance forecasts.	Solar irradiance	RMSE: 9.788 MSE: 9.721
Liu et al., 2023 [62]	The LSTM-based deep learning model is applied for solar power forecasts.	Time, wind speed, air pressure, humidity, temperature, wind direction, and pyranometer	MSE
Voyant et al., 2017 [52]	Machine learning models including SVM and regression trees are applied to predict solar radiation	Solar irradiance	RMSE
Torres et al., 2021 [50]	RNN-based deep learning model is applied for short-term PV power forecast	Weather inputs from IoT data set and historical PV generation	R^2: 0.988
Hacioğlu, 2017 [65]	Linear regression and Gaussian process regression-based machine learning models are utilized for solar irradiance forecasts.	Wind speed, temperature, humidity parameters, pressure and solar irradiance	MAE: 0.0166 RMSE: 0.0227
Wang et al., 2021 [66]	ANN-based machine learning model is applied to wind power forecast.	Temperature, pressure, wind direction	NMAE: 0.0044
Guo et al., 2010 [67]	ARMA model is utilized for wind speed forecast	Historical value of wind speed	MAE: 0.57
Ferreira et al., 2019 [68]	Hybrid time-series models are applied for short-term wind speed forecasts.	Air pressure, wind speed, wind direction	RMSE: 2.27
Santamaría-Bonfil et al., 2016 [69]	The SVM-based machine learning model is utilized for wind speed forecast	Wind speed, wind direction, humidity, solar radiation, temperature, atmospheric pressure, and heat radiation	NMAE: 0.15

3.1. Uncertainties Affecting the Generation

The outputs obtained from the aforementioned point forecast methods cannot be directly utilized for decision making by system operators due to their inherent limitations in providing sufficient information. In other words, instead of using only forecasting values to solve a deterministic problem, the literature applies a stochastic approach to ensure the robustness of the dispatch decision against all possible solar and wind energy generation uncertainties. As a result, outputs from the forecasting models are employed to generate probabilistic forecasts, which can be categorized as (1) probabilistic distribution; (2) distributional robust formulation; (3) independent identically distributed samples; and (4) robust uncertainty set [9]. The uncertainty parameter, which is the solar and wind generation forecasting error, is defined as δ in this paper. The actual solar and wind generation equals the forecast value plus the error, which is expressed in vector form.

$$P = \hat{P} + \delta \tag{2}$$

Assuming a normal distribution for the δ is most commonly used in the current literature for the sake of simplicity. However, the actual distribution is difficult to obtain in reality, and it might not perfectly follow a predefined distribution. Researchers in [70] propose using beta distribution to represent the wind forecast error instead of using the normal distribution. To overcome the drawback of defining a probabilistic distribution, a family of distributions in which the uncertainty parameter δ may fall is defined in [71]. The distributional robust formulation includes all possible distributions as the ambiguity set B. If we can access many possible realizations δ_s, e.g., historical data, and assume these samples are IID, these samples can represent the uncertainty parameter with the same probability of occurrence. The main drawback of IID is that an accurate probability distribution of δ may require the consideration of a large number of samples, and the historical data are normally not IID [72]. The uncertainty parameter δ can also be modelled without defining the possible distributions. Furthermore, we might lack adequate data to estimate the distribution of δ. Under this situation, the robust uncertainty set Δ can be defined using a set of scenarios δ_s which include all possible realizations of the uncertainty or allowing continuous variations of δ but restricting these fluctuations to remain within a predefined set, e.g., elliptical sets and box-constrained sets [73]. The scenarios for robust uncertainty sets can be generated using the MCMC model. Motivated by [74], the researchers in [75] propose a two-dimensional MCMC model assuming the current state of uncertainty parameter δ_t depends on the previous state of δ_{t-1}, the forecast value at t and a constant conditional probability. In this case, the properties, i.e., acf and pdf of the uncertainty parameter, can be maintained without explicitly defining them. However, due to the intermittency and the forecast being unable to reach 100% accuracy, the solar and wind generation uncertainties cannot be entirely tackled by relying only on the forecast methods.

3.2. Wildfires and Rainfall

A crucial motivation for enhancing grid resilience is the prediction of the grid state [76], representing how the grid or its components perform for various weather events, e.g., heavy rainfall and wildfire, and it can influence the grid-operating conditions. The grid state can be presented by the binary variable I as below:

$$I = \begin{cases} -1; \; Outage\ state \\ \;\;1; \; Normal\ state \end{cases} \tag{3}$$

The grid state serves as an essential input factor for the decision-making process of power system planning and operation. Before implementing measures to mitigate the impact of weather events, various strategies can be employed to estimate the potential grid state based on prevailing weather conditions in each region. Data-driven methods, particularly ML models, offer an effective means to model the intricate correlation between the grid state and the associated weather conditions. These methods demonstrate efficiency in terms of time and resource requirements, as they do not rely on a physical model. For instance, a predictive approach based on linear regression models was introduced in [77], aiming to forecast the electrical infrastructure damages caused by heavy rain and storm events.

The reliability, i.e., failure rate, of transmission and distribution facilities is significantly affected by the environment where they are located. As the weather prediction methods have been illustrated above, in this subsection, we analyzed the correlated characteristics between weather patterns and outage distributions to characterize both rainfall and bushfire events. A novel method is introduced to establish a correlation between the severe weather risk index and the power network model, incorporating geographical information about the transmission lines [78]. However, it is worth noting that a long transmission line may traverse multiple weather regions, potentially encountering different weather conditions in each region.

For instance, lines 7–8 and 15–16 cross two different weather regions in Figure 5. Consequently, the influence of weather conditions on a single line will differ from one region to another. Precisely defining the weather-affected failure rate of high-component components is crucial for modeling the impact of weather on the entire power infrastructure. Two common approaches exist for defining the failure rate of a transmission line crossing multiple regions. The first approach involves employing the weighted-average method, where the summation of the point-specific failure rates along each line segmentation is calculated based on an IMFR [79]. In the second category, the failure rate for the entire transmission line was determined by selecting the highest failure rate from any single point of a line [80].

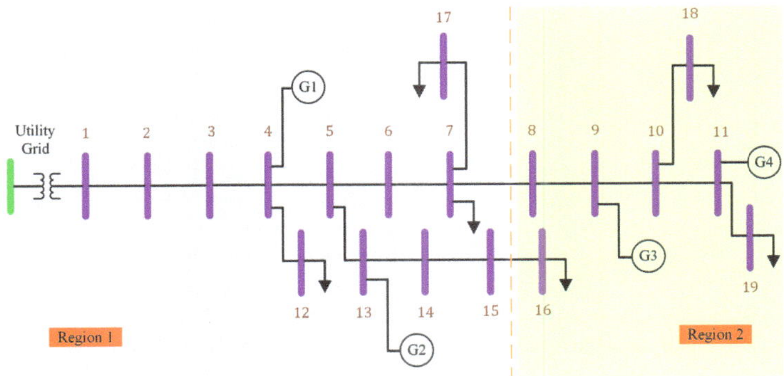

Figure 5. A 19-bus power system across two regions.

To depict the correlation between electric component failure rate and weather information, the risk for wildfire and rainfall needs to be estimated. The rating systems for wildfires are emphasized in this subsection, where a different standard is selected for each country. The Canadian Forest Service FWI is employed to assess the distribution of wildfire risk across a power grid. Meanwhile, in Australia, the FFDI is used widely to indicate the likelihood of wildfire in various regions. The FFDI encompasses measures of dryness based on rainfall and evaporation, temperature, wind speed, and humidity without considering the potential risk from fuel management and lightning [53]. According to Figure 6, the risk of wildfire in a certain region can be classified into six categories including low–moderate, high, very high, severe, extreme, and catastrophic based on the Australian standard.

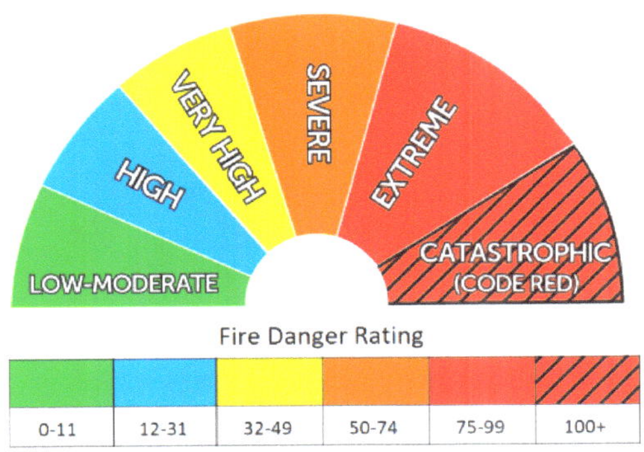

Figure 6. Fire danger ratings defined by the Country Fire Authority [81].

The impact of wildfire on the electricity supply of the transmission lines crossing the corresponding weather region can be quantified by constructing the fragility curve of the sub-transmission line failure rate with the weather conditions within the region. The fragility curve for the line failure rate and restoration time under varying risk levels can be achieved by a parametric method assuming that p_0 is the nominal line failure rate. The line failure rate remains constant when the FFDI is below 11, and the slope of the fragility curve is s when the FFDI is greater than 11. Therefore, the fragility curve of the line failure rate can be formulated as [81]:

$$p(FFDI) = \begin{cases} p_0, & FDDI \leq 11 \\ \min\{s(FFDI-11)p_0 + p_0, 1\}, & FDDI > 11 \end{cases} \tag{4}$$

3.3. Asset Management for Aging Assets

CBRM can be employed to analyze the health conditions of assets within the network and predict the potential risk to renew in optimal time. It operates as a structured framework that amalgamates data, engineering knowledge, and experience, creating a transparent, robust, and repeatable decision-support tool. The CBRM process unfolds through a series of sequential steps [82]: (1) define HI to indicate the asset conditions; (2) derive the correlation between HI and the POF of the assets; (3) employ the knowledge of degradation process to estimate the future condition and performance of assets, where HI and operation conditions are two main factors at this step; (4) assess COF to evaluate the consequences in specific categories, including safety, environmental and financial; (5) establish a risk model that combines POF and COF to quantify the risk and give priority to each asset; and (6) evaluate the optimal interventions in terms of risk and recalculate the new POF and COF after the risk reduction process. Overall, Steps 1–3 pertain to asset condition and performance, offering a systematic process to identify the relationship between the condition and POF. Steps 4–6 provide the assessment of COF and enable the quantification of risk when combined with POF. The relationship between the defined HI and POF at Step 2 is depicted in Figure 7, where the curve is constructed by curve-fitting technology given the function of the calibration curve. HI is affected by normal asset life, current age and conditions, and the degradation processes of the asset, and a higher health index indicates a lower condition.

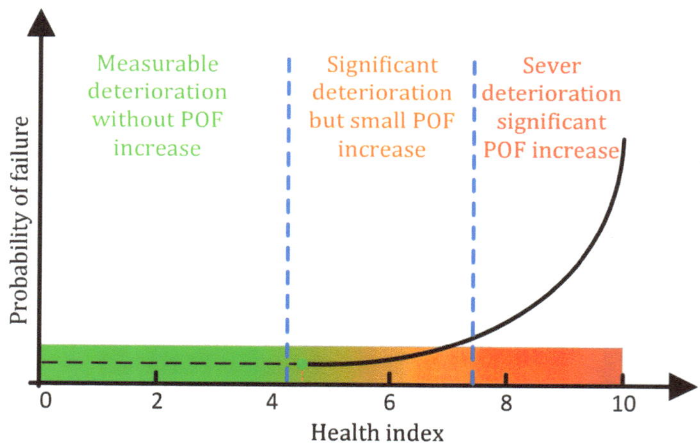

Figure 7. The increase in the probability of failure correlates with the rise in the health index [83].

3.4. Cyber-Attacks Detection

Cyber-attacks can influence critical power system operations, including state estimation, automatic generation control, voltage control, and energy market processes.

- Attacks on AGC: The purpose of AGC is to adjust the power output of the generators within an area to maintain the frequency and power exchange via the tie line. Given that the time resolution of AGC control signals operates on a time scale of seconds, it cannot afford to utilize elaborate data-validation algorithms [84].
- Attacks on state estimation: State estimation is crucial for operational decisions for a smart grid. The motivation for leading attacks on state estimation varies from causing blackouts to gaining financial profit from the market.
- Attacks on the energy market: Cyber-attacks on the energy market can be achieved by denial of service or injecting wrong data, e.g., jamming of price signals.
- Attacks on voltage control: Compared to the system frequency, voltage stability is a major concern in long distribution networks. LTC transformers are widely applied to

improve voltage stability in response to load variation. The control topology for LTC makes it vulnerable to man-in-the-middle attacks.

Various methods have been employed to detect the abnormal conditions for each of the above-mentioned system operations. Most research works feature ML algorithms, e.g., K-means, Q-learning, and GANs to generate malicious data for attacks. In [85], SVM-based technology and statistical anomaly detection are applied to identify the false data injection, which outperforms the statistical approach. Researchers in [86] propose a CDBN detection strategy that extracts temporal features from the measurements of distributed sensors, which is robust against different environment noise levels. A dynamic Bayesian network is applied in [87] to detect unobservable attacks; i.e., too many measurements are discarded in the system. In [88], a physics-data-based detection method is proposed to detect the cyber-attack on transformers to ensure solar farm security.

4. Mitigation Approaches

This paper first introduces several common mitigation approaches based on the operational procedures for each type of uncertainty mentioned in this study and subsequently focuses on the most emerging technologies including microgrid, demand-side management, and battery control strategies, potentially offering alternatives to extensive powerline reliance in high-risk environments. Power system follows a hierarchical control structure, where each dispatch step is decoupled based on the control horizon. In this paper, we focus on control resolutions of 5 min and above, addressing the steady state of the power system, while excluding control decisions within seconds, such as AGC. For more advanced load frequency control techniques under disturbances, refer to [89–91], which is out of the scope of this paper. It is worth noting that apart from the operational approaches discussed in this paper, hardware updates can be considered, e.g., undergrounding the distribution and transmission lines, to increase the robustness of the components to severe weather conditions.

4.1. Uncertainties Affecting the Generation

The objective of this paper is not to offer a comprehensive tutorial on any specific method. Instead, we explore the essential attributes and potential drawbacks of various optimization methods and developed mitigation techniques, providing readers with the knowledge necessary to make informed choices based on their specific context. The intermittent nature of wind and solar energy presents a demanding constraint for power system operators, as they must continually balance supply and demand in real time. In this review paper, strategies for addressing uncertainty in solar and wind generation encompass various mitigation methods. These methods involve fundamental optimization under uncertainty techniques and the developed market and operation frameworks including (1) microgrid strategies, (2) ESS operation strategies, and (3) the D-FCAS framework. However, incentivizing the customers to rely on an energy mix for power supply rather than individual resources is not the focus of this study.

To address the uncertainty parameter modeled in the previous subsection, stochastic, robust and chance-constrained optimization can be selected. As the stochastic optimization assumes the known probability distribution of δ which is difficult to obtain in reality and the robust optimization is too conservative which ensures the performance under the worst-case conditions, this paper focuses on the implementation of chance-constrained optimization denoted as Equations (5) and (6) to address the uncertainty in power system operations.

$$\min_{x,\, y_\delta} f(x) + R[g(x, y_\delta, \delta)] \tag{5}$$

$$s.t.\;\; P_\delta\{a(x, y_\delta, \delta) = 0,\; b(x, y_\delta, \delta) \leq 0\} \;\geq 1 - \varepsilon \tag{6}$$

x represents the variables that remain the same as the uncertainty parameter varies, whereas y_δ indicates the variables that vary when δ changes. R represents the cost of an additional reserve required to balance the power system. a and b are the equality and inequality constraints, respectively, which should be guaranteed with at least a probability of $1-\varepsilon$. According to [9,71], probabilistic distribution and a distributional robust formulation of δ can be addressed via DRCC optimization:

$$\inf_{P_\delta \in B} P_\delta\{a(x, y_\delta, \delta) = 0,\ b(x, y_\delta, \delta) \leq 0\} \geq 1-\varepsilon \quad (7)$$

The ambiguity set B is assumed to contain all possible candidates of P_δ. The model aims to minimize the objective function under the worst-case distribution of P_δ within B. The CVaR can be applied to make tractable approximations for the ambiguous chance constraints in Equation (7). For independent identically distributed samples and robust uncertainty set defined by a set of scenarios, SCC optimization is applied. It should be noted that at least a number of N scenarios need to be used to define the uncertainty set Δ, where $\delta \in \Delta$, to ensure the probability of constraint satisfaction [74].

$$N \geq \frac{1}{\varepsilon}\frac{e}{e-1}(ln\frac{1}{\beta} + N_x - 1) \quad (8)$$

β is the confidence factor and N_x represents the number of decision variables. Among the contemporary operational frameworks for addressing the challenges of integrating solar and wind power into the grid, microgrids have demonstrated their suitability and stability, as evidenced by pioneering projects.

4.2. Uncertainties Affecting the Network Assets

The potential risks caused by the uncertainties defined in this paper can be mitigated both in the planning phase and in real-time emergency management. For real-time emergency response, several strategies can be considered:

- If the local network is connected to the grid, the power capacity shortage can be compensated by emergency power imports from the grid given the fact that the transmission feeder has available capacity beyond the scheduled import power amount [92].
- Minimization of potential load shedding during the power capacity shortage can be achieved by the emergency discharge and rescheduling of ESSs for both grid-connected and off-grid local networks [93].
- Implementing various scheduling lead times, e.g., 5 min and 30 min ahead, to determine the charging and discharging profiles of ESSs based on generation and demand forecasts.
- Implementing the real-time network reconfiguration which dynamically adjusts the configuration of the electrical network to optimize its performance and respond to the uncertainty events. This process typically occurs in real time to enhance the overall reliability and efficiency of the power system.

A Monte Carlo simulation engine is widely used in the planning phase, which samples different outage scenarios of the system components including the outage of generators due to weather-related uncertainties, substations, and transmission feeders during high bushfire risk periods. Based on the controllable devices of the system and the proposed emergency response strategy, the Monte Carlo simulation engine can be coordinated with the system dispatch engine in various ways to examine the reliability of the system under uncertainties [53].

For aging infrastructure, wildfire and rainfall that can directly impact the transmission system, thus affecting the electricity supply. The following common mitigation approaches can be implemented [94]:

- Annual line and easement inspections: Qualified technicians will conduct thorough inspections of all lines, assets, and easements following the outlined process and evaluation criteria in "Lines Practices and Procedures".
- Tower-climbing inspections: Scheduled at intervals of three, six, or nine years, these inspections are tailored to the probability of asset failure in specific areas, considering factors like bushfires.
- Tower corrosion monitoring: Tower legs and lattice members which might be influenced by previous weather-related events are monitored. Corrosion assessment and grading adhere to established standards with results recorded for subsequent maintenance scheduling and replacement activities.
- Targeted asset replacement: Combining data from condition monitoring and line inspections, the identification of necessary targeted asset replacements, such as insulators, conductors, and ground wires, is carried out.

As the potential risk of aging assets is modelled, this subsection focuses on identifying the optimal preventive maintenance plan for a system to mitigate the associated risk. An MINLP problem is proposed in [95] to determine the optimal preventive maintenance scheduling for enhancing the resilience of power distribution systems. The objective of the presented MINLP problem is to minimize the total expected number of power outages throughout the entire planning time horizon while adhering to a total budget limit and considering various levels of periodic budget constraints. This optimization directly enhances the resilience of the power system in the face of various uncertainties by reducing the overall likelihood of power outages. In [83,96–98], reliability ranking is achieved based on the determined POF. Predictive maintenance identifies assets with the greatest impact on network reliability and the poorest health condition, posing operational risks that warrant immediate replacement. In contrast, assets with minimal impact on network reliability and a satisfactory relative condition do not need to take immediate action. It is noteworthy that assets with minimal contribution to system reliability may still be in poor technical condition, while assets in good technical condition may significantly influence the relative contribution to system reliability. The key advantage of utilizing this type of approach lies in its capacity to concurrently visualize the relative condition of a given unit and its contribution to system reliability. Establishing a connection between asset HI, POF and COF is the most crucial step, providing a means to estimate future conditions and performance. According to [99], individual asset risks are determined using criticality factors corresponding to consequence categories, adjusting consequences based on each asset's operating context. Criticality, expressed as a multiplication factor, ranges from c for an average asset to 0 for less crucial and c^{max} for more pivotal assets. Factors consider network performance, safety, and financial considerations.

4.3. Uncertainties Affecting the Communication Link

As illustrated in Figure 8, the cyber-attack on power grid and industrial control systems can be mitigated by implementing cyber and physical security measures, e.g., control centers from unauthorized access, improving the privacy of customers, and deploying the defense system that monitors network and system activities for abnormal behavior and potential cyber-attacks. The privacy of customers in the smart grid and industrial control system must be protected and cannot be shared with other parties without permission, as sensitive information leakage is increasingly important in modern power systems. After knowing the vulnerabilities of CPS, several defense techniques must be deployed to prevent cyber-attacks. Detailed mitigation approaches can be found in the following literature [100–104]. This subsection focuses on the prevention strategies, whereas Section 4.4 focuses on designing a fault tolerance network with resilient-preventive and protective measures.

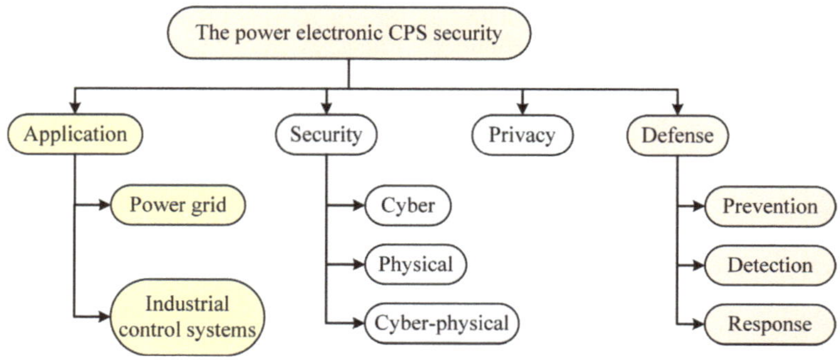

Figure 8. Security classification for the power CPS [48].

4.4. Emerging Technologies

This subsection offers insights into emerging technologies that could serve as ultimate solutions for addressing the uncertainties mentioned in this paper. It is important to note that this paper provides a review of these techniques rather than focusing on assessing their effectiveness. Microgrids, often described as localized energy ecosystems, offer an ingenious approach to managing the variability of solar and wind generation. These self-contained networks enable intelligent distribution, generation, and storage, providing flexibility in times of fluctuation [105]. They have shown remarkable effectiveness in enhancing energy resilience and sustainability [106]. The hierarchical control structure of a DMS with MGs is depicted in Figure 9. The microgrid includes an MCC, multiple LCs, and MSCs which coordinate the RESs with the MG [107].

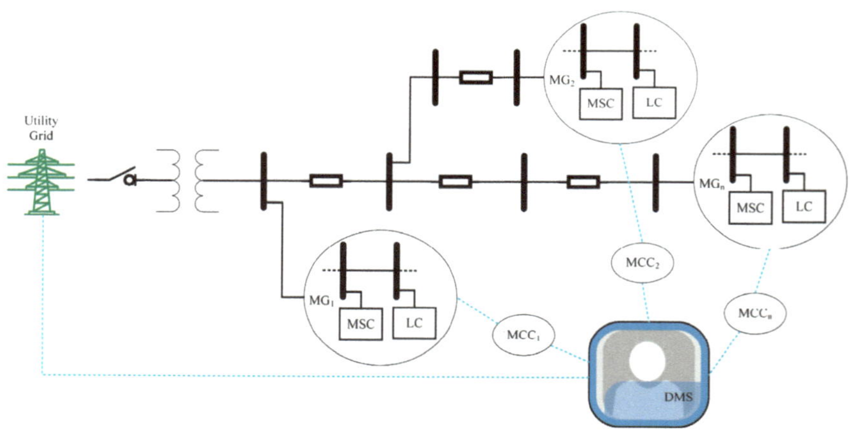

Figure 9. MG control diagram.

Energy storage, particularly in the form of batteries, is anticipated to revolutionize future power systems, significantly enhancing their resilience [54]. Depending on the energy capacity and charge/discharge capabilities, batteries can contribute to resilience through long-duration applications, such as reducing network flow congestions or supporting the security of supply in isolated areas. Additionally, batteries play a crucial role in short-duration applications like regulation control, responding quickly to weather events' impacts within seconds to minutes [60]. This distinction categorizes energy storage into energy applications (long duration) and power applications (short duration). Furthermore, energy storage includes bulk storage, employing large-scale units like PHS and CAES, as well as distributed energy storage within load centers utilizing smaller units [33].

As ESS technology advances, marked by increased energy density and reduced costs, these systems have become pivotal components in the quest for a stable and efficient energy landscape. Effective ESS operation strategies optimize the storage and release of energy, harmonizing supply and demand while cushioning the impact of intermittent generation [108]. Figure 10 demonstrates the expanding estimated range of the energy capacity, i.e., charging and discharging of the ESSs, and SOC as the control time horizon increases [109]. It can be seen that the uncertainty of the ESS SOC and required capacity propagates through the control horizon. In other words, the variance of uncertainties increases over time as the distance from the current time step increases. The estimation of the SOC becomes more difficult as the prediction horizon increases. Therefore, it is crucial to implement a tractable approach to manage the SOC of ESSs under solar and wind generation uncertainties.

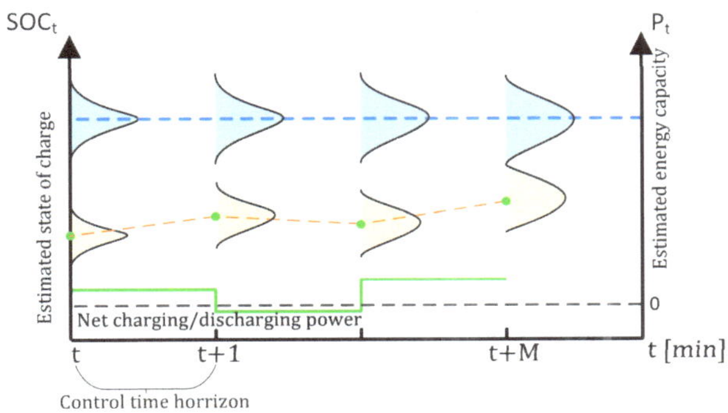

Figure 10. Estimated range of SOC (yellow) and energy capacity (blue) of ESS.

Figure 11 demonstrates an example of the change in SOC within an operation hour, where the blue curve is the reference trajectory of SOC for the scenario without any uncertainty and the green curve is the SOC for a scenario when the secondary reserves and redispatch mechanism, which are presented as SR and RD, respectively, are both activated. At the start of the hour, secondary reserves are deployed to compensate for the uncertainty within the first 15 min, and the ESS is required to store the excess generation in the system. The redispatch mechanism is not activated in the first interval as the actual SOC matches the reference value but is activated in the second interval, i.e., 15 to 30 min when there is an SOC mismatch. At 45 min, SOC drops due to the energy deficiency caused by uncertainty, resulting in the activation of redispatch to cover the SOC drop compared to the forecast trajectory. In this way, the SOC can be maintained the same as the forecast in real-time operation and prevent the propagation of uncertainty through the time horizon.

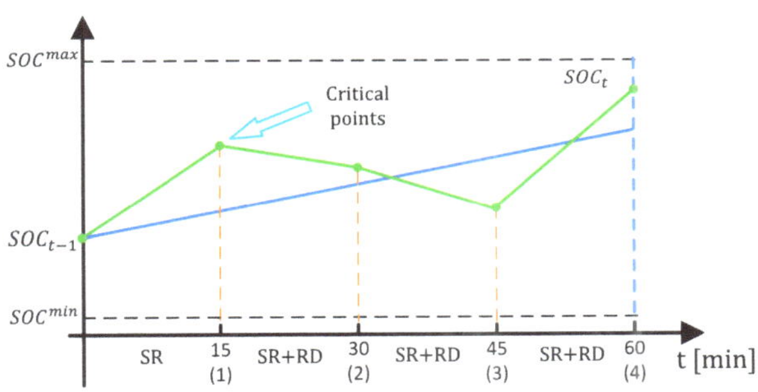

Figure 11. Reference trajectory of SOC (blue) and an example SOC curve (green) for a specific scenario during one operation hour.

The last technique included in this review to tackle the uncertainty is the D-FCAS framework, which invites consumers to play an active role in grid management. This approach engages demand-side response to balance grid frequency and delivers essential ancillary services, further fortifying the grid against the vagaries of renewable generation. A detailed survey of the existing literature using the methods is given in Table 2.

Table 2. Uncertainty mitigation approaches.

Category	Brief Description	Reference
MGs	The report discusses the use of transition matrices to model the probability distribution of solar generation and demand at different intervals of the day. It presents heat maps of transition matrices for solar power generation, showing the likelihood of transition from one state to another.	[53]
	The paper introduces a two-stage operation strategy for IMGs. In the initial stage, day-ahead scheduling is employed to forecast the electricity consumption baseline and regulation capacity for the subsequent day. The second stage focuses on real-time power consumption control, utilizing RegD signals. This second stage consists of two layers: the upper layer manages demand response signals and facilitates electricity exchange among microgrids through an energy-sharing mechanism, while the lower layer executes real-time power consumption control for each individual microgrid.	[62]
	The decentralized control approach divides the distribution system into intelligent small grids called microgrids, which can operate autonomously. In island mode, microgrids use voltage and frequency droop control characteristics to share the load automatically without the need for communication systems. In order to reduce the complexity of the network, a decentralized approach using microgrids is suggested.	[110]
	This document is a review of microgrid control techniques, specifically focusing on controlling microgrids with distributed RESs in island mode.	[107]
ESS control	The paper proposes explicit and implicit decision methods to address the scheduling problem with a focus on solution robustness and nonparticipative qualities. The explicit decision method assumes affine policies linking decision variables and uncertainty realizations, whereas the implicit decision method explores secure ranges of thermal unit outputs and SOC levels to ensure the feasibility of future economic dispatch solutions.	[111]
	The paper proposes a risk-based chance-constrained control strategy to optimize the dispatch of energy-constrained ESSs, taking into account the uncertainty associated with estimating the SOC and capacity of the ESSs. The controller coordinates the ESSs to minimize the unscheduled participation of generators and overcome ramp-rate limitations for balancing variability from renewable generation. The paper also introduces a temperature-based DLR approach to integrate ESSs and increase renewable generation.	[109]
	The authors introduce an innovative two-stage robust optimization approach that effectively captures the operation of storage devices, accounting for the anticipatory nature of the two-stage setting. The resultant robust counterpart constitutes a mixed-integer trilevel program featuring lower-level binary variables. To tackle the nonconvexity of the problem, the authors suggest employing an exact nested column-and-constraint generation algorithm.	[112]

Table 2. *Cont.*

Category	Brief Description	Reference
D-FCAS	This paper proposes a coordinated control strategy for a VPP aiming to enhance load frequency control. The VPP coordinates the allocation of energy and regulation signals among BESSs and HPWHs, which were determined by distribution coefficients derived through multi-objective optimization.	[113]
	The paper highlights the importance of demand response in enhancing the operational flexibility of power systems and the advantages of industrial loads in providing such a response. However, the discrete power changes in these loads restrict them from offering valuable ancillary services. To address this constraint, the document suggests techniques that empower these loads to offer regulation or load following with the assistance of an onsite energy storage system. The coordination between industrial loads and energy storage is established through a model predictive control approach.	[114]
	Researchers in this paper propose an optimization strategy that includes day-ahead scheduling and frequency regulation service to maximize profits and ensure real-time load-following performance. The paper presents a case study that demonstrates the cost-effectiveness and load-following capability of the proposed method compared to industrial loads equipped with only on-site ESS or passive use of solar energy.	[54]

5. Limitations and Possible Future Directions

5.1. Limitation for MCMC

The limitations of the aforementioned mitigation approaches can be reviewed from the technical perspective considering the computational burden of the operating system and from the perspective of the market and operation framework. The scenarios generated by the MCMC model for the robust uncertainty set and the input scenarios for the Monte Carlo engine can add computational burden to the operation system. Therefore, researchers have either investigated the scenario reduction methods or adjusted their decision-making models to accommodate a wider range of input scenarios. For instance, a simultaneous backward model from [40] is adopted to reduce the number of scenarios and maintain as much information as possible. Considering a set of scenarios δ with a number of Ns, the Euclidean distance between any scenario pair is defined as $D_{i,j} = \sqrt{\sum_{t=1}^{T}\left(\delta_t^i - \delta_t^j\right)^2}$ in Figure 12.

Another way to improve the computational speed is the boundary-tightening approach documented in [115], which is shown in Figure 13 below.

In ref. [116], the decisions are decoupled at various stages of the timeline. For day-ahead optimization, DC power flow is selected instead of the nonconvex AC power flow method to enhance computational efficiency while considering network constraints. During operation hours, where the number of decision variables is lower, a more accurate SOCP relaxed power flow model is employed for network constraints. Future research directions involve finding a balance between the performance of operating systems and the required computational time.

The limitations in the current market and operation framework include the following:

- Regulatory Framework: Regulatory frameworks and market designs may not fully account for the unique characteristics of RESs, such as their variability. Existing market structures and mechanisms may not incentivize the desired response to manage uncertainty.
- Transmission and Grid Constraints: Limited transmission capacity and grid constraints can exacerbate the challenges of integrating RESs, leading to uncertainty in power

flows and frequency control. Future improvements require the information exchange between DNSPs and RESs operators.

- Artificial Intelligence: Many AI models, especially deep learning models, operate as black boxes, making it challenging to understand how they arrive at specific decisions. This lack of transparency can hinder trust and make it difficult to explain data-driven outcomes.

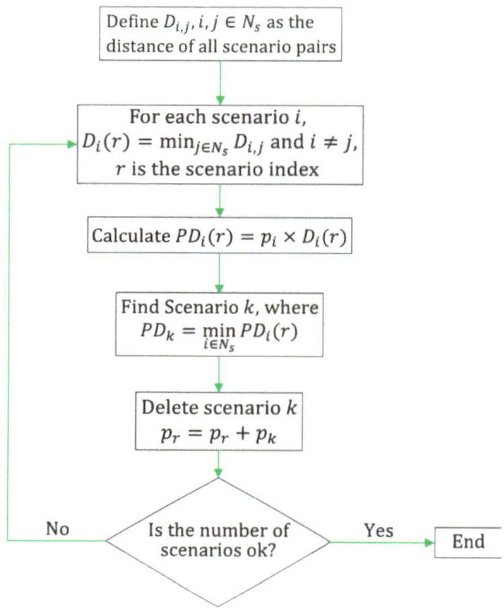

Figure 12. Simultaneous backward scenario reduction method.

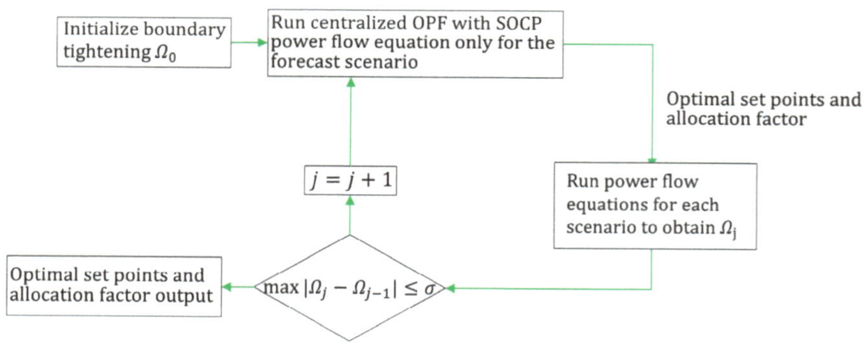

Figure 13. Boundary-tightening method.

5.2. Limitations for Defining the Weather Regions

The assumption underlying the methods mentioned for the modeling of wildfire and rainfall is that a weather condition concludes at the boundary separating two geographical weather regions. The network's segmentation into weather regions relies on this assumption. However, in reality, there are no distinct borders between weather regions. Instead, weather events cross the network with different densities, speeds, and durations, which will affect the accuracy of the estimated line failure rate under wildfire or heavy rainfall, thus impacting the reliability of the power system.

5.3. Limitations for Asset Management

CBRM offers valuable insights into asset condition and risk, but it also comes with certain limitations. It is recommended to combine CBRM with other risk management methods to mitigate these limitations. Some of the key limitations are listed below.

- Data Accuracy and Availability: CBRM heavily relies on accurate and reliable data for assessing asset conditions. Inaccuracies or insufficient data can compromise the effectiveness of the risk management process.
- Complexity of Degradation Processes: Predicting future conditions and performance is challenging due to the complexity of degradation processes. Some degradation mechanisms may not follow predictable patterns, making it difficult to precisely estimate asset deterioration.
- Assumption Sensitivity: CBRM involves making assumptions about the relationships between HI, POF, and COF. The accuracy of results is sensitive to the validity of these assumptions, and deviations can impact the reliability of risk assessments.
- Limited Predictive Ability for Catastrophic Events: CBRM may not be well suited for predicting rare but catastrophic events. Extreme events, such as natural disasters, may have consequences that are challenging to quantify accurately, leading to potential underestimation of risks.

6. Conclusions

The challenges faced by the electric grid, stemming from various sources, carry profound implications for reliability, stability, and economic efficiency. Notably, Australia's vulnerability to severe weather events amplifies the significance of addressing power network vulnerabilities, particularly in remote areas. Despite these challenges, the existing literature lacks a thorough examination of uncertainties in contemporary power systems, including weather-related events, cyber threats, and asset management, along with an exploration of mitigation approaches and their limitations. To address this gap, the review delves into both conventional robust control methods and modern probabilistic, data-driven approaches for modeling uncertainty events and their correlation to the state of the grid, facilitating optimal decision making. The exploration extends to the development of robust and scenario-based operations, control technologies for MGs and ESSs, and D-FCAS and reserve provision for frequency regulation. These advancements aim to design a power system robust against uncertainties.

Additionally, the conclusion underscores the current prevalence of deep learning models in weather prediction within the power system. The trend of utilizing deep learning in power system decision-making processes, such as fault restoration and predictive maintenance for aging assets, is evident in the existing literature. However, a primary challenge lies in the extensive data requirements for data-driven methods and the computational challenges for Monte Carlo engines. A potential future direction involves the reduction in input data through preprocessing and the implementation of realistic approximations to obtain accurate results promptly. Furthermore, the increasing intensity and frequency of severe weather conditions caused by global climate change necessitate the development of emerging technologies to enhance the reliability and stability of the current smart grid. As the penetration of AMI for the smart grid is anticipated to reach 100% in the future, cybersecurity measures need to be improved to prevent cyber-attacks. In addition, redundant generation and communication links must be applied to make the smart grid more robust against potential cyber threats. This exploration contributes insights to guide the understanding and management of uncertainties in evolving power systems, paving the way for informed decision making and resilient grid architectures.

Author Contributions: Conceptualization, H.H. and S.S.Y.; Methodology, H.H., S.S.Y. and H.T.; Writing—original draft preparation, H.H. and S.S.Y.; Writing—review and editing, S.S.Y. and H.T.; Supervision, S.S.Y. and H.T. All authors have read and agreed to the published version of the manuscript.

Funding: This research was supported in part by the Australian Research Council, IC210100021.

Data Availability Statement: No new data was created in this review.

Conflicts of Interest: The authors declare no conflict of interest.

Abbreviations

AMI	Advanced metering infrastructure
D-FCAS	Demand-side frequency control ancillary service
RES	Renewable energy resource
LMP	Locational marginal price
AI	Artificial intelligence
PV	Photovoltaic
MG	Microgrids
VPP	Virtual power plant
ERCOT	Electric Reliability Council of Texas
SCADA	Supervisory control and data acquisition
AR	Autoregressive
MA	Moving-average
ARMA	Autoregressive moving average model
SARIMA	Seasonal autoregressive integrated moving average
ARIMA	Autoregressive integrated moving average
LR	Linear regression
SVM	Support vector machine
BPN	Backpropagation network
LSTM	Long short-term memory
RNN	Recurrent neural network
ANN	Artificial neural network
MSE	Mean square error
RMSE	Root mean square error
MAE	Mean absolute error
MAPE	Mean absolute percentage error
NMAE	Normalized mean absolute error
IID	Independent and identically distributed
MCMC	Markov chain Monte Carlo
ML	Machine learning
IMFR	Incremental multiplier of failure rate
FWI	Fire weather index
FFDI	Forest fire danger index
CBRM	Condition-based risk management
HI	Health index
POF	Probability of failure
COF	Consequences of failure
AGC	Automatic generation control
LTC	Load tap changing
CDBN	Conditional deep belief network
ESS	Energy storage system
DRCC	Distributionally robust chance constrained
SCC	Scenario-based chance constrained
MINLP	Mixed integer nonlinear programming
CPS	Cyber–physical system
DMS	Distribution management system
MCC	Microgrid central controller
PHS	Pumped hydro storage
CAES	Compressed air energy storage
SOC	State of charge
DLR	Dynamic line rating
DC	Direct current

AC	Alternating current
SOCP	Second-order cone programming
DFIG	Doubly fed induction generators
CNN	Convolutional neural network
RPS	Renewable portfolio standard
acf	Autocorrelation function
pdf	Probability distribution function
GAN	Generative adversarial network
CVaR	Conditional Value-at-Risk
LC	Load controller
MSC	Micro-source controller
IMG	Interconnected microgrid
RegD	Dynamic regulation
BESS	Battery energy storage system
HPWH	Heat pump water heater
DNSP	Distributed network service provider

References

1. Schroeder, P.W. The nineteenth century system: Balance of power or political equilibrium? *Rev. Int. Stud.* **1989**, *15*, 135–153. [CrossRef]
2. Liu, Z. *Global Energy Interconnection*; Academic Press: Cambridge, MA, USA, 2015.
3. Carrasco, J.M.; Franquelo, L.G.; Bialasiewicz, J.T.; Galván, E.; Portillo, R.; Prats, M.M.; León, J.I.; Moreno-Alfonso, N. Power-electronic systems for the grid integration of renewable energy sources: A survey. *IEEE Trans. Ind. Electron.* **2006**, *53*, 1002–1016. (In English) [CrossRef]
4. Bublitz, A.; Keles, D.; Zimmermann, F.; Fraunholz, C.; Fichtner, W. A survey on electricity market design: Insights from theory and real-world implementations of capacity remuneration mechanisms. *Energy Econ.* **2019**, *80*, 1059–1078. (In English)
5. Rice, E.B.; AlMajali, A. Mitigating The Risk of Cyber Attack on Smart Grid Systems. *Procedia Comput. Sci.* **2014**, *28*, 575–582. (In English) [CrossRef]
6. Inquiry into the National Electricity Market. Available online: https://www.accc.gov.au/system/files/Inquiry%20into%20the%20National%20Electricity%20Market%20-%20May%202022%20report(13880615.1).pdf (accessed on 11 November 2023).
7. AlMajali, A.; Rice, E.; Viswanathan, A.; Tan, K.; Neuman, C. A Systems Approach to Analysing Cyber-Physical Threats in the Smart Grid. In Proceedings of the 2013 IEEE International Conference on Smart Grid Communications (SmartGridComm), Vancouver, BC, Canada, 21–24 October 2013; pp. 456–461. (In English).
8. Zhang, Y.; Wang, J.X.; Wang, X.F. Review on probabilistic forecasting of wind power generation. *Renew. Sustain. Energy Rev.* **2014**, *32*, 255–270. (In English)
9. Roald, L.A.; Pozo, D.; Papavasiliou, A.; Molzahn, D.K.; Kazempour, J.; Conejo, A. Power systems optimization under uncertainty: A review of methods and applications. *Electr. Power Syst. Res.* **2023**, *214*, 108725. (In English)
10. Borunda, M.; Rodríguez-Vázquez, K.; Garduno-Ramirez, R.; de la Cruz-Soto, J.; Antunez-Estrada, J.; Jaramillo, O.A. Long-Term Estimation of Wind Power by Probabilistic Forecast Using Genetic Programming. *Energies* **2020**, *13*, 1885. (In English)
11. Chandramowli, S.N.; Felder, F.A. Impact of climate change on electricity systems and markets—A review of models and forecasts. *Sustain. Energy Technol. Assess.* **2014**, *5*, 62–74. (In English)
12. Sathaye, J.; Shukla, P.R.; Ravindranath, N.H. Climate change, sustainable development and India: Global and national concerns. *Curr. Sci. India* **2006**, *90*, 314–325. (In English)
13. Climate Change Impacts on Power Systems. Available online: https://www.preventionweb.net/news/climate-change-impacts-power-systems (accessed on 13 October 2023).
14. How Does Weather Affect the Electrical Grid. Available online: https://marketscale.com/industries/building-management/how-does-weather-affect-the-electrical-grid/ (accessed on 13 October 2023).
15. Handayani, K.; Filatova, T.; Krozer, Y. The Vulnerability of the Power Sector to Climate Variability and Change: Evidence from Indonesia. *Energies* **2019**, *12*, 3640. (In English) [CrossRef]
16. Anadon, L.D.; Baker, E.; Bosetti, V. Integrating uncertainty into public energy research and development decisions. *Nat. Energy* **2017**, *2*, 17071.
17. Bhusal, N.; Gautam, M.; Abdelmalak, M.; Benidris, M. Modeling of Natural Disasters and Extreme Events for Power System Resilience Enhancement and Evaluation Methods. In Proceedings of the 2020 International Conference on Probabilistic Methods Applied to Power Systems (PMAPS), Liege, Belgium, 18–21 August 2020. (In English).
18. Stock, S.; Babazadeh, D.; Becker, C. Applications of Artificial Intelligence in Distribution Power System Operation. *IEEE Access* **2021**, *9*, 150098–150119. (In English) [CrossRef]
19. Brusentsev, V.; Vroman, W. *Disasters in the United States: Frequency, Costs, and Compensation*; WE Upjohn Institute: Kalamazoo, MI, USA, 2017.

20. Increasing Frequency and Intensity of Extreme Weather Events. Available online: https://www.epa.nsw.gov.au/your-environment/climate-change/trends/extreme-weather-events (accessed on 13 December 2023).
21. Hart, D.; Birson, K. *Deployment of Solar Photovoltaic Generation Capacity in the United States*; Office of Energy Policy and Systems Analysis, U. S. Department of Energy: Washington DC, USA, 2016.
22. Nazma Yeasmin, S. Performance of a Low Cost Circular Arc Horizontal Axis Wind Turbine. Master's Thesis, Bangladesh University of Engineering and Technology, Dhaka, Bangladesh, 2009. Available online: http://lib.buet.ac.bd:8080/xmlui/bitstream/handle/123456789/861/full%20Thesis.pdf?sequence=1&isAllowed=y (accessed on 13 October 2023).
23. Renewable Energy. Available online: https://ourworldindata.org/renewable-energy#citation (accessed on 23 October 2023).
24. 2022 Integrated System Plan (ISP). Available online: https://aemo.com.au/energy-systems/major-publications/integrated-system-plan-isp/2022-integrated-system-plan-isp (accessed on 23 October 2023).
25. Holttinen, H.; Meibom, P.; Orths, A.; Hulle, F.v.; Ensslin, C.; Hofmann, L.; McCann, J.; Pierik, J.; Tande, J.O.; Estanqueiro, A. Design and Operation of Power Systems with Large Amounts of Wind Power, first results of IEA collaboration. In Proceedings of the EWEC'2006-European Wind Energy Conference & Exhibition, Athens, Greece, 27 February–2 March 2006; pp. 1–15.
26. Pieper, C.; Rubel, H. Electricity storage: Making large-scale adoption of wind and solar energies a reality. In *Balanced Growth: Finding Strategies for Sustainable Development*; Springer: Berlin/Heidelberg, Germany, 2011. (In English)
27. Beier, J.; Thiede, S.; Herrmann, C. Energy flexibility of manufacturing systems for variable renewable energy supply integration: Real-time control method and simulation. *J. Clean. Prod.* **2017**, *141*, 648–661. (In English) [CrossRef]
28. Anvari, M.; Lohmann, G.; Wächter, M.; Milan, P.; Lorenz, E.; Heinemann, D.; Tabar, M.R.R.; Peinke, J. Short term fluctuations of wind and solar power systems. *New J. Phys.* **2016**, *18*, 063027. (In English)
29. Kahwash, F.; Maheri, A.; Mahkamov, K. Integration and optimisation of high-penetration Hybrid Renewable Energy Systems for fulfilling electrical and thermal demand for off-grid communities. *Energy Convers. Manag.* **2021**, *236*, 114035. (In English)
30. Smith, J.C.; Milligan, M.R.; DeMeo, E.A.; Parsons, B. Utility wind integration and operating impact state of the art. *IEEE Trans. Power Syst.* **2007**, *22*, 900–908. (In English)
31. Albadi, M.H.; El-Saadany, E.F. Overview of wind power intermittency impacts on power systems. *Electr. Power Syst. Res.* **2010**, *80*, 627–632. (In English) [CrossRef]
32. Mladenov, V.; Chobanov, V.; Georgiev, A. Impact of Renewable Energy Sources on Power System Flexibility Requirements. *Energies* **2021**, *14*, 2813. (In English)
33. Zipf, M.; Möst, D. Impacts of volatile and uncertain renewable energy sources on the German electricity system. In Proceedings of the 2013 10th International Conference on the European Energy Market (EEM), Stockholm, Sweden, 27–31 May 2013; pp. 1–8.
34. Towoju, O.A.; Oladele, O.A. Electricity Generation from Hydro, Wind, Solar and the Environment. *Eng. Technol. J.* **2021**, *39*, 1392–1398.
35. Rikos, E.; Tselepis, S.; Hoyer-Klick, C.; Schroedter-Homscheidt, M. Stability and Power Quality Issues in Microgrids Under Weather Disturbances. *IEEE J. Sel. Top. Appl. Earth Obs. Remote Sens.* **2008**, *1*, 170–179. (In English) [CrossRef]
36. Steen, D.; Goop, J.; Göransson, L.; Nursbo, S.; Brolin, M.; Sarasini, S.; Kåberger, T. Challenges of integrating solar and wind into the electricity grid. In *Systems Perspectives on Renewable Power*; Chalmers University of Technology: Gothenburg, Sweden, 2014; pp. 94–107.
37. Cost Pass through Application 2019–2020 Bushfire Disaster Event. Available online: https://www.aer.gov.au/system/files/Endeavour%20Energy%20-%202019-20%20bushfire%20disaster%20event%20cost%20pass%20through%20-%20Application%20-%20August%202020.pdf (accessed on 1 December 2023).
38. Serrano, R.; Carvalho, M.R.; Araneda, J.C.; Alamos, O.; Barroso, L.; Bayma, D.; Ferreira, R.; Moreno, R. Fighting against Wildfires in Power Systems: Lessons and Resilient Practices From the Chilean and Brazilian Experiences. *IEEE Power Energy Mag.* **2022**, *20*, 38–51. (In English)
39. Clements, D.; Mancarella, P. Systemic modelling and integrated assessment of asset management strategies and staff constraints on distribution network reliability. *Electr. Power Syst. Res.* **2018**, *155*, 164–171. (In English) [CrossRef]
40. Moseidjord, A. The California electricity crisis. *Energy J.* **2003**, *24*, 143–145. (In English)
41. Cleveland, K. "SIGNIFICANT BREAKING WORSE" The Fukushima Nuclear Crisis as a Moral Panic. *Crit. Asian Stud.* **2014**, *46*, 509–539. (In English) [CrossRef]
42. Avizienis, A.; Laprie, J.C.; Randell, B.; Landwehr, C. Basic concepts and taxonomy of dependable and secure computing. *IEEE Trans. Dependable Secur. Comput.* **2004**, *1*, 11–33. (In English) [CrossRef]
43. Lessons from the Massive United Kingdom Power Outage. Available online: https://www.govtech.com/blogs/lohrmann-on-cybersecurity/lessons-from-the-massive-uk-power-outage.html#:~:text=According%20to%20another%20source%20(Euronews,Great%20Britain%20said%20on%20Saturday (accessed on 15 December 2023).
44. Shair, J.; Xie, X.R.; Wang, L.P.; Liu, W.; He, J.B.; Liu, H. Overview of emerging subsynchronous oscillations in practical wind power systems. *Renew. Sust. Energy Rev.* **2019**, *99*, 159–168. (In English)
45. Ghafouri, M.; Karaagac, U.; Ameli, A.; Yan, J.; Assi, C. A Cyber Attack Mitigation Scheme for Series Compensated DFIG-Based Wind Parks. *IEEE Trans. Smart Grid* **2021**, *12*, 5221–5232. (In English) [CrossRef]
46. Haque, N.I.; Shahriar, M.H.; Dastgir, M.G.; Debnath, A.; Parvez, I.; Sarwat, A.; Rahman, M.A. Machine learning in generation, detection, and mitigation of cyberattacks in smart grid: A survey. *arXiv* **2020**, arXiv:2010.00661.

47. Ten, C.W.; Liu, C.C.; Manimaran, G. Vulnerability Assessment of Cybersecurity for SCADA Systems. *IEEE Trans. Power Syst.* **2008**, *23*, 1836–1846. (In English)
48. Amin, M.; El-Sousy, F.F.M.; Aziz, G.A.A.; Gaber, K.; Mohammed, O.A. CPS Attacks Mitigation Approaches on Power Electronic Systems With Security Challenges for Smart Grid Applications: A Review. *IEEE Access* **2021**, *9*, 38571–38601. (In English) [CrossRef]
49. Meenal, R.; Binu, D.; Ramya, K.C.; Michael, P.A.; Kumar, K.V.; Rajasekaran, E.; Sangeetha, B. Weather Forecasting for Renewable Energy System: A Review. *Arch. Comput. Methods Eng.* **2022**, *29*, 2875–2891. (In English)
50. Torres, J.F.; Hadjout, D.; Sebaa, A.; Martínez-alvarez, F.; Troncoso, A. Deep Learning for Time Series Forecasting: A Survey. *Big Data* **2021**, *9*, 3–21. (In English) [CrossRef] [PubMed]
51. Chang, W.-Y. A literature review of wind forecasting methods. *J. Power Energy Eng.* **2014**, *2*, 161. [CrossRef]
52. Voyant, C.; Notton, G.; Kalogirou, S.; Nivet, M.-L.; Paoli, C.; Motte, F.; Fouilloy, A. Machine learning methods for solar radiation forecasting: A review. *Renew. Energy* **2017**, *105*, 569–582.
53. Project 7: Microgrid Impact Study. Available online: https://c4net.com.au/wp-content/uploads/2022/08/49.07-Final_report_P7_v3.pdf (accessed on 25 October 2023).
54. Chau, T.K.; Yu, S.S.; Fernando, T.; Iu, H.H.C. Demand-Side Regulation Provision From Industrial Loads Integrated With Solar PV Panels and Energy Storage System for Ancillary Services. *IEEE Trans. Ind. Inform.* **2018**, *14*, 5038–5049. (In English) [CrossRef]
55. Dubey, A.K.; Kumar, A.; Garcia-Diaz, V.; Sharma, A.K.; Kanhaiya, K. Study and analysis of SARIMA and LSTM in forecasting time series data. *Sustain. Energy Technol. Assess.* **2021**, *47*, 101474. (In English)
56. Meenal, R.; Selvakumar, A.I. Assessment of SVM, empirical and ANN based solar radiation prediction models with most influencing input parameters. *Renew. Energy* **2018**, *121*, 324–343.
57. Livieris, I.E.; Pintelas, E.; Pintelas, P. A CNN-LSTM model for gold price time-series forecasting. *Neural Comput. Appl.* **2020**, *32*, 17351–17360. (In English)
58. Shamshirband, S.; Rabczuk, T.; Chau, K.W. A Survey of Deep Learning Techniques: Application in Wind and Solar Energy Resources. *IEEE Access* **2019**, *7*, 164650–164666. (In English) [CrossRef]
59. Carneiro, T.C.; Rocha, P.A.C.; Carvalho, P.C.M.; Fernandez-Ramirez, L.M. Ridge regression ensemble of machine learning models applied to solar and wind forecasting in Brazil and Spain. *Appl. Energy* **2022**, *314*, 118936. (In English)
60. Díaz-Vico, D.; Torres-Barrán, A.; Omari, A.; Dorronsoro, J.R. Deep Neural Networks for Wind and Solar Energy Prediction. *Neural Process Lett.* **2017**, *45*, 829–844. (In English)
61. Golder, A.; Jneid, J.; Zhao, J.Y.; Bouffard, F. Machine Learning-Based Demand and PV Power Forecasts. In Proceedings of the 2019 IEEE Electrical Power and Energy Conference (EPEC), Montreal, QC, Canada, 16–18 October 2019. (In English).
62. Liu, J.Q.; Yu, S.M.S.; Hu, H.J.; Zhao, J.B.; Trinh, H.M. Demand-Side Regulation Provision of Virtual Power Plants Consisting of Interconnected Microgrids Through Double-Stage Double-Layer Optimization. *IEEE Trans. Smart Grid* **2023**, *14*, 1946–1957. (In English)
63. Akash, R.; Rangaraj, A.G.; Meenal, R.; Lydia, M. Machine Learning Based Univariate Models For Long Term Wind Speed Forecasting. In Proceedings of the 5th International Conference on Inventive Computation Technologies (Icict-2020), Coimbatore, Tamilnadu, 26–28 February 2020. (In English).
64. Alzahrani, A.; Shamsi, P.; Ferdowsi, M.; Dagli, C. Solar Irradiance Forecasting Using Deep Recurrent Neural Networks. In Proceedings of the 2017 IEEE 6th International Conference on Renewable Energy Research and Applications (ICRERA), San Diego, CA, USA, 5–8 November 2017. (In English).
65. Hacioğlu, R. Prediction of solar radiation based on machine learning methods. *J. Cogn. Syst.* **2017**, *2*, 16–20.
66. Wang, Y.; Zou, R.M.; Liu, F.; Zhang, L.J.; Liu, Q.Y. A review of wind speed and wind power forecasting with deep neural networks. *Appl. Energy* **2021**, *304*, 117766. (In English) [CrossRef]
67. Guo, Z.H.; Dong, Y.; Wang, J.Z.; Lu, H.Y. The Forecasting Procedure for Long-Term Wind Speed in the Zhangye Area. *Math. Probl. Eng.* **2010**, *2010*, 684742. (In English) [CrossRef]
68. Ferreira, M.; Santos, A.; Lucio, P. Short-term forecast of wind speed through mathematical models. *Energy Rep.* **2019**, *5*, 1172–1184. (In English) [CrossRef]
69. Santamaría-Bonfil, G.; Reyes-Ballesteros, A.; Gershenson, C. Wind speed forecasting for wind farms: A method based on support vector regression. *Renew. Energy* **2016**, *85*, 790–809. [CrossRef]
70. Bludszuweit, H.; Dominguez-Navarro, J.A.; Llombart, A. Statistical analysis of wind power forecast error. *IEEE Trans. Power Syst.* **2008**, *23*, 983–991. (In English) [CrossRef]
71. Yuan, Z.P.; Xia, J.; Li, P. Two-Time-Scale Energy Management for Microgrids With Data-Based Day-Ahead Distributionally Robust Chance-Constrained Scheduling. *IEEE Trans. Smart Grid* **2021**, *12*, 4778–4787. (In English)
72. Riaz, M.; Ahmad, S.; Hussain, I.; Naeem, M.; Mihet-Popa, L. Probabilistic Optimization Techniques in Smart Power System. *Energies* **2022**, *15*, 825. (In English) [CrossRef]
73. Ben-Tal, A.; Nemirovski, A. Robust solutions of uncertain linear programs. *Oper. Res. Lett.* **1999**, *25*, 1–13. (In English)
74. Campi, M.C.; Garatti, S.; Prandini, M. The scenario approach for systems and control design. *Annu. Rev. Control* **2009**, *33*, 149–157. [CrossRef]
75. Hu, H.; Vrakopoulou, M.; Farokhi, F. Co-Optimization of Frequency Regulation Capacity and Performance Offered by Microgrid under Uncertainty. Master's Thesis, University of Melbourne, Melbourne, Australia, 2023.

76. Das, L.; Munikoti, S.; Natarajan, B.; Srinivasan, B. Measuring smart grid resilience: Methods, challenges and opportunities. *Renew. Sust. Energy Rev.* **2020**, *130*, 109918. (In English) [CrossRef]
77. Jufri, F.H.; Oh, S.; Jung, J.; Choi, M.-H. A method to forecast storm-caused distribution grid damages using cost-sensitive regression algorithm. In Proceedings of the 2019 IEEE Innovative Smart Grid Technologies-Asia (ISGT Asia), Chengdu, China, 21–24 May 2019; IEEE: Piscataway, NJ, USA, 2019; pp. 3986–3990.
78. Panteli, M.; Mancarella, P. Modeling and Evaluating the Resilience of Critical Electrical Power Infrastructure to Extreme Weather Events. *IEEE Syst. J.* **2017**, *11*, 1733–1742. (In English) [CrossRef]
79. Yang, W.J.; Sparrow, S.N.; Ashtine, M.; Wallom, D.C.H.; Morstyn, T. Resilient by design: Preventing wildfires and blackouts with microgrids. *Appl. Energy* **2022**, *313*, 118793. (In English) [CrossRef]
80. Panteli, M.; Mancarella, P. Influence of extreme weather and climate change on the resilience of power systems: Impacts and possible mitigation strategies. *Electr. Power Syst. Res.* **2015**, *127*, 259–270. (In English) [CrossRef]
81. Forest Fire Danger Index. Available online: https://research.csiro.au/bushfire/assessing-bushfire-hazards/hazard-identification/fire-danger-index/ (accessed on 16 October 2023).
82. Blackmore, P.; Leeprechanon, N. Improving Network Reliability Through Effective Asset Management. In Proceedings of the 2015 IEEE Innovative Smart Grid Technologies—Asia (ISGT ASIA), Bangkok, Thailand, 3–6 November 2015. (In English).
83. Fickler, R.; Cheim, L.; Lam, B. Application of Probabilistic Techniques in Asset Management. In Proceedings of the 2009 IEEE/PES Power Systems Conference and Exposition, Seattle, WA, USA, 15–18 March 2009; Volume 1–3. (In English).
84. Chatterjee, K.; Padmini, V.; Khaparde, S.A. Review of Cyber Attacks on Power System Operations. In Proceedings of the 2017 IEEE Region 10 Symposium (TENSYMP), Cochin, India, 14–16 July 2017. (In English).
85. Esmalifalak, M.; Liu, L.C.; Nguyen, N.; Zheng, R.; Han, Z. Detecting Stealthy False Data Injection Using Machine Learning in Smart Grid. *IEEE Syst. J.* **2017**, *11*, 1644–1652. (In English) [CrossRef]
86. He, Y.B.; Mendis, G.J.; Wei, J. Real-Time Detection of False Data Injection Attacks in Smart Grid: A Deep Learning-Based Intelligent Mechanism. *IEEE Trans. Smart Grid* **2017**, *8*, 2505–2516. (In English) [CrossRef]
87. Karimipour, H.; Dehghantanha, A.; Parizi, R.M.; Choo, K.K.R.; Leung, H. A Deep and Scalable Unsupervised Machine Learning System for Cyber-Attack Detection in Large-Scale Smart Grids. *IEEE Access* **2019**, *7*, 80778–80788. (In English) [CrossRef]
88. Zhang, J.N.; Guo, L.L.; Ye, J. Cyber-Attack Detection for Photovoltaic Farms Based on Power-Electronics-Enabled Harmonic State Space Modeling. *IEEE Trans. Smart Grid* **2022**, *13*, 3929–3942. (In English) [CrossRef]
89. Hu, J.Y.; Lanzon, A. Distributed Finite-Time Consensus Control for Heterogeneous Battery Energy Storage Systems in Droop-Controlled Microgrids. *IEEE Trans. Smart Grid* **2019**, *10*, 4751–4761. (In English) [CrossRef]
90. Shangguan, X.C.; Zhang, C.A.K.; He, Y.; Jin, L.; Jiang, L.; Spencer, J.; Wu, M. Robust Load Frequency Control for Power System Considering Transmission Delay and Sampling Period. *IEEE Trans. Ind. Inform.* **2021**, *17*, 5292–5303. (In English) [CrossRef]
91. Khan, S.A.; Ojha, K.; Prasad, S. Mitigation Law against Time-delay based Cyber Intrusion Uncertainty in Power Systems. In Proceedings of the 2023 5th International Conference on Power, Control & Embedded Systems (ICPCES), Allahabad, India, 6–8 January 2023; pp. 1–6.
92. Dileep, G. A survey on smart grid technologies and applications. *Renew. Energy* **2020**, *146*, 2589–2625. (In English)
93. Hannan, M.A.; Faisal, M.; Ker, P.J.; Begum, R.A.; Dong, Z.Y.; Zhang, C. Review of optimal methods and algorithms for sizing energy storage systems to achieve decarbonization in microgrid applications. *Renew. Sust. Energy Rev.* **2020**, *131*, 110022. (In English) [CrossRef]
94. Sharafi, D.; Dowdy, A.; Landsberg, J.; Bryant, P.; Ward, D.; Eggleston, J.; Liu, G.C. Wildfires Down Under: Impacts and Mitigation Strategies for Australian Electricity Grids. *IEEE Power Energy Mag.* **2022**, *20*, 52–63. (In English)
95. Dehghani, N.L.; Darestani, Y.M.; Shafieezadeh, A. Optimal Life-Cycle Resilience Enhancement of Aging Power Distribution Systems: A MINLP-Based Preventive Maintenance Planning. *IEEE Access* **2020**, *8*, 22324–22334. (In English) [CrossRef]
96. Gunantara, N. A review of multi-objective optimization: Methods and its applications. *Cogent Eng.* **2018**, *5*, 1502242. (In English) [CrossRef]
97. Dehghanian, P.; Fotuhi-Firuzabad, M.; Bagheri-Shouraki, S.; Kazemi, A.A.R. Critical Component Identification in Reliability Centered Asset Management of Power Distribution Systems Via Fuzzy AHP. *IEEE Syst. J.* **2012**, *6*, 593–602. (In English) [CrossRef]
98. Alvarez-Alvarado, M.S.; Donaldson, D.L.; Recalde, A.A.; Noriega, H.H.; Khan, Z.A.; Velasquez, W.; Rodríguez-Gallegos, C.D. Power System Reliability and Maintenance Evolution: A Critical Review and Future Perspectives. *IEEE Access* **2022**, *10*, 51922–51950. (In English)
99. Awadallah, S.K.E.; Milanovic, J.V.; Jarman, P.N.; Wang, Z.D. Probabilistic Indicators for Assessing Age- and Loading-Based Criticality of Transformers to Cascading Failure Events. *IEEE Trans. Power Syst.* **2014**, *29*, 2558–2566. (In English) [CrossRef]
100. Rajkumar, V.S.; Tealane, M.; Stefanov, A.; Presekal, A.; Palensky, P. Cyber Attacks on Power System Automation and Protection and Impact Analysis. In Proceedings of the 2020 IEEE PES Innovative Smart Grid Technologies Europe (ISGT-Europe), The Hague, The Netherlands, 26–28 October 2020. (In English).
101. Barreto, C.; Eghtesad, T.; Eisele, S.; Laszka, A.; Dubey, A.; Koutsoukos, X. Cyber-attacks and mitigation in blockchain based transactive energy systems. In Proceedings of the 2020 IEEE Conference on Industrial Cyberphysical Systems (ICPS), Tampere, Finland, 9–12 June 2020; IEEE: Piscataway, NJ, USA, 2020; pp. 129–136.
102. Sridhar, S.; Hahn, A.; Govindarasu, M. Cyber-Physical System Security for the Electric Power Grid. *Proc. IEEE* **2012**, *100*, 210–224. (In English)

103. Ten, C.W.; Manimaran, G.; Liu, C.C. Cybersecurity for Critical Infrastructures: Attack and Defense Modeling. *IEEE Trans. Syst. Man. Cybern.-Part A* **2010**, *40*, 853–865. (In English) [CrossRef]
104. Zografopoulos, I.; Hatziargyriou, N.D.; Konstantinou, C. Distributed Energy Resources Cybersecurity Outlook: Vulnerabilities, Attacks, Impacts, and Mitigations. *IEEE Syst. J.* **2023**, *17*, 6695–6709. (In English)
105. Hernández-Mayoral, E.; Madrigal-Martínez, M.; Mina-Antonio, J.D.; Iracheta-Cortez, R.; Enríquez-Santiago, J.A.; Rodríguez-Rivera, O.; Martínez-Reyes, G.; Mendoza-Santos, E. A Comprehensive Review on Power-Quality Issues, Optimization Techniques, and Control Strategies of Microgrid Based on Renewable Energy Sources. *Sustainability* **2023**, *15*, 9847. (In English) [CrossRef]
106. Nair, R.P.; Kanakasabapathy, P. PR controller-based droop control strategy for AC microgrid using Ant Lion Optimization technique. *Energy Rep.* **2023**, *9*, 6189–6198. (In English) [CrossRef]
107. Borazjani, P.; Wahab, N.I.A.; Hizam, H.B.; Soh, A.B.C. A review on microgrid control techniques. In Proceedings of the 2014 IEEE Innovative Smart Grid Technologies—Asia (ISGT ASIA), Kuala Lumpur, Malaysia, 20–23 May 2014.
108. Blanco, H.; Faaij, A. A review at the role of storage in energy systems with a focus on Power to Gas and long-term storage. *Renew. Sustain. Energy Rev.* **2018**, *81*, 1049–1086. [CrossRef]
109. Amini, M.; Almassalkhi, M. Optimal Corrective Dispatch of Uncertain Virtual Energy Storage Systems. *IEEE Trans. Smart Grid* **2020**, *11*, 4155–4166. [CrossRef]
110. Reddy, Y.J.; Dash, S.; Ramsesh, A.; Kumar, Y.V.P.; Raju, K.P. Monitoring and Control of Real Time Simulated Microgrid with Renewable Energy Sources. In Proceedings of the 2012 IEEE Fifth Power India Conference, Murthal, India, 19–22 December 2012. (In English).
111. Zhou, Y.Z.; Zhai, Q.Z.; Wu, L. Multistage Transmission-Constrained Unit Commitment With Renewable Energy and Energy Storage: Implicit and Explicit Decision Methods. *IEEE Trans. Sustain. Energy* **2021**, *12*, 1032–1043. (In English)
112. Cobos, N.G.; Arroyo, J.M.; Alguacil, N.; Wang, J.H. Robust Energy and Reserve Scheduling Considering Bulk Energy Storage Units and Wind Uncertainty. *IEEE Trans. Power Syst.* **2018**, *33*, 5206–5216. (In English) [CrossRef]
113. Oshnoei, A.; Kheradmandi, M.; Blaabjerg, F.; Hatziargyriou, N.D.; Muyeen, S.M.; Anvari-Moghaddam, A. Coordinated control scheme for provision of frequency regulation service by virtual power plants. *Appl. Energy* **2022**, *325*, 119734. (In English) [CrossRef]
114. Zhang, X.; Hug, G.; Kolter, J.Z.; Harjunkoski, I. Demand Response of Ancillary Service From Industrial Loads Coordinated With Energy Storage. *IEEE Trans. Power Syst.* **2018**, *33*, 951–961. (In English) [CrossRef]
115. Karagiannopoulos, S.; Roald, L.; Aristidou, P.; Hug, G. Operational planning of active distribution grids under uncertainty. In Proceedings of the IREP 2017, X Bulk Power Systems Dynamics and Control Symposium, Espinho, Portugal, 27 August 2017.
116. Naughton, J.; Wang, H.; Cantoni, M.; Mancarella, P. Co-optimizing virtual power plant services under uncertainty: A robust scheduling and receding horizon dispatch approach. *IEEE Trans. Power Syst.* **2021**, *36*, 3960–3972. [CrossRef]

designs

MDPI

Article

Design and Application of Low-Temperature Geothermal Thermoelectric Power Generation (Lotemg–TPG) in Sari Ater Hot Spring, Ciater, Subang, West Java, Indonesia

Harapan Marpaung [1], Supriyadi [1,2], Ni Ketut Lasmi [1], Alamta Singarimbun [1] and Wahyu Srigutomo [1,*]

1 Earth Physics and Complex System, Physics Department, Faculty of Mathematics and Natural Science, Bandung Institute of Technology, Jl. Ganesha 10, Bandung 40132, Indonesia; mharapan887@gmail.com (H.M.); supriyadii@unej.ac.id (S.); tjeku17@gmail.com (N.K.L.); alamtasingarimbung60@gmail.com (A.S.)

2 Physics Department, Faculty of Mathematics and Natural Sciences, Universitas Jember, Jl. Kalimantan 37, Jember 68121, Indonesia

* Correspondence: srigutomo@itb.ac.id

Abstract: The use of surface geothermal manifestations in Indonesia is still very limited as a tourist attraction. Solid-state thermoelectric generator technology is an alternative to converting electrical energy directly from a heat source in the form of low-temperature geothermal manifestation. Low-temperature geothermal thermoelectric power generation (Lotemg–TPG) was designed, manufactured, and tested to take advantage of this opportunity. It was also applied to the Sari Ater Hot Spring, Ciater. The Lotemg–TPG unit comprises seven M8T modules in two frame blocks equipped with hot- and cold-water circulation channels. The M8T module is the main part of the Lotemg–TPG, which consists of eight TEG elements of type TEG1-241-1.4-1.2, flanked by a hot-side radiator and a cold-side radiator. The measurement results showed that at the temperature difference between the hot-side T_h and the cold-side T_c of ΔT 17.38 °C, one module can produce 1.30 W of power, so the total power of the Lotemg–TPG unit is around 9.10 W. This result is quite good considering that the heat source is obtained for free, and the device can operate to produce stable electrical power.

Keywords: geothermal manifestation; low temperature; Sari Ater Hot Spring; Lotemg–TPG; temperature difference, ΔT; electrical power

Citation: Marpaung, H.; Supriyadi; Lasmi, N.K.; Singarimbun, A.; Srigutomo, W. Design and Application of Low-Temperature Geothermal Thermoelectric Power Generation (Lotemg–TPG) in Sari Ater Hot Spring, Ciater, Subang, West Java, Indonesia. *Designs* **2024**, *8*, 60. https://doi.org/10.3390/designs8030060

Academic Editor: Min-Hwi Kim

Received: 2 May 2024
Revised: 3 June 2024
Accepted: 13 June 2024
Published: 17 June 2024

1. Introduction

Geothermal is a renewable energy source that can be used to meet electrical needs. Geothermal energy in Indonesia is spread across Sumatra Island, Java Island, Sulawesi, and the Nusa Tenggara Islands. Indonesia's geothermal potential is 40% of the world's geothermal potential, but only 3.01% of national electricity is used [1]. Bina et al. [2] have developed a geothermal classification in Indonesia based on the exergy concept, but researchers in Indonesia still tend to use a classification based on temperature. The geothermal potential is categorized into high temperature (>225 °C), medium temperature (125–225 °C), and low temperature (<125 °C) [3]. According to DiPippo [4], most geothermal power plants utilize high and medium temperatures because they are considered more economical, while utilizing low-temperature geothermal potential is still limited. In geothermal systems, surface appearances are usually characterized by thermal manifestations. Most geothermal manifestations, especially hot springs, are only used for tourism because they generally have low rates. Low-temperature geothermal sources, such as hot springs, can be utilized to produce electricity using thermoelectric generators.

Ciater Hot Springs are located in Subang, West Java, Indonesia. They are a part of North of Tangkuban Parahu (NTP), the geothermal manifestation zone [5]. The hot springs have a discharge of more than 50 L/s. They are situated at 1000 m.a.s.l, around 3 km northeast of Mt. Tangkuban Parahu. Recent research indicates two hot springs in Ciater

Village: Ciater 1 Hot Spring, with a temperature of 40.8 °C, and Ciater 2 Hot Spring, with a temperature of 42.8 °C [6]. They are considered low-temperature geothermal heat sources with limited utilization potential. Nevertheless, these hot springs have been developed into tourist attractions, i.e., Sari Ater Hot Spring in Ciater Village, Ciater District, Subang Regency, West Java Province, Indonesia. Using these hot springs to generate electricity without disrupting their function as tourist attractions is also possible.

According to Rana et al. [7], the Ciater Hot Springs' potential for geothermal heat is low due to their limited utilization and low-grade temperature, less than 230 °C. However, Remeli et al. [8] suggested that waste heat with a temperature of less than 140 °C can be used as a source of electricity. Technologies such as mechanical systems, semiconductor materials (solid state), or a combination of both can utilize waste heat. However, the mechanical system has several weaknesses, which has resulted in the slow development of technology to utilize low-temperature geothermal heat as an electricity generator [9,10]. One alternative power generation technology that has emerged is a thermoelectric generator (TEG) that works based on the temperature difference between two sides. The thermoelectric phenomenon was first discovered by T.J. Seebeck in 1820 [11]. In 1822, H.C. Oersted explained that an electric current appears and flows in a circuit if there is a temperature difference between two metal junctions. The potential difference, V, resulting from the temperature difference, ΔT, in the two metal junctions is expressed by Equation (1) [12]:

$$V = \alpha(\Delta T), \tag{1}$$

where α is the Seebeck coefficient. Most metals have a relatively small Seebeck coefficient value, ranging from 1 to 10 μV/K. Therefore, in order to obtain larger values, the use of semiconductor technology is necessary.

In 1956, A.F. Ioffe significantly advanced thermoelectric technology by utilizing semiconductor materials, which increased the efficiency by 4%. The theory of thermoelectric-semiconductor materials was documented in books, and it continues to serve as a reference for researchers even today [13]. Using n-type and p-type semiconductor pairs in thermoelectric technology can significantly enhance the Seebeck coefficient up to hundreds of times. Figure 1a depicts a semiconductor TEG schematic diagram, and Figure 1b shows a commercial TEG.

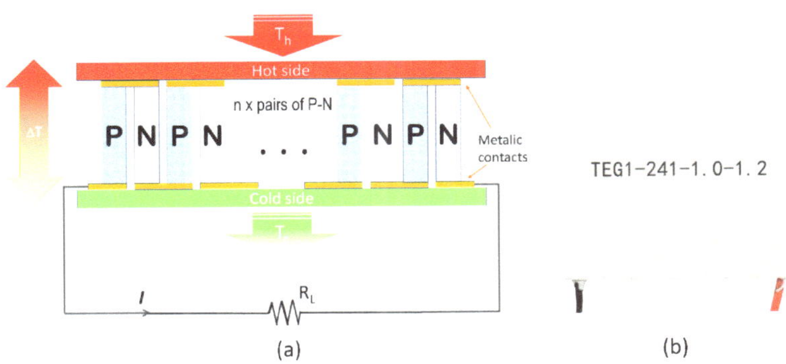

Figure 1. (**a**) A diagram showing the structure of a TEG created using semiconductors, modified from Jaziri et al. [14], and (**b**) an example of a commercial TEG.

The TEG performance of semiconductor materials is determined by a dimensionless quantity, called the figure of merit, ZT [13–17]:

$$ZT = \frac{\alpha^2 \cdot \sigma}{k} T, \tag{2}$$

where α is the Seebeck coefficient in $V \cdot K^{-1}$, σ is the electrical conductivity in $S \cdot m^{-1}$, k is the thermal conductivity in $W \cdot m^{-1} \cdot K^{-1}$, and T is the temperature in K. Thermoelectric materials with a high Seebeck coefficient, high electrical conductivity, and low heat conductivity will also have a high ZT value.

TEG efficiency, η, is expressed by Equation (3) [11,16–19]:

$$\eta = \frac{\Delta T}{Th} \frac{\sqrt{1 + ZT} - 1}{\sqrt{1 + ZT} + \frac{Tc}{Th}}, \tag{3}$$

where ΔT is the temperature difference between the hot-side temperature, Th, and the cold-side temperature, Tc, in K. Based on Equation (3), TEG efficiency is directly proportional to the temperature difference, ΔT. Many commercial TEGs are based on the Bi_2Te_3 semiconductor material and have a ZT value of between 0.6 and 0.8 and an efficiency range of 3–5%. Twaha et al. [20] even noted that commercial TEGs operating below 373.15 K have efficiencies of only 1–3%.

The output power, P, of each TEG element is given by Equation (4) [14]:

$$P = V_{oc}^2 \frac{R_L}{(R_{TEG} + R_L)^2}, \tag{4}$$

where R_L and R_{TEG} are the TEG element's external loads and internal resistance, respectively, and V_{OC} is the open-circuit voltage of the TEG element. If the load resistance, R_L, matches the internal resistance, R_{TEG}, then the maximum output power, P_{max}, is expressed by Equation (5) [14,20,21]:

$$P_{max} = \frac{V_{oc}^2}{4R_{TEG}}, \tag{5}$$

As an electrical energy conversion device based on semiconductor materials, TEG can directly convert heat energy into electricity if there is a temperature difference between the hot side (Th) and the cold side (Tc) [22]. The low efficiency is an obstacle to TEG development. Nevertheless, researchers are still interested in developing TEG applications because of their advantages. This system has no mechanical parts and is compact, quiet, reliable, and environmentally friendly, so it can be operated for a long time with minimal maintenance [7,23–29]. In addition, the low efficiency of TEG can be compensated by utilizing freely obtained heat sources or waste heat. Until now, using TEG as an electricity generator has primarily utilized waste heat from vehicles/machines and solar heat sources (solar thermoelectric generator (STEG)). Some researchers also focus on TEG applications with low-temperature heat sources, such as Maneewan and Chindaruksa [30] with biomass heat sources, Hsu et al. [31] with vehicle exhaust waste heat sources, and Liu et al. [24] and Rana et al. [7], who focused on arranging and constructing TEG as a power generator. TEG elements capable of operating at low temperatures are available on the market and are discussed in a review paper by Zulkepli et al. [16]. The use of TEG in power generation applications with low-temperature geothermal sources is still rare, and portable hardware still needs to be developed. A few researchers have developed TEG applications in geothermal fields, i.e., Niu et al. [32], Suter et al. [33], Liu et al. [24], and Li et al. [34], but they are still limited to experimental and modeling studies with a temperature difference of 72.2–120 °C. The review paper on TEG applications conducted by Zoui et al. [17] does not even mention the application of TEG to geothermal energy. The application of TEG for low-temperature geothermal heat was found in a review paper by Lee et al. [35], which focused on discussing the commercialization of low-temperature TEGs. The paper explained that Taiwan, where most of its geothermal potential has low-temperature types in the form of hot springs, has developed a large-scale TEG-based power generation consisting of six sub-systems, where each system consists of 128 TEG elements. This generating unit is large, so it needs to be developed to a more compact size. The latest development of a thermoelectric generator for medium- and low-temperature geothermal applications was carried out by Xie et al. [36]

with advanced fabrication, accompanied by advanced measurement and control systems, and using 768 TEG elements. Thermoelectric generator systems on a large scale (many TEG elements) are expensive and challenging. The initial development of a portable thermoelectric generator consisting of only 20 TEG elements and working at a low-temperature difference of 67 °C has been carried out by Ahiska and Mamur [23,37], but has yet to be applied to geothermal fields. Another study on applying a TEG-based power generation that utilizes hot steam from geothermal wells directly in Bottle Rock, California, was carried out by Li et al. [38]. Another researcher, Catalan et al. [39], developed and focused on TEG applications in hot dry rock fields (geothermal manifestation) requiring unique heat exchanger designs. The two above studies were not adopted here because this study focuses more on low-temperature geothermal energy, especially hot springs. After conducting a literature review, it appeared possible to create a TEG-based power generation device that is small, portable, low cost, and capable of running on low-temperature heat sources in water, steam, or a combination of the two phases with low-temperature. The product is expected to be applied to geothermal manifestations, low-temperature geothermal energy, or waste heat in existing geothermal power plants in Indonesia.

The availability of low-temperature heat sources in the form of hot springs, which have only been used for tourism purposes, opens up opportunities for them to produce electrical energy. This research aims to design, produce, and test a portable, small-scale TEG-based power generation system that can operate on a low-temperature heat source in either the water or steam phase. The design and production of this power generation unit went through a series of tests using data collectors to guarantee its performance. The Sari Ater Hot Springs, Ciater, part of the manifestation of the Tangkuban Parahu geothermal system, was chosen as the test site for the power generation unit. The resulting device is expected to be an alternative for utilizing renewable energy, namely, low-temperature geothermal energy, and can be operated to support tourism needs. Applying this power generation unit in other geothermal manifestations is possible, considering that Indonesia has many low-temperature geothermal systems, especially those that have not been exploited economically and have yet to be reached by the electricity network.

2. Methods

The research was approached with a direct experimental method, which included designing, manufacturing, testing, and applying the low-temperature geothermal thermoelectric power generation (Lotemg–TPG) unit. Tests were carried out at the laboratory and field scales. Tests and applications in the field were carried out at the Sari Ater Hot Springs, Ciater, Subang, West Java, Indonesia.

The parameters used as a reference in the Lotemg–TPG unit test included cold-side temperature, Tc, hot-side temperature, Th, electric voltage, V, electric current, I, measurement time, t, and load resistance, R_L. These parameters were presented as a temperature difference curve, ΔT, and an output power curve, P, versus time t. Analysis was carried out on these curves to understand and evaluate the performance of each component and generator module. The performance of the Lotemg–TPG was reviewed based on the electrical power produced versus the temperature of the heat source and its stability during operation.

2.1. Lotemg–TPG Test and Application Site

The Lotemg–TPG unit test was performed in stages, including laboratory and field application tests. The design, manufacture, and laboratory-scale tests of the Lotemg–TPG were carried out in a private workshop with adequate facilities and equipment. Meanwhile, field-scale tests were carried out at the Sari Ater Hot Spring, Ciater, Subang, West Java, Indonesia (Figure 2). The hot springs in this area have been developed as a tourist attraction in the form of a bathing pool.

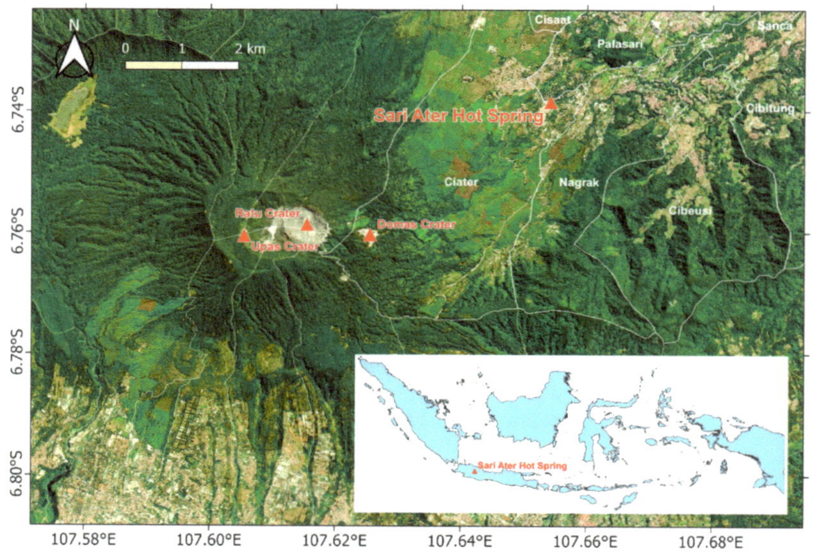

Figure 2. Field test location for the Lotemg–TPG unit, namely, the Sari Ater Hot Spring, Ciater, Subang, West Java, Indonesia.

2.2. Research Stages

In general, the research stages were carried out in the sequence described in Figure 3, in the form of design, manufacture, and test, so that the Lotemg–TPG unit was realized, tested, and could be applied to low-temperature geothermal heat.

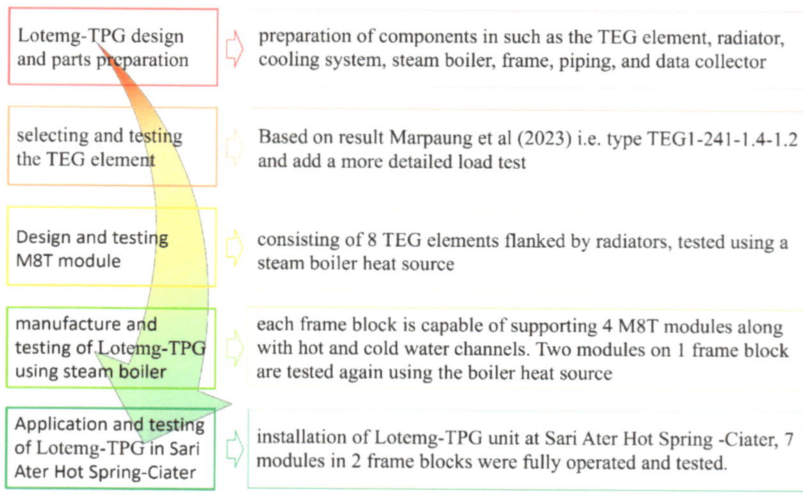

Figure 3. Research stages illustration [40].

2.3. Device Manufacture and Performance Test of Lotemg–TPG

The primary basis of this research was manufacturing the Lotemg–TPG unit. The main component of the Lotemg–TPG unit is the TEG element type TEG1-241-1.4-1.2 (Figure 1b). This type of TEG element has previously been tested and had the best performance [40]. Some essential parameters of TEG type TEG1-241-1.4-1.2 are dimensions of 54.4 mm × 54.4 mm

× 3.4 mm, efficiency of 4%, and operating temperature of 200 °C. To adjust the hot side and cold side of the TEG element to reach the specified temperature, a radiator is needed, which was attached to side of TEG element using thermal paste and reinforced with a holder. The hot-side and cold-side radiators were made of aluminum with dimensions of 25 cm × 12.5 cm × 1.4 cm (Figure 4, top). The radiator was made of aluminum and had many channels, so heat transfer occurred optimally. The arrangement of the hot-side radiator, with 8 TEG elements, and the cold-side radiator, glued together with thermal paste and a holder, is called the M8T module (Figure 4).

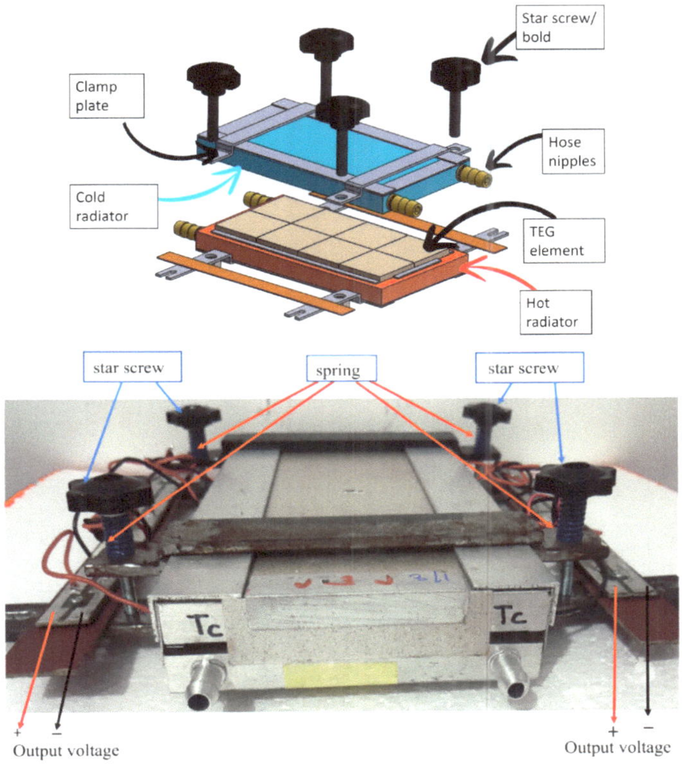

Figure 4. M8T module consisting of 8 TEG elements flanked by a hot-side radiator and a cold-side radiator: schematic diagram (**top**) and real unit (**bottom**).

The power generation block frame was made of iron, which supported the main components of the Lotemg–TPG, which were packaged in the form of an M8T module and supporting components, i.e., hot-water and cold-water circulation systems. The hot-water and cold-water circulation systems were made similarly in the form of 1″ iron pipes connected to each radiator of the M8T module via a hose. The hoses connecting the hot pipes and hot-side radiators were made from heat-resistant materials. This block frame measured 40 cm long, 28 cm wide, and 50 cm high, weighing around 20 kg. Each block could be filled with 4 M8T modules (Figure 5).

Figure 5. Seven M8T layouts on two block frames of Lotemg–TPG units (schematic and real unit).

Other supporting components in this research were for the purpose of laboratory tests and measurement equipment. For the purpose of laboratory tests, namely, tests on the M8T module and the Lotemg–TPG unit, a heat source provided by a steam boiler and a cold-water circulation system equipped with a pump were used. The steam boiler was designed and built with a height of 1 m and a diameter of 22 cm. The bottom of the steam boiler until was 20 cm high, and the diameter was made wider at 30 cm to maximize the heating process (Figure 6). This boiler can produce hot steam, hot water, or a combination of both at temperatures up to 170.0 °C (±2.5 °C) and pressures up to 5.0 kg/cm^2 (±0.1 kg/cm^2), with a heat-up time of approximately 45 min. The cooling system for the M8T module test and the Lotemg–TPG unit in the lab used an active system supported by a 12V DC submersible pump to circulate water between the cold-side radiator and the reservoir. During the application and field test of the Lotemg–TPG unit, hot-water circulation came from the bathtub faucet, and cold-water circulation came from the garden faucet at Sari Ater Hot Springs, Ciater.

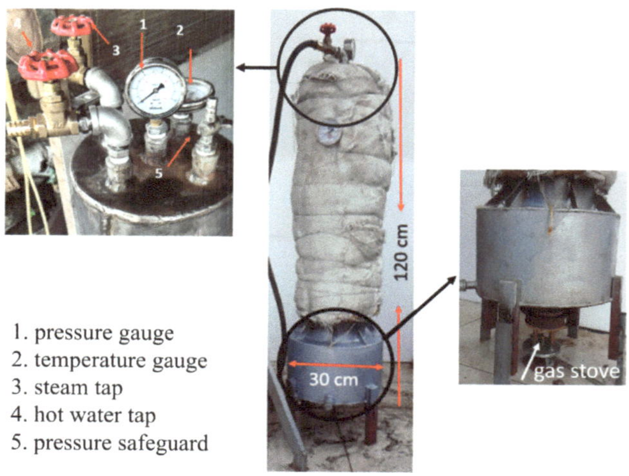

Figure 6. Control system on top of the steam boiler (**left**) and vertical tubular steam boiler with a larger-diameter bottom (**right**).

A critical supporting component in this research was the integrated data collector, used as a test measurement tool. This integrated data collector was deliberately designed and made specifically to measure test parameters in this research. The function of each button, switch, display, and I/O socket is shown in Figure 7 and Table 1. During the measurements, external load resistance in the form of a cermet film resistor set was also used when testing the power characteristics of TEG type TEG1-241-1.4-1.2, the cermet film resistor set for the M8T module test, and the Lotemg–TPG unit test in the laboratory, as well as a 12 W 5 V LED lamp for the test at Sari Ater Hot Spring. Measurements were documented with the help of a recording camera.

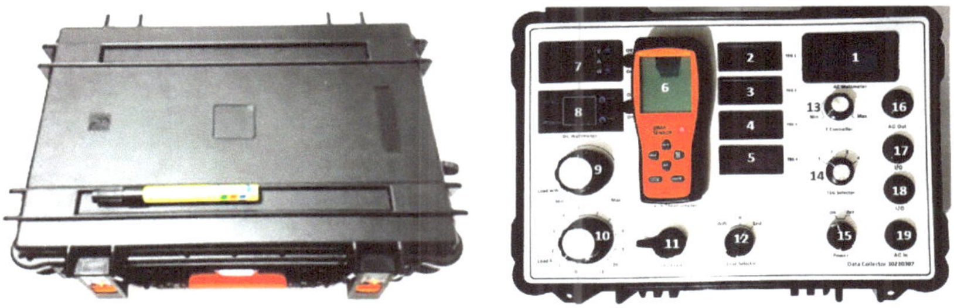

(**a**) (**b**)

Figure 7. Integrated data collector (**a**) box and (**b**) user interface, i.e., switch, display, and I/O sockets.

Table 1. Display functions, control switches, and I/O sockets for data collection.

Number of Panels	Description/Specification
2, 3, 4, and 5	DC voltage meter/4-digit voltmeter, range 0–100 V, accuracy: ±0.3%
6	4-Channel digital thermometer (T_c and T_h measurements)/ Smart Sensor Thermocouple 4-Channel AS887, range −200~1372 °C, accuracy: ±0.1% + 0.6 °C
7 and 8	DC power meter/DC Meter 3in1 Volt Ampere Watt 4 digits, accuracy: 1% (±1 digit)
9	Variable potentiometers for R load (0–30 Ω)
10	Rotary switch for R load (0–80 Ω)
11	LED light load sockets
12	Load selector switch
14	TEG power generation output line selector switch
15	On/off switch
17 and 18	Input/output sockets (in/out socket)

3. Results and Discussion

3.1. TEG Element Type TEG1-241-1.4-1.2 and M8T Module

The TEG elements were a crucial component in preparing the Lotemg–TPG unit. Different types of TEG elements were considered based on their operational temperature, efficiency, and output power. In a previous publication [40], tests were conducted on five types of TEG elements, including TEG1-241-1.4-1.2, TEG1-287, TEG1-199, SP1848, and TEC1-12706. The results showed that TEG1-241-1.4-1.2 produced the largest output power at various temperature differences, such as 4.9 W at ΔT = 100 °C. TEG1-241-1.4-1.2 can operate up to a temperature of 200 °C and has an efficiency of 4%. However, the output power, P, versus load resistance, R_L, presented by Marpaung et al. [40] did not clearly show the maximum power, as it was limited to a maximum load resistance of <16 Ω. To obtain a larger load resistance range, repeated measurements using the same tool settings were conducted, and the results are presented in Figure 8. The load resistance range was obtained by setting the rotary switch for load resistance, R_L, on the data collector (switch No. 12 in Figure 7b) and adding an external load in the form of a cermet film resistor set. The heat source was obtained from a 200 W/220 V electric heating element to maintain a stable hot-side temperature, *Th*, at 165 °C. The cooling system consisted of a radiator connected to a water tank, and circulation was assisted by a DC pump. The temperature of the cold side of the TEG element increased slightly and then stabilized at *Tc* of 28 °C.

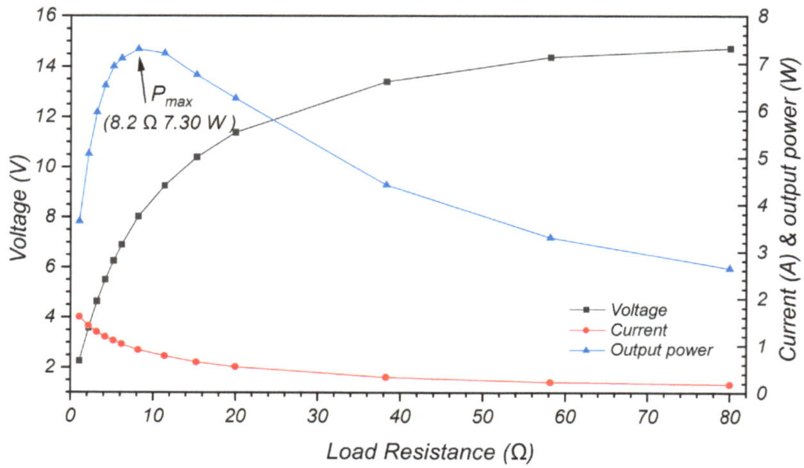

Figure 8. Voltage (left axis), current, and output power (right axis) characteristics curve of TEG element type TEG1-241-1.4-1.2 over a wide range of load resistance.

Based on the curve in Figure 8, by Ohm's law, when the load was added slowly, the current decreased and the voltage increased. In the load range of 0–80 Ω, the maximum output power, P_{max}, was observed to be 7.30 W, obtained when the load resistance was 8.2 Ω. This is in accordance with Equation (4), which states that maximum output power is obtained when the load resistance value approaches the internal resistance of TEG type TEG1-241-1.4-1.2, namely, 7.8 Ω. Increasing the maximum output power of the TEG element by means of impedance matching, namely, equalizing the load resistance to the internal resistance of the TEG element, was also carried out and explained in research by Ahiska and Mamur [23], Twaha et al. [20], Zhang et al. [41], Morais et al. [42], and Jaziri et al. [14]. Morais et al. [42] further explained that the arrangement of TEG elements in the design of a thermoelectric-based power plant influences the total internal resistance of the TEG. The arrangement of TEG elements in a series will produce a large output voltage and internal resistance. On the other hand, the arrangement of TEG elements in parallel will produce a large current and small internal resistance.

After obtaining the TEG1-241-1.4-1.2 type as the TEG element with the best performance, the next step was to make the M8T module. The construction of this module consists of eight TEG elements arranged between the hot-side radiator and the cold-side radiator (Figure 4). In detail, the eight TEG elements were arranged in two back-to-back paths, making it easier to arrange the wiring. The hot-side and cold-side radiators were also arranged with the input–output channels back-to-back to facilitate the piping line and avoid temperature influences between the two. The documentation for testing this module using a heat source from a steam boiler is presented in Figure 9. This module utilized an active cooling system, where the circulation of cold water between the water tank and the cold-side radiator was assisted by a pump. The schematic arrangement of the eight TEG elements in the M8T module is shown in Figure 10, top. Assuming the hot-side radiator and cold-side radiator were able to distribute heat uniformly on both sides of the TEG element, measurements of temperature and power characteristics over time were only carried out on TEG element 1 (V1), and the curves are presented in Figure 10, bottom.

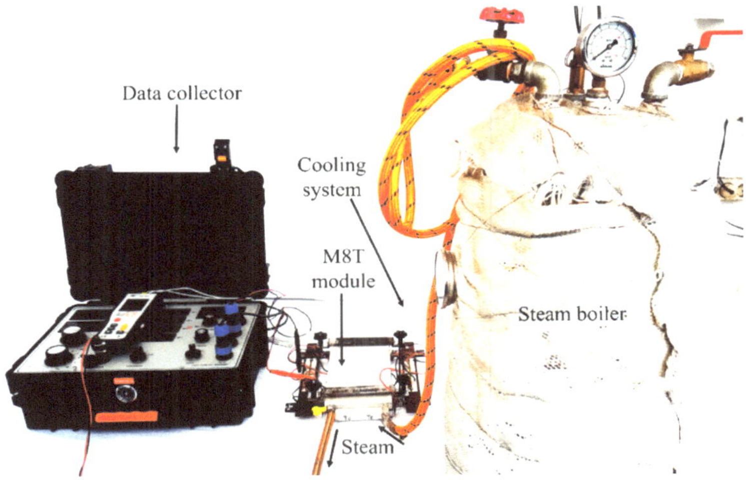

Figure 9. Equipment arrangement for testing the M8T module.

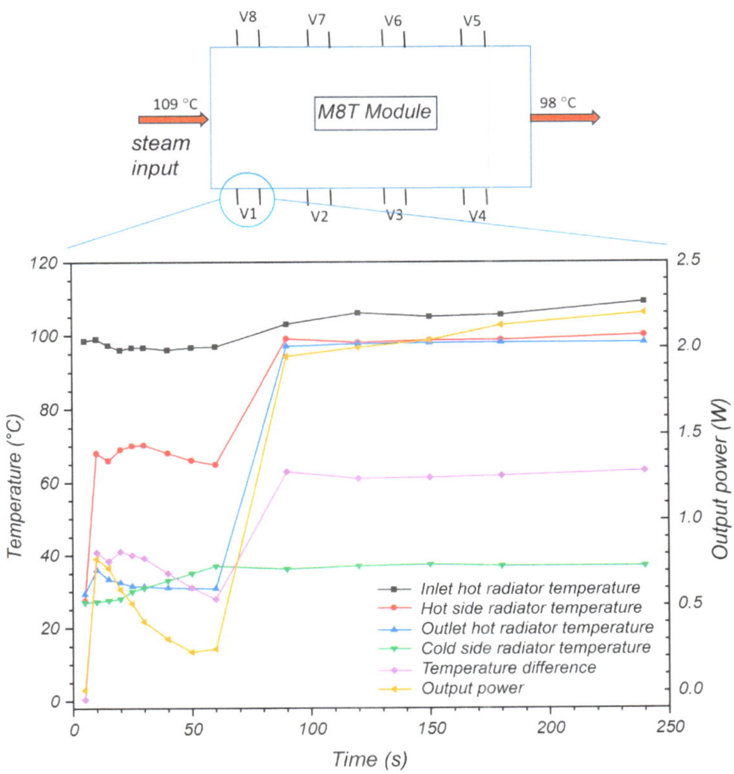

Figure 10. Schematic arrangement of the eight TEG elements on the M8T module (**top**) and temperature (left axis) and output power (right axis) of TEG element 1 (V1) on the M8T module (**bottom**).

During the testing of the M8T module, the steam boiler temperature was set to 135.0 °C (±2.5 °C) with a pressure of 4.0 kg/cm^2 (±0.1 kg/cm^2) and was directed to the hot-side radiator inlet via a heat-resistant hose. At these temperature and pressure levels, the temperature at the inlet hot radiator, T_3, remained relatively constant, reaching an average temperature of 105.70 °C after 90 s. The heat from the hot steam heated up the hot side of the radiator, which then passed on to the TEG elements to produce an electric potential. Based on the curve in Figure 10, during the first 60 s, the temperature on the hot side of the radiator remained unstable and even decreased slightly due to the cold water trapped inside, left over from the previous experiment. After reaching a stable condition, the average temperature of the hot-side radiator was Th = 98.86 °C, with the outlet hot radiator temperature still relatively high at T_4 = 97.72 °C. The temperature of the cold-side radiator experienced a slight increase as a result of the heat transfer from the hot steam through the TEG element, and then remained relatively constant at an average temperature of 36.92 °C. This indicates that the cooling system was working well. Under stable conditions, an average temperature difference, ΔT, of around 61.94 °C was obtained, which produced an output power of around 2.06 W at TEG element 1 (V1), with a load resistance of 7.8 Ω. In this M8T module test, the current, voltage, and output power measurements were taken only on TEG element 1. Assuming that the eight TEG elements were in the same condition (with the same temperature difference), the measurements on TEG element 1 can represent the other TEG elements. This output power was slightly lower than the output power resulting from the TEG type TEG1-241-1.4-1.2 characteristic test conducted by Marpaung et al. [40] on the same ΔT. Assuming that all TEG elements in the M8T module were in the same condition, they would produce a total output power of 16.48 W. The eight TEG elements in the M8T module can be connected in series, parallel, or any combination to meet output voltage requirements for future tests and applications.

3.2. Lotemg–TPG Unit Block Frame and Laboratory Test

The Lotemg–TPG unit is an integrated power generation system consisting of seven M8T modules in two blocks with a solid frame equipped with hot-side and cold-side circulation systems and TEG element output wiring lines. The first block consists of three M8T modules, and the second block consists of four M8T modules. The radiators on each M8T module in the blocks were interconnected with the central piping system, hot-water hoses, and cold-water hoses. The front and top views, accompanied by the finished product size of the Lotemg–TPG unit, are shown in Figure 11, bottom. The connection between the radiator and the central pipe through the hose is shown in Figure 12. The blue arrow illustrates the direction of cooling water flow, distributed through a hose from a 1″ pipe to each Block 1 and Block 2 module. The hot fluid flows through the tap, illustrated by the red arrow line, and is distributed from the 1″ pipe to each module through a heat-resistant hose up to 260 °C.

The design of the wiring between TEG elements and M8T modules in Lotemg–TPG is flexible. The eight TEG elements in each M8T module can be connected in series or parallel to adjust the voltage and current output, as desired. When a large potential difference is needed, the TEG elements are connected in series. Conversely, when a large current is needed, the TEG elements are connected in parallel. To obtain certain voltage and current values, combinations of TEG elements in series and parallel can also be used. The same applies to the arrangement between M8T modules in Lotemg–TPG. Some M8T modules can be connected in series or parallel to achieve the desired output. According to the explanation of Morais et al. [42], the combination of circuits between TEG elements and between M8T modules is also useful for obtaining maximum output power. The electrical energy produced by Lotemg–TPG is in the form of direct current (DC) electricity.

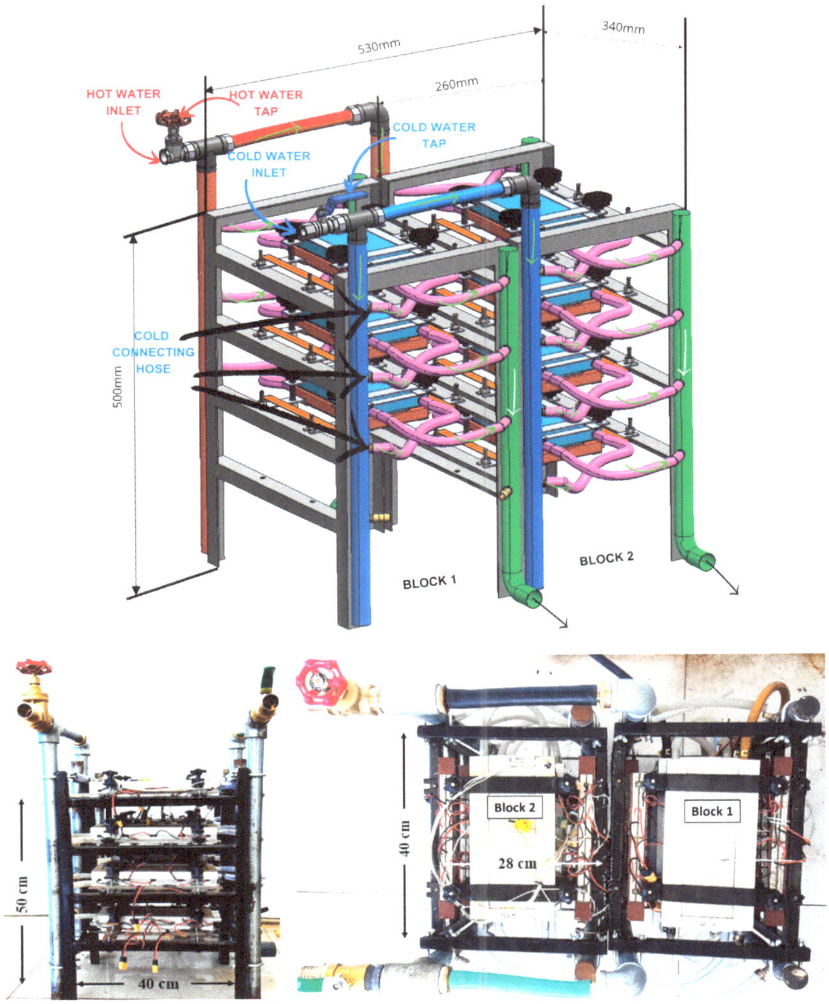

Figure 11. Schematic diagram of Lotemg–TPG (**top**) and the real unit of Lotemg–TPG seen from the front and top (**bottom**), with information on the size of the block frame.

The steam boiler previously used to test the M8T module was repurposed to test the performance of the Lotemg–TPG unit. However, there was limited hot-water steam flow capacity to supply all seven M8T modules on the Lotemg–TPG. Therefore, during the Lotemg–TPG unit test, using a heat source from the steam boiler, only Module 1 and Module 2 in one of the blocks were operated, as shown in Figure 13. Hot steam from the steam boiler was directed to the hot-side radiator of Module 1. The hot-side radiator output from Module 1 then flowed back to the Module 2 radiator. The cold-side radiators in Module 1 and Module 2 were connected to the cooling system with the help of a pump.

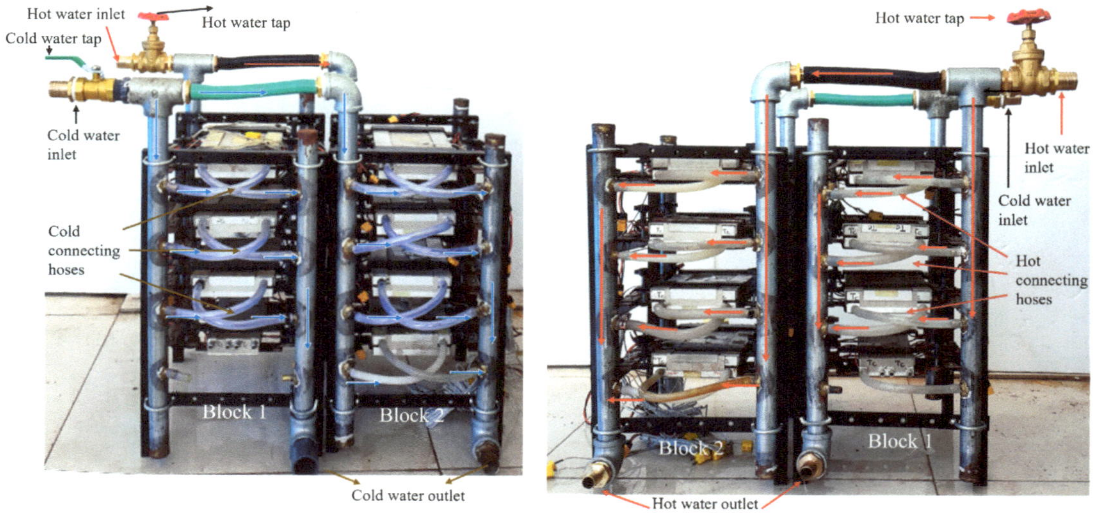

(a) (b)

Figure 12. Lotemg–TPG unit seen from (**a**) the right side, with an illustration of the cooling fluid flow in the form of a blue arrow, and (**b**) the left side, with an illustration of the hot fluid flow in the form of a red arrow.

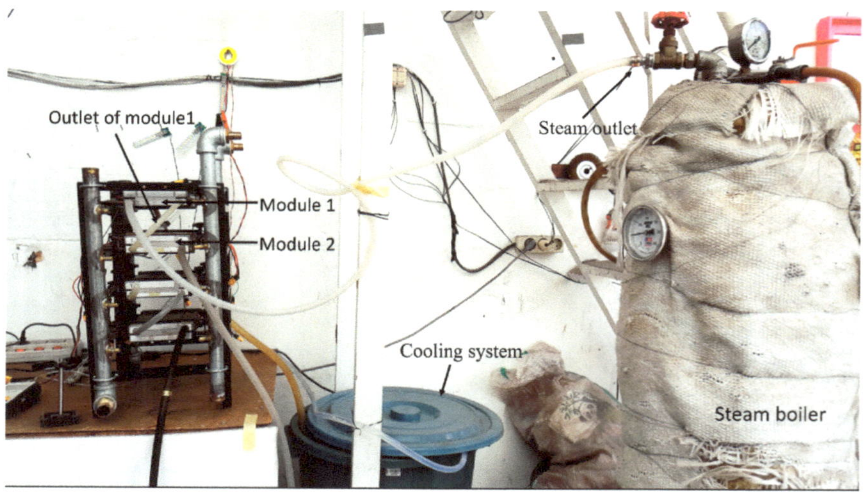

Figure 13. Connection of radiator Module 1 and Module 2 on Lotemg–TPG to the steam boiler and cooling system.

Out of eight TEG elements in Module 1 and Module 2, only four TEG elements at one line side (see Figure 4, top and Figure 10, top) were measured. In order to achieve maximum power, an impedance matching technique was used, which has been explained by Morais et al. [42]. A series-parallel combination of four TEG elements was used, as shown in Figure 14. The results of the calculations showed that the internal resistance of the measured elements was the same as the internal resistance of the TEG1-241-1.4-1.2 element type, which was 7.8 Ω. To achieve maximum power, the load resistance, R_L, should be set close to 7.8 Ω. In this test, the load resistance value was approximated by a parallel array of 4 pieces of 33 Ω resistors to obtain 8.25 Ω. This load value was also relatively close to the load resistance, which produced maximum output power, P_{max}, in Figure 8, which was 8.2 Ω.

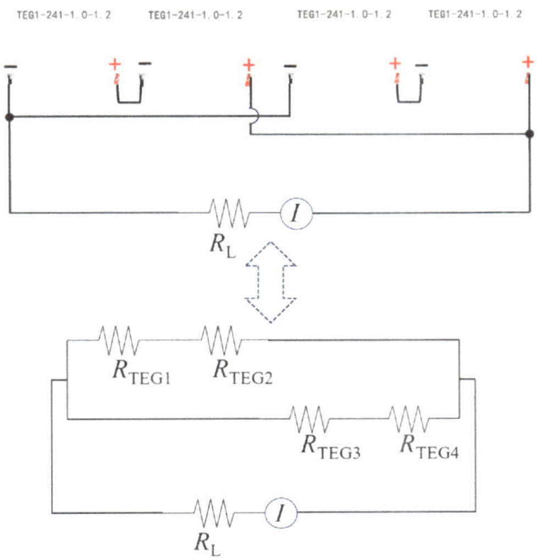

Figure 14. Series-parallel connection of four TEG elements in Module 1 and Module 2.

The results of the Lotemg–TPG test measurements with a heat source as a steam boiler on four TEG elements in Module 1 and Module 2 are presented in the form of curves of temperature difference (left axis) and output power (right axis) versus operating time, shown in Figure 15. With a heat source in the form of hot steam from a steam boiler and the previous radiator conditions emptied, the temperature in the radiator quickly rose and reaches a stable condition in <60 s. The output power of Module 1 (blue line) was much greater than that of Module 2 (green line) because the heat was received first by Module 1, while Module 2 only received the residual heat. The average output power in Module 1 in a stable state was around 9.48 W at an average temperature difference, ΔT, of around 69.21 °C for the four TEG elements, so that for each TEG, the output power was 2.37 W. This output power was also smaller than the output power of the characteristic test of TEG type TEG1-241-1.4-1.2 by Marpaung et al. [40] at the same ΔT, which was probably due to the load resistance (8.25 Ω) being more significant than the internal resistance of the four-element TEG circuit (7.8 Ω). Meanwhile, the four TEG elements at Module 2, in a stable state with an average temperature difference, ΔT, of 58.56 °C, still produced an average power of 4.12 W. These values were relatively constant in the time range of 60 to 300 s (stable state), which indicates that the modules in the Lotemg–TPG unit worked well. When the flow of hot steam from the steam boiler was turned off at 305 s, the temperature difference, ΔT, and output power, P, decreased slowly due to the residual heat remaining in the hot-side radiator. The test results on Module 1 were assumed to be representative of the other modules in Lotemg–TPG, considering that all M8T modules were made uniformly and placed in the generating frame with similar conditions, especially for the hot-side temperature, Th, and cold-side temperature, Tc.

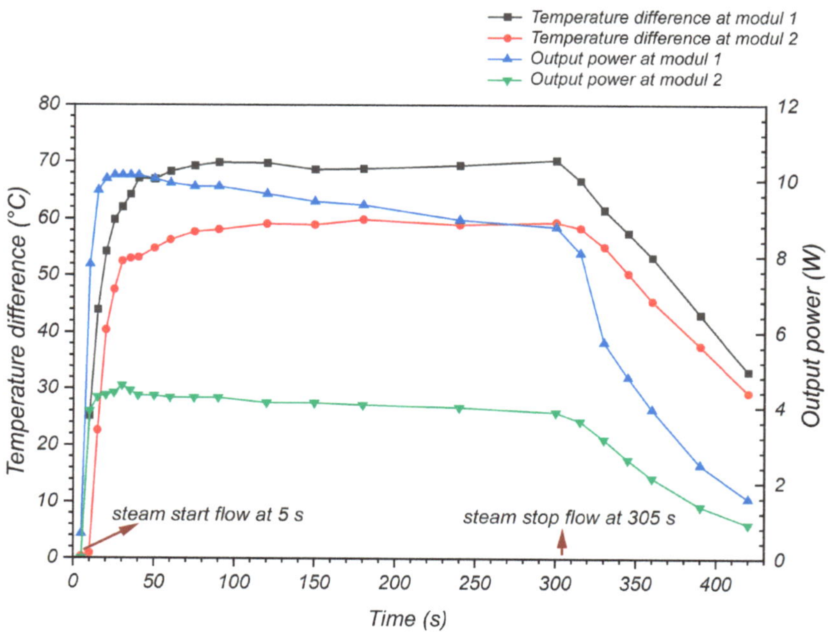

Figure 15. Curves of temperature difference (left axis) and output power (right axis) versus operating time of four TEG elements in Module 1 and Module 2 of the Lotemg–TPG, with a heat source from a steam boiler.

3.3. Application and Field Test of Lotemg–TPG Unit at Sari Ater Hot Springs

Sari Ater Hot Spring is a natural tourist attraction managed by PT Sari Ater and located in Ciater Village, Ciater District, Subang Regency, Wes Java Province, Indonesia. It is situated on the northeast slope of Mt. Tangkuban Parahu and is part of the geothermal manifestation system. The temperature of the hot spring ranges from 43 to 46 °C, according to direct measurements. These results are consistent with research by Nasution et al. [5] and Sentosa et al. [6], considering the dynamics of the Mt. Tangkuban Parahu geothermal system. The relatively low temperature of this hot spring was utilized for electricity production through TEG-based energy conversion equipment previously created and tested.

The Lotemg–TPG unit tested at the Sari Ater hot springs consisted of two blocks (Figure 16). The hot-water source was taken from the soak room tap and channeled to the power generation system via a 6 m hose. Measurements indicated that the hot-water discharge from the soaking room tap was 7.74 L per minute at 45.50 °C. At the main inlet tap of the power generation system, the hot-water temperature dropped to 43 °C. However, the hot-water channel in the form of an ordinary hose with a length of up to 6 m was less than ideal due to heat radiation occurring along the hose. After supplying heat to the hot-side radiator for seven M8T modules, the hot water from the power generation system was still 41 °C. This means the power generation system can only absorb 2 °C of heat from hot water at 43 °C and a flow rate of 7.74 L per minute. The cold-water circulation utilized tap water available in the park, which was channeled to the power generation system via a 10 m hose.

Figure 16. (**a**) Hot water from the bathroom tap and (**b**) cold water from the garden tap. (**c**,**d**) Lotemg–TPG was connected to a hot-water and cold-water source via a hose.

The Lotemg–TPG unit was tested with a heat source from a Sari Ater Hot Spring. The tests involved measuring the temperature of the hot-side radiator (Th) and cold side (Tc), current (I), and output voltage (V) of the M8T module. The measurements were conducted using the data collector on Module 1 in Block 1 and Module 4 in Block 2. Due to the relatively low temperature of the hot spring, it will generate a low voltage. Therefore, the eight TEG elements in Module 1 and Module 4 were wired in a series connection to deliver a voltage capable of supplying LED lights. The results are presented in curves of output power, P, and temperature difference, ΔT, versus time, t in Figure 17. P_1 is the output power on Module 1 in Block 1 and P_2 is the power on Module 4 in Block 2. According to the temperature difference, ΔT, versus time, t, it took around 60 s for the Lotemg–TPG unit to reach a stable condition, with Th_1 around 40.63 °C and Th_2 around 40.40 °C. With a passive cooling system, where cold water flows directly from the tap in the garden without the help of a pump, the temperature of the cold-side radiators Tc_1 and Tc_2 only experienced a slight increase and then stabilized, with Tc_1 around 23.25 °C and Tc_2 around 23.42 °C. The difference in Th_1 and Th_2 as well as Tc_1 and Tc_2 caused a difference in ΔT_1 (17.38 °C) and ΔT_2 (16.98 °C), which affected the output power. The output power from Module 1 in Block 1 was slightly greater than that of Module 4 in Block 2. This was due to the limited flow of hot water from the tap, so the hot-water supply to Block 2 was not as optimal as that in in Block 1, where the hot water passed first. After reaching a stable condition, the average electrical power produced by Module 1 was 1.30 W and that by Module 4 was 1.29 W. The Lotemg–TPG unit was capable of producing a total power of around 9.10 W with seven M8T modules installed. Taking the average output power in Modules 1 and 4, each TEG element produced 0.16 W of output power. From the results of the recent study of TEG applications in low-temperature geothermal energy conducted by Liao et al. [43], each TEG element was able to produce 0.96 W at a temperature difference of 73 °C. Some factors that were presumed to cause the results of this study to be lower than those of Liao et al. were the low temperature of the hot springs, the low flow rate of the hot water, and the use of a far simpler and cheaper heat exchanger. The power output of the Lotem–TPG with a heat source in the form of the Sari Ater Hot Spring, Ciater, was still in the form of direct current (DC), where each module produced an average output voltage of 3.50 V and could be used directly to supply 5 V 12 W LED light sticks with a relatively bright flame. Although the power output was relatively small when compared to the results of the Lotemg–TPG unit test using a steam boiler heat source, the unit is still worth developing. This is because the heat source from hot springs is available for free, and the unit can operate stably. Several improvements to the hot-water circulation system by adding insulators can increase the output power of the Lotemg–TPG unit. The addition of an external power regulator is also needed so that the Lotemg–TPG unit can produce maximum power and safely supply electronic equipment.

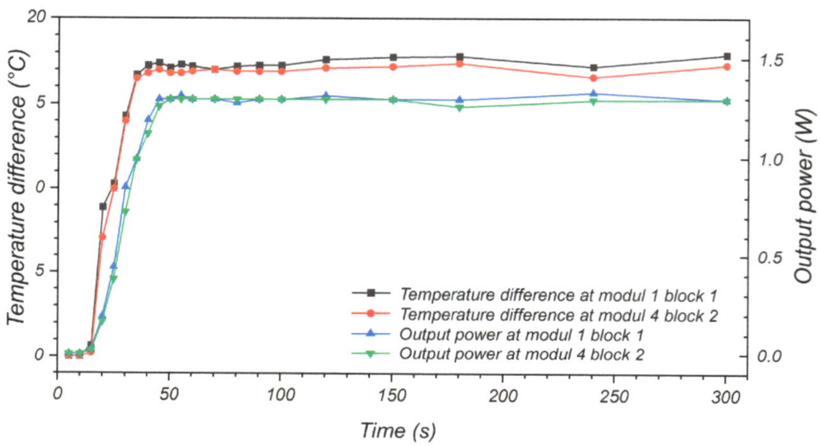

Figure 17. Temperature difference (left axis) and output power (right axis) versus time in the Lotemg–TPG unit test at Sari Ater Hot Spring.

4. Conclusions

The Lotemg–TPG unit was successfully manufactured and tested in this research. The Lotemg–TPG unit is a solid-state TEG-based energy conversion device that directly converts low-temperature heat sources into electrical energy. This unit comprised seven M8T modules in two frame blocks equipped with hot-water and cold-water circulation channels. The M8T module is the main part of the Lotemg–TPG, which consists of eight TEG elements of type TEG1-241-1.4-1.2, flanked by a hot-side radiator and a cold-side radiator. A more detailed load test on TEG type TEG1-241-1.4-1.2 obtained a maximum power of 7.30 W at a load of 8.2 Ω. The test results of the M8T module with a steam boiler heat source showed that each TEG element could produce 2.06 W of power at ΔT 61.94 °C. After being assembled in the Lotemg–TPG unit with a steam boiler heat source, the test results showed that the series-parallel combination of four TEG elements could produce 2.37 W of power at ΔT 69.21 °C. Installing the Lotemg–TPG unit in the Sari Ater Hot Spring was relatively easy, namely, by connecting the hot-water inlet to the tap at the heat source in the soaking room and the cold-water inlet to the water tap in the garden. The measurement results showed that at ΔT 17.38 °C, Module 1 with all TEG elements in series connection could produce 1.30 W of power, so the total power of the Lotemg–TPG unit was around 9.10 W. This result is quite good considering that the heat source was obtained for free and the device can operate to produce stable power.

Author Contributions: Conceptualization, A.S. and W.S.; methodology, N.K.L. and W.S.; software, S. and W.S.; validation, N.K.L., A.S. and W.S.; formal analysis, H.M., S. and W.S.; investigation, N.K.L., A.S. and W.S.; resources, H.M.; data curation, H.M. and S.; writing—original draft preparation, H.M. and S.; writing—review and editing, S., N.K.L., A.S. and W.S.; visualization, H.M. and S.; supervision, N.K.L., A.S. and W.S.; project administration, H.M. and W.S.; funding acquisition, H.M. All authors have read and agreed to the published version of the manuscript.

Funding: This research received no external funding.

Data Availability Statement: Data are contained within the article.

Acknowledgments: The authors would like to thank PT Sari Ater, which manages the Sari Ater Hot Springs Tourist Attraction, for allowing this research to take place.

Conflicts of Interest: The authors declare no conflicts of interest.

References

1. Rizaty, M.A. Kapasitas PLTP Indonesia Terbesar Kedua Di Dunia Pada 2022. Available online: https://dataindonesia.id/energi-sda/detail/kapasitas-pltp-indonesia-terbesar-kedua-di-dunia-pada-2022 (accessed on 13 March 2023).
2. Bina, S.M.; Jalilinasrabady, S.; Fujii, H.; Pambudi, N.A. Classification of Geothermal Resources in Indonesia by Applying Exergy Concept. *Renew. Sustain. Energy Rev.* **2018**, *93*, 499–506. [CrossRef]
3. Hochstein, M.P. *Classification and Assessment of Geothermal Resources*; Dickson, M.H., Fanelli, M., Eds.; UNITAR/UNDP: Rome, Italy, 1990; pp. 31–57.
4. *Ronald DiPippo Geothermal Power Plants: Principles, Applications, Case Studies and Environmental Impact*; Elsevier: Amsterdam, The Netherlands, 2015; ISBN 9780081008799.
5. Nasution, A.; Kartadinata, M.N.; Kobayashi, T.; Siregar, D.; Sutaningsih, E.; Hadisantono, R.; Kadarstia, E. Geology, Age Dating and Geochemistry of the Tangkuban Parahu Geothermal Area, West Java, Indonesia. *J. Geotherm. Res. Soc. Jpn.* **2004**, *26*, 285–303. [CrossRef]
6. Sentosa, R.A.; Haryanto, N.S.A.D.; Gentana, D. Land Surface Temperature Pada Manifestasi Permukaan Panasbumi: Studi Kasus Gunung Tangkuban Parahu, Jawa Barat. *Padjadjaran Geosci. J.* **2018**, *2*, 376–386.
7. Rana, S.; Orr, B.; Iqbal, A.; Ding, L.C.; Akbarzadeh, A.; Date, A. Modelling and Optimization of Low-Temperature Waste Heat Thermoelectric Generator System. *Energy Procedia* **2017**, *110*, 196–201. [CrossRef]
8. Remeli, M.F.; Tan, L.; Date, A.; Singh, B.; Akbarzadeh, A. Simultaneous Power Generation and Heat Recovery Using a Heat Pipe Assisted Thermoelectric Generator System. *Energy Convers. Manag.* **2015**, *91*, 110–119. [CrossRef]
9. Hettiarachchi, H.D.M.; Golubovic, M.; Worek, W.M.; Ikegami, Y. Optimum Design Criteria for an Organic Rankine Cycle Using Low-Temperature Geothermal Heat Sources. *Energy* **2007**, *32*, 1698–1706. [CrossRef]
10. Yu, K.; Zhou, Y.; Liu, Y.; Liu, F.; Hu, L.; Ao, W.; Zhang, C.; Li, Y.; Li, J.; Xie, H. Near-Room-Temperature Thermoelectric Materials and Their Application Prospects in Geothermal Power Generation. *Geomech. Geophys. Geo-Energy Geo-Resour.* **2020**, *6*, 12. [CrossRef]
11. Velmre, E. Thomas Johann Seebeck (1770–1831). *Est. J. Eng.* **2007**, *13*, 276. [CrossRef]
12. Rowe, D.M. Review, Thermoelectric Waste Heat Recovery as a Renewable Energy Source. *Int. J. Innov. Energy Syst. Power* **2006**, *1*, 13–23.
13. Snyder, G.J. Small Thermoelectric Generators. *Electrochem. Soc. Interface* **2008**, *17*, 54–56. [CrossRef]
14. Jaziri, N.; Boughamoura, A.; Müller, J.; Mezghani, B.; Tounsi, F.; Ismail, M. A Comprehensive Review of Thermoelectric Generators: Technologies and Common Applications. *Energy Rep.* **2020**, *6*, 264–287. [CrossRef]
15. Abdel-Motaleb, I.M.; Qadri, S.M. Thermoelectric Devices: Principles and Future Trends. *arXiv* **2017**, arXiv:1704.07742.
16. Zulkepli, N.; Yunas, J.; Mohamed, M.A.; Hamzah, A.A. Review of Thermoelectric Generators at Low Operating Temperatures: Working Principles and Materials. *Micromachines* **2021**, *12*, 734. [CrossRef] [PubMed]
17. Zoui, M.A.; Bentouba, S.; Stocholm, J.G.; Bourouis, M. A Review on Thermoelectric Generators: Progress and Applications. *Energies* **2020**, *13*, 3606. [CrossRef]
18. Ravindra, N.M.; Jariwala, B.; Bañobre, A.; Maske, A. *Thermoelectrics*; SpringerBriefs in Materials; Springer International Publishing: Cham, Switzerland, 2019; ISBN 978-3-319-96339-6.
19. Northwestern_University Thermoelectric Properties of Materials. Available online: https://thermoelectrics.matsci.northwestern.edu/thermoelectrics/index.html (accessed on 22 November 2021).
20. Twaha, S.; Zhu, J.; Yan, Y.; Li, B. A Comprehensive Review of Thermoelectric Technology: Materials, Applications, Modelling and Performance Improvement. *Renew. Sustain. Energy Rev.* **2016**, *65*, 698–726. [CrossRef]
21. Lee, H. *Thermoelectrics: Design and Materials*, 1st ed.; John Wiley & Sons, Inc.: Hokoben, NJ, USA; Western Michigan University: Kalamazoo, MI, USA, 2017; ISBN 9781118848951.
22. Rowe, D.M. *Thermoelectrics Handbook Macro to Nano*; CRC Press: Boca Raton, FL, USA, 2006; ISBN 9781315220390.
23. Ahiska, R.; Mamur, H. Design and Implementation of a New Portable Thermoelectric Generator for Low Geothermal Temperatures. *IET Renew. Power Gener.* **2013**, *7*, 700–706. [CrossRef]

24. Liu, C.; Chen, P.; Li, K. A 500 W Low-Temperature Thermoelectric Generator: Design and Experimental Study. *Int. J. Hydrogen Energy* **2014**, *39*, 15497–15505. [CrossRef]
25. Ziapour, B.M.; Saadat, M.; Palideh, V.; Afzal, S. Power Generation Enhancement in a Salinity-Gradient Solar Pond Power Plant Using Thermoelectric Generator. *Energy Convers. Manag.* **2017**, *136*, 283–293. [CrossRef]
26. Kumar, S.; Roy, D.; Ghosh, S. Thermodynamic Assessment of TEG-ORC Combined Cycle Powered by Solar Energy. *Int. J. Renew. Energy Technol.* **2017**, *8*, 346. [CrossRef]
27. Zare, V.; Palideh, V. Employing Thermoelectric Generator for Power Generation Enhancement in a Kalina Cycle Driven by Low-Grade Geothermal Energy. *Appl. Therm. Eng.* **2018**, *130*, 418–428. [CrossRef]
28. Gholamian, E.; Habibollahzade, A.; Zare, V. Development and Multi-Objective Optimization of Geothermal-Based Organic Rankine Cycle Integrated with Thermoelectric Generator and Proton Exchange Membrane Electrolyzer for Power and Hydrogen Production. *Energy Convers. Manag.* **2018**, *174*, 112–125. [CrossRef]
29. Khanmohammadi, S.; Saadat-Targhi, M.; Al-Rashed, A.A.A.A.; Afrand, M. Thermodynamic and Economic Analyses and Multi-Objective Optimization of Harvesting Waste Heat from a Biomass Gasifier Integrated System by Thermoelectric Generator. *Energy Convers. Manag.* **2019**, *195*, 1022–1034. [CrossRef]
30. Maneewan, S.; Chindaruksa, S. Thermoelectric Power Generation System Using Waste Heat from Biomass Drying. *J. Electron. Mater.* **2009**, *38*, 974–980. [CrossRef]
31. Hsu, C.-T.; Huang, G.-Y.; Chu, H.-S.; Yu, B.; Yao, D.-J. Experiments and Simulations on Low-Temperature Waste Heat Harvesting System by Thermoelectric Power Generators. *Appl. Energy* **2011**, *88*, 1291–1297. [CrossRef]
32. Niu, X.; Yu, J.; Wang, S. Experimental Study on Low-Temperature Waste Heat Thermoelectric Generator. *J. Power Sources* **2009**, *188*, 621–626. [CrossRef]
33. Suter, C.; Jovanovic, Z.R.; Steinfeld, A. A 1kWe Thermoelectric Stack for Geothermal Power Generation—Modeling and Geometrical Optimization. *Appl. Energy* **2012**, *99*, 379–385. [CrossRef]
34. Li, K.; Garrison, G.; Moore, M.; Zhu, Y.; Liu, C.; Horne, R.; Petty, S. An Expandable Thermoelectric Power Generator and the Experimental Studies on Power Output. *Int. J. Heat Mass Transf.* **2020**, *160*, 120205. [CrossRef]
35. Lee, K.-T.; Lee, D.-S.; Chen, W.-H.; Lin, Y.-L.; Luo, D.; Park, Y.-K.; Bandala, A. An Overview of Commercialization and Marketization of Thermoelectric Generators for Low-Temperature Waste Heat Recovery. *iScience* **2023**, *26*, 107874. [CrossRef]
36. Xie, H.; Gao, T.; Long, X.; Sun, L.; Wang, J.; Xia, E.; Li, S.; Li, B.; Li, C.; Gao, M.; et al. Design and Performance of a Modular 1 Kilowatt-Level Thermoelectric Generator for Geothermal Application at Medium-Low Temperature. *Energy Convers. Manag.* **2023**, *298*, 117782. [CrossRef]
37. Ahiska, R.; Mamur, H. Development and Application of a New Power Analysis System for Testing of Geothermal Thermoelectric Generators. *Int. J. Green Energy* **2016**, *13*, 672–681. [CrossRef]
38. Li, K.; Garrison, G.; Zhu, Y.; Moore, M.; Liu, C.; Hepper, J.; Bandt, L.; Horne, R.; Petty, S. Thermoelectric Power Generator: Field Test at Bottle Rock Geothermal Power Plant. *J. Power Sources* **2021**, *485*, 229266. [CrossRef]
39. Catalan, L.; Aranguren, P.; Araiz, M.; Perez, G.; Astrain, D. New Opportunities for Electricity Generation in Shallow Hot Dry Rock Fields: A Study of Thermoelectric Generators with Different Heat Exchangers. *Energy Convers. Manag.* **2019**, *200*, 112061. [CrossRef]
40. Marpaung, H.; Singarimbun, A.; Srigutomo, W.; Lasmi, N.K. Thermoelectric Application of TEG1-241-1.4-1.2 for Power Generation from Low Temperature Geothermal Fluid. *Adv. Sci. Technol.* **2023**, *126*, 67–77. [CrossRef]
41. Zhang, Z.; Wu, Y.; Li, W.; Xu, D. Performance of a Solar Thermoelectric Power-Harvesting Device Based on an All-Glass Solar Heat Transfer Pipe and Gravity-Assisted Heat Pipe with Recycling Air Cooling and Water Cooling Circuits. *Energies* **2020**, *13*, 947. [CrossRef]
42. Morais, F.; Carvalhaes-Dias, P.; Duarte, L.; Spengler, A.; De Paiva, K.; Martins, T.; Cabot, A.; Dias, J.S. Optimization of the TEGs Configuration (Series/Parallel) in Energy Harvesting Systems with Low-Voltage Thermoelectric Generators Connected to Ultra-Low Voltage DC-DC Converters. *Energies* **2020**, *13*, 2297. [CrossRef]
43. Liao, J.; Xie, H.; Wang, J.; Sun, L.; Long, X.; Li, C.; Gao, T.; Xia, E. Effect of Operating Conditions on the Output Performance of a Compact TEG for Low-Grade Geothermal Energy Utilization. *Appl. Therm. Eng.* **2024**, *236*, 121878. [CrossRef]

designs

MDPI

Review

A Systematic Literature Review on AC Microgrids

Marcos Gomez-Redondo [1,2,3], Marco Rivera [2,4,*], Javier Muñoz [5] and Patrick Wheeler [4]

1 Doctorado en Sistemas de Ingeniería, Faculty of Engineering, University of Talca, Campus Curicó, Curicó 3344158, Chile; marcos.gomez@utalca.cl or marcosgomez@fiuna.edu.py
2 Laboratory of Energy Conversion and Power Electronics, Direction of Research, Universidad de Talca, Curicó 3341717, Chile
3 Laboratory of Power and Control Systems, Faculty of Engineering, Universidad Nacional de Asunción, Luque 2060, Paraguay
4 Power Electronics, Machines and Control (PEMC) Group, 15 Triumph Rd, Lenton, Nottingham NG7 2GT, UK; pat.wheeler@nottingham.ac.uk
5 Department of Electrical Engineering, Faculty of Engineering, University of Talca, Campus Curicó, Curicó 3344158, Chile; jamunoz@utalca.cl
* Correspondence: marco.rivera@nottingham.ac.uk or marcoriv@utalca.cl; Tel.: +44-7871-099302

Abstract: The objective of this work is to analyze and compare AC microgrid (ACMG) solutions to introduce the topic to new researchers. The methodology used to achieve this goal is a systematic literature review using five questions: (1) How have ACMGs evolved in five years? (2) What are the standards for ACMGs? (3) What are the different schemes for connecting MGs to the utility grid? (4) What are the different control schemes in ACMGs? (5) What is an appropriate way to compare results when working with ACMGs? The articles were published in Q1/Q2 journals as based on either the Scimago Journal Rank (SJR) and/or the Journal Citation Report (JCR) between 2018 and 2022 and were from three databases: (1) Web of Science (WoS), (2) Scopus, and (3) IEEE Xplore. Publications not describing pure ACMGs, review papers, publications not related to the questions, and papers describing work that did not meet a quality assessment were excluded, resulting in 34 articles being included in this review. Results show: (1) the energy sources and AC bus nature of microgrids over five years, (2) the identification and quantification of cited standards for microgrids, (3) the pros and cons of different schemes for connecting an AC microgrid to the main grid, (4) the control schemes, classified in a hierarchical control structure, and (5) the simulation tools and experimental benches used in microgrids. Most studies considered a generic energy source and a low-voltage three-phase AC bus, 16 standards were found, and the most cited standard was IEEE Standard 1547. The most common connection scheme to the utility grid was a direct connection, most of the works proposed a modification to a hierarchical control system scheme, and the most common simulation tool was MATLAB. The preferred experimental setup consisted of parallel inverters for testing a control scheme, a prototype when proposing a power electronic system, and a laboratory microgrid for testing fault detection methods.

Keywords: AC microgrid; experimental microgrid; grid-tied microgrid; microgrid control systems; microgrid standards; simulation tools; systematic literature review

Citation: Gomez-Redondo, M.; Rivera, M.; Muñoz, J.; Wheeler, P. A Systematic Literature Review on AC Microgrids. *Designs* **2024**, *8*, 77. https://doi.org/10.3390/designs8040077

Academic Editor: Mohammad Hassan Khooban

Received: 17 April 2024
Revised: 28 July 2024
Accepted: 30 July 2024
Published: 5 August 2024

1. Introduction

With the advent of renewable energy, the paradigm in electrical systems has shifted from centralized to distributed generation. This has given birth to the concept of microgrids. Although the concept of a microgrid varies between different authors and has been changing over the years, a good definition of a microgrid is "a group of interconnected loads and distributed energy resources within clearly defined electrical boundaries that acts as a single controllable entity with respect to the grid. A microgrid can connect and disconnect from the grid to enable it to operate in both grid-connected or island mode". Microgrids can enable grid modernization, allow the integration of renewable energies, reduce peak loads and losses by locating generation near demand, ensure power availability for critical loads, and may support the main grid [1].

Microgrids are often classified according to the nature of the common bus to which the generators, loads, and storage elements are connected. If the bus works in alternating current (AC), the microgrid can be called an AC microgrid, if the bus is direct current (DC), the microgrid is known as DC microgrid, and if it has both AC and DC buses, it is known as a hybrid microgrid. Microgrids can be connected to the main electrical system (so-called grid-connected MGs) or they can be independent of the grid, so-called off-grid MGs. Nevertheless, the imcrogrid must be able to operate independently of the grid, which it is known as island mode [2]. This last requirement makes it necessary to include systems that allow continuous and controlled generation of power. Controlled generation sources are generally based on non-renewable and polluting energy, so they should be avoided if possible. However, the variability of renewable energy production usually requires operation with the addition of an appropriate storage system. Currently, bidirectional converter technologies are crucial for efficient energy management in microgrids, as they allow users to store renewable energy and release it when necessary [3,4].

Matrix converters are bidirectional AC/AC converters that have been studied for over 50 years, during which time researchers have acquired mature knowledge for applications in grid operations, sustainable transportation, and electrical drives [5]. Papers have proposed the inclusion of matrix converters in microgrids, but they are usually focused on the modulation or control scheme and do not consider distributed generation [6,7]. The motivation of the work described in this paper is to consider the state of development of microgrids to evaluate the inclusion of matrix converters, taking into account the current problems of microgrids and their required functionalities. Given the nature of matrix converters, which are bidirectional AC/AC converters, only AC microgrids (ACMGs) are considered. Matrix converters have applications in grid-tied microgrids as the interface between the microgrid (MG) and the main grid [5,8] for bidirectional power flow, so grid-connected microgrids receive more points than islanded microgrids in the evaluation process.

There are related reviews on ACMGs, but they tend to be very specialized, whereas more general approaches are presented in this paper. Some review articles discuss protection schemes for ACMGs [9–13]; other works center on control and management systems [14,15]. In addition, none of the related papers developed a systematic literature review (SLR). The main objective of this work is to summarize the current information on ACMGs in order to introduce the topic to new researchers in a simple way.

This paper is organized into five sections. This section introduced the paper. Section 2 shows the material and methods employed for developing the review. Section 3 shows the findings of this review. Section 4 discusses the results, and Section 5 presents conclusions.

2. Materials and Methods

This systematic literature review (SLR) follows the structure suggested by the PRISMA 2020 statement [16] as much as possible, considering this journal's requirements. The workflow starts by formulating the following five research questions (RQs):

1. Research question 1 (RQ1): How have AC microgrids (ACMGs) evolved over five years? This question aims to find the most common structures of microgrids (MGs) and the tendency for the next years. It is too wide, so just two characteristics were considered: the nature of every distribution generation unit (DGU) employed and the nature of the AC bus.
2. Research question 2 (RQ2): What are the standards for ACMGs? This question is intended to compile the current requirements of ACMGs.
3. Research question 3 (RQ3): What are the different schemes for connecting MGs to the utility grid? This focuses on the method for grid connection. It may be direct with possible connecting relays or through different power converters.

4. Research question 4 (RQ4): What are the different control schemes in ACMGs? The control system design depends on the source of energy, filters, and power converter employed.
5. Research question 5 (RQ5): What is an appropriate way to show results when working with ACMGs? This question will be helpful in investigating ACMGs, so it can introduce the equipment required to lead an investigation.

These questions are the basis of the work, and every decision made during the development of the work has been taken to answer these questions successfully. A general scheme of the paper selection procedure is shown in Figure 1. The different steps in the process are detailed in the following subsections.

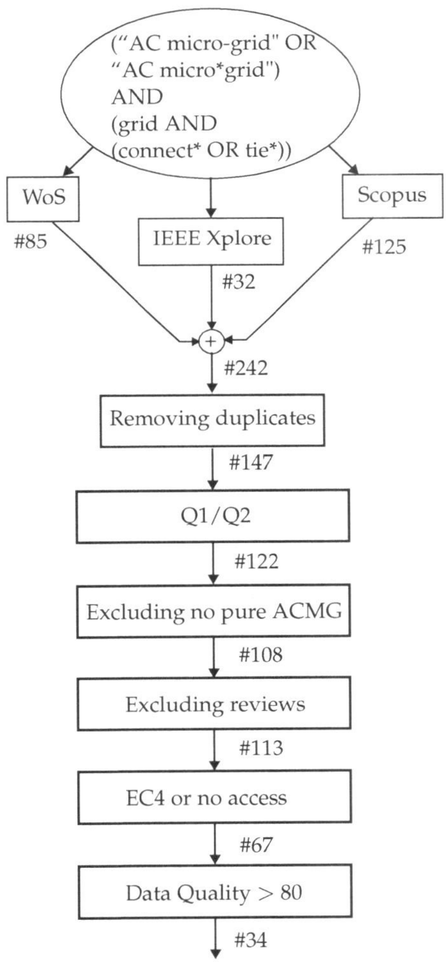

Figure 1. Selection process.

Tools employed to organize and evaluate the papers were:

- Mendeley reference manager: for reading, taking notes, and organizing papers along the process;
- Microsoft (MS) Excel: for arranging, evaluating, and extracting data.

2.1. Inclusion and Exclusion Criteria

2.1.1. Inclusion Criterion (IC)

- IC1: articles from journals written in English and that included ACMGs, especially if they are grid-tied, published between 2018 and 2022.

2.1.2. Exclusion Criteria (EC)

- EC1: not being in a Q1/Q2 journal—this is a guarantee of quality;
- EC2: abstract showing that it does not study a pure ACMG;
- EC3: the paper is a review—only primary sources were considered;
- EC4: lack of information to answer more than three research questions. The objective of this criterion is to reduce the number of articles included and to keep those with more information related to the study. It was included for three main reasons: (1) publications that include a real system or at least an experimental MG are considered more important to answer RQ1 than those that only have theoretical formulations (equations) or simulations of non-real systems; (2) RQ5 only makes sense when the study answers RQ1 and RQ3 or RQ4; (3) most of the works are high-quality papers, so they get a high score even when not answering any question.

2.2. Information Sources

The databases employed for the research were Web of Science (WoS) and Scopus due to their reputation and reliability. IEEE Xplore was also consulted to recover newer papers that might not yet be available in the aforementioned databases.

2.3. Search Strategy

The main query used for the research was: `(''AC micro-grid'' OR ''AC micro*grid'') AND (grid AND (connect* OR tie*))`. The search query was obtained in an iterative process, as we tried to obtain only works related to the research questions. This search command is understandable in all three databases, simplifying the reproducibility of the work. Other concepts such as "active distribution network", "parallel converters", and "distributed generation" can also be intimately related to MGs, but the authors preferred a short, simple query instead of trying to cover the whole universe. As initial search queries such as "microgrid" resulted in thousands of works, it was not considered necessary to capture more papers; instead, our purpose was to reduce the number of resulting papers and to keep those that could help answer the research questions. As the term "microgrid" involves not only the most popular DC MGs but also other topics such as heat systems, policies, economic aspects, stability studies, and communication systems, the employed term "AC microgrid" could filter the results more successfully. It was not an easy task to establish a search query because the research questions were too general and the results too numerous. Moreover, the following filters were considered:

- Only journal papers, which have passed through a more rigorous revision process than other sources;
- Only articles published between 2018 and 2022 to ensure that the information used in the study was up to date;
- Only papers in English because English is considered to be the more extended language for scientific research;
- In Scopus, the query was applied to the title, abstract, or keywords to reduce the high number of articles presented without this restriction (3768 articles).

2.4. Selection Process

2.4.1. Q1/Q2 filtering

The search process was performed by the first author, who also defined the eligibility criteria, so the work might be biased based on his criteria. However, the rest of this section tries to explain in great detail all the reasons for the different choices the author made during the review process. To guarantee trusted sources, only articles in the first or second quartile (Q1/Q2) were included, reducing the number to 122 studies from 45 different journals, which may be found in Appendix A. Two rankings were considered: SJR (Scientific Journal Ranking) in the categories *Energy* or *Engineering* and JCR (Journal Citation Report) in the category *Engineering, Electrical and Electronic*. These categories were chosen in order for us to focus on research on the electrical and electronic aspects of MGs. Nevertheless, in practice, the categories chosen in SJR were much more inclusive than the categories in JCR. As SJR has more journals, most of the papers that appear in JCR also appear in SJR with a better percentile. Only the journals: *Electronics*, *IEEE Transactions on Smart Grid*, and *IET Renewable Power Generation* were included by the JCR side and not by SJR. The first exclusion criterion (EC1) is: not belonging to a journal that has been Q1 or Q2 in at least one of the aforementioned rankings in the included five years.

2.4.2. Title and Abstract Screening

Titles and abstracts were screened, and some articles were excluded:

- EC2: abstract shows that the study does not include a pure ACMG but instead studies a DC MG, hybrid MG, or other type of MG [17–30].
- EC3: the article is a review—review articles are excluded as they are not primary sources [31,32].

Thereafter, exclusion criterion 4 (EC4) (explained in the Section 2.4.3) was applied.

2.4.3. Excluding Criterion 4 (EC4)

To reduce the number of works studied while keeping the more relevant ones for answering the research questions, it was decided to exclude works that do not satisfy at least two of the three following requirements:

- Includes a grid-tied MG;
- Has real or experimental results (not just theoretical or simulation results);
- Answers three or more research questions.

This rule reduced the number of works to 67. In this step, one article could not be verified because access was not possible, so it was also excluded [33]. Emulation platforms were considered experimental setups. The electrolyzer found in one article is only used for hydrogen generation, so it does not count as a distribution generation unit (DGU), and its scheme is not considered an MG [34].

2.4.4. Quality Evaluation

Finally, these 67 works were evaluated with the criteria shown in Table 1, resulting in 34 articles included in this review. Articles with a score greater than or equal to 80 points out of 100 were included. The evaluation was done in an MS Excel sheet.

Table 1. Quality evaluation criteria.

Paper		Weight
Title	<15 words	3
	Keywords in the title	3
Abstract	Presents a logic structure	3
Introduction	Context	3
	Implicit or explicit hypothesis	3
	Problem	3
	Explicit objective	3
Theoretical framework	State-of-the-art in a logical order	3
	Appropriate content	3
	Detailed methodology	3

Table 1. *Cont.*

Paper		Weight
Results	Available data	3
	Results match objectives	3
	Shows results with standardized metrics	3
	Information in figures complements the text	3
Discussion	Findings related to the objectives	3
	Results are compared to those from the state-of-the-art	3
Conclusions	Correspond to the objectives	3
	Show future work	3
References	References match	3
	Complete references	3
EC4	Includes a grid-tied MG	14
	Real or experimental application	13
	Answer three or more research questions	13
	Total	100

The first 60 points relate to the quality of the article, while the last 40 points relate to the research questions. The threshold of 80 points was selected to reduce the sample size to a manageable number.

2.5. Data Collection Process

The first author gathered data. Some data were tabulated in the same MS Excel Version 2496 sheet used for evaluation, and some information was directly inserted into this document.

2.6. Data Items

Included studies are summarized in Table 2, where excluded works and the exclusion reasons are also shown. Many areas are included, such as the design of a power converter for solving a specific problem, the simulation and emulation of control systems for ACMGs, optimization of MG parameters, techniques for noise reduction, and fault detection and protection systems.

Table 2. Included and excluded works.

	Articles
Included	[35–68]
Excluded—reviews	[31,32]
Excluded—no pure AC	[17–30]
Excluded—EC4	[33,69–106]
Excluded—<80	[34,107–138]

2.7. Study Risk of Bias Assessment

As for all search queries, evaluation and data extraction were performed by one author; the authors recognize the possibility of introducing bias in the process. However, as this is not a quantitative study, it is difficult to evaluate bias, and the best that can be done is to show the work and explain the method as transparently as possible.

2.8. Synthesis Methods

2.8.1. RQ1a

For RQ1a, the number of articles using each type of DGU was counted. Only DGUs used in the main work were counted. Energy storage systems were also counted, as they can act as DGUs.

2.8.2. RQ1b

The information was searched in the text and figures.

2.8.3. RQ2

The requirement was to explicitly include the name of the standard in the document. On the first read of the articles, the standard was remarked on if found. Then, in a second phase, the word *standard* was searched for with the text finder in Mendeley. In this way, some standards included in the references were also found. A problem this last search may have is the separation at the end of the line, as was found in some articles, so the first compilation was kept, and the second procedure was only to complement and could not replace the full reading of an article.

2.8.4. RQ3

The connection scheme of the ACMG to the main grid was first searched for in the figures of the developed work. The text was then read if any information was missing. When not all the necessary information was found, the work was excluded based on.

2.8.5. RQ4

The control scheme was first searched for in the figures, and then the text was read again for missing information. As was found during the work, there are different approaches to controlling a microgrid. The first time, all control systems were extracted, including their function and the level of hierarchy control (if this existed). In a second round, after gaining more knowledge on the topic, it was decided to classify the different control schemes to answer this question. The modulation stage is not considered a control system. In HMGs, only the control system involved in AC bus control was considered, but sometimes it was difficult to separate the control system, especially when the system had interleaved converters that worked between the AC and DC buses.

2.8.6. RQ5

The simulation and experimental sections were read to find this information. The results consisted of simulations or experimental results, so all the simulation tools and experimental platforms or benches were compiled.

3. Results

3.1. Research Question 1 (RQ1): Evolution of AC Microgrids (ACMGs) over Five Years

This research question is very wide, so in this study, two aspects are considered: the source of energy of the distributed generation units and the characteristics of the AC bus—in other words, the level of voltage, number of phases, and frequency of the AC bus.

3.1.1. RQ1a: Evolution of Distributed Generation (DG) in ACMGs over Five Years

To answer this question, the number of articles that contained each DG source was counted and tabulated. Data were drawn using MATLAB and are shown in Figure 2.

Most of the works used a hierarchical control scheme for microgrids. Defining the type of energy source was mainly necessary for the control of the tertiary level for optimal dispatch [62]. Modeling the different sources of energy is another case when the nature of the DG unit is important [35]. Furthermore, designing a specific system based on one DG unit is another case in which the energy source of the DG unit is mentioned [43,44,61].

In other cases, such as proposing solid state transformers or interlinking devices for AC and DC buses [64], modifying the primary level [53,56,59,60,65,68], dealing with the secondary level [58], seamless operation [39,46,48,55], grid feeding [38,41], grid support [45,50,54,57,63], detecting faults [36,47,49,67], dynamic response enhancement [51], measurements [52], and communication improvements [66], energy sources were not important.

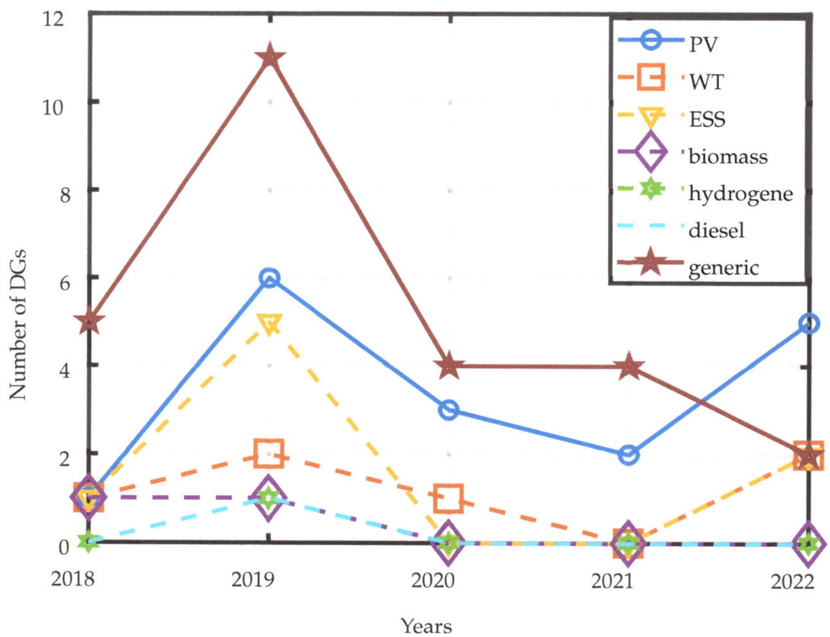

Figure 2. Evolution of the number of sources of DG included in the literature over five years (RQ1). We can see a decline in research on the topic after the peak in 2019. Numbers show that photovoltaic (PV) systems are the most popular DG source. The data also reflect that energy storage systems (ESSs) are one of the most popular topics, given the problem of the variability of renewable energy sources (RESs). Sometimes, this variability is compensated for with other sources of energy, such as diesel or biomass, and, of course, the main grid supply. Nevertheless, most of the works do not care about the energy source and use a generic DC voltage to feed the inverters, given that they work on another issue such as a control scheme, a power converter, or solving an optimization problem.

DG units (DGUs) are assumed to use generic DG when DG is mentioned but not used in the main work [57,58]. Superconducting magnetic energy storage is considered an ESS [44]. PV emulators used in the experimental setup were also considered as PV DGUs [37,40,61]. In [62], the DDGs mentioned were considered generic DGs, as they can came from diesel, hydro, thermal, or other sources. The difference the author claims is that DDG power is programmable, whereas renewable sources such as PV or WT are not always available. In [64], the source was considered to be a PV system, because even though it uses an ideal DC link, the power-decoupled current source inverter (PD-CSI) was designed taking into account PV systems.

3.1.2. RQ1b: Evolution of AC buses in ACMGs over Five years

The results for the different AC buses found are summarized in Table 3. The nature of the AC bus is classified as: low-voltage single-phase and high-frequency (LV 1-ϕ HF), low-voltage single-phase (LV 1-ϕ), low-voltage three-phase (LV 3-ϕ), and medium-voltage three-phase (MV 3-ϕ). The works are arranged by publication year. For the voltage level, we considered:

- Low-voltage (LV): up to 1000 V;
- Medium-voltage (MV): between 1000 V and 45 kV.

Table 3. AC bus scheme in the included works (RQ1).

	2018	2019	2020	2021	2022	Sum
LV 1-ϕ HF					[37]	1
LV 1-ϕ	[52]	[35,57,63]	[59]	[64]		6
LV 3-ϕ	[44–46,48]	[40–43,50,54–56,61]	[36,66]	[38,47,53,65,67,68]	[39,62]	23
MV 3-ϕ		[60]	[49,51]			3
Sum	5	13	5	7	3	33

Some remarks regarding the data extraction include: the voltage operation level is not mentioned, but it was assumed to be low-voltage due to its experimental setup [45]. The voltage level is not shown, but this is assumed to be MV because of the power levels [51]. An article included an unusual AC bus frequency of 400 Hz, so it was separated from the others [37]. The article shows a low voltage, but it is not clear whether it is single-phase or three-phase, so it was not included in answering this part of the question [58]. The voltage level is not mentioned, but it is assumed to be low-voltage because the power levels are within the range of generators with low voltages [62]. The MG does not have an AC bus but is connected to a single-phase grid [63].

3.2. Research Question 2 (RQ2): Standards for ACMGs

Standards found in the included works are tabulated in Table 4. Many works do not mention any standard [35–37,41,42,45,46,48,49,53,55,56,62,63,65–67]. Some of them refer to *IEEE Standards* [48,53,54] or *IEC Standards* [54] but do not mention anything specific. It is important to note that VDE 0126-1-1 is a German standard for which the current status is "withdrawn".

Table 4. Standards for microgrids (RQ2).

Standard	Name	Mentioned by
IEEE Standard 1547-2018 [139]	Standard for interconnection and interoperability of distributed energy resources with associated electric power system interfaces	[17,44,51,58–60]
IEEE Standard 519-2014 [140]	IEEE recommended practices and requirements for harmonic control in electric power systems	[39,52,57,61]
EN-50160 [141]	Voltage characteristics of electricity supplied by public electricity networks	[38,50]
IEEE Std C37.118.1 [142]	IEEE standard for synchrophasor measurements for power systems	[38]
IEEE 1159-1995 [143]	IEEE recommended practices for monitoring electric power quality	[50]
IEEE 2030.7-2017 [144]	IEEE standard for the specifications of microgrid controllers	[40]
IEEE 1459-2010 [145]	IEEE standard definitions for the measurement of electric power quantities under sinusoidal, nonsinusoidal, balanced, and unbalanced conditions	[43]
EN 61000-2-2[146]	Electromagnetic compatibility (EMC)–part 2-2: environment—compatibility levels for low-frequency conducted disturbances and signaling in public low-voltage power supply systems	[50]
EN IEC 61000-3-2 [147]	Electromagnetic compatibility (EMC)–part 3-2: limits—limits for harmonic current emissions (equipment input current <= 16 A per phase)	[50]
IEC 61727 [148]	Photovoltaic (PV) systems—characteristics of the utility interface	[51]

Table 4. *Cont.*

Standard	Name	Mentioned by
IEC 60831-1 [149]	Shunt power capacitors of the self-healing type for AC systems having a rated voltage up to and including 1 kV–part 1: general—performance, testing, and rating safety requirements guide for installation and operation	[52]
IEC 60831-2 [150]	Shunt power capacitors of the self-healing type for AC systems with a rated voltage up to and including 1kV–part 2: aging test, self-healing test, and destruction test	[52]
IEC 60255 [151]	Measuring relays and protection equipment–part 1: common requirements	[47]
IEC 61850 [152]	Communication protocols for intelligent electronic devices at electrical substations	[47]
IEC 62040-2 [153]	Uninterruptible power systems (UPSs)–part 2: electromagnetic compatibility (EMC) requirements	[68]
VDE 0126-1-1 [154]	Automatic disconnection device between a generator and the public low-voltage grid	[64]

3.3. Research Question 3 (RQ3): Schemes for Connecting MGs to the Utility Grid

The direct connection scheme can be through a filter, a circuit breaker, or a combination of both. The schemes we found for connecting ACMGs to the main grid with their respective pros and cons are tabulated in Table 5. The information extracted is only from the main work the papers include and does not include information about grid connections that is present in the introduction.

There were some considerations when collecting information. The article involving the high-frequency AC bus is considered an islanded solution [37]. Another article included two schemes, but the laboratory grid scheme uses an autotransformer [41]. Some MGs are considered islanded because the work does not show a clear connection of the AC bus to the main grid; also, the work focuses on distributed generation and does not consider the grid connection [56]. A photovoltaic unified power quality conditioner (PV-UPQC) connects to the grid at two points through an isolation transformer—on the primary and the secondary—which is why it is put into another category [43]. One article does not have only one AC bus: instead, three AC buses with their AC sources are interconnected through inverters connected to the same DC bus [50]. A paper used TRIAC to connect the AC bus to the grid, but this is considered to be a type of direct connection [55].

Table 5. Pros and cons of different schemes for grid connection (RQ3).

Connection Scheme	Pros	Cons	Works
Direct	Easy connection and low cost	Little control	[35,38,41,44–46,48,51–53,55,57–60,66,67]
Isolation transformer	Galvanic isolation	High cost	[40,42,47,49,50,62]
Autotransformer	Efficiency and cost	No isolation	[39,41]
Grid-tied inverter	Simple, quality of the AC bus voltage	High cost	[54]
PV-UPQC system	Power quality of the grid	Depends on PV system	[43]
Energy storage system	Palliates the intermittence of RESs	Highest cost	[63]
H-bridge, active power decoupler	Power quality	Complex control and high cost	[64]
Islanded			[36,37,56,61,65,68]

3.4. Research Question 4 (RQ4): Control Schemes in ACMGs

Before answering this question, note that there are many control system schemes for different purposes among the different subsystems inside a microgrid. Commonly controlled variables can be power (active or reactive), voltage (magnitude and frequency), and current. Some works pretend to improve the quality of the electrical variables with their control system by controlling reactive power, reducing harmonics, and seeking seamless transitions between modes of operation. Other problems must also be resolved with control systems, such as the capability of working in an islanded and grid-tied mode and delays in the communication systems of centralized or distributed control.

Taking into account the results, hierarchical control is a promising approach to controlling a microgrid; it defines three levels for controlling different variables. Primary-level control is responsible for controlling active and reactive power, and this level is usually decentralized. Secondary-level control is responsible for correcting the steady-state deviation of the primary level and can be centralized, decentralized, or distributed. Finally, tertiary-level control is in charge of energy dispatching, i.e., establishing the power references. Again, control schemes can be found for a seamless transition from islanded to grid-tied mode or vice versa. Decentralized droop-based control schemes are the most popular at the primary level for power sharing. Secondary control may be centralized, decentralized, or distributed. Some authors prefer to avoid communication and propose decentralized secondary control, while others prefer to include a communication system to have more information in a centralized or distributed secondary control system. Tertiary control is centralized. Taking into account the results, the authors recommend using hierarchical control for a matter of organization. The authors consider that the most organized way to divide this multipurpose control system is hierarchical control, so the control schemes we found were classified into the hierarchical scheme structure when possible, as shown in Table 6. Some works did not explicitly show a control system [47,49,67], and some used only pulse width modulation [52].

After this classification, it can be seen that most of the works accept the hierarchical control scheme, as this scheme facilitates the design of a specific part of the control system. This classification shows the (highest) level that each control system occupies (or might occupy) in a hierarchical control scheme. For example, a control system occupying the secondary level may replace the secondary control or both the secondary and primary levels. The first 0-level is for control systems that could be employed in the inner loops of the primary level for controlling the voltages and currents.

Table 6. Control schemes in ACMGs (RQ4).

0-Level	PI-resonant for current control [36], PI-resonant for voltage control [36], PI-resonant for voltage control [36], model predictive current control [41], active damping and LCL output current control [41], PR for voltage control [37], negative-sequence voltage elimination [38], negative-sequence current sharing [38], deadbeat current control [50], PI current control [51,55], PI voltage control [55], and hysteresis current controller [61]
Primary level	Classic droop [48,58], PI for reactive power control [41], bilinear PI controller based on passivity-based formulations [44], bilevel functional-rotation-based active damping control [45], improved droop controller [53], repetitive and state feedback control combined with droop control [56], optimal direct control method (FCS-MPC current and voltage control for active power filter) [57], classic droop modification [59], linear quadratic Gaussian control [60], efficiency-prioritized droop control strategy [65], model-predictive-control-based virtual synchronous generator (VSG-MPC) [103], 3-phase improved-magnitude phase-locked-loop control [61]
Secondary level	Microgrid central controller [40], distributed droop-based [46] distributed leader–follower control, fuzzy multitask secondary controller [58], droop [62], decentralized passive dynamic PI controllers [42]
Tertiary level	MPC for optimal dispatch [35], supervisory control [46], master–slave configuration [39], PI power control [51,58]
Hard to classify in hierarchical control schemes	Current and voltage control of PV-unified power quality conditioner [43], quasi-proportional-resonant-integral (PRI) current controller (grid-tied inverter with MPPT for PV) [63], quasi-proportional-resonant (PR) current controller (grid-tied inverter with MPPT for PV) [64], composite controller (internal model controller + quasi-PR controller with multiple resonance compensation) for grid-tied inverter [66]

3.5. Research Question 5 (RQ5): Tools and Experimental Setups for ACMGs

The answer to this question is classified into two categories: simulation and experimental results. The proposal of each article can be also arranged into five groups, and the results shown depend on this proposal. Proposals found among the works are (1) control schemes, (2) systems involving power converters, (3) fault detection methods, (4) optimization methods, and (5) power quality monitoring indexes. Depending on these classifications, the results and experiments, as well as the tools employed, were different. The information on the simulation tools used is summarized in Table 7. Most of the works involving MG control systems employ simulation tools to test their proposed schemes. Some of the works go further, with real-time simulation of MGs. Works proposing power converters for a specific application in MGs usually test their platform experimentally [36]. Some works do not specify the simulation tool [42,46,53]. In some of the articles, the plots seem to be made in MATLAB, but it is not clear, so they are not included in the table. MATLAB/Simulink was the most popular tool for simulations. PLECS can be used in a MATLAB environment [65], which suggests that authors perform simulations in MATLAB before PLECS implementation.

The experimental results were also classified with the same organization, but in this case, instead of the simulation tools, the type of experimental bench is considered, as shown in Table 8. The most common experimental method to test the control in MGs is through parallel converters. Only five works employed full experimental MGs with emulated or real generation units [39,49,61,67]. Another interesting approach is emulating an MG in real-time [40,41,53].

Table 7. Results shown and simulation tools.

Proposal	Results	Simulation Tool	Works
Control scheme	Works commonly show figures of merit and figures of physical variables to show the tracking capability and response time	MATLAB PLECS PSCAD PSIM	[39,41,44,48,50,54,59,66,68] [65] [56] [55]
Power electronic system	Experimental prototype of their proposal	MATLAB PLECS	[21,43,62,124] [37]
Fault detection method	Works show tables of delay times for fault detection	MATLAB PSCAD	[47,49] [67]
Optimization method	Figures of physical variables	MATLAB	[51]
Power quality monitoring index	Comparison to other indexes	MATLAB	[52]

Table 8. Experimental results.

Proposal	Experimental Setup	Works
Control scheme	2-parallel VSI 3-parallel VSI 1-VSI OPAL-RT simulated MG Control in FPGA + dSPACE RT simulated scheme Typhoon HIL emulated MG	[39,54,58,60,61,65,68] [38,45,66] [40,50,56] [40] [41] [53]
Power electronic system	Solid state transformer Other converter	[43] [36,37,63,64]
Fault detection method	Laboratory MG	[49,67]

4. Discussion

4.1. Limitation of Evidence

In general, the search query could have been more extensive given other synonyms found in the current literature: nanogrid, picogrid, minigrid, smart grid, distributed generation, and parallel inverter; however, the term microgrid (MG) is the most common. Similarly, not including hybrid microgrids in the search query omits related information, as AC microgrids are present in hybrid microgrids. Nonetheless, limiting the search query to ACMGs is a price that must be paid to obtain a more selective sample, as the ultimate goal was not to consult all the literature but to answer the research questions. Distributed generation and parallel inverters are general terms that may or may not describe microgrids. Distributed generation is the basis for the development of microgrids and typically is related to renewable energy sources (RESs) in an MG; likewise, parallel inverters typically function in MGs. The main difference between the other concepts is the levels of power they manage, and only MGs are normally connected to the grid [155], so the employed term is considered correct. However, choosing only grid-connected MGs was an important cut of information, but it was one of the purposes of the work to review the different connection schemes.

More databases could have been included, but the choice of these databases does not represent a significant risk of bias in the posterior quality assessment because it is considered that the more qualified works and journals are present in these databases.

Taking into account the SLR technique, the research questions were too general, and they were modified during the process. They could have been defined more precisely if the knowledge of AC microgrids (ACMGs) was deeper. It is necessary to have closely related research questions, maybe fewer, to define a unique search query; otherwise, it could be better to define multiple search queries, but again, multiple queries can result in multiple articles with fewer questions. Moreover, the systematic literature review (SLR) technique was developed for resolving conflicts (different numerical results) between different sources, so PRISMA 2020 recommendations are best suited for solving conflicts or contradictory statements between different sources.

4.2. Interpretation

4.2.1. Research Question 1 (RQ1)

It can be seen that the distributed generation units (DGUs) are not of importance for many authors, given that they use generic DGUs in their studies. In any case, the most studied DGUs are photovoltaic (PV) systems along with a proper energy storage system, which should be of no surprise for DC or hybrid microgrids (HMGs), but they also occur in ACMGs.

PV systems as renewable energy sources have gained more interest than others given their high scalability and easy installation. PV efficiency has also increased over the years. Another reason PV systems are interesting is that solar irradiance is less unpredictable than other energy sources. These factors have produced a fast decline in costs, which makes PV systems less expensive than all of the other renewable energy sources. Evidence suggests that PV systems will continue to be the fastest growing renewable energy source for the coming years [156].

Concerning the AC bus, the most popular is the three-phase low-voltage. Works including medium-voltage buses were rarely included because they are difficult to implement experimentally. The fact that low-power distributed generation units usually work at low-voltage promotes the preference for low-voltage ACMGs. Another factor that drives the preference for low-voltage microgrids is local consumption, which does not require the voltage elevation that is necessary to reduce transmission losses. Three-phase systems are preferred over single-phase systems because the power is constant: they can transmit three times more power with just one more conductor, and they can supply three-phase loads as well as single-phase loads. One article presented an unusually high-frequency AC bus [37], and more works of this type are expected because this can help in high-frequency applications and allows reducing the number of components required for the filters; however, their usage is expected for specific islanded microgrids, like space applications, and not for the main utility grid. In conclusion, current facts and evidence imply that low-voltage three-phase AC buses will be preferred in the future.

In particular, the studied aspects were not the most interesting, but the work unveiled others that could be considered for the evolution of ACMGs. It is considered that the most relevant topics given the change of paradigm from centralized to distributed generation in ACMGs will be: (1) Control systems need to be modified, given that distributed generation has changed the usual control schemes and different standards and protocols have different requirements. (2) Energy storage systems need to be developed due to the variability of the energy sources. This will consist of enhancing the energy and power capacity of the energy storage systems. It includes appropriate energy conversion systems for storing and recovering electrical energy (fuel cells, flywheels, etc.) as well as the power electronics for adequate electrical energy delivery [3,4]. (3) Interconnection systems can help to connect DC and AC buses in microgrids and also interconnect different microgrids in the electrical system. (4) Communication systems can reduce package loss and allow faster and more reliable data transmission. (5) Measurement systems can provide appropriate power management and control. (6) Protection systems need to be updated, as current devices are not designed for distributed generation. (7) Optimization in terms of costs, losses, efficiency, etc., needs to be performed, and self-adaptive tuning methods for control parameters need to be formulated given the current developments in artificial intelligence and self-learning and self-adapting systems.

It is also shown that the number of included publications decreased since 2019; however, the first set of 147 articles did not show this decrease. Other search queries like "microgrid" show an increasing number of publications over time. This means that the selection process is likely to be responsible for the observed decrease. This could imply that the ACMG tied to the grid, which is the studied system, is less relevant for researchers than other topics like energy storage systems or DC and hybrid microgrids. In the future, one may expect a reduction in ACMGs and a dominance of DC microgrids since energy storage systems (ESSs), PV systems, and most household devices work with DC voltage. Today, HMGs lead this transition.

4.2.2. Research Question 2 (RQ2)

Standards from the IEEE Standard Association, which is an American institution, and the International Electrotechnical Commission (IEC) from Europe are the most popular. The standard to be adopted depends on the specific topic of study in the MG and regional and national regulations. As MGs are diverse, many standards on different topics were found. The reader should define the problem and select the standard according to their region and purpose.

The most frequently mentioned standard was IEEE Standard 1547, which is a series of standards for the connection of distributed energy resources with associated electric power system interfaces, with IEEE Standard 1547-2018 being the basis of the series [139]. Other standards of the series complement the first in other aspects like details, schematics, cybersecurity, island mode operation, secondary network systems, distribution impact studies for distributed generation interconnection, and storage systems. This standard has been fully adopted in 8 states and 25 utilities in the USA. It is also in the process of being adopted by other USA states and utilities [157]. The IEEE 519-2014 standard is the second-most cited standard and defines the voltage and current harmonic distortion criteria for the design of electrical systems [140].

The IEC is an international standards organization that prepares and publishes international standards for all electrical and electronic technologies; the IEC consists of 170 countries: 60 full members, 23 associate members, and 87 affiliates. IEC standards cover a vast range of technologies from power generation, transmission, distribution to home appliances and office equipment, semiconductors, fiber optics, batteries, solar energy, nanotechnology, and marine energy, as well as many others. EN 50160 is a standard from the IEC that defines the voltage characteristics for the electricity supplied by public electricity networks. IEC 61000-4-30 is another standard of interest that defines measurement procedures for applying the previous standard correctly.

Standards reference other standards, and nowadays, efforts are put into making them all compatible. For example, IEEE 1547 refers to IEEE Standard 519 and IEC 61000.4.15 [158] regarding flicker and fluctuation. Another example is IEEE Standard 519, which was purely an IEEE standard with no references to the IEC standards in its version from 1992, but its updated version from 2014 includes all of the measurement methods referenced to IEC standards.

Information on potential gaps and areas for improvement of standards has not been found in the included literature. However, standards will continue evolving, with every revision searching for global compatibility (as far as possible) for electric specifications; IEEE 1547 is a good option to adopt given its current state of development, its current permeation in the utilities, and its periodic revisions and improvements every 10 years by the IEEE Standard Association.

4.2.3. Research Question 3 (RQ3)

The most usual option is direct connection: letting the AC bus be directly connected to the main grid, perhaps with a current filter and a circuit breaker for protection. Transformers or auto-transformers are also based on passive elements and are also simple and more secure, but they are more expensive. One implication of using these passive schemes is that the AC bus voltage inside the microgrid is regulated by the main grid. Another implication is that microgrids change their mode of operation from island mode to grid-connected mode when the microgrid is connected to the main grid. This is how real-world microgrids are normally connected to the main grid [159,160].

Other, more complex systems like solid-state transformers are attractive as they allow a controlled interconnection of the MG and provide ancillary services. Practical implications include a more complex and expensive interface. They are preferred in DC or hybrid microgrids since they provide a DC stage, which serves as a DC bus [17]. However, a real-world application of this configuration has not been found. In the future, we expect more research on bidirectional grid/MG power interfaces, such as solid-state transformers, and devices that can provide ancillary services to this interface, such as unified power quality conditioners, active power filters, dynamic voltage restorers, and static synchronous series compensators.

One thing to bear in mind is the concept of MGs that different authors may have. Some works considered a grid-tied AC bus that only considered the grid-tied mode of operation, which lacks the autonomy of MGs. In this context, the stability and inertia of the MG are not taken into account, and some other problems, such as power quality or optimal dispatch, are approached. Some authors may consider a grid-tied electrical system with one generator alone as an MG, and some others define an MG as an autonomous electrical system with distributed generation that may or not be connected to the grid. We prefer the definition of IEEE Standard 1547 [139].

4.2.4. Research Question 4 (RQ4)

Control systems in microgrids are too wide, but in general, hierarchical control is taking over and promises to be the first option in the coming years as it provides an organized framework to establish a multipurpose control scheme. Central control is more effective but is not scalable like decentralized control [42,60]. Centralized control also requires good and fast communication, which represents more cost. However, central control can be applied to critical systems that do not change over time. Distributed control occupies a midpoint between centralized and decentralized control: being scalable but requiring a communication system among all (or some) of the converters [46,161].

The most popular recent trend in ACMG control systems is to look for a more robust control system and to improve the stability by appropriate voltage and frequency control systems [162–167]. Taking this trend into account, potential future directions would be to test and validate non-linear control as well as to include artificial intelligence in the control system. Another innovative trend is handling cyberattacks in the control system [168–170]. Other topics included in ACMG include the optimization of the energy management system (EMS) [171] and the seamless transition between grid-connected and island modes of operation [39,55]. Considering the need for energy storage systems, appropriate control systems for further developing energy storage technologies are also expected.

4.2.5. Research Question 5 (RQ5)

Many papers describing different proposals validated their work through simulations. MATLAB/Simulink has proven to be the first choice for simulation and is complemented by equipment like dSPACE for experimental applications. Other options are not very common. Moreover, MATLAB allows for an easy way to plot results and is widely used to show results.

The experimental microgrids found are mainly laboratory prototypes, and the results show only one-time operation, which does not tell about continuous operation. Most of the included works propose a control scheme for the MG, and the best choice for testing a control scheme is to have parallel inverters connected to the same bus. However, these experiments should include, if possible, the generation unit to test the controller response to the variability of renewable energy sources. When proposing a converter or system, it is almost a rule that an experimental prototype supports the proposal.

5. Conclusions

This work presents a systematic literature review on AC microgrids (ACMGs) based on five research questions, all of which have been addressed and discussed. The article serves as an introductory overview of ACMGs; it focuses on five key aspects and acts as a hub for accessing related research papers. These points include: (1) the evolution of ACMGs, (2) standards applicable to ACMGs, (3) grid connection schemes, (4) control systems within ACMGs, and (5) findings from articles on ACMGs.

According to the literature reviewed, the typical ACMG configuration involves distributed generation units connected via power converters to an AC bus that supplies AC loads. When connected to the grid, this AC bus is typically linked to the main grid through breakers and passive components. ACMGs commonly employ hierarchical control systems, facilitating operation in both grid-connected and island modes.

It is noteworthy that the development of microgrids is closely linked to the advancement of standards, with IEEE Standard 1547-2018 being particularly prominent.

Considering the trend towards autonomy achieved through batteries and the rising popularity of DC loads alongside the dominance of photovoltaic systems, the adoption of hybrid microgrids appears more pragmatic than relying solely on ACMGs.

Author Contributions: Conceptualization, M.G.-R. and M.R.; methodology, M.G.-R.; software, M.G.-R.; validation, M.G.-R. and M.R.; formal analysis, M.G.-R.; investigation, M.G.-R.; resources, M.G.-R., M.R. and J.M.; data curation, M.G.-R.; writing—original draft preparation, M.G.-R. and M.R.; writing—review and editing, M.G.-R., M.R., J.M. and P.W.; visualization, M.G.-R.; supervision, M.R., J.M. and P.W.; project administration, M.R., J.M. and P.W.; funding acquisition, M.R., J.M. and P.W. All authors have read and agreed to the published version of the manuscript.

Funding: This research was funded by the Agencia Nacional de Investigación y Desarrollo (ANID) ANID-Subdirección de Capital Humano/Doctorado Nacional/2022-21220504, FONDECYT regular grant number 1220556, ANID FOVI230169, Fondap SERC 1523A0006, and IRCF project 24932270 from the University of Nottingham.

Data Availability Statement: Data sharing is not applicable.

Conflicts of Interest: All authors declare they do not have competing interests.

Abbreviations

The following abbreviations are used in this manuscript:

APF	Active power filter
AC	Alternating current
ACMG	AC microgrid
DDG	Dispatchable distributed generation
DG	Distributed generation
DGU	Distributed generation unit
EC	Exclusion criterion
ESS	Energy storage system
FCS-MPC	Model predictive control with finite control set
FPGA	Field programmable gate array
HF	High frequency
HIL	Hardware in the loop
HMG	Hybrid microgrid
IC	Inclusion criterion
JCR	Journal Citation Report
LV	Low voltage
MG	Microgrid
MPPT	Maximum power point tracking
MS	Microsoft
MV	Medium voltage
PD-CSI	Power decoupled current source inverter
PI	Proportional-integral
PR	Proportional-resonant
PRI	Proportional-resonant-integral
PV	Photovoltaic
PV-UPQC	Photovoltaic unified power quality conditioner
PWM	Pulse width modulation
SG	Synchronous generator
SJR	Scimago Journal Rank
RES	Renewable energy source
RQ	Research question
VSG-MPC	Model-predictive-control-based virtual synchronous generator
VSI	Voltage source inverter
WoS	Web of Science
WT	Wind turbine

Appendix A

A summary of the journals included in the study is presented in Table A1.

Table A1. Journals (Q1/Q2) in SJR or JCR and the number of works obtained from each one.

Journal	Article
Applied Sciences (Switzerland)	3
Computers and Electrical Engineering	2
Electric Power Systems Research	4
Electricity Journal	1
Electronics	5
Electronics Letters	1
Energies	9
Energy Conversion and Management	1
Energy Reports	3
Heliyon	1
IEEE Access	7

Table A1. *Cont.*

Journal	Article
IEEE Canadian Journal of Electrical and Computer Engineering	1
IEEE Journal of Emerging and Selected Topics in Power Electronics	3
IEEE Systems Journal	4
IEEE Transactions on Circuits and Systems I: Regular Papers	1
IEEE Transactions on Control Systems Technology	1
IEEE Transactions on Industrial Electronics	7
IEEE Transactions on Industrial Informatics	2
IEEE Transactions on Industry Applications	6
IEEE Transactions on Power Delivery	1
IEEE Transactions on Power Electronics	5
IEEE Transactions on Power Systems	2
IEEE Transactions on Smart Grid	11
IET Generation, Transmission and Distribution	4
IET Power Electronics	2
IET Renewable Power Generation	3
International Journal of Electrical Power and Energy Systems	9
International Journal of Energy Research	1
International Journal of Hydrogen Energy	1
International Review of Electrical Engineering	1
International Transactions on Electrical Energy Systems	6
Inventions	1
Iranian Journal of Science and Technology-Transactions of Electrical Engineering	1
Journal of Energy Storage	1
Journal of Engineering	1
Journal of Intelligent and Fuzzy Systems	1
Journal of Modern Power Systems and Clean Energy	1
Protection and Control of Modern Power Systems	1
Renewable and Sustainable Energy Reviews	2
Renewable Energy Focus	1
Sustainable Cities and Society	1
Sustainable Energy Technologies and Assessments	1
Sustainable Energy, Grids, and Networks	1
Systems and Control Letters	1
TOTAL	122

Appendix A.1. Registration and Protocol

This review is not registered and no protocol was prepared for this work.

Appendix A.2. Additional Data from This Work

The first author can supply more data related to this work.

References

1. Ton, D.T.; Smith, M.A. The U.S. Department of Energy's Microgrid Initiative. *Electr. J.* **2012**, *25*, 84–94. [CrossRef]
2. Justo, J.J.; Mwasilu, F.; Lee, J.; Jung, J.W. AC-microgrids versus DC-microgrids with distributed energy resources: A review. *Renew. Sustain. Energy Rev.* **2013**, *24*, 387–405. [CrossRef]
3. Min, J.; Ordonez, M. Bidirectional Resonant CLLC Charger for Wide Battery Voltage Range: Asymmetric Parameters Methodology. *IEEE Trans. Power Electron.* **2021**, *36*, 6662–6673. [CrossRef]
4. Min, J.; Ordonez, M. Unified Bidirectional Resonant Frequency Tracking for CLLC Converters. *IEEE Trans. Power Electron.* **2022**, *37*, 5637–5649. [CrossRef]
5. Bento, A.; Paraíso, G.; Costa, P.; Zhang, L.; Geury, T.; Pinto, S.F.; Silva, J. On the potential contributions of matrix converters for the future grid operation, sustainable transportation and electrical drives innovation. *Appl. Sci.* **2021**, *11*, 4597. [CrossRef]

6. Malekjamshidi, Z.; Jafari, M.; Zhu, J.; Rivera, M. Design, Implementation, and Stability Analysis of a Space Vector Modulated Direct Matrix Converter for Power Flow Control in a More Reliable and Sustainable Microgrid. *Sustainability* **2020**, *12*, 8591. [CrossRef]
7. Gontijo, G.; Soares, M.; Tricarico, T.; Dias, R.; Aredes, M.; Guerrero, J. Direct Matrix Converter Topologies with Model Predictive Current Control Applied as Power Interfaces in AC, DC, and Hybrid Microgrids in Islanded and Grid-Connected Modes. *Energies* **2019**, *12*, 3302. [CrossRef]
8. Malekjamshidi, Z.; Jafari, M.; Zhu, J.; Xiao, D. Bidirectional power flow control with stability analysis of the matrix converter for microgrid applications. *Int. J. Electr. Power Energy Syst.* **2019**, *110*, 725–736. [CrossRef]
9. Memon, A.A.; Kauhaniemi, K. A critical review of AC Microgrid protection issues and available solutions. *Electr. Power Syst. Res.* **2015**, *129*, 23–31. [CrossRef]
10. Patnaik, B.; Mishra, M.; Bansal, R.C.; Jena, R.K. AC microgrid protection—A review: Current and future prospective. *Appl. Energy* **2020**, *271*, 115210. [CrossRef]
11. Sarangi, S.; Sahu, B.K.; Rout, P.K. Review of distributed generator integrated AC microgrid protection: Issues, strategies, and future trends. *Int. J. Energy Res.* **2021**, *45*, 14117–14144. [CrossRef]
12. Bui, D.M.; Chen, S.L.; Lien, K.Y.; Chang, Y.R.; Lee, Y.D.; Jiang, J.L. Investigation on transient behaviours of a uni-grounded low-voltage AC microgrid and evaluation on its available fault protection methods: Review and proposals. *Renew. Sustain. Energy Rev.* **2017**, *75*, 1417–1452. [CrossRef]
13. Dagar, A.; Gupta, P.; Niranjan, V. Microgrid protection: A comprehensive review. *Renew. Sustain. Energy Rev.* **2021**, *149*, 111401. [CrossRef]
14. Rajesh, K.; Dash, S.; Rajagopal, R.; Sridhar, R. A review on control of ac microgrid. *Renew. Sustain. Energy Rev.* **2017**, *71*, 814–819. [CrossRef]
15. Mohammed, A.; Refaat, S.S.; Bayhan, S.; Abu-Rub, H. AC Microgrid Control and Management Strategies: Evaluation and Review. *IEEE Power Electron. Mag.* **2019**, *6*, 18–31. [CrossRef]
16. Page, M.J.; McKenzie, J.E.; Bossuyt, P.M.; Boutron, I.; Hoffmann, T.C.; Mulrow, C.D.; Shamseer, L.; Tetzlaff, J.M.; Akl, E.A.; Brennan, S.E.; et al. The PRISMA 2020 statement: An updated guideline for reporting systematic reviews. *Int. J. Surg.* **2021**, *88*, 105906. [CrossRef]
17. Agrawal, A.; Nalamati, C.S.; Gupta, R. Hybrid DC–AC Zonal Microgrid Enabled by Solid-State Transformer and Centralized ESD Integration. *IEEE Trans. Ind. Electron.* **2019**, *66*, 9097–9107. [CrossRef]
18. Che, Y.; Zhou, J.; Lin, T.; Li, W.; Xu, J. A simplified control method for tie-line power of DC micro-grid. *Energies* **2018**, *11*, 933. [CrossRef]
19. Eisapour-Moarref, A.; Kalantar, M.; Esmaili, M. Power sharing in hybrid microgrids using a harmonic-based multidimensional droop. *IEEE Trans. Ind. Inform.* **2020**, *16*, 109–119. [CrossRef]
20. Jin, X.; Shen, Y.; Zhou, Q. A systematic review of robust control strategies in DC microgrids. *Electr. J.* **2022**, *35*, 107125. [CrossRef]
21. Khosravi, N.; Abdolmohammadi, H.R.; Bagheri, S.; Miveh, M.R. Improvement of harmonic conditions in the AC/DC microgrids with the presence of filter compensation modules. *Renew. Sustain. Energy Rev.* **2021**, *143*, 110898. [CrossRef]
22. Li, Y.; He, L.; Liu, F.; Li, C.; Cao, Y.; Shahidehpour, M. Flexible Voltage Control Strategy Considering Distributed Energy Storages for DC Distribution Network. *IEEE Trans. Smart Grid* **2019**, *10*, 163–172. [CrossRef]
23. Lin, B. Modular soft-switching converter in DC micro-grid system applications. *Electron. Lett.* **2018**, *54*, 649–651. [CrossRef]
24. Liu, Q.; Caldognetto, T.; Buso, S. Flexible Control of Interlinking Converters for DC Microgrids Coupled to Smart AC Power Systems. *IEEE Trans. Ind. Electron.* **2019**, *66*, 3477–3485. [CrossRef]
25. Nalamati, C.; Agrawal, A.; Gupta, R. Multiple parallel-connected dab-based solid-state transformer for hybrid dc/ac microgrid system. *IET Gener. Transm. Distrib.* **2020**, *14*, 6359–6370. [CrossRef]
26. Nie, J.; Yuan, L.; Wen, W.; Duan, R.; Shi, B.; Zhao, Z. Communication-Independent Power Balance Control for Solid State Transformer Interfaced Multiple Power Conversion Systems. *IEEE Trans. Power Electron.* **2020**, *35*, 4256–4271. [CrossRef]
27. Sadeghi, M.; Khederzadeh, M. Hybrid multi-DC-AC MG based on multilevel interlinking converter. *IET Power Electron.* **2019**, *12*, 1187–1194. [CrossRef]
28. Sowmmiya, U.; Govindarajan, U. Control and power transfer operation of WRIG-based WECS in a hybrid AC/DC microgrid. *IET Renew. Power Gener.* **2018**, *12*, 359–373. [CrossRef]
29. Tricarico, T.; Gontijo, G.; Neves, M.; Soares, M.; Aredes, M.; Guerrero, J. Control design, stability analysis and experimental validation of new application of an interleaved converter operating as a power interface in hybrid microgrids. *Energies* **2019**, *12*, 437. [CrossRef]
30. Tricarico, T.; Gontijo, G.; Aredes, M.; Dias, R.; Guerrero, J. New hybrid-microgrid topology using a bidirectional interleaved converter as a robust power interface operating in grid-connected and islanded modes. *IET Renew. Power Gener.* **2020**, *14*, 134–144. [CrossRef]
31. Adineh, B.; Keypour, R.; Davari, P.; Blaabjerg, F. Review of Harmonic Mitigation Methods in Microgrid: From a Hierarchical Control Perspective. *IEEE J. Emerg. Sel. Top. Power Electron.* **2021**, *9*, 3044–3060. [CrossRef]
32. Zhang, H.; Xiang, W.; Lin, W.; Wen, J. Grid Forming Converters in Renewable Energy Sources Dominated Power Grid: Control Strategy, Stability, Application, and Challenges. *J. Mod. Power Syst. Clean Energy* **2021**, *9*, 1239–1256. [CrossRef]

33. Prompinit, K.; Khomfoi, S. A battery energy storage system control technique with ramp rate and C-rate parameter consideration for AC microgrid applications. *Int. Rev. Electr. Eng.* **2018**, *13*. [CrossRef]
34. Combe, Q.; Abasian, A.; Pierfederici, S.; Weber, M.; Dufour, S. Control of a Three-Phase Current Source Rectifier for H2 Storage Applications in AC Microgrids. *Energies* **2022**, *15*, 2436. [CrossRef]
35. Acevedo-Arenas, C.Y.; Correcher, A.; Sánchez-Díaz, C.; Ariza, E.; Alfonso-Solar, D.; Vargas-Salgado, C.; Petit-Suárez, J.F. MPC for optimal dispatch of an AC-linked hybrid PV/wind/biomass/H2 system incorporating demand response. *Energy Convers. Manag.* **2019**, *186*, 241–257. [CrossRef]
36. Antunes, H.; Silva, S.; Brandao, D.; Machado, A.; Ferreira, R. A fault-tolerant grid-forming converter applied to AC microgrids. *Int. J. Electr. Power Energy Syst.* **2020**, *121*, 106072. [CrossRef]
37. Barzegarkhoo, R.; Farhangi, M.; Lee, S.S.; Aguilera, R.P.; Siwakoti, Y.P.; Pou, J. Nine-Level Nine-Switch Common-Ground Switched-Capacitor Inverter Suitable for High-Frequency AC-Microgrid Applications. *IEEE Trans. Power Electron.* **2022**, *37*, 6132–6143. [CrossRef]
38. Borrell, Á; Velasco, M.; Miret, J.; Camacho, A.; Martí, P.; Castilla, M. Collaborative Voltage Unbalance Elimination in Grid-Connected AC Microgrids with Grid-Feeding Inverters. *IEEE Trans. Power Electron.* **2021**, *36*, 7189–7201. [CrossRef]
39. Buduma, P.; Das, M.; Naayagi, R.; Mishra, S.; Panda, G. Seamless Operation of Master-Slave Organized AC Microgrid with Robust Control, Islanding Detection and Grid Synchronization. *IEEE Trans. Ind. Appl.* **2022**, *58*, 6724–6738. [CrossRef]
40. Charalambous, A.; Hadjidemetriou, L.; Zacharia, L.; Bintoudi, A.; Tsolakis, A.; Tzovaras, D.; Kyriakides, E. Phase balancing and reactive power support services for microgrids. *Appl. Sci.* **2019**, *9*, 5067. [CrossRef]
41. Chhor, J.; Sourkounis, C. Optimal voltage control strategy for grid-feeding power converters in AC microgrids. *Electr. Power Syst. Res.* **2019**, *176*, 105945. [CrossRef]
42. Firuzi, M.F.; Roosta, A.; Gitizadeh, M. Stability analysis and decentralized control of inverter-based ac microgrid. *Prot. Control Mod. Power Syst.* **2019**, *4*, 6. [CrossRef]
43. Campanhol, L.B.G.; Da Silva, S.A.O.; De Oliveira, A.A.; Bacon, V.D. Power Flow and Stability Analyses of a Multifunctional Distributed Generation System Integrating a Photovoltaic System with Unified Power Quality Conditioner. *IEEE Trans. Power Electron.* **2019**, *34*, 6241–6256. [CrossRef]
44. Gil-González, W.; Montoya, O. Passivity-based PI control of a SMES system to support power in electrical grids: A bilinear approach. *J. Energy Storage* **2018**, *18*, 459–466. [CrossRef]
45. Guo, Y.; Lu, X.; Chen, L.; Zheng, T.; Wang, J.; Mei, S. Functional-Rotation-Based Active Dampers in AC Microgrids with Multiple Parallel Interface Inverters. *IEEE Trans. Ind. Appl.* **2018**, *54*, 5206–5215. [CrossRef]
46. Hou, X.; Sun, Y.; Lu, J.; Zhang, X.; Koh, L.H.; Su, M.; Guerrero, J.M. Distributed Hierarchical Control of AC Microgrid Operating in Grid-Connected, Islanded and Their Transition Modes. *IEEE Access* **2018**, *6*, 77388–77401. [CrossRef]
47. Hussain, N.; Khayat, Y.; Golestan, S.; Nasir, M.; Vasquez, J.; Guerrero, J.; Kauhaniemi, K. Ac microgrids protection: A digital coordinated adaptive scheme. *Appl. Sci.* **2021**, *11*, 7066. [CrossRef]
48. Issa, W.R.; Khateb, A.H.E.; Abusara, M.A.; Mallick, T.K. Control Strategy for Uninterrupted Microgrid Mode Transfer during Unintentional Islanding Scenarios. *IEEE Trans. Ind. Electron.* **2018**, *65*, 4831–4839. [CrossRef]
49. Jarrahi, M.; Samet, H.; Ghanbari, T. Novel Change Detection and Fault Classification Scheme for AC Microgrids. *IEEE Syst. J.* **2020**, *14*, 3987–3998. [CrossRef]
50. Jarwar, A.; Soomro, A.; Memon, Z.; Odhano, S.; Uqaili, M.; Larik, A. High dynamic performance power quality conditioner for AC microgrids. *IET Power Electron.* **2019**, *12*, 550–556. [CrossRef]
51. Jumani, T.; Mustafa, M.; Rasid, M.; Memon, Z. Dynamic response enhancement of grid-tied ac microgrid using salp swarm optimization algorithm. *Int. Trans. Electr. Energy Syst.* **2020**, *30*, e12321. [CrossRef]
52. Kaushal, J.; Basak, P. A Novel Approach for Determination of Power Quality Monitoring Index of an AC Microgrid Using Fuzzy Inference System. *Iran. J. Sci. Technol.-Trans. Electr. Eng.* **2018**, *42*, 429–450. [CrossRef]
53. Kulkarni, S.; Gaonkar, D. Improved droop control strategy for parallel connected power electronic converter based distributed generation sources in an Islanded Microgrid. *Electr. Power Syst. Res.* **2021**, *201*, 107531. [CrossRef]
54. Li, H.; Li, S.; Lu, J.; Qu, Y.; Guo, C. A Novel Strategy Based on Linear Active Disturbance Rejection Control for Harmonic Detection and Compensation in Low Voltage AC Microgrid. *Energies* **2019**, *12*, 3982. [CrossRef]
55. Lim, J.U.; Kwon, I.S.; Kim, H.W.; Cho, K.Y. Seamless transfer algorithm of AC microgrid inverter compensating load current for weak grid. *Energies* **2019**, *12*, 728. [CrossRef]
56. Ma, W.; Ouyang, S. Control strategy for inverters in microgrid based on repetitive and state feedback control. *Int. J. Electr. Power Energy Syst.* **2019**, *111*, 447–458. [CrossRef]
57. Naderi, Y.; Hosseini, S.; Zadeh, S.; Mohammadi-Ivatloo, B.; Savaghebi, M.; Guerrero, J. An optimized direct control method applied to multilevel inverter for microgrid power quality enhancement. *Int. J. Electr. Power Energy Syst.* **2019**, *107*, 496–506. [CrossRef]
58. Neves, R.V.A.; Machado, R.Q.; Oliveira, V.A.; Wang, X.; Blaabjerg, F. Multitask Fuzzy Secondary Controller for AC Microgrid Operating in Stand-Alone and Grid-Tied Mode. *IEEE Trans. Smart Grid* **2019**, *10*, 5640–5649. [CrossRef]
59. Panjaitan, S.D.; Kurnianto, R.; Sanjaya, B.W. Flexible Power-Sharing Control for Inverters-Based Microgrid Systems. *IEEE Access* **2020**, *8*, 177984–177994. [CrossRef]

60. Pérez-Ibacache, R.; Yazdani, A.; Silva, C.; Agüero, J. Decentralized Unified Control for Inverter-Based AC Microgrids Subject to Voltage Constraints. *IEEE Access* **2019**, *7*, 157318–157329. [CrossRef]
61. Sharma, R.; Kewat, S.; Singh, B. Robust 3IMPL control algorithm for power management of SyRG/PV/BES-Based distributed islanded microgrid. *IEEE Trans. Ind. Electron.* **2019**, *66*, 7765–7777. [CrossRef]
62. Vilaisarn, Y.; Moradzadeh, M.; Abdelaziz, M.; Cros, J. An MILP formulation for the optimum operation of AC microgrids with hierarchical control. *Int. J. Electr. Power Energy Syst.* **2022**, *137*, 107674. [CrossRef]
63. Wang, M.H.; Yang, T.B.; Tan, S.C.; Hui, S. Hybrid Electric Springs for Grid-Tied Power Control and Storage Reduction in AC Microgrids. *IEEE Trans. Power Electron.* **2019**, *34*, 3214–3225. [CrossRef]
64. Wang, M.H.; He, Y.; Jia, Y.; Xu, Z. A Power-Decoupled Current-Source Inverter for PV Energy Harvest and Grid Voltage Regulation. *IEEE Trans. Ind. Electron.* **2021**, *68*, 9540–9549. [CrossRef]
65. Yuan, W.; Wang, Y.; Liu, D.; Deng, F.; Chen, Z. Efficiency-Prioritized Droop Control Strategy of AC Microgrid. *IEEE J. Emerg. Sel. Top. Power Electron.* **2021**, *9*, 2936–2950. [CrossRef]
66. Zhang, H.; Ke, C.; Zhi, N.; Cheng, Y. Compound control method for overcoming transmission delay impact on networked control inverter for AC microgrid. *J. Intell. Fuzzy Syst.* **2020**, *38*, 7875–7892. [CrossRef]
67. Zheng, X.; Zeng, Y.; Zhao, M.; Venkatesh, B. Early Identification and Location of Short-Circuit Fault in Grid-Connected AC Microgrid. *IEEE Trans. Smart Grid* **2021**, *12*, 2869–2878. [CrossRef]
68. Zheng, C.; Dragičević, T.; Blaabjerg, F. Model Predictive Control-Based Virtual Inertia Emulator for an Islanded Alternating Current Microgrid. *IEEE Trans. Ind. Electron.* **2021**, *68*, 7167–7177. [CrossRef]
69. Abhinav, S.; Schizas, I.; Lewis, F.; Davoudi, A. Distributed noise-resilient networked synchrony of active distribution systems. *IEEE Trans. Smart Grid* **2018**, *9*, 836–846. [CrossRef]
70. Alam, M. Overcurrent protection of AC microgrids using mixed characteristic curves of relays. *Comput. Electr. Eng.* **2019**, *74*, 74–88. [CrossRef]
71. Alfergani, A.; Khalil, A.; Rajab, Z. Networked control of AC microgrid. *Sustain. Cities Soc.* **2018**, *37*, 371–387. [CrossRef]
72. Ali, S.; Aamir, M.; Jafri, A.; Subramaniam, U.; Haroon, F.; Waqar, A.; Yaseen, M. Model predictive control—Based distributed control algorithm for bidirectional interlinking converter in hybrid microgrids. *Int. Trans. Electr. Energy Syst.* **2021**, *31*, e12817. [CrossRef]
73. Aref, M.; Mossa, M.; Lan, N.; Quynh, N.; Oboskalov, V.; Ali, A. Improvement of Fault Current Calculation and Static Security Risk for Droop Control of the Inverter-Interfaced DG of Grid-Connected and Isolated Microgrids. *Inventions* **2022**, *7*, 52. [CrossRef]
74. Cai, H.; Hu, G. Distributed Nonlinear Hierarchical Control of AC Microgrid via Unreliable Communication. *IEEE Trans. Smart Grid* **2018**, *9*, 2429–2441. [CrossRef]
75. Cai, H.; Hu, G. Distributed Robust Hierarchical Power Sharing Control of Grid-Connected Spatially Concentrated AC Microgrid. *IEEE Trans. Control Syst. Technol.* **2019**, *27*, 1012–1022. [CrossRef]
76. Cai, C.; Liu, H.; Tao, Y.; Deng, Z.; Dai, W.; Chen, J. Microgrid Equivalent Modeling Based on Long Short-Term Memory Neural Network. *IEEE Access* **2020**, *8*, 23120–23133. [CrossRef]
77. Cao, W.; Ma, Y.; Wang, F.; Tolbert, L.; Xue, Y. Low-Frequency Stability Analysis of Inverter-Based Islanded Multiple-Bus AC Microgrids Based on Terminal Characteristics. *IEEE Trans. Smart Grid* **2020**, *11*, 3662–3676. [CrossRef]
78. Dutta, R.; Samantaray, S. Assessment of impedance based fault locator for AC micro-grid. *Renew. Energy Focus* **2018**, *26*, 1–10. [CrossRef]
79. Eyisi, C.; Li, Q. Load sharing scheme incorporating power security margins for parallel operation of voltage source inverters. *Energies* **2021**, *14*, 5825. [CrossRef]
80. Gupta, S.; Maulik, A.; Das, D.; Singh, A. Coordinated stochastic optimal energy management of grid-connected microgrids considering demand response, plug-in hybrid electric vehicles, and smart transformers. *Renew. Sustain. Energy Rev.* **2022**, *155*, 111861. [CrossRef]
81. Habibi, S.; Bidram, A. Unfalsified Switching Adaptive Voltage Control for Islanded Microgrids. *IEEE Trans. Power Syst.* **2022**, *37*, 3394–3407. [CrossRef]
82. Islam, M.; Yang, F.; Ekanayek, C.; Amin, M. Grid power fluctuation reduction by fuzzy control based energy management system in residential microgrids. *Int. Trans. Electr. Energy Syst.* **2019**, *29*, e2758. [CrossRef]
83. Iyakaremye, J.D.D.; Nyakoe, G.N.; Wekesa, C.W. MPC-Based Arctan Droop Control Strategy of the Parallel Inverter System in an Islanded AC Microgrid. *J. Eng.* **2021**, *2021*, 1870590. [CrossRef]
84. Jayachandran, M.; Ravi, G. Predictive power management strategy for PV/battery hybrid unit based islanded AC microgrid. *Int. J. Electr. Power Energy Syst.* **2019**, *110*, 487–496. [CrossRef]
85. John, B.; Ghosh, A.; Goyal, M.; Zare, F. A DC Power Exchange Highway Based Power Flow Management for Interconnected Microgrid Clusters. *IEEE Syst. J.* **2019**, *13*, 3347–3357. [CrossRef]
86. Kabalan, M.; Singh, P.; Niebur, D. A Design and Optimization Tool for Inverter-Based Microgrids Using Large-Signal Nonlinear Analysis. *IEEE Trans. Smart Grid* **2019**, *10*, 4566–4576. [CrossRef]
87. Kryonidis, G.; Kontis, E.; Chrysochos, A.; Oureilidis, K.; Demoulias, C.; Papagiannis, G. Power Flow of Islanded AC Microgrids: Revisited. *IEEE Trans. Smart Grid* **2018**, *9*, 3903–3905. [CrossRef]
88. Lenz, E.; Pagano, D.J.; Pou, J. Bifurcation Analysis of Parallel-Connected Voltage-Source Inverters with Constant Power Loads. *IEEE Trans. Smart Grid* **2018**, *9*, 5482–5493. [CrossRef]

89. Memon, A.; Kauhaniemi, K. An adaptive protection for radial AC microgrid using IEC 61850 communication standard: Algorithm proposal using offline simulations. *Energies* **2020**, *13*, 5316. [CrossRef]
90. Mohammadi, J.; Ajaei, F.B.; Stevens, G. Grounding the AC microgrid. *IEEE Trans. Ind. Appl.* **2019**, *55*, 98–105. [CrossRef]
91. Mohammadshahi, S.; Boulaire, F.; Love, J.; Gorji, S.; Mackinnon, I. A flexible analytical model for operational investigation of solar hydrogen plants. *Int. J. Hydrogen Energy* **2022**, *47*, 782–808. [CrossRef]
92. Moreira, A.; Paredes, H.; Souza, W.D.; Marafao, F.; Silva, L.D. Intelligent expert system for power quality improvement under distorted and unbalanced conditions in three-phase AC microgrids. *IEEE Trans. Smart Grid* **2018**, *9*, 6951–6960. [CrossRef]
93. Nayak, A.; Maulik, A.; Das, D. An integrated optimal operating strategy for a grid-connected AC microgrid under load and renewable generation uncertainty considering demand response. *Sustain. Energy Technol. Assessments* **2021**, *45*, 101169. [CrossRef]
94. Perez-Ibacache, R.; Cedeno, A.; Silva, C.; Carvajal, G.; Aguero, J.; Yazdani, A. Decentralized Model-Based Predictive Control for der Units Integration in AC Microgrids Subject to Operational and Safety Constraints. *IEEE Trans. Power Deliv.* **2021**, *36*, 2479–2489. [CrossRef]
95. Rahman, A.; Syed, I.; Ullah, M. Small signal stability of a balanced three-phase AC microgrid using harmonic linearization: Parametric-based analysis. *Electronics* **2019**, *8*, 12. [CrossRef]
96. Saldarriaga-Zuluaga, S.; López-Lezama, J.; Muñoz-Galeano, N. Adaptive protection coordination scheme in microgrids using directional over-current relays with non-standard characteristics. *Heliyon* **2021**, *7*, e06665. [CrossRef]
97. Shaker, Y.; Yousri, D.; Osama, A.; Al-Gindy, A.; Tag-Eldin, E.; Allam, D. Optimal Charging/Discharging Decision of Energy Storage Community in Grid-Connected Microgrid Using Multi-Objective Hunger Game Search Optimizer. *IEEE Access* **2021**, *9*, 120774–120794. [CrossRef]
98. Soundarya, G.; Sitharthan, R.; Sundarabalan, C.K.; Balasundar, C.; Karthikaikannan, D.; Sharma, J. Design and Modeling of Hybrid DC/AC Microgrid with Manifold Renewable Energy Sources. *IEEE Can. J. Electr. Comput. Eng.* **2021**, *44*, 130–135. [CrossRef]
99. Srivastava, A.; Parida, S. Data driven approach for fault detection and Gaussian process regression based location prognosis in smart AC microgrid. *Electr. Power Syst. Res.* **2022**, *208*, 107889. [CrossRef]
100. Torabi-Farsani, K.; Vafamand, N.; Razavi-Far, R.; Saif, M. Secure Frequency Regulation of Electric Vehicle-connected Microgrid System under Multiple Latency Attacks. *Comput. Electr. Eng.* **2022**, *101*, 108008. [CrossRef]
101. Yang, T.; He, Y.; Liu, G.P. Distributed Voltage Restoration of AC Microgrids Under Communication Delays: A Predictive Control Perspective. *IEEE Trans. Circuits Syst. I Regul. Pap.* **2022**, *69*, 2614–2624. [CrossRef]
102. Zholbaryssov, M.; Fooladivanda, D.; Domínguez-García, A. Resilient distributed optimal generation dispatch for lossy AC microgrids. *Syst. Control Lett.* **2019**, *123*, 47–54. [CrossRef]
103. Zhang, Y.; Wei, W. Decentralised coordination control strategy of the PV generator, storage battery and hydrogen production unit in islanded AC microgrid. *IET Renew. Power Gener.* **2020**, *14*, 1053–1062. [CrossRef]
104. Zolfaghari, M.; Abedi, M.; Gharehpetian, G.B. Power Flow Control of Interconnected AC–DC Microgrids in Grid-Connected Hybrid Microgrids Using Modified UIPC. *IEEE Trans. Smart Grid* **2019**, *10*, 6298–6307. [CrossRef]
105. Zolfaghari, M.; Abedi, M.; Gharehpetian, G.B. Robust Nonlinear State Feedback Control of Bidirectional Interlink Power Converters in Grid-Connected Hybrid Microgrids. *IEEE Syst. J.* **2020**, *14*, 1117–1124. [CrossRef]
106. Zolfaghari, M.; Abedi, M.; Gharehpetian, G.B.; Guerrero, J.M. Flatness-Based Decentralized Control of Bidirectional Interlink Power Converters in Grid-Connected Hybrid Microgrids Using Adaptive High-Gain PI-Observer. *IEEE Syst. J.* **2021**, *15*, 478–486. [CrossRef]
107. Ahmad, S.; Mekhilef, S.; Mokhlis, H.; Karimi, M.; Pourdaryaei, A.; Ahmed, T.; Jhuma, U.; Afzal, S. Fuzzy logic-based direct power control method for pv inverter of grid-tied ac microgrid without phase-locked loop. *Electronics* **2021**, *10*, 3095. [CrossRef]
108. Amirkhan, S.; Radmehr, M.; Rezanejad, M.; Khormali, S. A robust control technique for stable operation of a DC/AC hybrid microgrid under parameters and loads variations. *Int. J. Electr. Power Energy Syst.* **2020**, *117*, 105659. [CrossRef]
109. Anand, A.; Affijulla, S. Ensemble empirical mode decomposition based differential protection scheme for islanded and grid-tied AC microgrid. *IET Gener. Transm. Distrib.* **2020**, *14*, 6674–6681. [CrossRef]
110. Antunes, H.; Silva, S.; Brandao, D.; Machado, A.; Filho, B. A new multifunctional converter based on a series compensator applied to AC microgrids. *Int. J. Electr. Power Energy Syst.* **2018**, *102*, 160–170. [CrossRef]
111. Aragon, C.A.; Guzman, R.; de Vicuña, L.G.; Miret, J.; Castilla, M. Constrained Predictive Control Based on a Large-Signal Model for a Three-Phase Inverter Connected to a Microgrid. *IEEE Trans. Ind. Electron.* **2022**, *69*, 6497–6507. [CrossRef]
112. Awad, H.; Bayoumi, E.; Soliman, H.; Santis, M.D. Robust tracker of hybrid microgrids by the invariant-ellipsoid set. *Electronics* **2021**, *10*, 1794. [CrossRef]
113. Barr, N.; Li, S.; Fu, X. Control of grid-connected inverters for circulating current suppression using artificial neural network and conventional control methods. *Int. Trans. Electr. Energy Syst.* **2021**, *31*. [CrossRef]
114. Chakraborty, S.; Das, S. Communication-less protection scheme for AC microgrids using hybrid tripping characteristic. *Electr. Power Syst. Res.* **2020**, *187*, 106453. [CrossRef]
115. Deng, F.; Yao, W.; Zhang, X.; Mattavelli, P. A Decentralized Current Sharing Strategy for Islanded Resistive Microgrids Based on Iterative Virtual Impedance Regulation. *IEEE Trans. Ind. Inform.* **2022**, *18*, 3958–3969. [CrossRef]
116. Dheer, D.K.; Vijay, A.S.; Kulkarni, O.V.; Doolla, S. Improvement of Stability Margin of Droop-Based Islanded Microgrids by Cascading of Lead Compensators. *IEEE Trans. Ind. Appl.* **2019**, *55*, 3241–3251. [CrossRef]

117. Dong, M.; Li, L.; Wang, L.; Song, D.; Liu, Z.; Tian, X.; Li, Z.; Wang, Y. A distributed secondary control algorithm for automatic generation control considering EDP and automatic voltage control in an AC microgrid. *Energies* **2018**, *11*, 932. [CrossRef]
118. Ebrahim, M.A.; Ayoub, B.A.A.; Nashed, M.N.F.; Osman, F.A.M. A Novel Hybrid-HHOPSO Algorithm Based Optimal Compensators of Four-Layer Cascaded Control for a New Structurally Modified AC Microgrid. *IEEE Access* **2021**, *9*, 4008–4037. [CrossRef]
119. Ebrahim, M.; Aziz, B.; Nashed, M.; Osman, F. Optimal design of controllers and harmonic compensators for three-level cascaded control in stationary reference frame for grid-supporting inverters-based AC microgrid. *Energy Rep.* **2022**, *8*, 860–877. [CrossRef]
120. Heredero-Peris, D.; Chillón-Antón, C.; Pagès-Giménez, M.; Montesinos-Miracle, D.; Santamaría, M.; Rivas, D.; Aguado, M. An enhancing fault current limitation hybrid droop/V-f control for grid-tied four-wire inverters in AC microgrids. *Appl. Sci.* **2018**, *8*, 1725. [CrossRef]
121. Jmii, H.; Abbes, M.; Meddeb, A.; Chebbi, S. Centralized VSM control of an AC meshed microgrid for ancillary services provision. *Int. J. Electr. Power Energy Syst.* **2020**, *115*, 105450. [CrossRef]
122. Kumar, A.; Jha, B.K.; Dheer, D.K.; Singh, D.; Misra, R.K. Nested backward/forward sweep algorithm for power flow analysis of droop regulated islanded microgrids. *IET Gener. Transm. Distrib.* **2019**, *13*, 3086–3095. [CrossRef]
123. Lenz, E.; Pagano, D.J.; Ruseler, A.; Heldwein, M.L. Two-Parameter Stability Analysis of Resistive Droop Control Applied to Parallel-Connected Voltage-Source Inverters. *IEEE J. Emerg. Sel. Top. Power Electron.* **2020**, *8*, 3318–3332. [CrossRef]
124. Li, T.; Su, C.; Zhang, Y. Multi-source residual capacity collaborative dynamic compensation strategy based on robust residual generator. *Energy Rep.* **2021**, *7*, 3318–3332. [CrossRef]
125. Mahmud, K.; Sahoo, A.; Ravishankar, J.; Dong, Z. Coordinated Multilayer Control for Energy Management of Grid-Connected AC Microgrids. *IEEE Trans. Ind. Appl.* **2019**, *55*, 535–548. [CrossRef]
126. Mi, Y.; Yan, L.; Cai, H.; Fu, Y.; Tian, S.; Jin, C. Reactive power sharing control for islanded AC microgrid based on fuzzy adaptive compensation method. *IET Gener. Transm. Distrib.* **2020**, *14*, 585–593. [CrossRef]
127. Patankar, P.P.; Munshi, M.M.; Deshmukh, R.R.; Ballal, M.S. A Modified Control Method for Grid Connected Multiple Rooftop Solar Power Plants. *IEEE Trans. Ind. Appl.* **2021**, *57*, 3306–3316. [CrossRef]
128. Pompodakis, E.; Kryonidis, G.; Demoulias, C.; Alexiadis, M. A Generic Power Flow Algorithm for Unbalanced Islanded Hybrid AC/DC Microgrids. *IEEE Trans. Power Syst.* **2021**, *36*, 1107–1120. [CrossRef]
129. Souraki, H.P.; Radmehr, M.; Rezanejad, M. Distributed energy storage system-based control strategy for hybrid DC/AC microgrids in grid-connected mode. *Int. J. Energy Res.* **2019**, *43*, 6283–6295. [CrossRef]
130. souraki, H.P.; Radmehr, M.; Rezanejad, M. Lypunov theory combined with small signal linearization for regulated operation of a hybrid DC/AC microgrid. *Int. Trans. Electr. Energy Syst.* **2020**, *30*. [CrossRef]
131. Roncero-Clemente, C.; Gonzalez-Romera, E.; Barrero-González, F.; Milanés-Montero, M.I.; Romero-Cadaval, E. Power-Flow-Based Secondary Control for Autonomous Droop-Controlled AC Nanogrids with Peer-to-Peer Energy Trading. *IEEE Access* **2021**, *9*, 22339–22350. [CrossRef]
132. Sahu, P.; Mishra, S.; Prusty, R.; Panda, S. Improved -salp swarm optimized type-II fuzzy controller in load frequency control of multi area islanded AC microgrid. *Sustain. Energy, Grids Netw.* **2018**, *16*, 380–392. [CrossRef]
133. Serrano-Fontova, A.; Azab, M. Development and performance analysis of a multi-functional algorithm for AC microgrids: Simultaneous power sharing, voltage support and islanding detection. *Int. J. Electr. Power Energy Syst.* **2022**, *135*, 107341. [CrossRef]
134. Sreelekshmi, R.; Lakshmi, R.; Nair, M. AC microgrid with battery energy storage management under grid connected and islanded modes of operation. *Energy Rep.* **2022**, *8*, 350–357. [CrossRef]
135. Talapur, G.; Suryawanshi, H. A modified control scheme for power management in an AC microgrid with integration of multiple nanogrids. *Electronics* **2019**, *8*, 490. [CrossRef]
136. Tang, X.; Zhang, D.; Xiao, D.; Li, M. Modeling and Stability Analysis of a Novel Voltage-Oriented Power Coordination Controlled Constant-Frequency AC Microgrid System. *Electronics* **2021**, *10*, 1935. [CrossRef]
137. Holari, Y.T.; Taher, S.; Mehrasa, M. Distributed energy storage system-based nonlinear control strategy for hybrid microgrid power management included wind/PV units in grid-connected operation. *Int. Trans. Electr. Energy Syst.* **2020**, *30*, e12237. [CrossRef]
138. Xie, B.; Liu, Y.; Ji, Y.; Wang, J. Two-stage battery energy storage system (BESS) in AC microgrids with balanced state-of-charge and guaranteed small-signal stability. *Energies* **2018**, *11*, 322. [CrossRef]
139. *IEEE Std 1547-2018*; IEEE Standard for Interconnection and Interoperability of Distributed Energy Resources with Associated Electric Power Systems Interfaces. (Revision of IEEE Std 1547-2003). IEEE Standards Association: Piscataway, NJ, USA, 2018; pp. 1–138. [CrossRef]
140. *IEEE Std 519-2014*; IEEE Recommended Practice and Requirements for Harmonic Control in Electric Power Systems. (Revision of IEEE Std 519-1992). IEEE Standards Association: Piscataway, NJ, USA, 2014; pp. 1–29. [CrossRef]
141. *EN 50160*; Voltage Characteristics of Electricity Supplied by Public Electricity Networks. CENELEC (European Committee for Electrotechnical Standardization): Brussels, Belgium, 2022; pp. 1–34. [CrossRef]
142. *IEEE Std C37.118.1-2011*; IEEE Standard for Synchrophasor Measurements for Power Systems. (Revision of IEEE Std C37.118-2005). IEEE Standards Association: Piscataway, NJ, USA, 2011; pp. 1–61. [CrossRef]

143. *IEEE Std 1159-1995;* IEEE Recommended Practice for Monitoring Electric Power Quality. IEEE Standards Association: Piscataway, NJ, USA, 1995; pp.1–80. [CrossRef]
144. *IEEE Std 2030.7-2017;* IEEE Standard for the Specification of Microgrid Controllers. IEEE Standards Association: Piscataway, NJ, USA, 2018; pp. 1–43. [CrossRef]
145. *IEEE Std 1459-2010;* IEEE Standard Definitions for the Measurement of Electric Power Quantities Under Sinusoidal, Nonsinusoidal, Balanced, or Unbalanced Conditions. (Revision of IEEE Std 1459-2000). IEEE Standards Association: Piscataway, NJ, USA, 2010; pp. 1–50. [CrossRef]
146. *IEC 61000-2-2:2002;* Electromagnetic compatibility (EMC)—Part 2-2: Environment—Compatibility Levels for Low-Frequency Conducted Disturbances and Signalling in Public Low-Voltage Power Supply Systems. IEC (International Electrotechnical Commission): Geneva, Switzerland, 2002; pp. 1–57. Available online: https://webstore.iec.ch/en/publication/4133 (accessed on 21 May 2024).
147. *IEC 61000-3-2:2000;* Electromagnetic Compatibility (EMC)—Part 3-2: Limits—Limits for Harmonic Current Emissions (Equipment Input Current <= 16 A per Phase). IEC (International Electrotechnical Commission): Geneva, Switzerland, 2000; pp. 1–55. Available online: https://webstore.iec.ch/en/publication/18724 (accessed on 21 May 2024).
148. *IEC 61727:2004;* Photovoltaic (PV) Systems—Characteristics of the Utility Interface. IEC (International Electrotechnical Commission): Geneva, Switzerland, 2004; pp. 1–23. Available online: https://webstore.iec.ch/en/publication/5736 (accessed on 21 May 2024).
149. *IEC 60831-1:2014;* Shunt Power Capacitors of the Self-Healing Type for a.c. Systems Having a Rated Voltage Up to and Including 1000 V—Part 1: General—Performance, Testing and Rating—Safety Requirements—Guide for Installation and Operation. IEC (International Electrotechnical Commission): Geneva, Switzerland, 2014; pp. 1–73. Available online: https://webstore.iec.ch/en/publication/22263 (accessed on 21 May 2024).
150. *IEC 60831-2:2014;* Shunt Power Capacitors of the Self-Healing Type for a.c. Systems Having a Rated Voltage Up to and Including 000 V—Part 2: Ageing Test, Self-Healing Test and Destruction Test. IEC (International Electrotechnical Commission): Geneva, Switzerland, 2014; pp. 1–22. Available online: https://webstore.iec.ch/en/publication/3610 (accessed on 21 May 2024).
151. *IEC 60255-1:2022;* Measuring Relays and Protection Equipment—Part 1: Common Requirements. IEC (International Electrotechnical Commission): Geneva, Switzerland, 2022; pp. 1–92. Available online: https://webstore.iec.ch/en/publication/59762 (accessed on 21 May 2024).
152. *IEC 61850-3:2013;* Communication Networks and Systems for Power Utility Automation—Part 3: General Requirements. IEC (International Electrotechnical Commission): Geneva, Switzerland, 2013; pp. 1–136. Available online: https://webstore.iec.ch/en/publication/6010 (accessed on 21 May 2024).
153. *IEC 62040-1:2017;* Uninterruptible Power Systems (UPS)—Part 2: Electromagnetic Compatibility (EMC) Requirements. IEC (International Electrotechnical Commission): Geneva, Switzerland, 2017; pp. 1–146. Available online: https://webstore.iec.ch/en/publication/31983 (accessed on 21 May 2024).
154. VDE VERLAG. *DIN VDE V 0126-1-1 VDE V 0126-1-1:2013-08.* Automatic Disconnection Device between a Generator and the Public Low-Voltage Grid. 2013. Available online: https://www.vde-verlag.de/standards/0100178/din-vde-v-0126-1-1-vde-v-0126-1-1-2013-08.html (accessed on 21 May 2024).
155. Ogg, F. Pico, Nano, Micro, Minigrids. 2015. Available online: https://www.researchgate.net/publication/282947314_Pico_nano_micro_minigrids (accessed on 21 May 2024).
156. IRENA. Renewable Power Generation Costs in 2022. *International Renewable Energy Agency, Abu Dhabi.* 2023. Available online: https://www.irena.org/Publications/2023/Aug/Renewable-Power-Generation-Costs-in-2022 (accessed on 21 May 2024).
157. IREC. IEEE 1547-2018 Adoption Tracker-Interstate Renewable Energy Council (IREC). Available online: https://irecusa.org/resources/ieee-1547-2018-adoption-tracker/ (accessed on 3 June 2024).
158. *IEC 61000-4-15:2010 RLV;* Electromagnetic Compatibility (EMC)—Part 4-15: Testing and Measurement Techniques—Flickermeter—Functional and Design Specifications. IEC (International Electrotechnical Commission): Geneva, Switzerland, 2010; pp. 1–190. Available online: https://webstore.iec.ch/en/publication/22269 (accessed on 3 June 2024).
159. Barnes, M.; Kondoh, J.; Asano, H.; Oyarzabal, J.; Ventakaramanan, G.; Lasseter, R.; Hatziargyriou, N.; Green, T. Real-World MicroGrids-An Overview. In Proceedings of the 2007 IEEE International Conference on System of Systems Engineering, San Antonio, TX, USA, 16–18 April 2007; pp. 1–8. [CrossRef]
160. Panora, R.; Gehret, J.E.; Furse, M.M.; Lasseter, R.H. Real-World Performance of a CERTS Microgrid in Manhattan. *IEEE Trans. Sustain. Energy* **2014**, *5*, 1356–1360. [CrossRef]
161. Zhang, X.; Zhang, Y.; Li, R. Distributed Event-Triggered Control for Power Sharing in Grid-Connected AC Microgrid. *IEEE Trans. Power Syst.* **2024**, *39*, 770–778. [CrossRef]
162. Arora, A.; Bhadu, M.; Kumar, A. Simultaneous Damping and Frequency Control in AC Microgrid Using Coordinated Control Considering Time Delay and Noise. *Trans. Inst. Meas. Control* **2024**, 01423312241230735. [CrossRef]
163. Kumar, A.; Bhadu, M.; Arabi, A.I.A.; Kamangar, S.; Bhutto, J.K.; Ali, M.A.; Kumar, S. Optimized robust control for improving frequency response of delay dependent AC microgrid with uncertainties. *Electr. Power Syst. Res.* **2024**, *229*, 110138. [CrossRef]
164. Rafia, A.N.A.; Jayaprakash, R. Fractional Order Controller For Power Control in AC Islanded PV Microgrid using Electric Vehicles. *Recent Adv. Electr. Electron. Eng.* **2024**, *17*, 137–146. [CrossRef]

165. Li, G.; Hu, T.; Liu, X. A Quasi-Disturbance Current-Based Droop Control Strategy for Enhancing Voltage Robustness of Islanded AC Microgrid. *IEEE Trans. Ind. Inform.* **2024**, 1–11. [CrossRef]
166. Jain, S.K.; Agrawal, S. Robust control of frequency considering operations of AC microgrid in islanded mode. *Int. J. Eng. Syst. Model. Simul.* **2024**, *15*, 1–10. [CrossRef]
167. Kiani, S.; Salmanpour, A.; Hamzeh, M.; Kebriaei, H. Learning Robust Model Predictive Control for Voltage Control of Islanded Microgrid. *IEEE Trans. Autom. Sci. Eng.* **2024**, 1–12. [CrossRef]
168. Ding, K.; Zhu, Q.; Huang, T. Partial-Information-Based Non-Fragile Intermittent Estimator for Microgrids with Semi-Aperiodic DoS Attacks: Gain Stochastic Float. *IEEE Trans. Power Syst.* **2024**, *39*, 2271–2283. [CrossRef]
169. Mahvash, H.; Taher, S.A.; Guerrero, J.M. Mitigation of severe false data injection attacks (FDIAs) in marine current turbine (MCT) type 4 synchronous generator renewable energy using promoted backstepping method. *Renew. Energy* **2024**, *222*, 120008. [CrossRef]
170. Kharchouf, I.; Mohammed, O.A. Controller Hardware-in-the-Loop Testbed of a Distributed Consensus Multi-Agent System Control under Deception and Disruption Cyber-Attacks. *Energies* **2024**, *17*, 1669. [CrossRef]
171. Akarne, Y.; Essadki, A.; Nasser, T.; Bhiri, B.E. Experimental Analysis of Efficient Dual-Layer Energy Management and Power Control in an AC Microgrid System. *IEEE Access* **2024**, *12*, 30577–30592. [CrossRef]

designs

MDPI

Article

Enhancing Energy Access in Rural Indonesia: A Holistic Assessment of a 1 kW Portable Power Generator Based on Proton-Exchange Membrane Fuel Cells (PEMFCs)

Handrea Bernando Tambunan [1], Reynolds Widhiyanurrochmansyach [2], Sabastian PraninditYo [2] and Jayan Sentanuhady [2,*]

[1] Power Generation Research Division, PT PLN (Persero), Jakarta 12760, Indonesia; handrea.bernando.t@gmail.com

[2] Department of Mechanical and Industrial Engineering, Gadjah Mada University, Yogyakarta 55284, Indonesia; reynolds2020@mail.ugm.ac.id (R.W.); sabastian.pranindityo@ugm.ac.id (S.P.)

* Correspondence: jayan@ugm.ac.id

Abstract: Hydrogen energy is a promising alternative to traditional fossil fuels, offering a clean and sustainable solution to address the challenges of climate change and environmental degradation. Fuel cells provide direct and environmentally friendly conversion of chemical energy from a fuel source into electrical energy, emitting only water vapor when utilizing hydrogen from renewable sources. This study delves into the design of a portable proton-exchange membrane fuel cell (PEMFC) device tailored for household use in rural areas. The research focuses on achieving a minimum peak power of 1000 W and a voltage of 220 VAC at 50 Hz for the fuel cell. Employing theoretical calculations derived from existing formulas and literature reviews, various fuel cell components are meticulously assessed, including real power, voltage drop, performance under current load, and pressure drop on the bipolar plate. Additionally, the study encompasses the selection of auxiliary components like converters, inverters, fans, and others. The resultant fuel cell design showcases a device capable of generating a peak power of 1132.32 W with an efficiency rating of 48.66%. Identifying suitable auxiliary components further contributes to developing a practical and efficient portable power solution for rural households.

Keywords: proton-exchange membrane; fuel cells; rural area; portable power generator

Citation: Tambunan, H.B.; Widhiyanurrochmansyach, R.; Pranindityo, S.; Sentanuhady, J. Enhancing Energy Access in Rural Indonesia: A Holistic Assessment of a 1 kW Portable Power Generator Based on Proton-Exchange Membrane Fuel Cells (PEMFCs). *Designs* **2024**, *8*, 117. https://doi.org/10.3390/designs8060117

Academic Editor: Dibin Zhu

Received: 6 September 2024
Revised: 16 October 2024
Accepted: 24 October 2024
Published: 7 November 2024

1. Introduction

Renewable energy has emerged as a pivotal solution to address the growing global demand for electrical power while mitigating the environmental impacts of traditional fossil fuel-based energy sources [1–3]. Unlike finite and environmentally taxing resources such as coal or oil, renewable energy derives from naturally replenishing sources, including sunlight, wind, water, biomass, and geothermal heat [4]. Harnessing these resources to generate electricity has become a cornerstone in transitioning towards a more sustainable and environmentally friendly energy landscape. Solar photovoltaic panels convert sunlight into electricity, wind turbines harness the wind's kinetic energy, hydropower systems utilize flowing water, biomass uses the energy from plants and plant-derived materials, and geothermal plants tap into the Earth's internal heat. Embracing renewable energy diversifies the energy mix and contributes to reducing greenhouse gas emissions [5,6], combating climate change [7], and fostering energy independence [8]. As advancements in technology and infrastructure continue to drive down costs [9] and improve efficiency [10], the widespread adoption of renewable energy promises to create a sustainable clean energy development [11], more resilient [12], and equitable energy future [13].

Hydrogen energy is a promising alternative to traditional fossil fuels, offering a clean and sustainable solution to address the challenges of climate change and environmental degradation [14–16]. Unlike conventional fossil fuels, hydrogen combustion produces only water vapor as a byproduct, making it a zero-emission energy carrier when generated from renewable sources [17]. The versatility of hydrogen lies in its potential applications, ranging from fueling vehicles and power generation to industrial processes [18–20]. When produced through electrolysis using renewable energy sources like wind or solar power, green hydrogen becomes an eco-friendly option, minimizing the carbon footprint associated with energy production [21,22]. Additionally, hydrogen's energy density and storability make it a viable candidate for overcoming intermittency issues in renewable energy systems, acting as an energy carrier and storage medium [23–25]. As research and technological advancements progress, hydrogen's role in the global energy transition is expected to grow, contributing to a more sustainable and resilient energy landscape [26–28]. However, challenges such as production costs [29], infrastructure development [30], and efficient storage methods [31] must be addressed to unlock the potential of hydrogen as a clean alternative to traditional fossil fuels.

Hydrogen fuel cells represent a cutting-edge technology with the potential to revolutionize the way we generate power for various applications. At its fundamental principles, a hydrogen fuel cell converts chemical energy directly into electrical energy through an electrochemical reaction between hydrogen and oxygen, producing electricity, water, and heat as byproducts [32–34]. This process is remarkably efficient and emits only water vapor, making it an environmentally friendly alternative to traditional combustion-based power systems. Hydrogen fuel cells are particularly promising for transportation, as they can power electric vehicles and emit water as exhaust [35,36]. Beyond transportation, fuel cells find applications in stationary power generation for residential [37], commercial [38], and industrial use [39]. The scalability and versatility of hydrogen fuel cells position them as a critical player in the transition to a sustainable energy future, offering a clean and efficient energy solution that can help reduce greenhouse gas emissions and dependence on fossil fuels [40].

Fuel cell technology encompasses various types, each with distinct features and applications. Alkaline fuel cells (AFCs) operate with an alkaline electrolyte, typically potassium hydroxide, offering efficiency and reliability [41]. AFCs have found applications in space missions and niche industrial uses [42]. On the other hand, solid oxide fuel cells (SOFCs) function at high temperatures and are adaptable to various fuels [28], including hydrogen and natural gas [43]. Known for their high efficiency and low emissions, SOFCs are employed in stationary power generation [44] and combined heat and power systems [45]. Proton-exchange membrane (PEM) fuel cells utilize a polymer electrolyte membrane, making them suitable for applications requiring quick start-up and variable power output, such as automotive and portable electronic devices [46,47]. Each type of fuel cell technology addresses specific needs and contributes to the evolving landscape of clean and sustainable energy solutions, driving innovation in diverse sectors.

One notable advantage is their ability to operate at lower temperatures, facilitating quick start-up times, which is crucial for meeting the unpredictable energy demands of rural settings [48]. The fuel cells are compact, lightweight, and have high power density, making them ideal for various portable electronic devices, camping equipment, and backup power systems. Operating at lower temperatures further enhances their suitability for portable applications, ensuring a quick response to changing energy needs. PEM fuel cells utilize hydrogen as a fuel source, which can be supplied from easily transportable cartridges or tanks. As a result, these fuel cells reduce reliance on traditional batteries and fossil fuel-powered generators, offering a cleaner and more sustainable power source for individuals on the move or in remote locations. Their scalability and efficiency make PEM fuel cells a promising option for addressing the growing demand for portable power in a world increasingly reliant on mobile and off-grid devices.

PEMFCs offer high electrical efficiency and zero emissions, with potential socio-economic benefits such as job creation and environmental improvement. However, in order to contextualize the use of PEMFC technology, it is essential to consider region-specific factors like the energy mix, renewable resources, and infrastructure challenges [1]. Reference [2] provides an example of PEMFC technology applied in an eco-neighborhood, demonstrating its feasibility when integrated with local housing and energy systems. This kind of localized analysis is crucial for understanding how PEMFC can contribute to regional energy policies and economic development. Additionally, comparing the environmental performance of PEMFC with other fuel cell technologies for the Brazilian context could further justify its selection by highlighting PEMFC's potential to reduce environmental impacts when combined with renewable energy resources [3].

Proton-exchange membrane (PEM) fuel cells hold significant promise for addressing the energy needs of rural areas. Their versatility and efficiency make them well-suited for decentralized power generation in off-grid or remote locations. PEM fuel cells can provide a reliable source of electricity for rural communities, offering a clean energy solution with minimal environmental impact. The modular nature of PEM technology allows for scalability, enabling systems to be tailored to the specific energy requirements of different communities. Moreover, locally sourced hydrogen can power PEM fuel cells, promoting energy independence and economic development in rural areas. As research and development efforts continue to optimize PEM fuel cell performance and reduce costs, the technology holds great potential to contribute to sustainable electrification and improve the quality of life in rural communities worldwide.

This study aims to assess the design process using several calculations of a 1 kW portable power generator employing proton-exchange membrane (PEM) fuel cell technology. The focus is on evaluating the potential of such a system to address the energy access challenges in rural areas of Indonesia. The portable nature of the power generator makes it a practical and versatile solution for off-grid and remote locations where reliable access to electricity is often limited. By leveraging PEM fuel cell technology, known for its efficiency, quick start-up times, and minimal environmental impact, this study aims to explore the feasibility of deploying such systems in Indonesia's diverse and challenging rural landscapes.

2. Materials and Methods

This article provides several design processes to determine the proposed product in the proton-exchange membrane fuel cell (PEMFC) stack design used as the portable power generator. Basic conceptual design and several performance calculations are generated in this article, which led to the further process of detailed engineering design. This research develops a portable 1 kW PEMFC system specifically designed to address the challenges of energy access in rural Indonesia. One of the primary contributions of this project is its focus on delivering a sustainable, off-grid power solution for remote and underserved areas where grid extension is often difficult or economically unfeasible. This portable fuel cell system supports PLN's (Indonesia's state electricity company) efforts to electrify rural regions by providing a reliable and sustainable energy source that can be easily deployed in areas lacking conventional infrastructure. Another significant contribution is the system's compact and portable design. With a power capacity of 1 kW, the design balances portability and the ability to meet household energy needs (lighting) in rural settings. The lightweight and easy-to-carry nature of the system makes it practical for deployment in remote locations where transportation and installation of traditional power solutions can be challenging. This portability aspect is critical for reaching the most isolated communities in Indonesia, where infrastructure limitations make conventional electrification difficult.

Moreover, the system's simplicity and reliability were central to the design. The decision to use a low-complexity, air-cooled thermal management system ensures that the PEMFC operates efficiently without resource-intensive cooling methods. By relying on forced air cooling, the system avoids the additional complexity and maintenance requirements of liquid cooling systems, which are unsuitable for rural environments with limited technical support. This design choice makes the system reliable and easy to maintain, ensuring long-term functionality in remote areas. The design provides a scalable solution for rural electrification. While the current focus is on a 1 kW PEMFC system, the design principles can be adapted to create higher-power versions for larger applications. This scalability makes the system versatile, enabling it to be adjusted to different power needs in various rural electrification contexts across Indonesia.

The first stage is the design concept, where the overarching principles and objectives of the fuel cell system are defined. As discussed in the introduction, the judgement of the PEM fuel cell as the chosen fuel cell technology will be discussed comprehensively in this study stage. This stage also involves several basic judgments, such as the basic architecture and outlining of the intended applications and performance requirements. Conceptual sketches and initial design parameters are developed during this phase to guide the subsequent stages. Our calculations are proposed in this stage using the governing equations specified in this section, including the fundamental electrochemical reactions, ideal voltage, and performance losses such as activation, ohmic, and concentration losses. These calculations lay the groundwork for assessing the efficiency and performance of the PEMFC system.

The second stage is data specification, which comprehensively analyses the specific data and parameters needed for the design. This includes the required power output, efficiency targets, operating conditions, material selection, and environmental or regulatory constraints. The collection and specification of accurate data are crucial for informing the design process and ensuring that the fuel cell system meets performance expectations and aligns with technical and regulatory standards. At this stage, key parameters such as fuel flow rates, power output, and reactant flow rates are carefully analyzed to ensure the system's operational effectiveness.

The third and final stage is engineering design, where the detailed design of the fuel cell system is carried out based on the established concept and specified data. This involves creating detailed engineering drawings and assessing the system's mechanical and electrical integration. However, this engineering design stage is not provided in this article, which is later discussed in another article.

By systematically progressing through these three stages—design concept, data specification, and engineering design—engineers can ensure a thorough and informed approach to developing a fuel cell system that meets the desired performance goals and applications. This structured process is essential for successful and efficient fuel cell design, ensuring the fuel cell's real-world applicability and long-term sustainability, particularly in portable power applications.

2.1. Electrochemical Reactions

In a PEM fuel cell, the critical electrochemical reactions occur at the anode and cathode, separated by a polymer electrolyte membrane. The most common fuel for PEM fuel cells is hydrogen gas (H_2). The hydrogen molecules split through electrolysis into protons (H^+) and electrons (e^-). This reaction is represented in Equation (1). The protons (H^+) move through the electrolyte to the cathode, creating an electrical current. Simultaneously, on the cathode side, oxygen (O_2) is introduced. Hydrogen is oxidized, meaning it loses electrons (e^-) that flow through an external circuit, creating an electrical current. The protons (H^+) pass through the polymer electrolyte membrane towards the cathode.

At the cathode, the protons (H^+) arriving from the anode combine with electrons (e^-) travelling through the external circuit. This reaction results in water formation (H_2O), as shown in Equation (2). This reaction represents the reduction process, where oxygen is reduced by gaining electrons. The overall reaction makes PEM fuel cells an environmentally friendly and sustainable technology for electricity generation, particularly in applications such as portable power generators.

$$2H_2 \rightarrow 4H^+ + 4e^- \quad (1)$$

$$O_2 + 4H^+ + 4e^- \rightarrow 2H_2O \quad (2)$$

2.2. Ideal Fuel Cell Voltage

A fuel cell's open circuit voltage (OCV) is a crucial parameter that reflects the thermodynamic efficiency of the cell under no load conditions, meaning when no current is drawn from the cell. The OCV is determined by the Gibbs free energy (ΔG) change in the electrochemical reactions occurring within the fuel cell. Gibbs free energy, denoted as G, is a thermodynamic potential that measures the maximum reversible work a system can perform at constant temperature and pressure. It is named after American physicist Josiah Willard Gibbs, who significantly contributed to thermodynamics. The Gibbs free energy is vital in chemical and physical processes as it helps determine the spontaneity of a reaction and provides insights into the system's capacity to perform work. The ideal voltage of a PEM fuel cell is derived from the Gibbs free energy change (ΔG), which represents the maximum amount of reversible work that the fuel cell can perform. Gibbs free energy (ΔG) is affected by the operating temperature (T), enthalpy difference (ΔH_T), and entropy difference (ΔS_T), as shown by Equation (3).

$$\Delta G = \Delta H_T - T\,\Delta S_T \tag{3}$$

The voltage (E) generated by Gibbs free energy (ΔG) in a fuel cell is significantly influenced by the operating temperature of the cell. This is because enthalpy (H) and entropy (S) are temperature-dependent thermodynamic properties. In standard conditions (298.15 K temperature and 1 atm pressure), the enthalpy and entropy of the system can be determined. However, when operating at different temperatures, the change in enthalpy (ΔH) and entropy (ΔS) can be calculated using the combined polynomial heat capacity function. The integrated polynomial function of heat capacity estimates the enthalpy and entropy differences at various temperatures, as shown in Equations (4) and (5). This function incorporates the temperature dependence of heat capacity and allows for estimating enthalpy and entropy changes over a range of temperatures. Understanding how these thermodynamic parameters change with temperature is essential for optimizing the performance and efficiency of fuel cells at different operational conditions. This knowledge aids in designing fuel cell systems that can function effectively across a spectrum of temperatures, contributing to enhanced energy conversion and overall efficiency.

$$\Delta H_T = h_{298.15} + \int_{298.15}^{T} C_p dT \tag{4}$$

$$\Delta S_T = s_{298.15} + \int_{298.15}^{T} \frac{C_p}{T} dT \tag{5}$$

The voltage (E) generated by a fuel cell as a function of temperature is calculated by combining the equations that describe the relationship between enthalpy change (ΔH), entropy change (ΔS), temperature (T), the number of electrons transferred in the cell reaction (n), and Faraday's constant (F = 96,485 C/mol) as shown in Equation (6).

$$E = -\frac{\Delta H_T - T\Delta S_T}{nF} \tag{6}$$

To incorporate the effect of operating pressure on the ideal voltage of a fuel cell, the governing equation is modified by including pressure-dependent terms. The second term on the right side of the equation accounts for the impact of operating pressure on the cell potential. The logarithmic term adjusts as pressure increases to reflect this change, providing a more accurate representation of the cell's behavior under varying pressure conditions. By applying Equation (7), which accounts for the impact of temperature and pressure on the ideal voltage of a fuel cell, researchers and engineers can gain a comprehensive understanding of the system's performance across a range of operating conditions. This information is crucial for optimizing fuel cell designs and ensuring efficient energy conversion in real-world applications.

$$E = -\frac{\Delta H_T - T\Delta S_T}{nF} - \frac{RT}{2F}\ln\left(\frac{(P_{H_2O})}{(P_{H_2})(P_{O_2})^{\frac{1}{2}}}\right) \tag{7}$$

The intrinsic efficiency (ε_i) is a metric used to quantify the fundamental efficiency of a chemical reaction in a fuel cell. It precisely measures the ratio of the reversible free energy that can be converted into electrical energy (ΔG) to the internal energy of the reactants (ΔH). This ratio provides insights into the inherent efficiency of the chemical reaction itself, excluding losses or inefficiencies external to the reaction, such as those related to heat dissipation or system losses. Mathematically, the intrinsic efficiency can be expressed as shown in Equation (8).

$$\varepsilon_i = \frac{\Delta G}{\Delta H} \tag{8}$$

The numerator, ΔG, represents the reaction's reversible energy that can be converted into electrical energy. This is the maximum practical work that can be obtained from the reaction under ideal conditions. The denominator, ΔH, represents the reactants' internal energy, indicating the system's total heat content. A higher intrinsic efficiency shows a more thermodynamically favorable reaction, meaning a more significant fraction of the internal energy of the reactants can be converted into useful electrical energy. However, it is important to note that achieving high intrinsic efficiency does not guarantee high overall efficiency in a real-world fuel cell system, as additional factors such as activation losses, ohmic losses, and mass transport losses contribute to overall system efficiency. In summary, intrinsic efficiency is a valuable metric for understanding the inherent efficiency of the chemical reaction within a fuel cell, providing a fundamental perspective on the thermodynamic potential for converting chemical energy into electrical energy.

2.3. Voltage Kinetic Losses

In practical fuel cell operation, electricity production involves various losses associated with the kinetics of electron transfer and other factors. Three main types of losses occur in fuel cells, collectively affecting the overall performance and efficiency. Activation losses (ΔV_{act}) occur at the electrochemical interfaces of the fuel cell, specifically at the anode and cathode. These losses are related to the activation energy required to initiate the electrochemical reactions. The activation energy barrier must be overcome for the reaction to proceed, and the extent of these losses is influenced by factors such as the nature of the catalyst, surface properties, and the kinetics of the chemical reactions. Activation losses increase with overpotential, the difference between the actual and thermodynamically ideal potential required for the reaction. Internal resistance losses (ΔV_{ohm}), also known as ohmic losses or resistance losses, are related to the resistance encountered by the flow of ions through the electrolyte and the electrical resistance in the electrodes and interconnects. These losses are influenced by the conductivity of the materials used in the fuel cell. Internal resistance losses contribute to voltage drops within the cell, reducing efficiency and increasing heat generation. Concentration losses (ΔV_{conc}), also referred to as mass transport losses, arise from limitations in transporting reactants (such as hydrogen and oxygen) to the electrodes and removing reaction products. Concentration polarization occurs when a gradient in reactant or product concentrations near the electrode surfaces reduces reaction rates and cell performance. Effective mass transport is crucial for maintaining uniform reactant/product concentrations throughout the cell. The total losses in a fuel cell, which impact the overall efficiency and voltage output, can be expressed as Equation (9). Several losses are derived from the Butler–Volmer equation, which determines the reacting current from the voltage difference.

$$E_{sel} = E_0 - \Delta V_{act} - \Delta V_{ohm} - \Delta V_{conc} \tag{9}$$

The fuel cell reaction experiences a voltage drop due to the activation energy required by the cell, as shown in Equation (10). Activation loss results in a sharp initial drop in the circuit with increased cell current. This activation loss (ΔV_{act}) is affected by the exchange current density (j_0), current density (j) and charge transfer coefficient (α_c).

$$\Delta V_{act} = \frac{RT}{\alpha_c F}\log(j_0) - \frac{RT}{\alpha_c F}\log(j) \quad (10)$$

The internal ionization resistance in the fuel cell's MEA (membrane electrode assembly) is shown in Equation (11). Ionization resistance is typically represented by the area-specific resistance (Ri), which eliminates the dependence on the active area of the fuel cell. This internal resistance loss (ΔV_{ohm}) is affected by current density (j) and area-specific resistance (Ri).

$$\Delta V_{ohm} = j \times R_i \quad (11)$$

This loss is caused by the decreased ability of the fuel cell to maintain reactant concentration due to the high current drawn from the fuel cell, shown in Equation (12). This concentration losses (ΔV_{conc}) is affected by the current density upper limit (j_L) and current density (j).

$$\Delta V_{conc} = \frac{RT}{nF} ln\left(\frac{j_L}{j_L - j}\right) \quad (12)$$

2.4. Fuel Cell Power Produced

Once the voltage of a fuel cell has been calculated, the power that the fuel cell stack can produce can be determined. The power output is directly related to the voltage and the current flowing through the stack. Multiple individual cells are connected in series to form a fuel cell stack. When cells are connected in series, the voltage across each cell adds up, resulting in a higher stack voltage. The fuel cell stack's power (P) can be calculated using the Equation (13). The stack voltage is the total voltage across all the cells connected in series within the stack. Since the cells are stacked in series, the voltage of each cell adds up, resulting in a higher overall stack voltage. This is beneficial for achieving higher power output from the fuel cell stack. The number of cells in the stack directly influences the stack voltage. As more cells are added to the stack, the voltage increases proportionally. This serial connection of cells allows for the convenient scaling of voltage to meet the power requirements of specific applications. This series of cell connections is a crucial aspect of designing fuel cell stacks to meet the desired power output for various applications.

$$P_{fc} = \left(N_{stack} \times E_{sel}\right) \times i \quad (13)$$

The energy conversion efficiency in a fuel cell system is a critical parameter that reflects how effectively chemical energy stored in reactants is converted into electrical energy. Various losses within the fuel cell influence the efficiency of the electrochemical processes. The system's overall efficiency can be determined by combining the intrinsic efficiency with the losses, as shown in Equation (14).

$$\varepsilon_i \times \varepsilon_e = \frac{\Delta G}{\Delta H} \times \frac{E}{E_0} \quad (14)$$

2.5. Reactant Flow Rate

The flow rate of reactants during fuel cell operation is crucial to ensure the continuous supply of fuel and oxidants necessary for the electrochemical reactions. The reactant flow rate is typically determined by the current drawn by the electrical load connected to the fuel cell. The molar flow rate (N_x) measures the number of moles of reactant (fuel or oxidant) passing through the fuel cell per unit of time. It is directly related to the current (i) flowing through the electrical circuit. The relationship between the molar flow rate, current, and the stoichiometric coefficient of the reactant in the electrochemical reaction can be expressed using Faraday's law of electrolysis in Equation (15). Volumetric flow rate ($\dot{V}_x$) is a measure of the volume of gas that passes through a given cross-sectional area per unit of time, as shown in Equation (16). It is commonly expressed in units such as liters per minute (L/min) or cubic meters per second (m^3/s).

$$\dot{N}_x = \frac{i}{nF} \times N_{stack} \tag{15}$$

$$\dot{V}_x = \frac{M_x}{\rho_x} \times \dot{N}_x \tag{16}$$

3. Results and Discussion

The outcomes of this study will illuminate various critical aspects of the fuel cell system. A comprehensive analysis of fuel cell performance will be presented, encompassing key metrics such as power output, efficiency, and voltage-current characteristics. The investigation will delve into the intricacies of the required reactant flow rate, shedding light on the optimal rates of hydrogen and oxygen necessary for the fuel cell's efficient operation. Additionally, the study will provide insights into the design and performance of the bipolar plate flow field, elucidating the configuration of channels crucial for the uniform distribution of reactants and efficient removal of byproducts. Purging strategies will also be detailed, essential for maintaining reactant purity and preventing undesired substance buildup.

Furthermore, the examination will extend to evaluating fuel cell secondary components, including end plates and current collector plates, which play pivotal roles in supporting the fuel cell stack's structural integrity and electrical connectivity. The study will explore Balance of Plants (BOP) devices, elucidating their contribution to temperature control, humidity management, and overall system functionality. Through this comprehensive approach, the research aims to provide a nuanced understanding of the fuel cell system's performance, design intricacies, and the role of essential components and auxiliary devices. Both processes are created as preliminary design processes, to be further examined to achieve the engineering design process.

3.1. Fuel Cell Technology Comparison and Judgement

As discussed in the introduction, the outcome of the fuel cell technology will generate electricity in rural areas. The chosen fuel cell technology was the PEM fuel cell, which has several advantages. The key points in choosing this technology will be discussed comprehensively in this section.

A comparison of several fuel cell technologies should be conceived to judge the most suitable fuel cell technology for this purpose. Several studies comparing fuel cell technology have been summarized in Table 1.

Table 1. Comparison of fuel cell technologies [32].

Type of Fuel Cell	Alkaline Fuel Cell (AFC)	Phosphoric Acid Fuel Cell (PAFC)	Solid Oxide Fuel Cell (SOFC)
Operating Temperature (°C)	90—-100	150–200	600–1000
Output System (kW)	10–100	50–1000	<1–3000
Electrical Efficiency (%)	60	>40	35–43
Combines Heat Power Efficiency (%)	>80	>85	<90
Application	Military, Aerospace, Distributed power generator	Distributed power generator, co-generation	High-scale power generator, co-generation

Table 1. *Cont.*

Type of Fuel Cell	Alkaline Fuel Cell (AFC)	Phosphoric Acid Fuel Cell (PAFC)	Solid Oxide Fuel Cell (SOFC)
Advantages	High power density, fast start-up	High efficiency, stable electrolyte characteristic	High combined heat power efficiency
Disadvantages	Expensive catalyst, fuel impurity intolerance	Corrosive electrolyte, fuel impurity intolerance	Expensive, slow start-up, sulfur intolerance
Type of Fuel Cell	Molten carbonate fuel cell (MCFC)	Proton-exchange membrane fuel cell (PEMFC)	Direct methanol fuel cell (DMFC)
Operating Temperature (°C)	600–700	50–100	60–200
Output System (kW)	<1–1000	<1–250	<1–100
Electrical Efficiency (%)	45–47	53–58	40
Combines Heat Power Efficiency (%)	>80	70–90	80
Application	High-scale power generator, co-generation	Transportation, backup power plant, small-scale power generator	Mobile energy storage
Advantages	High efficiency, stable electrolyte characteristic	High power density, fast start-up, non-corrosive electrolyte	Impurities tolerance
Disadvantages	Expensive, slow start-up, sulfur intolerance	Expensive catalyst, impurity intolerance	Low efficiency, low power density

For this purpose, the fuel cell technology for the rural area should be as compact as possible. As additional information, the rural area has minimum human resources during operation. The complexity of fuel cell technology is one of the considerations during judgment. The PEM fuel cell has these advantages since it has the highest power density. In addition, the PEM fuel cell has the lowest operating temperature compared with several other fuel cell technologies. The higher operating temperature will increase the complexity of the heat management system.

Besides complexity, efficiency is also considered during specification judgment. Table 1 demonstrates that the PEM fuel cell has a competitive advantage in terms of efficiency. Efficiency refers to the amount of output work generated compared to the amount of input added to the system. In fuel cell systems, efficiency is mainly measured by dividing the system's output by the amount of fuel (hydrogen) supplied. Higher efficiency means less fuel (hydrogen) is required to produce the same output.

3.2. Fuel Cell Performance

A crucial step in predicting the actual performance of the fuel cell involves calculating the three primary voltage losses as a function of current density. These losses encompass activation, internal resistance, and concentration losses, as shown in Figure 1. Each of these components contributes to the overall voltage losses experienced by the fuel cell during its operation. Activation losses relate to the energy required to initiate electrochemical reactions at the cell's electrodes, internal resistance losses account for the electrical resistance within the cell components, and concentration losses reflect the impact of reactant concentration gradients on the electrochemical processes. The analysis comprehensively explains how different operating conditions influence the fuel cell's performance by quantifying these losses as a function of current density. This predictive approach is essential for optimizing the fuel cell's efficiency and tailoring its operational parameters to achieve desired energy conversion outcomes.

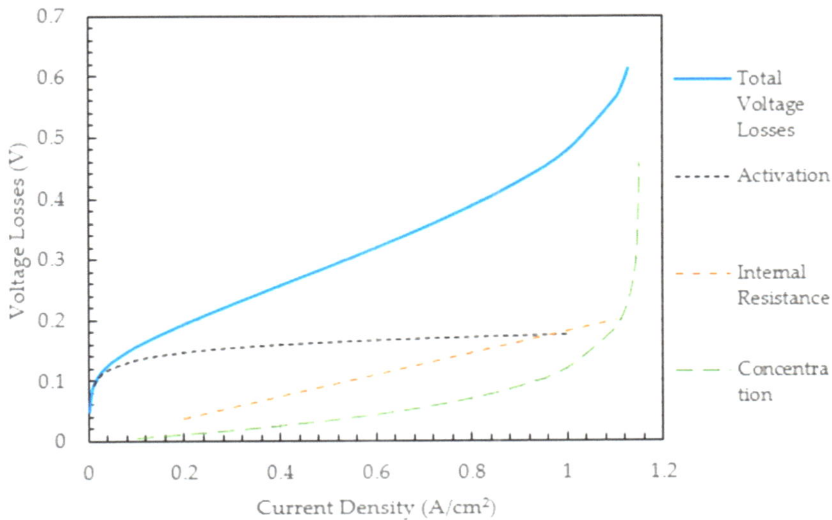

Figure 1. Total voltage loss of fuel cell.

After calculating the three main voltage losses as a function of current density, the obtained results are utilized to predict the performance of a single fuel cell. The prediction is typically visualized using a polarization curve representation, as shown in Figure 2. A fuel cell's performance is effectively characterized by this curve, which graphically illustrates the relationship between the voltage produced by the fuel cell and its corresponding current. The polarization curve comprehensively overviews the fuel cell's behavior under different operating conditions. It helps identify the optimal operating points where the fuel cell achieves maximum efficiency and power output. Analyzing the polarization curve is fundamental in evaluating and optimizing fuel cell performance, providing critical insights for designing and fine-tuning these energy conversion systems.

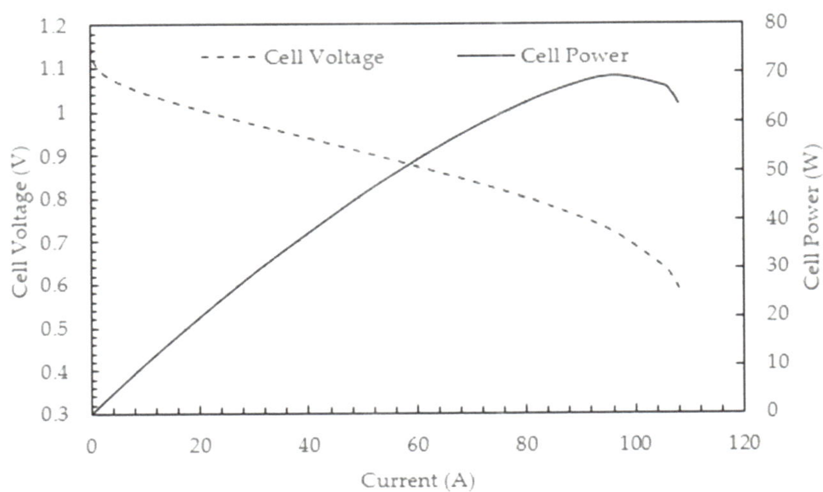

Figure 2. Single-cell polarization curve.

The polarization curve analysis reveals a crucial parameter for the performance of a single fuel cell: the peak current that can be drawn, determined to be 57.6 Amperes.

This specific value is strategically identified to avoid excessively high currents that may introduce unpredictability, considering the many factors that come into play. Operating at or below this peak current ensures a more predictable and stable performance, considering various influences such as activation losses, internal resistance, and concentration losses. As shown in Table 2, the peak characteristic obtained encapsulates the fuel cell's optimal operating point, highlighting the balance between maximizing current draw for power output and maintaining a controlled and manageable operational environment. This information is instrumental in designing and operating the fuel cell within reliable and efficient performance limits.

Table 2. Single-cell peak characteristic.

Power (W)	Current (A)	Voltage (V)
50.107	57.6	0.86992

Once the peak power output of a single cell is established, it becomes possible to determine the number of cells needed in a specific arrangement to achieve a target peak power output. In this case, the desired peak power is set at 1000 Watts. Given the known peak power of a single cell, a crucial parameter identified from the polarization curve analysis, the calculation reveals that 25 cells in one array would be sufficient to meet the 1000 Watts target.

However, it is recognized that the actual power available for external use may be influenced by additional components in the BOP that draw power from the fuel cell. To account for these power draws and ensure self-sustainment, the number of cells is prudently increased to 28, as shown in Table 3. This adjustment considers the broader system requirements and aims to provide a sufficient power margin to accommodate the needs of both the external load and the internal components of the BOP, ensuring the fuel cell system operates reliably and sustainably. The calculated performance of the fuel cell provides valuable insights into its efficiency and behavior, revealing an intrinsic efficiency of approximately 80.3%. This metric reflects the fundamental efficiency of the chemical reactions occurring within the fuel cell, considering losses inherent to the electrochemical processes.

Table 3. Stack peak characteristic.

Power (W)	Current (A)	Voltage (V)
1403.3	57.6	24.3578

Analyzing the fuel cell's polarization curve further elucidates its characteristics, as shown in Figure 3. Notably, as more current is drawn from the fuel cell, there is a corresponding decrease in the voltage it produces. This observation aligns with typical behavior seen in fuel cells, where higher current draw induces more significant losses in voltage due to factors like activation losses and internal resistance.

A notable phenomenon mentioned is the drastic voltage drop at the onset of the current draw, which is near the maximum current limit. This behavior is typical in fuel cells and is often attributed to activation losses dominating low currents and concentration losses becoming significant at high currents. Despite the continuous decline in voltage, the fuel cell's power output increases with increasing current draw, reaching a peak before starting to decline. This characteristic power peak results from the interplay between voltage and current, emphasizing the importance of understanding the polarization curve for optimizing the fuel cell's performance under varying operational conditions.

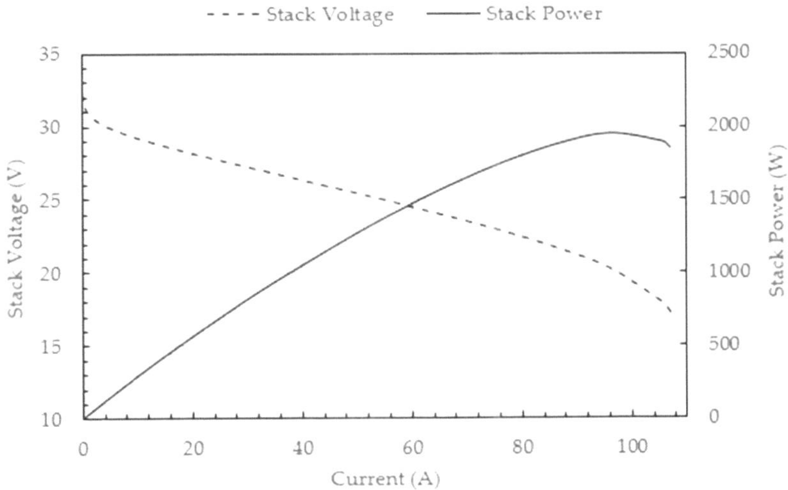

Figure 3. Fuel cell stack polarization curve.

3.3. *Required Reactant Flow Rate*

Determining the required reactant flow rate in a fuel cell is influenced by two key factors: the current drawn by the fuel cell and the number of cells in the arrangement. This relationship is mathematically expressed through a governing equation that considers these variables. The calculation results highlight a significant observation: the reactant flow rate exhibits a linear relationship with the current drawn by the fuel cell. As the current drawn by the fuel cell increases, there is a proportional increase in the required reactant flow rate, as shown in Figure 4. This linear relationship suggests that the reactant demand is directly tied to the electrical output of the fuel cell. Such insights are crucial for system design and optimization, enabling a precise control mechanism for adjusting reactant flow rates with variations in electrical demand. Understanding this linear relationship aids in efficiently managing reactant resources, contributing to the overall performance and stability of the fuel cell system.

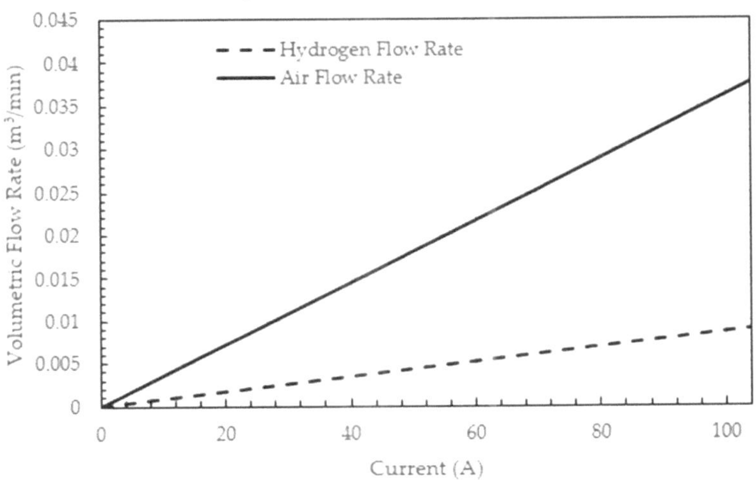

Figure 4. Reactant flow rate required.

3.4. Bipolar Plate Flow Field Design

The choice of graphite plates as the material for bipolar plates in this design is motivated by several favorable characteristics. Graphite offers notable corrosion resistance properties, ensuring the durability and longevity of the bipolar plates in the fuel cell system. Additionally, graphite is lightweight, providing an advantage over stainless steel. This lighter weight contributes to overall system efficiency and reduces the load on the fuel cell stack. Furthermore, graphite exhibits higher conductivity than carbon composites, enhancing electron transfer efficiency within the bipolar plates. This conductivity is a critical factor in optimizing the overall performance of the fuel cell by minimizing electrical resistance. Comparing specific masses, graphite plates have a particular mass of 2.63 kg/kWh, while stainless steel, a common alternative, has a higher mass of 3.85 kg/kWh. This lower specific mass of graphite contributes to a lighter overall system, which benefits applications where weight considerations are essential.

Despite these advantages, graphite plates are not recyclable. However, this design does not prioritize recycling, suggesting that other factors such as performance, weight, and corrosion resistance take precedence. This decision underscores the importance of considering a holistic set of criteria and trade-offs in material selection for bipolar plates based on the specific requirements and priorities of the fuel cell system design.

The design of the bipolar plate flow field, encompassing both the cathode and anode flow fields, is a critical aspect of optimizing the performance and efficiency of a fuel cell system. The bipolar plates serve as the structural backbone of the fuel cell stack while facilitating the distribution of reactants (oxygen and hydrogen) to the electrodes and expeditious removal of reaction products. The design of these flow fields influences mass transport, pressure distribution, and overall electrochemical performance. The cathode flow field is engineered to ensure a uniform distribution of oxygen across the electrode surface. This is achieved through a well-defined network of channels in the bipolar plate that facilitates the even flow of air. The design aims to prevent concentration gradients, enhance mass transport, and maintain a steady oxygen supply to support the electrochemical reactions at the cathode. The channels' geometry, size, and distribution pattern are crucial parameters that influence the performance of the cathode flow field.

Similarly, the anode flow field is meticulously designed to facilitate the uniform distribution of hydrogen to the anode electrode. The channels in the anode flow field are configured to optimize hydrogen distribution, prevent concentration gradients, and enhance mass transport to the anode. The anode flow field design considers factors such as pressure drop, flow uniformity, and avoidance of dead zones to maximize the utilization of the available catalyst and ensure efficient hydrogen oxidation.

3.4.1. Cathode Flow Field

The cathode flow supply method is a forced-convection open-cathode system. The air, which serves as the reactant at the cathode, is actively drawn into the system from the surrounding environment. The external mechanism, such as a fan or blower, enhances airflow over the cathode. The essential characteristic of this open-cathode system is that it does not restrict the air reactant source during the fuel cell's operation. In other words, the fuel cell freely draws air from its surroundings without conserving or limiting the supply. This contrasts with specific closed-cathode systems where the air supply may be more controlled and conserved. The design choice of a parallel cathode flow path type is facilitated by the nature of the forced-convection open-cathode setup, as shown in Figure 5. In a parallel flow path configuration, multiple channels or paths are available for the cathode reactant to flow. In this specific context, the parallel configuration is chosen because there is no necessity to conserve the air reactant. The system can allow for a more straightforward and efficient design, providing flexibility in managing the cathode flow paths to optimize mass transport and fuel cell performance. The optimal pressure drop is determined based on the flow field size in length (l), width (w), and height (h) as illustrated in Figure 6 and detailed in Table 4, to give the optimal pressure drop. The pressure drop formed by the field is calculated to be 5955 Pa. This approach offers simplicity and ease of operation advantages, as the fuel cell can draw in air continuously from the environment without the need for complex conservation measures during its normal functioning.

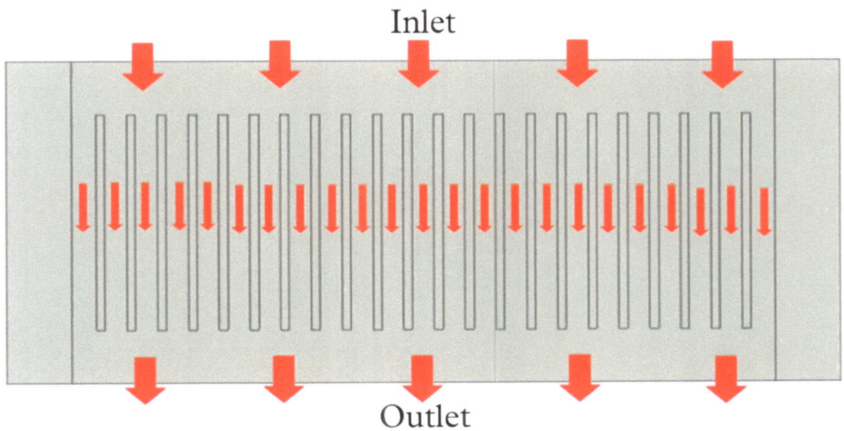

Figure 5. Parallel flow field design.

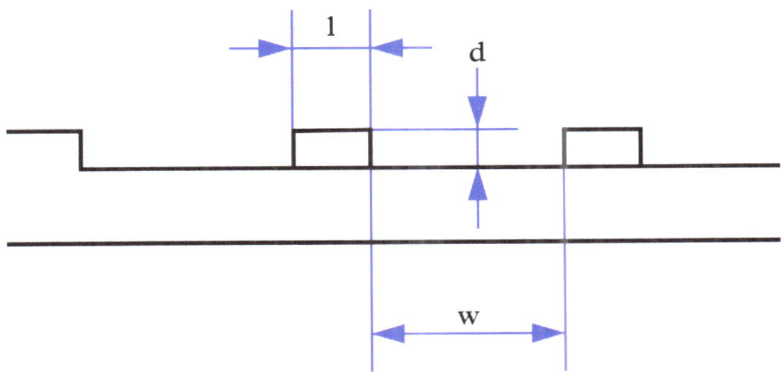

Figure 6. Field size parameters.

Table 4. Parallel flow field size.

w (mm)	d (mm)	w:l
1.3	1.5	1:0.7

3.4.2. Anode Flow Field

The chosen approach for achieving maximum hydrogen utilization in the fuel cell involves employing a dead-end anode flow configuration. In this configuration, the anode flow is closed except during specific activities such as draining or purging. This setup aims to enhance hydrogen utilization efficiency by allowing for a more thorough interaction of hydrogen with the anode electrode. However, specific challenges are associated with this dead-end configuration, particularly in managing pressure drops and preventing water flooding. To address these concerns, the pressure drop at the anode must be carefully controlled to ensure it accommodates the pressure required for the purging process. Purging removes inert gases and accumulated water from the anode flow path.

A serpentine flow design is chosen for the anode flow path to prevent water flooding, as shown in Figure 7. The serpentine configuration offers advantages in draining accumulated water compared to other flow path types like pin and parallel designs. The serpentine path allows for the effective removal of water through its winding channels. However, it is acknowledged that the serpentine flow design may increase the total pressure drop during the purging process. This trade-off is a consideration in the design process, where the benefits of improved water drainage must be weighed against the potential increase in pressure drop [49,50]. The selected design on the serpentine flow channel is defined in Table 5.

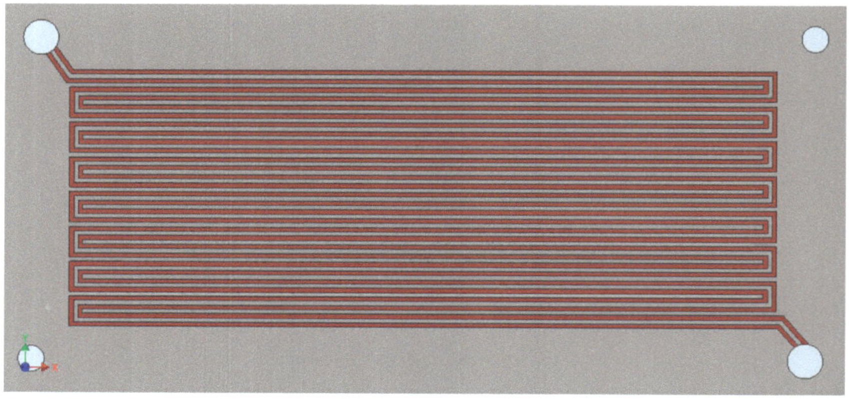

Figure 7. Serpentine flow field design.

Table 5. Serpentine anode flow field design.

w (mm)	d (mm)	Number of Channel
1	1	2

The design is based on several considerations, i.e., manufacturability, active area of reaction, and reaction byproduct management. For the manufacturability, the depth (d) of 1 mm is designed based on conventional manufacturing ability in Indonesia, which gives lower cost than the advanced process like etching.

3.5. Purging

Efficient purging in a fuel cell system involves carefully considering two key variables: purging frequency and duration. These parameters are crucial for maintaining optimal performance by addressing issues related to water accumulation in the anode flow. Purging frequency is defined as the time interval between successive purging events. Ideally, the purging frequency should be determined based on occurrences of hydrogen starvation resulting from water accumulation in the anode flow. The goal is to perform purging as infrequently as possible to minimize wasted hydrogen while ensuring it is completed before accumulated water negatively impacts the fuel cell's performance. This research uses a purging frequency that occurs approximately every 600 s after the previous purging event.

Moreover, purging duration refers to when the purge valve is open during the purging process. Longer purging times have the advantage of removing all accumulated water but may lead to hydrogen wastage and reduced overall efficiency. This research uses the optimal purge valve, which should be open for 28 milliseconds to perform drainage effectively and then closed again. This duration strikes a balance, ensuring efficient water removal without unnecessary hydrogen loss. Optimizing purging in a fuel cell system involves finding the right balance between purging frequency and duration. The goal is to minimize hydrogen wastage while effectively removing accumulated water to prevent adverse effects on fuel cell performance. The purging frequency and purge valve duration achieve this delicate balance and ensure the fuel cell operates efficiently over its operational life.

However, the purging frequency and duration of PEMFC have different characteristics in each fuel cell. The best practical value of purging frequency and duration can be obtained by practical research. Furthermore, several research studies that indicate the optimum purging rate and efficiency will be researched in our research schedule.

3.6. Fuel Cell Secondary Components

Secondary components in a fuel cell system ensure the system's structural integrity, electrical connectivity, and overall efficiency. The end plate and the current collector plate are two primary secondary components. The end plate is a structural component that encloses one side of the fuel cell stack. It provides mechanical support to the individual cells within the stack and helps maintain their alignment. Additionally, end plates contribute to sealing the fuel cell stack, preventing the leakage of reactants and reaction products. The design of end plates is critical for maintaining the uniform compression of the cell components, ensuring good electrical contact, and supporting the overall durability of the fuel cell stack. The current collector plate is an essential component responsible for collecting electrical current generated by individual cells in the stack. It is a conductive pathway, allowing electrons to flow from one cell's electrodes to the adjacent ones. The current collector plate helps distribute the electrical load evenly across the fuel cell stack, promoting uniform current collection and reducing resistive losses. Materials with high electrical conductivity, such as graphite or metal-coated graphite, are commonly used for current collector plates to enhance their efficiency in facilitating electron transport.

3.6.1. End Plate

The choice of aluminum as the material for the end plate in a fuel cell system is based on several advantageous characteristics. The selection of materials in fuel cell components is crucial to achieving optimal performance and durability. Aluminum possesses a relatively lightweight density compared to other metals. This characteristic is vital for applications where weight is a significant consideration, such as portable power generation. Using aluminum helps reduce the overall weight of the fuel cell system, contributing to improved efficiency and portability. Aluminum is widely available in the market, making it a convenient and accessible material for manufacturing. Its availability contributes to cost-effectiveness and ease of sourcing, facilitating the production of fuel cell systems on a larger scale. Specific mass is a critical parameter in fuel cell design, representing the mass of the material per unit of energy produced (kg/kWh). Compared to other metals commonly used in hydrogen handling, such as stainless steel, aluminum exhibits a lower specific mass. This characteristic is advantageous in achieving a favorable balance between material strength and weight, enhancing the overall efficiency of the fuel cell system. The type of aluminum chosen for this design is 5052, known for its high corrosion resistance. Corrosion resistance is essential in fuel cell applications, where exposure to moisture or other corrosive elements could compromise components' structural integrity and performance. Aluminum 5052 provides durability and longevity under varying environmental conditions. Aluminum offers ease of material engineering, particularly in thread manufacturing for fittings in the flow system. This characteristic simplifies manufacturing and creates precise and reliable connections within the fuel cell system.

3.6.2. Current Collector Plate

The design of the current collector plate in a fuel cell system is a critical consideration, considering the electrical resistance that occurs when an electric current flows through the plate. Electrical resistance in the current collector plate generates unwanted heat, and the design aims to minimize this resistance for efficient operation. The chosen material for the current collector plate is copper. Copper is selected for its relatively low electrical resistance, making it an excellent conductor of electricity. Low resistance is crucial in minimizing the heat generated during the flow of electric current through the plate. Additionally, copper is readily available in the field, contributing to the practicality and accessibility of the material for manufacturing processes. The primary goal of using copper is to exploit its low resistance, which allows for efficient current collection with minimal heat dissipation. A current collector plate with low resistance ensures that a significant portion of the electrical energy generated in the fuel cell is efficiently collected and utilized, contributing to the overall performance and efficiency of the fuel cell system. The design choice includes determining the thickness of the current collector plate to be 2 mm. This thickness is carefully selected to balance structural integrity and electrical performance. A relatively small cross-section helps manage the resistance and heat generation in the plate while maintaining the necessary mechanical strength. This design choice ensures that the current collector plate efficiently performs its role without unnecessary energy losses or excessive heat buildup.

3.7. System Design

Figure 8 illustrates the configuration and positioning of each component within the process. The diagram effectively delineates the system into two primary sections: the fuel line and the electrical line. The fuel line segment primarily comprises flow regulation and measurement devices essential for safely transporting and utilizing hydrogen. Key components within this section include the pressure regulator valve, the hand-operated valve, the hydrogen flow meter, and the solenoid purge valve. These devices collectively ensure the secure and efficient management of hydrogen flow throughout the system by regulating pressure, controlling the flow rate, and safely purging excess gas to maintain system balance and safety.

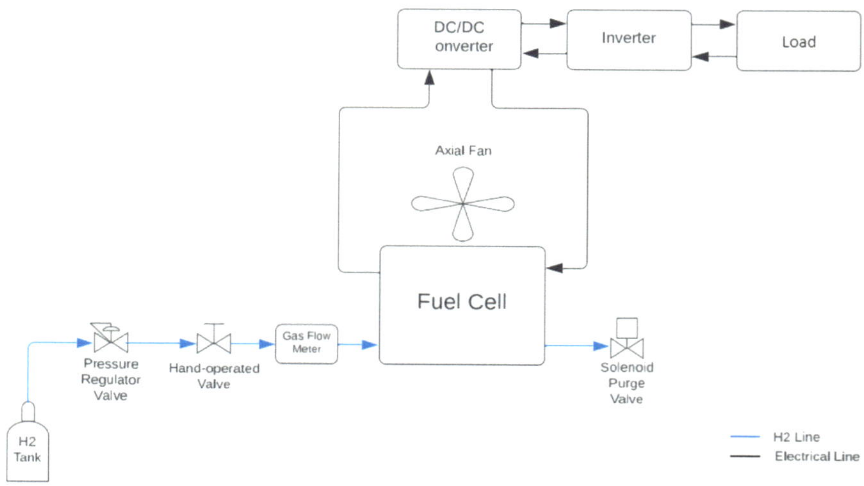

Figure 8. Schematic of PEMFC system.

On the other hand, the electrical line section consists of a DC/DC converter, an inverter, and a load. These components play a pivotal role in ensuring that the electricity generated by the fuel cell can be efficiently converted and utilized. The DC/DC converter adjusts the voltage output to the required level. At the same time, the inverter converts direct current (DC) produced by the fuel cell into alternating current (AC), making it suitable for household or industrial use. The load represents the electrical devices or systems powered by this AC output.

Additionally, an axial fan is integrated into the system to maintain optimal operating temperatures by providing necessary cooling. This ensures the fuel cell operates efficiently and prevents overheating during prolonged operation. Together, these components contribute to a cohesive system that supports the secure handling of hydrogen fuel and effectively utilizes the generated electrical power for various applications.

Figure 9 illustrates the overall assembly of the portable PEMFC system, showcasing how key components from the flow diagram are compactly integrated into a cohesive and efficient system. The design preserves the core elements, including the fuel line, flow regulation components, electrical converters, and safety mechanisms, while ensuring a compact form factor suitable for portable applications. The fuel line in this assembly includes several critical elements to manage the flow and pressure of hydrogen gas. Regulating valves precisely control the fuel flow rate and pressure, ensuring that the fuel cell receives the correct amount of hydrogen for efficient operation. These valves play a pivotal role in maintaining the balance between fuel consumption and performance, preventing potential over-pressurization or under-supply that could negatively impact the fuel cell's operation.

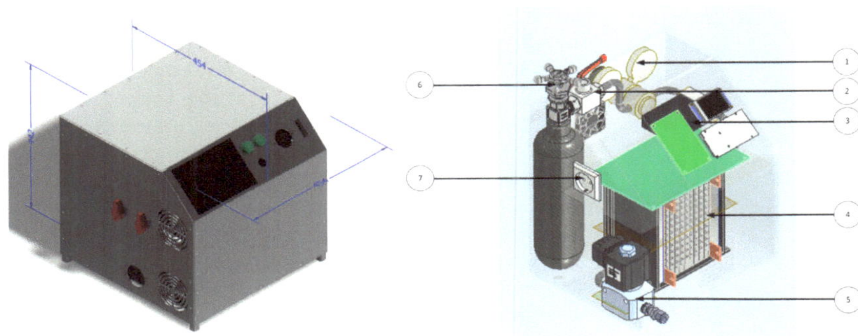

1. Pressure Meter
2. Hand-Operated Ball Valve
3. Flow Meter
4. Fuel Cell
5. Purging Valve
6. Inlet Valve
7. Load

Figure 9. Design of PEMFC power generator.

In addition to the regulating valves, the system features a flow meter, which continuously monitors the hydrogen feed. Real-time hydrogen flow monitoring is essential for ensuring that the fuel cell operates at peak efficiency and provides accurate data on fuel consumption, enabling the operator to manage fuel resources effectively in real-world conditions. The system is equipped with a purging valve on the anode side of the fuel cell. This component is responsible for the critical function of purging inert gases and excess water that may accumulate during the electrochemical reactions inside the fuel cell. Inert gases like nitrogen and water can reduce the fuel cell's efficiency if allowed to build up, and the purging valve helps maintain a clean, reactive environment in the anode. Proper purging ensures that the hydrogen reacts efficiently within the cell, improving the fuel cell's performance and longevity.

The cooling system, consisting of axial fans, is another critical component of this assembly. These fans ensure the fuel cell stack operates within the optimal temperature range by preventing overheating. Adequate cooling is vital for maintaining the chemical stability of the membrane and other internal components, directly impacting the system's durability and operational efficiency. In terms of electrical management, the system incorporates a DC/DC converter and inverter, essential for conditioning the power output from the fuel cell. The DC/DC converter regulates the voltage level from the fuel cell to match the load requirements. At the same time, the inverter converts the direct current (DC) generated by the fuel cell into alternating current (AC), which is necessary for powering standard electrical appliances and devices.

Finally, the assembly also integrates several safety features, including pressure regulators, sensors, and feedback systems that monitor the overall status of the fuel cell.

These components work together to ensure the fuel cell operates within safe parameters, automatically adjusting flow rates or initiating purging when necessary. In summary, the assembly in Figure 9 brings together a well-organized arrangement of critical components, including fuel regulation devices, cooling systems, power conditioning units, and safety mechanisms. Each element works in unison to create a portable, efficient, and safe fuel cell system designed to meet the energy needs of remote or off-grid environments. This comprehensive integration ensures the system can deliver reliable power while maintaining high operational standards, durability, and ease of use in real-world applications.

3.8. Water Management System

The water management system is a critical design element in PEMFCs. Proper water management is essential for the efficient operation of the PEMFC, as it ensures the membrane's hydration level is optimal for proton conduction, directly impacting overall performance. The hydration of the PEMFC membrane is closely related to the surrounding humidity, which plays a role in the membrane's ability to conduct protons effectively. Insufficient hydration can lead to higher resistances within the membrane, reducing efficiency, while excessive water can cause flooding, leading to performance degradation.

One commonly used method for membrane hydration in PEMFCs is the self-humidification technique. In this approach, the membrane utilizes the water produced as a byproduct of the electrochemical reaction to keep itself hydrated. This method works well under high-power loads, where the water production is sufficient to maintain adequate hydration levels. However, this type of self-humidification becomes less effective under low-power conditions, where the water production rate is reduced. As a result, at lower current densities, the membrane may not receive enough water to remain hydrated solely through self-humidification. External water injection into the fuel supply line becomes necessary to ensure consistent membrane hydration. This method, while effective, introduces additional complexities and will require further investigation in future research, as several design considerations must be addressed to optimize the water injection system for varying operating conditions.

Flooding is another significant issue associated with the water management system in PEMFCs. Under high-load conditions, the water production rate can exceed the membrane's hydration needs, accumulating excess water. This excess water can block the flow of reactants, such as hydrogen and oxygen, which are essential for the fuel cell's operation. When the reactants are obstructed, the overall power output of the PEMFC drops, resulting in a lower power density. This phenomenon, known as flooding, poses a considerable challenge to the consistent performance of the fuel cell. The flooding phenomena are also related to the bipolar design, explained in Sections 3.4.1 and 3.4.2. The technique to address flooding is the anode purging method, where excess water is expelled from the system through the anode side. Another approach is liquid separation, which involves physically removing the excess water to prevent blockages. The anode purging method is part of this model's simple design and judgment. However, the time and period of anode purging are critical aspects of the design model.

3.9. Thermal Management System

The thermal management methods primarily used in PEMFCs are air-cooled and liquid-cooled models. The air-cooled model is often utilized in lower-power PEMFC systems. In this configuration, the air supply serves a dual purpose: it provides the necessary oxygen for the electrochemical reaction and simultaneously aids in cooling the fuel cell. The design leverages the natural air flow to dissipate heat generated during operation, making it a straightforward and cost-effective solution for systems with lower power demands. This approach is particularly advantageous in applications with a preference for simplicity and minimal additional infrastructure.

On the other hand, the liquid-cooled model is employed in higher-power PEMFC systems, where more significant thermal management is required due to the increased heat generation. This system uses a separate cooling circuit, with a coolant—typically a water-glycol mixture—circulated through channels or passages within the fuel cell stack. This method allows for more efficient and precise temperature control by directly transferring heat away from critical components. The liquid-cooled model is essential for high-power applications, such as larger stationary power systems or advanced automotive applications, where maintaining optimal operating temperatures is vital for performance and reliability. Both thermal management methods are designed to ensure that the PEMFC operates within its optimal temperature range, enhancing efficiency and extending the lifespan of the fuel cell components. The choice between air-cooled and liquid-cooled systems depends on the specific power requirements and operational conditions of the PEMFC application.

In the current design model, the thermal management system employs forced airflow, the exact mechanism for supplying air to the cathode side of the PEMFC. This design choice was carefully made based on several factors, primarily focusing on the power output of the PEMFC, which is approximately 1000 watts. Given this relatively low power output, it was determined that a more complex cooling system, such as a liquid-cooled thermal management system, would be unnecessary and overcomplicate the design.

The forced air-cooling method is a more suitable solution for this power range, as it can efficiently dissipate the heat generated during the fuel cell's operation without adding significant weight, cost, or complexity to the system. Forced air cooling systems are more straightforward to design and implement, and they can be easily integrated into the existing air supply infrastructure used for the cathode, minimizing the need for additional components. This dual-purpose approach—using the same airflow for cooling and oxygen supply—streamlines the overall design, making it more compact and efficient.

In contrast, a liquid-cooled thermal management system, though highly effective for high-power fuel cells or systems with greater heat loads, would be excessive for a PEMFC producing only 1000 watts. Liquid cooling systems involve pumps, heat exchangers, and additional plumbing, increasing the system's weight, cost, and complexity without providing proportional benefits. Moreover, maintaining and servicing a liquid cooling system requires more effort and resources, which would not be justified for a fuel cell of this power capacity. Therefore, the design team concluded that forced air cooling offers a more balanced solution, providing adequate thermal management while keeping the system lightweight and cost-effective.

Additionally, this decision aligns with optimizing the PEMFC system for smaller-scale applications where simplicity, reliability, and ease of operation are prioritized. By avoiding using an overdesigned liquid cooling system, the overall efficiency of the fuel cell can be maximized without unnecessary trade-offs. Future design iterations could consider alternative cooling methods if the power output increases, but for the current model, forced air cooling is deemed the most practical and efficient solution.

3.10. Balance of Plants (BOP) Devices

In a fuel cell system, the BOP encompasses various auxiliary components and devices essential for the overall operation and control of the system. Two crucial BOP devices are converters and inverters, along with self-sustainable components, which play distinct roles in managing electrical power within the system. A converter in a fuel cell system is responsible for converting the direct current (DC) generated by the fuel cell stack into alternating current (AC). This conversion is crucial for compatibility with AC-based electrical systems, as many applications and power grids utilize AC power. The converter ensures that the electrical output from the fuel cell stack is suitable for use in various devices and applications that require AC power. Conversely, an inverter performs the opposite function. It converts AC power from the grid or an external source into DC power. This functionality is necessary when external power sources, such as the electrical grid, need to supplement or support the fuel cell system. The inverter ensures that the external AC power is converted into DC power compatible with the fuel cell stack.

The power consumption of BOP devices is an essential consideration in the overall efficiency of the fuel cell system. These devices consume electrical power during the conversion process, and minimizing power losses is crucial for optimizing the system's efficiency. Efficient BOP devices contribute to maximizing the net electrical output of the fuel cell system. The interplay between components of these devices is carefully managed to ensure seamless integration with the electrical grid, efficient utilization of electrical power, and compatibility with the requirements of various applications. The design and selection of BOP devices are crucial for achieving a well-balanced and high-performance fuel cell system. Engineers aim to optimize these components to minimize energy losses, enhance efficiency, and provide a reliable and adequate power generation solution.

3.10.1. Converter and Inverter

In a fuel cell system, both DC/DC converters and inverters play crucial roles in converting the fuel cell's output into usable electricity for various applications. The DC/DC converter is employed to stabilize the electricity output from the fuel cell. Fuel cells can exhibit fluctuating outputs, and the DC/DC converter is essential for ensuring a consistent and stable electrical output. Dynamic resistance control is a crucial feature of the DC/DC converter, enabling it to adapt to changes in the fuel cell's output conditions. This control mechanism helps optimize the electrical output to meet the requirements of connected devices or systems. The inverter, in contrast, is responsible for converting the direct current (DC) output from the fuel cell or the DC/DC converter into alternating current (AC). AC is commonly used in household appliances and electrical grids. By converting the fuel cell's DC output to AC, the inverter facilitates plug-and-play usage for various household devices, making it compatible with existing electrical systems. The DC/DC converter and the inverter introduce some efficiency loss during the conversion process, as shown in Figure 10.

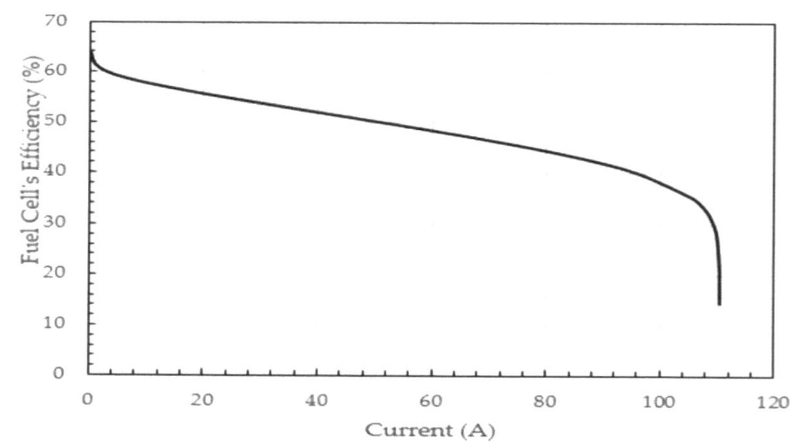

Figure 10. Fuel cell theoretical efficiency according to current load.

Efficiency measures how effectively the devices perform their functions without unnecessary energy losses. The efficiency of these devices impacts the overall electrical output of the fuel cell system. It is important to note that the fuel cell's output will experience a further decrease due to the efficiency losses in the converter and inverter. This decrease is a result of the energy lost during the conversion process. Engineers aim to design and select efficient converters and inverters to minimize these losses and maximize the net electrical output available for consumption.

The overall efficiency of a fuel cell system is determined by combining the efficiencies of the DC/DC converter and the inverter with the intrinsic efficiency of the fuel cell itself.

The efficiency of the DC/DC converter and the inverter refers to the effectiveness of these devices in converting and managing electrical power with minimal losses. These efficiencies are typically expressed as percentages and represent the ratio of helpful output power to the input power. For example, if the DC/DC converter has an efficiency of 90%, 90% of the input power is converted into useful output power, and 10% is lost as heat.

The intrinsic efficiency of the fuel cell reflects its fundamental efficiency in converting chemical energy into electrical energy. It is based on the electrochemical processes occurring within the fuel cell and is a crucial determinant of the overall system efficiency. The overall efficiency of the fuel cell system is calculated by multiplying the efficiencies of the DC/DC converter and the inverter with the intrinsic efficiency of the fuel cell. The efficiency of the fuel cell system can vary depending on operating conditions, as shown in Figures 11 and 12. During peak power demand, when the fuel cell operates at its maximum output, the overall efficiency is calculated to be around 48.664%. On the other hand, during the open circuit voltage (OCV) state, where no electrical power is drawn from the fuel cell, the efficiency is higher, measured at around 66%. The OCV state reflects the fuel cell's ability to generate voltage without delivering power to external loads.

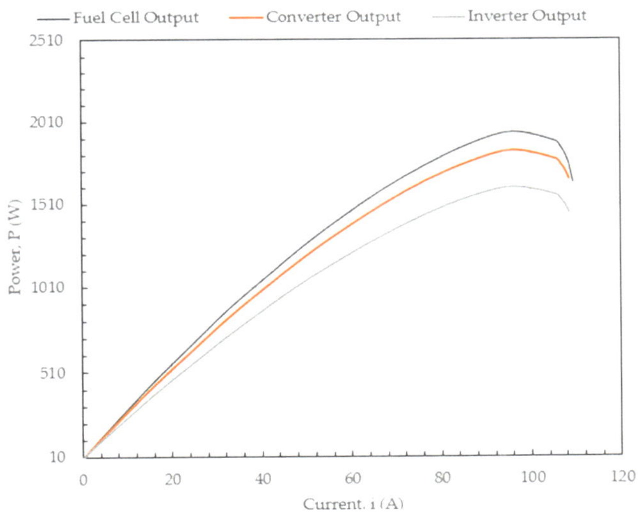

Figure 11. Fuel cell power output with converter and inverter efficiency.

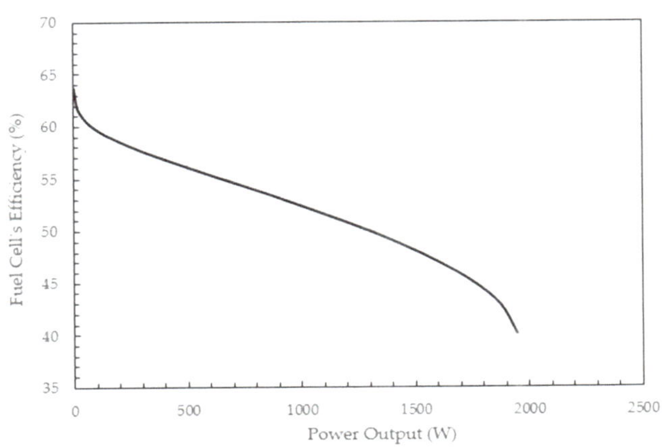

Figure 12. Fuel cell efficiency is determined according to the power drawn.

3.10.2. Power Consumption

Incorporating various BOP components in a fuel cell system introduces a degree of self-sustainability by redirecting a portion of the generated power to fulfil the energy needs of these auxiliary components. This self-sustaining approach ensures the proper functioning of critical BOP devices. The BOP components requiring power from the fuel cell include cathode fans, solenoid valves, hydrogen sensors, display, pressure sensor, current sensor, and voltage sensor. The cathode fans are crucial for maintaining the appropriate airflow on the cathode side of the fuel cell. These fans contribute to cooling and ensuring optimal operating conditions, enhancing the overall efficiency and performance of the fuel cell. The solenoid valve regulates the flow of reactants into the fuel cell. It controls the hydrogen or air supply, ensuring proper functioning and safety in the fuel cell system. The hydrogen sensor is a safety component that monitors the concentration of hydrogen in the system. It plays a vital role in detecting and alerting potential leaks or unsafe conditions, contributing to the overall safety of the fuel cell device. The display is a user interface providing real-time information about the fuel cell system's status, performance, and relevant parameters. It enhances the user experience and facilitates monitoring and control. The pressure sensor measures the pressure levels within the fuel cell system. It helps maintain optimal pressure conditions, ensuring efficient operation and safety. The current sensor monitors the electrical current flowing within the system. It plays a role in assessing and optimizing the electrical performance of the fuel cell. The voltage sensor measures the electrical potential difference within the fuel cell. It aids in monitoring and regulating the electrical output to meet system requirements.

By utilizing a portion of the generated power to meet the energy demands of these BOP components, the fuel cell system achieves a degree of autonomy and self-sustainability. This design strategy contributes to the reliability and continuous operation of the fuel cell, ensuring that critical components receive the necessary power for proper functioning. The specific power requirements and specifications of each BOP component are tailored to align with the fuel cell system's overall design and operational needs, as shown in Table 6.

Table 6. BOP power consumption.

Devices	Power (W)
Axial Fan (2 Unit)	3.3
Hydrogen Valve Sensor	2.5
Solenoid Valve	6.48
Display	5
Pressure Sensor	0.32
Current Sensor	0.3
Voltage Sensor	0.3
Total	18.2

4. Conclusions

This study addresses the critical issue of energy access in rural Indonesia by comprehensively assessing a 1 kW portable power generator based on PEMFC. The design, rooted in a meticulous series of calculations, has successfully yielded specifications for the fuel cell system. The results indicate that the fuel cell, configured with 28 cells, achieves a peak output of 1403 W and 24.3578 VDC, translating into usable electrical power of 1133.32 W at 220 VAC and 50 Hz. The fuel cell's efficiency at peak power is measured at 48.66%. Additionally, the study revealed non-linear relationships between the power generated and current load, emphasizing the influence of current load magnitude on voltage and power values. Notably, the fuel cell's efficiency varies with changes in the current load, reaching 66% at the open-circuit voltage (OCV) state and recording 48.66% at peak power. This holistic assessment provides valuable insights into the design, performance characteristics, and efficiency dynamics of the PEMFC-based portable power generator, contributing to the advancement of sustainable and accessible energy solutions in rural contexts.

One exploration avenue involves optimizing the PEMFC system's design parameters to enhance efficiency and power output further. Investigating advanced materials for fuel cell components, such as catalysts and membranes, could improve performance and durability. Additionally, integrating energy storage systems and smart grid technologies would better manage fluctuating power demands in off-grid environments. The current research focuses on the theoretical design and assessment of the portable PEMFC system, with performance metrics such as peak power, efficiency, voltage, and current being evaluated through direct analytical calculations. However, experimental validation is planned for the next phase of this research, where we will develop a physical prototype of the portable PEMFC and conduct tests under controlled conditions to validate the theoretical findings. Furthermore, future research could focus on upscaling to 5 kW, field trials, and real-world implementations of the designed portable power generator in diverse rural settings, as well as assessing its adaptability, reliability, and socio-economic impacts. This holistic approach would provide valuable insights for refining the technology and tailoring it to the specific needs of rural communities.

Author Contributions: Conceptualization, H.B.T., J.S. and R.W.; methodology, J.S.; software, R.W.; validation, H.B.T. and J.S.; formal analysis, J.S.; investigation, R.W.; resources, R.W.; data curation, R.W.; writing—original draft preparation, R.W. and S.P.; writing—review and editing, H.B.T.; visualization, R.W.; supervision, J.S.; project administration, H.B.T.; funding acquisition, H.B.T. All authors have read and agreed to the published version of the manuscript.

Funding: This research received no external funding.

Data Availability Statement: Not applicable.

Acknowledgments: The completion of this research project was possible through the invaluable collaboration with UGM and generous financial support provided by PT PLN (Persero). The funding played a crucial role in enabling new advancements and contributing to understanding hydrogen fuel cell research. This research is a testament to the positive outcomes that can arise from effective collaboration between industry and academia.

Conflicts of Interest: The authors declare no conflicts of interest.

References

1. Anwar, A.; Siddique, M.; Eyup, D.; Sharif, A. The Moderating Role of Renewable and Non-Renewable Energy in Environment-Income Nexus for ASEAN Countries: Evidence from Method of Moments Quantile Regression. *Renew. Energy* **2021**, *164*, 956–967. [CrossRef]
2. Shahzadi, I.; Yaseen, M.R.; Iqbal Khan, M.T.; Amjad Makhdum, M.S.; Ali, Q. The Nexus between Research and Development, Renewable Energy and Environmental Quality: Evidence from Developed and Developing Countries. *Renew. Energy* **2022**, *190*, 1089–1099. [CrossRef]
3. Asongu, S.A.; Iheonu, C.O.; Odo, K.O. The Conditional Relationship between Renewable Energy and Environmental Quality in Sub-Saharan Africa. *Environ. Sci. Pollut. Res.* **2019**, *26*, 36993–37000. [CrossRef] [PubMed]
4. Halkos, G.E.; Gkampoura, E.C. Reviewing Usage, Potentials, and Limitations of Renewable Energy Sources. *Energies* **2020**, *13*, 2906. [CrossRef]
5. Chien, F.; Hsu, C.C.; Ozturk, I.; Sharif, A.; Sadiq, M. The Role of Renewable Energy and Urbanization towards Greenhouse Gas Emission in Top Asian Countries: Evidence from Advance Panel Estimations. *Renew. Energy* **2022**, *186*, 207–216. [CrossRef]
6. Khan, M.T.I.; Yaseen, M.R.; Ali, Q. Nexus between Financial Development, Tourism, Renewable Energy, and Greenhouse Gas Emission in High-Income Countries: A Continent-Wise Analysis. *Energy Econ.* **2019**, *83*, 293–310. [CrossRef]
7. Owusu, P.A.; Asumadu-Sarkodie, S. A Review of Renewable Energy Sources, Sustainability Issues and Climate Change Mitigation. *Cogent Eng.* **2016**, *3*, 1167990. [CrossRef]
8. Zou, C.; Pan, S.; Hao, Q. On the Connotation, Challenge and Significance of China's "Energy Independence" Strategy. *Pet. Explor. Dev.* **2020**, *47*, 449–462. [CrossRef]
9. Elia, A.; Kamidelivand, M.; Rogan, F.; Ó Gallachóir, B. Impacts of Innovation on Renewable Energy Technology Cost Reductions. *Renew. Sustain. Energy Rev.* **2021**, *138*, 110488. [CrossRef]
10. Rasoulinezhad, E.; Taghizadeh-Hesary, F. Role of Green Finance in Improving Energy Efficiency and Renewable Energy Development. *Energy Effic.* **2022**, *15*, 14. [CrossRef]

11. Jaiswal, K.K.; Chowdhury, C.R.; Yadav, D.; Verma, R.; Dutta, S.; Jaiswal, K.S.; Sangmesh, B.; Karuppasamy, K.S.K. Renewable and Sustainable Clean Energy Development and Impact on Social, Economic, and Environmental Health. *Energy Nexus* **2022**, *7*, 100118. [CrossRef]
12. Parag, Y.; Ainspan, M. Sustainable Microgrids: Economic, Environmental and Social Costs and Benefits of Microgrid Deployment. *Energy Sustain. Dev.* **2019**, *52*, 72–81. [CrossRef]
13. Carley, S.; Konisky, D.M. The Justice and Equity Implications of the Clean Energy Transition. *Nat. Energy* **2020**, *5*, 569–577. [CrossRef]
14. Bauer, C.; Treyer, K.; Antonini, C.; Bergerson, J.; Gazzani, M.; Gencer, E.; Gibbins, J.; Mazzotti, M.; McCoy, S.T.; McKenna, R.; et al. On the Climate Impacts of Blue Hydrogen Production. *Sustain. Energy Fuels* **2021**, *6*, 66–75. [CrossRef]
15. Kovač, A.; Paranos, M.; Marciuš, D. Hydrogen in Energy Transition: A Review. *Int. J. Hydrogen Energy* **2021**, *46*, 10016–10035. [CrossRef]
16. Megia, P.J.; Vizcaino, A.J.; Calles, J.A.; Carrero, A. Hydrogen Production Technologies: From Fossil Fuels toward Renewable Sources. A Mini Review. *Energy Fuels* **2021**, *35*, 16403–16415. [CrossRef]
17. Akinyele, D.; Olabode, E.; Amole, A. Review of Fuel Cell Technologies and Applications for Sustainable Microgrid Systems. *Inventions* **2020**, *5*, 42. [CrossRef]
18. Hosseini, S.E.; Butler, B. An Overview of Development and Challenges in Hydrogen Powered Vehicles. *Int. J. Green Energy* **2020**, *17*, 13–37. [CrossRef]
19. Griffiths, S.; Sovacool, B.K.; Kim, J.; Bazilian, M.; Uratani, J.M. Industrial Decarbonization via Hydrogen: A Critical and Systematic Review of Developments, Socio-Technical Systems and Policy Options. *Energy Res. Soc. Sci.* **2021**, *80*, 102208. [CrossRef]
20. Liu, W.; Zuo, H.; Wang, J.; Xue, Q.; Ren, B.; Yang, F. The Production and Application of Hydrogen in Steel Industry. *Int. J. Hydrogen Energy* **2021**, *46*, 10548–10569. [CrossRef]
21. Gambou, F.; Guilbert, D.; Zasadzinski, M.; Rafaralahy, H. A Comprehensive Survey of Alkaline Electrolyzer Modeling: Electrical Domain and Specific Electrolyte Conductivity. *Energies* **2022**, *15*, 3452. [CrossRef]
22. Kumar, K.; Alam, M.; Dutta, V. Energy Management Strategy for Integration of Fuel Cell-Electrolyzer Technologies in Microgrid. *Int. J. Hydrogen Energy* **2021**, *46*, 33738–33755. [CrossRef]
23. Furfari, S.; Clerici, A. Green Hydrogen: The Crucial Performance of Electrolysers Fed by Variable and Intermittent Renewable Electricity. *Eur. Phys. J. Plus* **2021**, *136*, 5. [CrossRef]
24. Cloete, S.; Hirth, L. Flexible Power and Hydrogen Production: Finding Synergy between CCS and Variable Renewables. *Energy* **2020**, *192*, 116671. [CrossRef]
25. Ruggles, T.H.; Dowling, J.A.; Lewis, N.S.; Caldeira, K. Opportunities for Flexible Electricity Loads Such as Hydrogen Production from Curtailed Generation. *Adv. Appl. Energy* **2021**, *3*, 100051. [CrossRef]
26. Arsalis, A.; Georghiou, G.E.; Papanastasiou, P. Recent Research Progress in Hybrid Photovoltaic–Regenerative Hydrogen Fuel Cell Microgrid Systems. *Energies* **2022**, *15*, 3512. [CrossRef]
27. Zhang, B.; Zhang, S.-X.; Yao, R.; Wu, Y.-H.; Qiu, J.-S. Progress and Prospects of Hydrogen Production: Opportunities and Challenges. *J. Electron. Sci. Technol.* **2021**, *19*, 100080. [CrossRef]
28. Singh, M.; Zappa, D.; Comini, E. Solid Oxide Fuel Cell: Decade of Progress, Future Perspectives and Challenges. *Int. J. Hydrogen Energy* **2021**, *46*, 27643–27674. [CrossRef]
29. Yukesh Kannah, R.; Kavitha, S.; Preethi; Parthiba Karthikeyan, O.; Kumar, G.; Dai-Viet, N.V.; Rajesh Banu, J. Techno-Economic Assessment of Various Hydrogen Production Methods—A Review. *Bioresour. Technol.* **2021**, *319*, 124175. [CrossRef]
30. Wang, Y.; Pang, Y.; Xu, H.; Martinez, A.; Chen, K.S. PEM Fuel Cell and Electrolysis Cell Technologies and Hydrogen Infrastructure Development—A Review. *Energy Environ. Sci.* **2022**, *15*, 2288–2328. [CrossRef]
31. Kaur, M.; Pal, K. Review on Hydrogen Storage Materials and Methods from an Electrochemical Viewpoint. *J. Energy Storage* **2019**, *23*, 234–249. [CrossRef]
32. Mekhilef, S.; Saidur, R.; Safari, A. Comparative Study of Different Fuel Cell Technologies. *Renew. Sustain. Energy Rev.* **2012**, *16*, 981–989. [CrossRef]
33. Emmanuel, B.O.; Barendse, P.; Chamier, J. Effect of Anode and Cathode Reletive Humidity Variance and Pressure Gradient on Single Cell PEMFC Performance. In Proceedings of the 2018 IEEE Energy Conversion Congress and Exposition, ECCE 2018, Portland, OR, USA, 23–27 September 2018; pp. 3608–3615. [CrossRef]
34. Singla, M.K.; Nijhawan, P.; Oberoi, A.S. Hydrogen Fuel and Fuel Cell Technology for Cleaner Future: A Review. *Environ. Sci. Pollut. Res.* **2021**, *28*, 15607–15626. [CrossRef]
35. Foorginezhad, S.; Mohseni-Dargah, M.; Falahati, Z.; Abbassi, R.; Razmjou, A.; Asadnia, M. Sensing Advancement towards Safety Assessment of Hydrogen Fuel Cell Vehicles. *J. Power Sources* **2021**, *489*, 229450. [CrossRef]
36. Li, Y.; Taghizadeh-Hesary, F. The Economic Feasibility of Green Hydrogen and Fuel Cell Electric Vehicles for Road Transport in China. *Energy Policy* **2022**, *160*, 112703. [CrossRef]
37. Babatunde, O.M.; Munda, J.L.; Hamam, Y. Hybridized Off-Grid Fuel Cell/Wind/Solar PV /Battery for Energy Generation in a Small Household: A Multi-Criteria Perspective. *Int. J. Hydrogen Energy* **2022**, *47*, 6437–6452. [CrossRef]
38. Baroutaji, A.; Wilberforce, T.; Ramadan, M.; Olabi, A.G. Comprehensive Investigation on Hydrogen and Fuel Cell Technology in the Aviation and Aerospace Sectors. *Renew. Sustain. Energy Rev.* **2019**, *106*, 31–40. [CrossRef]

39. Olabi, A.G.; Wilberforce, T.; Abdelkareem, M.A. Fuel Cell Application in the Automotive Industry and Future Perspective. *Energy* **2021**, *214*, 118955. [CrossRef]
40. Ghenai, C.; Bettayeb, M.; Brdjanin, B.; Hamid, A.K. Hybrid Solar PV/PEM Fuel Cell/Diesel Generator Power System for Cruise Ship: A Case Study in Stockholm, Sweden. *Case Stud. Therm. Eng.* **2019**, *14*, 100497. [CrossRef]
41. Ferriday, T.B.; Middleton, P.H. Alkaline Fuel Cell Technology—A Review. *Int. J. Hydrogen Energy* **2021**, *46*, 18489–18510. [CrossRef]
42. Pu, Z.; Zhang, G.; Hassanpour, A.; Zheng, D.; Wang, S.; Liao, S.; Chen, Z.; Sun, S. Regenerative Fuel Cells: Recent Progress, Challenges, Perspectives and Their Applications for Space Energy System. *Appl. Energy* **2021**, *283*, 116376. [CrossRef]
43. Bicer, Y.; Khalid, F. Life Cycle Environmental Impact Comparison of Solid Oxide Fuel Cells Fueled by Natural Gas, Hydrogen, Ammonia and Methanol for Combined Heat and Power Generation. *Int. J. Hydrogen Energy* **2020**, *45*, 3670–3685. [CrossRef]
44. Blum, L.; Fang, Q.; Groß-Barsnick, S.M.; de Haart, L.G.J.; Malzbender, J.; Menzler, N.H.; Quadakkers, W.J. Long-Term Operation of Solid Oxide Fuel Cells and Preliminary Findings on Accelerated Testing. *Int. J. Hydrogen Energy* **2020**, *45*, 8955–8964. [CrossRef]
45. Emadi, M.A.; Chitgar, N.; Oyewunmi, O.A.; Markides, C.N. Working-Fluid Selection and Thermoeconomic Optimisation of a Combined Cycle Co-generation Dual-Loop Organic Rankine Cycle (ORC) System for Solid Oxide Fuel Cell (SOFC) Waste-Heat Recovery. *Appl. Energy* **2020**, *261*, 114384. [CrossRef]
46. Yodwong, B.; Guilbert, D.; Phattanasak, M.; Kaewmanee, W.; Hinaje, M.; Vitale, G. Proton Exchange Membrane Electrolyzer Modeling for Power Electronics Control: A Short Review. *C* **2020**, *6*, 29. [CrossRef]
47. Pan, M.; Pan, C.; Li, C.; Zhao, J. A Review of Membranes in Proton Exchange Membrane Fuel Cells: Transport Phenomena, Performance and Durability. *Renew. Sustain. Energy Rev.* **2021**, *141*, 110771. [CrossRef]
48. Jiao, K.; Xuan, J.; Du, Q.; Bao, Z.; Xie, B.; Wang, B.; Zhao, Y.; Fan, L.; Wang, H.; Hou, Z.; et al. Designing the next Generation of Proton-Exchange Membrane Fuel Cells. *Nature* **2021**, *595*, 361–369. [CrossRef]
49. Masaeli, N.; Afshari, E.; Baniasadi, E.; Baharlou-Houreh, N. Performance Studies of a Membrane-Based Water and Heat Exchanger Using Serpentine Flow Channels for Polymer Electrolyte Membrane Fuel Cell Application. *Appl. Therm. Eng.* **2023**, *222*, 119950. [CrossRef]
50. Rostami, L.; Haghshenasfard, M.; Sadeghi, M.; Zhiani, M. A 3D CFD Model of Novel Flow Channel Designs Based on the Serpentine and the Parallel Design for Performance Enhancement of PEMFC. *Energy* **2022**, *258*, 124726. [CrossRef]

designs

MDPI

Article

Parameter Identification in Triple-Diode Photovoltaic Modules Using Hybrid Optimization Algorithms

Dhiaa Halboot Muhsen *, Haider Tarish Haider and Yaarob Al-Nidawi

Department of Computer Engineering, Mustansiriyah University, Baghdad 14022, Iraq; haiderth@uomustansiriyah.edu.iq (H.T.H.); yaarob.al_nidawi@uomustansiriyah.edu.iq (Y.A.-N.)
* Correspondence: dhiaahm@uomustansiriyah.edu.iq

Abstract: Identifying the parameters of a triple-diode electrical circuit structure in PV modules is a critical issue, and it has been regarded as an important research area. Accordingly, in this study, a differential evolution algorithm (DEA) is hybridized with an electromagnetism-like algorithm (EMA) in the mutation stage to enhance the reliability and efficiency of the DEA. A new formula is presented to adapt the control parameters (mutation factor and crossover rate) of the DEA. Seven different experimental data sets are used to improve the performance of the proposed differential evolution with an integrated mutation per iteration algorithm (DEIMA). The results of the proposed PV modeling method are evaluated with other state-of-the-art approaches. According to different statistical criteria, the DEIMA demonstrates superiority in terms of root mean square error and main bias error by at least 5.4% and 10%, respectively, as compared to other methods. Furthermore, the DEIMA has an average execution time of 27.69 s, which is less than that of the other methods.

Keywords: differential evolution; electromagnetism-like; photovoltaic; triple-diode model; optimization; solar irradiance; mutation factor

Citation: Muhsen, D.H.; Haider, H.T.; Al-Nidawi, Y. Parameter Identification in Triple-Diode Photovoltaic Modules Using Hybrid Optimization Algorithms. *Designs* **2024**, *8*, 119. https://doi.org/10.3390/designs8060119

Academic Editors: Surender Reddy Salkuti and Dibin Zhu

Received: 21 August 2024
Revised: 29 October 2024
Accepted: 5 November 2024
Published: 12 November 2024

1. Introduction

Air pollution and the fluctuation of fossil fuel prices due to political and economic crises around the world are the main problems that affect energy demand. In addition to the aforementioned drawbacks of conventional energy sources, fossil fuels are not abundant or sustainable [1]. Furthermore, energy demand has dramatically increased over time due to urbanized lifestyles. Therefore, finding a new sustainable energy source is a crucial research topic. Solar energy represents one of the most important and promising alternative energy sources [2,3]. Photovoltaic (PV) technology plays an important role in converting solar irradiance into direct-current electricity. The long lifecycle of PV modules and their environmentally friendly features are the main advantages of this technology [4]. In addition, the price of PV modules decreases over time [4]. The main drawback of PV modules is the low efficiency of converting solar energy into electricity [5]. This issue makes the output of PV modules limited, and it also suffers from weather fluctuations. Thus, the modeling of PV modules has to be more accurate to enable project administrators to use an appropriate number of PV modules in order to achieve a reliable and cost-effective PV system. Modeling PV modules significantly depends on the process of estimating the procedure of identifying unknown parameters for equivalent electrical circuits. These parameters are unknown and sensitive to weather conditions. Generally, the estimation of PV module parameters is performed via three approaches: analytical, numerical, and artificial intelligence (AI)-based methods [6]. In analytical methods, a relationship is established between the parameters of the PV module under a standard test condition (STC) and other weather conditions using manufacturer data [7]. The importance of analytical approaches is that they allow the fast and simple calculation of parameters [8]. The main issue with analytical methods is the significant deviation between experimental and simulated performance due to the impact of geographical situations on PV module parameters.

Numerical methods are used to overcome the drawbacks of analytical methods. In numerical methods, all the points of the I-V characteristic curve are used to extract the parameters of a PV module based on an iteration method [9]. Numerical methods offer a more accurate estimation of PV module parameters than analytical methods. The drawback of numerical methods is their requirement of big I-V data curves, and the accuracy of the results is affected by the assumed initial conditions of the parameters [10]. Furthermore, numerical methods need great computational resources. A combination of numerical and analytical methods (a compound method) was presented in [11–13] to obtain the benefits of the aforementioned methods. The drawbacks of both analytical and numerical methods still exist in the compound method.

The third type of PV modeling method is based on artificial intelligence (AI). Many research works adopted artificial neural networks (ANNs) for modeling PV modules [14]. ANN-based PV modeling can be represented as a black box that requires a large amount of data on I-V characteristic points. Moreover, it represents a complex, location-dependent modeling method. Due to the reliability and efficiency of metaheuristic algorithms, the latter are extensively used for modeling PV modules [15]. The I-V characteristics of PV modules are nonlinear, so the metaheuristic is an appropriate choice for handling the modeling problem. In the last decade, great research efforts have been devoted to estimating the parameters of the single-diode model (SDM) and double-diode model (DDM) of PV modules using various metaheuristic algorithms. On the other hand, few research works have been devoted to determining the parameters of the triple-diode model (TDM) of PV modules. In [16], the authors utilized an iterative process using the PSO algorithm to estimate PV model parameters and fitting the measured I-V curves to the calculated I-V curves. The series resistance parameter was considered to vary linearly with the load current through the device. The TDM proposed in [16] is compared to the two-diode model, and the findings indicate that the TDM's performance is better than that of the DDM. Meanwhile, nine unknown parameters of the TDM PV module were extracted via a novel implementation of the coyote optimization algorithm (COA) in [17]. The obtained ideal design variables of the presented COA-TDM were studied against the ideal variables achieved through a whale optimization algorithm (WOA)-based TDM PV model, a genetic algorithm (GA), and simulated annealing (SA) for both modules (KC200GT and MSX-60). The proposed COA-TDM showed optimal design variables that were quite close to those achieved by applying other metaheuristic optimization algorithms for two commercial PV modules.

In [18], a compound of an analytical method and an improved differential evolution algorithm called the IDEA was presented to identify the parameters of the DDM and TDM of PV modules. For both the TDM and DDM, the parameters were partially extracted via an analytical process (seven TDM parameters and five DDM parameters) and by using optimization techniques (both the TDM and DDM have two variables). In [19], a slime mold algorithm (SMA) was presented according to slime mold's natural oscillation. Based on a unique mathematical expression, the SMA was introduced, and it adapts the weights to collect negative and positive feedback of the slime mold propagation wave. According to the results of [19], the SMA showed better performance for the TDM, DDM, and SDM as compared to other heuristic methods. Transient search optimization (TSO) was proposed in [20]. It is based on a novel competent metaheuristic optimizer and aims to estimate a TDM PV module's optimal nine parameters. Different companies have applied TSO for a given objective function to identify three PV modules. Accordingly, PV I-V characteristics have been validated by the measured data with regards to different solar radiation values and temperatures. The TSO algorithm has proven its effectiveness as compared to other models, as indicated through the convergence curves. The authors of [21] customized the collected data and mathematical representations of a PV model for different diode numbers (SDM/DDM/TDM). In their study, the optimal parameters of the studied approach were determined according to variants of the fractional chaotic ensemble particle swarm optimizer (FC-EPSO) and other models. The root mean square error (RMSE) is also one of the data sets that has been calculated, assessed, and adapted as an objective function by the proposed algorithm. To justify the superiority of the presented approach, it has been tested against other approaches in the literature. The outcomes in [21] showed the least deviation between the estimated and measured curves with the fastest convergence.

Integrating both computational processes and the Harris hawk optimization (HHO) algorithm is another approach that was presented by the authors in [22]. The target of this model is to determine the parameters of the TDM regarding a PV module. In this work, the authors utilized the standard test condition (STC) datasheet values of the PV modules with a normal operating cell temperature (NOCT). This process is required to analytically examine four parameters while finding the remaining five parameters through the HHO model. Seeking to estimate the parameters, two commercial PV panels were used as the mono-crystalline CS6K280M and multi-crystal KC200GT. The results in [22] proved the efficacy of the presented model by depending on the datasheet values only. Accordingly, it can simply be implemented to find the electrical parameters of any commercial PV panel. In [23], a new optimization method, namely the interval branch and bound algorithm, was introduced and validated for three different parameter estimation models of PV cells (SDM, DDM, and TDM). The outcomes were justified against other available outcomes in the literature regarding the same data set. The behavior of the presented model was examined with regards to convergence speed and resulted in variability as a comparison to metaheuristics. The estimated features of the tested cells for both *P-V* and I-V have shown to be very similar to the experimental data, and the obtained findings are quite close to other efficient algorithms. In [24], an improved wind-driven optimization (IWDO) model was presented and applied, which calculated the triple-diode parameters of the PV cell model. In order to evaluate the proposed model, the IWDO model was implemented on three PV model techniques, poly-crystalline, mono-crystalline, and thin film. Accordingly, the obtained results were compared with other findings that were collected from other contributions to validate their accuracy. According to the results in [24], the presented algorithm showed superiority over other models in terms of accuracy and convergence speed. The authors in [25] relied on manta ray foraging optimization (MRFO) to find the unknown parameters of PV cells. MRFO was adapted to extract the optimal PV parameters regarding each of the single-, double-, and three-diode algorithms. Based on comparative results among different metaheuristic algorithms and MRFO, the findings proved that MRFO showed a better balance between the experimental and calculated I–V curves.

The authors in [26] proposed an enhanced LSHADE optimization technique called the Chaotic LSHADE (CLSHADE) model. The model was traced to the Lambert W-function that estimates both the TDM and DDM parameters of PV equivalent circuits. The outcomes indicated that the accuracy of the CLSHADE model could be enhanced through adapting the presented solar cell current expressions. The RMSE was calculated according to both the TDM and DDM. The authors in [27] proposed an improved spherical evolution technique that is based on a novel dynamic sine cosine mechanism (DSCSE). The experimental findings indicated the superiority of the DSCSE over other comparative approaches with different models and provided superior outcomes regardless of different temperatures and light intensities. Based on the DE (HDE), a novel heterogeneous mechanism was presented in [28] to identify PV model parameters. Parameter fitting for the DDM, SDM, STP6-120/36 MM, TDM, Photowatt-PWP201 MM, and STM6-40/36 MM were determined by the HDE. The findings indicated the superiority of the HDE regarding the majority of PV techniques. Moreover, the HDE also takes little execution time to conduct its task. Although the HDE showed necessary features, its behavior regarding a few PV models like the TDM and DDM can be further improved. An enhanced model of the slime mold technique was presented in [29]. This model is based on the Lambert W-function (ImSMA_LW) to extract each of the SDM, DDM, and TDM parameters regarding a PV module. According to [29], ImSMA_LW offered compromised results under various scenarios with different conditions.

The contributions of this article can be presented as follows:

1. Nine unknown parameters of the TDM are accurately extracted by using a new differential evolution with integrated mutation per iteration algorithm (DEIMA). The proposed DEIM hybridized the conventional differential evolution algorithm (DEA) and the electromagnetism-like algorithm (EMA). This integration is required to overcome the drawbacks of these algorithms. The attraction–repulsion concept of EMA is utilized to synergize the mutation strategy of the conventional DE algorithm. In the DEIMA, both the EMA and DEA mutation strategies are applied within the same iteration.
2. A novel adaptive formula is proposed to realize an adaptive mutation factor with crossover rate values of the DEA's mutation strategy. The adaptive technique in this formula is based on the evolution of the fitness function.
3. The results of the proposed PV model are evaluated with regards to the DEIM algorithm and using realistic experimental I-V data of seven different operation conditions. In addition, the outcomes of the proposed DEIMA are compared with other related contributions in the literature using various statistical indices.

The manuscript is structured as follows: the Introduction is presented in Section 1. In Section 2, the triple-diode PV module is clarified and the PV module parameter estimation process is discussed. The proposed DEIMA is highlighted in Section 3. Next, the evaluation criteria that are utilized for evaluating the DEIMA results are presented in Section 4. The results of the proposed DEIMA-based PV method are presented, discussed, and evaluated in Section 5. The conclusions of this contribution and suggestions for future work are drawn in Section 6.

2. PV Modeling Method

This section demonstrates the mathematical representation of the triple-diode PV module. Furthermore, it discusses the parameter estimation process of the TDM-PV module, where it is considered as an optimization problem.

2.1. Triple-Diode PV Module Model

The TDM of the PV module comprises three diodes that are connected in parallel, as shown in Figure 1. The third diode is added into an equivalent electrical circuit of the PV module to simulate the recombination in grain sites and defected regions. The PV current output can be formulated by [22]

$$I = I_{ph} - I_{o1}\left[exp\left(q\frac{V+IR_S}{a_1BT_C}\right)-1\right] - I_{o2}\left[exp\left(q\frac{V+IR_S}{a_2BT_C}\right)-1\right] - I_{o3}\left[exp\left(q\frac{V+IR_S}{a_3BT_C}\right)-1\right] - \frac{V+IR_S}{R_P} \quad (1)$$

where I and I_{ph} are the PV output current (A) and photogenerated current (A), respectively. I_{o1}, I_{o2}, and I_{o3} are the saturation currents (A) that flow in the first, second, and third diodes, respectively. V is the PV output voltage (V); a_1, a_2, and a_3 are the ideality factors of the first, second, and third diodes, respectively; q, B, and T_C are the electron charge (1.60217646E-19C), Boltzman constant (1.3806503E-23J/K), and cell temperature (K), respectively. R_S is the series resistor, and R_P is the parallel resistor (Ω).

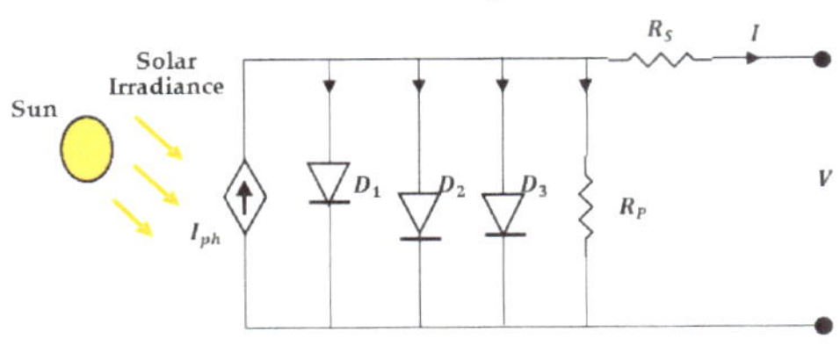

Figure 1. Electrical equivalent circuit of triple-diode PV module model.

According to Equation (1), there are nine indefinite parameters that should be precisely estimated. As it was stated previously, these parameters are related to weather conditions, specifically the ambient temperature and solar irradiance.

2.2. Problem Formulation

The issue of estimating nine indefinite parameters in the TDM-PV module can be formulated as an optimization problem with nine decision variables and an objective function, as follows:

$$OF = \sqrt{\frac{1}{n}\left(\sum_{i=1}^{n} f(I_e, V, \delta)^2\right)} \quad (2)$$

where

$$f(I_e, V, \delta) = I_e - I \quad (3)$$

I_e is the experimental output current (A) of the PV module. I refers to the PV module's computed current according to the estimated parameters that were described by vector $\delta = \left[I_{ph},\ I_{o1},\ I_{o2},\ I_{o3},\ a_1,\ a_2,\ a_3,\ R_S,\ R_P\right]$. Variable n represents the total number of points in the experimental I-V characteristic curve.

3. DEIMA Optimization Algorithm

The proposed PV modeling method is based on the DEIMA. In addition, this section highlights the strategy of deriving an adaptive formula for both the mutation and crossover rate control parameters of the DEIMA. The DEIMA is a stochastic metaheuristic optimization algorithm, which is initiated by hybridizing the DEA and EMA. The attraction–repulsion concept of the EMA is synergically utilized with the conventional mutation stage of the DEA. The four steps of the DEIMA will be discussed in detail in the following subsections.

3.1. Initialization

The first step of the DEIMA is initializing the $D_P \times N_P$ population, where D_P is the number of decision variables and N_P refers to the number of individual candidate vectors. It is worth mentioning that each individual vector comprises D_P decision variables. The population (pop) can be described as follows:

$$pop = \begin{pmatrix} x_{11} & x_{12} & \cdots & x_{1N_P} \\ x_{21} & x_{22} & \cdots & x_{2N_P} \\ \vdots & \vdots & \ddots & \vdots \\ x_{D_P1} & x_{D_P2} & \cdots & x_{D_PN_P} \end{pmatrix} \quad (4)$$

The D_P parameters of the ith individual vector are randomly initiated using Equation (5) and uniformly distributed within the interval of the jth decision variable $[XL_j,\ XH_j]$, where XL_j is the lower boundary and XH_j is the upper bounds of the search space.

$$X_{ji}^0 = XL_{ji} + rand \times (XH_{ji} - XL_{ji}) \quad (5)$$

where $rand$ is a random number, which is selected randomly in the range of (0, 1).

3.2. Mutation

The proposed DEIMA utilizes two different types of mutation, (M_e and M_d). Both types are used in each iteration and based on the following criterion:

$$Mutation\ operation = \begin{cases} M_e & if\ \theta_l^G < \varepsilon_1 \theta_l^0 \\ M_d & otherwise \end{cases} \quad (6)$$

where θ_l^0 and θ_l^G refer to the lth standard deviation vectors of the row vectors of the pop population in the initial and G generations, respectively; l is a random number, which is randomly chosen within the $[1,\ D_P]$ interval; and ε_1 is a constant control parameter with a value belonging to the interval [0, 1] [30]. The ε_1 controls the switch between M_e and M_d mutation operations within the population. The M_d mutation operation is achieved by computing the mutant vector $\hat{X}_i^G$, as follows:

$$\hat{X}_i^G = \hat{X}_{r1}^G + F\left(\hat{X}_{r2}^G - \hat{X}_{r3}^G\right) \tag{7}$$

where $\hat{X}_{r1}^G$, $\hat{X}_{r2}^G$, and $\hat{X}_{r3}^G$ are three distinct individual vectors randomly selected from the population. In other words, r_1, r_2, and r_3 are distinct indices and belong to the period $[1, N_P]$. F is the mutation control factor having values within [0.5, 1] [31]. It is worth mentioning that the control parameter F and crossover control rate (CR) in the crossover step are adaptive in each iteration and are being set by using a novel formula, as follows:

$$F,\ CR = d\left(\frac{L}{1 + \exp(-k(w - w_0))} + b\right) \tag{8}$$

The above formula is based on a sigmoid function, where the maximum value is labeled as the L of a sigmoid function curve and k and w_0 are the steepness of the curve and midpoint of the x-axis sigmoid function, respectively. In the current research work, L, k, and w_0 are set to 1, 12, and 0 values according to many experiments; b and d are constants, which are chosen to maintain F and CR within [0.5, 1], where b is set to 1 and d equals 0.5; and w refers to the random weighted difference between the best objective function value of w and is computed by

$$w = \left[OF\left(X_{best}^G\right) - OF\left(X_{best}^{G-1}\right)\right] * rand \tag{9}$$

As it was stated previously, the second part of the mutation is M_e. The last operation utilized the total force exerted on X_{r1}^G by X_{r2}^G and X_{r3}^G to compute the mutant vector X_i^G. The force exerted by any individual vector on another one depends on the charge between them, which is calculated as shown below:

$$q_{r1r2}^G = \frac{OF(X_{r1}^G) - OF(X_{r2}^G)}{OF(X_{worst}^G) - OF(X_{best}^G)} \tag{10}$$

$$q_{r1r3}^G = \frac{OF(X_{r1}^G) - OF(X_{r3}^G)}{OF(X_{worst}^G) - OF(X_{best}^G)} \tag{11}$$

where $OF(X^G)$ refers to the objective function value of individual vector X in the Gth iteration. X_{worst}^G and X_{best}^G are the worst and best solution in the Gth iteration, respectively. Consequently, the exerted forces by X_{r2}^G and X_{r3}^G on the individual vector X_{r1}^G are described by [32]

$$F_{r1r2}^G = \left(X_{r2}^G - X_{r1}^G\right) * q_{r1r2}^G \tag{12}$$

$$F_{r1r3}^G = \left(X_{r3}^G - X_{r1}^G\right) * q_{r1r3}^G \tag{13}$$

Then, the resultant force exerted by X_{r2}^G and X_{r3}^G on X_{r1}^G is calculated by

$$F_{r1}^G = F_{r1r2}^G + X_{r1r3}^G \tag{14}$$

Next, the M_e mutation operation produces the X_i^G mutant vector in the Gth iteration and is corresponding to X_i^G as follows:

$$\hat{X}_i^G = X_{r1}^G + F_\alpha^G \tag{15}$$

3.3. Crossover

In this step, the trial vector of Gth iteration $y_{j,i}^G$ is initiated by utilizing both the corresponding target X_{r1}^G or the mutant $\hat{X}_i^G$ vector based on the following formula:

$$Y_{j,i}^G = \begin{cases} \hat{X}_{j,i}^G & if\ rand \le CR\ or\ j = I_i \\ X_{j,i}^G & otherwise \end{cases} \tag{16}$$

where I_i is an index number, which is randomly selected from the interval $[1,\ D_P]$. As was stated in previous subsections, CR is the crossover control rate, which has an adaptive value that is computed by Equation (8). At the end of the crossover step, the element of the trial vector (decision variable) will be checked to determine if one of them violates the boundaries of the corresponding search space. Equation (5) is used to initiate the element of the trial vector when it is an unphysical value.

3.4. Selection

The last step of the DEIMA is the selection step, which chooses between the trial vector and the corresponding individual vector for the Gth iteration, and is described as

$$X_i^{G+1} = \begin{cases} Y_i^G & if\ OF\left(Y_i^G\right) < OF\left(X_i^G\right) \\ X_i^G & otherwise \end{cases} \tag{17}$$

At the end of the selection step, the individual vector of generation $(G+1)$ is generated to constitute the new population. Figure 2 illustrates the proposed DEIMA method for estimating the unknown nine parameters of the TDM-PD module.

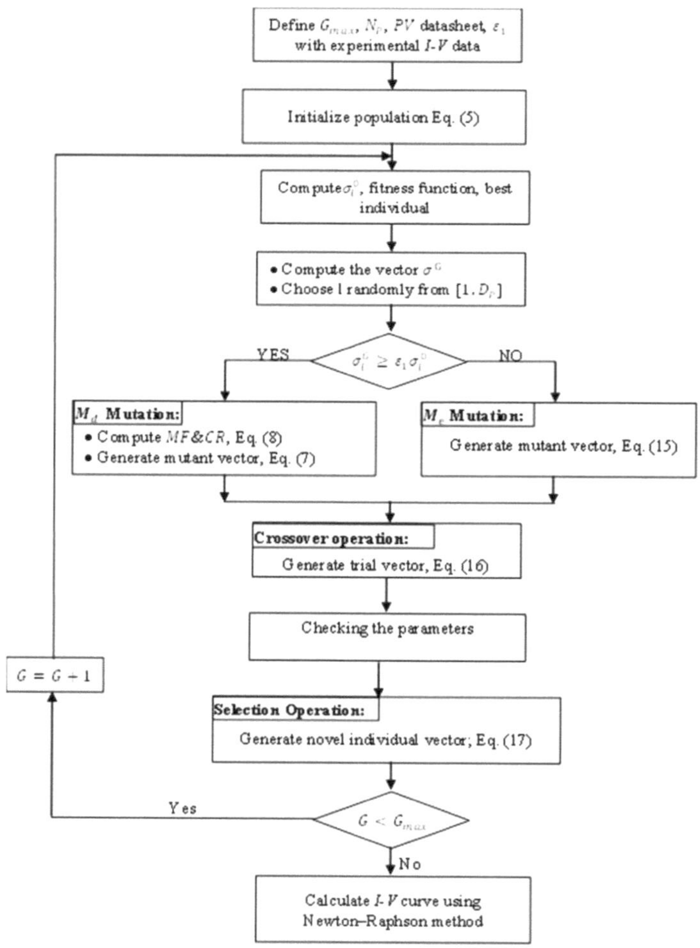

Figure 2. DEIMA-based PV modeling method.

4. Evaluation Criteria

This research work adapts seven criteria, which are required in the assessment step of the introduced PV modeling method and are compared to other methods. More details about the criteria evaluation are presented in the next subsection.

4.1. Root Mean Square Error (RMSE)

The $RMSE$ criterion presents the deviation between the computed and experimented points of the I-V characteristic curve along the n-data set, as depicted below:

$$RMSE = \sqrt{\frac{1}{n}\sum_{t=1}^{n}(I_{et} - I_{ct})^2} \tag{18}$$

where I_{et} and I_{ct} are the experimental and computed currents at the ith point of the I-V curve.

4.2. Mean Bias Error (MBE)

The overall bias error between the computed and experimented I-V characteristic curve is measured by using the MBE formula, as presented below:

$$MBE = \frac{1}{n}\sum_{t=1}^{n}(I_{et} - I_{ct})^2 \tag{19}$$

4.3. Coefficient of Determination (R^2)

R^2 is another criterion that is utilized to assess the accuracy and performance of the proposed PV modeling method. This coefficient explains the degree of ability that a computed I-V curve has to follow the experimental one. It is worth mentioning that the better value of the R^2 criterion should be close to 1. The R^2 is computed by

$$R^2 = 1 - \frac{\sum_{t=1}^{n}(I_{ct} - I_{et})^2}{\sum_{t=1}^{n}(I_{et} - \overline{I_e})^2} \tag{20}$$

where

$$\overline{I_e} = \frac{1}{n}\sum_{t=1}^{n}I_{et} \tag{21}$$

4.4. Average Absolute Error (AAE)

Absolute error shows the absolute difference between the experimental and calculated currents. The average value of absolute error is computed by finding the average of absolute error over all the points of the I-V curve. The AAE is calculated by

$$AAE = \frac{1}{n}\sum_{t=1}^{n}|I_{ct} - I_{et}| \tag{22}$$

4.5. Deviation of RMSE (d_m)

The evaluation criterion d_m refers to the $RMSE$ deviation of each solar irradiance from the mean $RMSE$ as regards the seven operation conditions, and is formulated as

$$d_m = RMSE_m - \overline{RMSE} \tag{23}$$

where $RMSE_m$ refers to the $RMSE$ of the mth solar irradiance (operation condition) and $\overline{RMSE}$ is the arithmetic mean of $RMSE$ regarding all operation conditions ($G_1 - G_7$). $\overline{RMSEE}$ is computed by

$$\overline{RMSE} = \frac{1}{r}\sum_{t=1}^{r}RMSE_t \tag{24}$$

where r is the total number of operation conditions.

4.6. Standard Test Deviation (STD)

Another statistical criterion, STD, is used to evaluate the performance of the proposed PV modeling method. The STD value explains the average deviation in the $RMSE$ for the seven operation conditions. The STD is computed by

$$STD = \sqrt{\frac{1}{(r-1)} \sum_{m=1}^{r} (d_m)^2} \tag{25}$$

4.7. CPU Execution Time (CET)

The required CPU time for executing the DEIMA is the last criterion utilized by the result evaluation process.

5. Results and Discussion

A multicrystalline Kyocera KC120-1 with a 120 Wp capacity PV module is used in this paper for testing the proposed modeling method. The specifications of the aforementioned PV module are tabulated in Table 1.

Table 1. The specifications of Kyocera KC120-1 PV module [32].

Characteristics	Value
Number of cells connected in series	**36**
Short-circuit current (I_{sc})	**7.45 A**
Open-circuit voltage (V_{oc})	**21.5 V**
Current at maximum power point (I_{mp})	**7.1 A**
Voltage at maximum power point (V_{mp})	**16.9 V**
Maximum power at STC (P_{max})	**120 W_P**
Temperature coefficient of I_{sc} (α)	**1.325 mA/K**
Temperature coefficient of V_{oc} (β)	**−77.5 mV/K**

Seven different operation conditions are used as experimental data to extract the parameters of the PV module. These operation conditions are denoted by G1 to G7, and they are tabulated in Table 2. The first and second columns of Table 2 comprise the solar irradiance and cell temperature of various operation conditions. It is worth mentioning that each operation condition includes various length of experimental I-V data points that were explained in the fourth column of Table 2. It should be noted that the weather condition and operation condition terms are interchangeably used throughout this paper.

Table 2. Seven operation conditions and related solar irradiance and cell temperatures.

Weather Condition	Length of Data (n)	Solar Radiance (W/m^2)	Cell Temperature (K)
G_1	22	118.28	318.32
G_2	24	148	321.25
G_3	50	306	327.7
G_4	91	711	324.21
G_5	92	780	329.1
G_6	101	840	331.42
G_7	102	978	328.56

A different evolution with an integrated mutation per each iteration is adopted to estimate the unknown TDM parameters. Since the dimension of the PV module optimization problem is 9, then there are 9 decision variables (D_P). Accordingly, the number of individual solutions will be $10D$ [33,34]. The maximum number of iterations (G_{max}) for the DEIMA and other adapted methods used for comparison is proposed to be 500 as a typical value. This number is based on several trial-and-error tests. According to many operations, the switching control parameter (ε_1) finds that the value of 0.28 is considered as the best performance of the DEIMA. Meanwhile, the mutation factor and crossover rate of the DEIMA are adaptive according to the proposed formula, which was discussed previously.

The search space range of the nine parameters regarding the TDM is based on some addressed values in the literature. The photocurrent, R_S, R_P, diode saturation currents (I_{o1}, I_{o2}, and I_{o3}), and diode ideality factors (a_1, a_2, and a_3) are chosen to be [1, 8] A, [0.1, 2] Ω, [100, 5000] Ω, [10^{-12}, 10^{-5}] A, and [1, 2], respectively [35].

The nine parameters of the TDM PV module are estimated according to the DEIMA and are explained in Table 3. According to Table 3, the value of the photocurrent (I_{ph}) increases when the solar irradiance increases, which supports the direct relationship between them. Based on the estimated parameters and Newton–Raphson method, the I-V and P-V curves of the TDM PV module under seven different operation conditions can be obtained as illustrated in Figures 3 and 4, respectively. According to Figures 3 and 4, it can be visually concluded that the I-V and P-V curves obtained by the DEIMA-based PV modeling method are closer to the experimental ones. It is worth mentioning that the experimental I-V and P-V curves of the PV module are irregular due to the error in the I-V generator, which is used in the field to collect the experimental data. It should be noticed that most of the deviation between the computed and experimental currents is concentrated near the vicinity of the maximum power point (MPP), especially at high solar irradiance levels. This happened because the predicted current is higher than the experimental current. Furthermore, the experimental I-V curves of the PV module are irregular.

Table 3. The nine estimated parameters of the TDM of the PV module.

Parameter	G1	G2	G3	G4	G5	G6	G7
a_1	1.00004	1.40544	1.97468	1.99608	1.64065	1.53082	1.49762
a_2	1.97660	1.45038	1.06383	1.00095	1.49554	1.44084	1.49754
a_3	1.99144	1.35464	1.04220	1.31040	1.45000	1.45671	1.49766
R_S	1.90299	0.13195	0.75172	0.53455	0.23549	0.16636	0.18106
R_P	100.00554	125.38801	195.59178	169.22920	100.00108	4798.1068	100.00002
I_{ph}	1.00000	1.00000	1.95796	4.42801	5.05995	5.18610	6.27363
I_{o1}	1.59×10^{-8}	6.695×10^{-10}	8.974×10^{-7}	3.610×10^{-6}	9.966×10^{-6}	9.996×10^{-6}	1.00×10^{-5}
I_{o2}	9.97×10^{-6}	5.555×10^{-6}	2.591×10^{-9}	6.494×10^{-9}	9.948×10^{-6}	9.779×10^{-6}	1.00×10^{-5}
I_{o3}	9.96×10^{-6}	4.180×10^{-7}	1.232×10^{-7}	2.915×10^{-6}	9.692×10^{-6}	9.994×10^{-6}	1.00×10^{-5}

Figure 5 shows the evolution of the objective function within all of the generations under the seven operation conditions for the DEIMA, which are required to calculate the TDM parameters of the PV module. The presented DEIMA shows fast convergence and minimum objective function values for all operation conditions. It should be noted that the proposed modeling method is based on the DEIMA and exhibits a stable objective function value at the first 20 iterations, which is around 1.1076 s as the average convergence time. The proposed novel formula for adapting F and CR values manage to have fast DEIMA converges and realizes stable performance. The F and CR values are chosen based on the evolution of the objective function, and the F and CR values converge to 0.5 with the progress of DEIMA generations and make the latter work in the exploitation phase. Due to the distinctive exploitation of the DEIMA, the proposed model predicts the optimal parameter values with minimum objective function value. The accurate estimation of nine parameters of the TDM PV module have enabled the production of both I-V and P-V points with more accuracy and closer to the experimental data. Figure 6 shows the effectiveness values of F and CR according to w values.

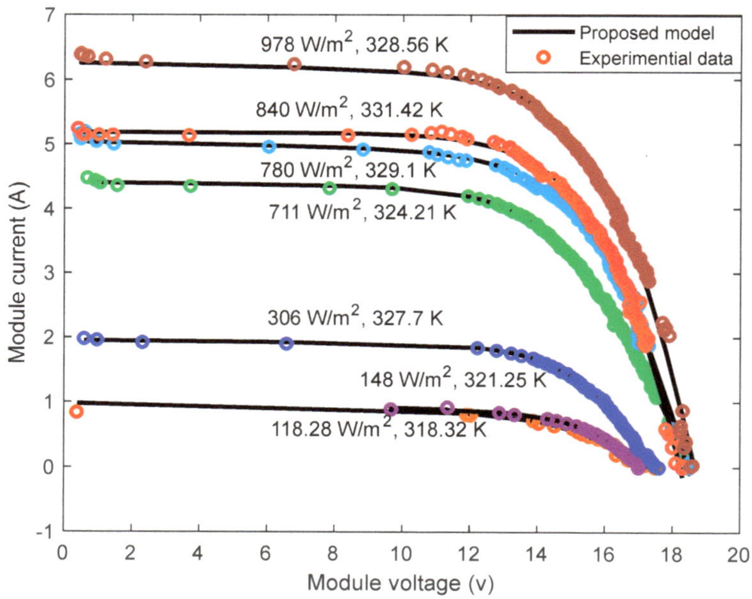

Figure 3. I-V characteristics of TDM of PV module under seven distinct operation conditions (different colors are experimental I-V data with different solar irradiance and cell temperature).

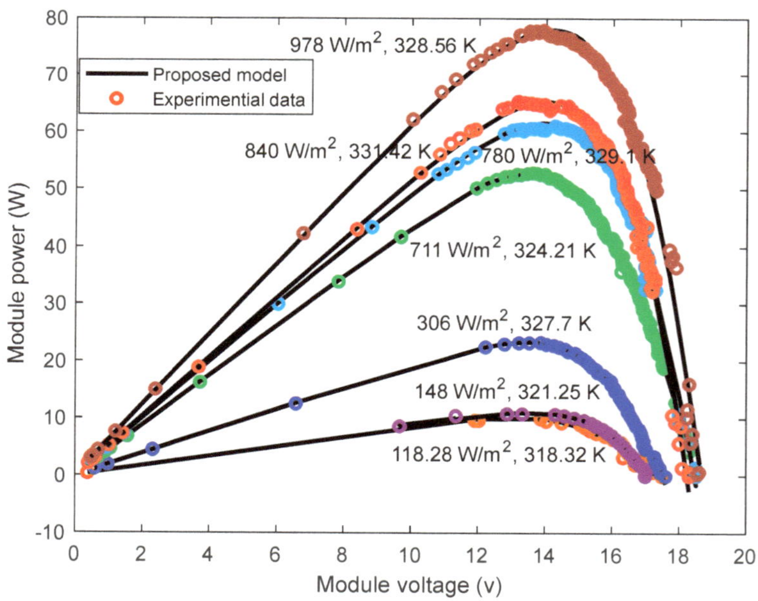

Figure 4. P-V characteristics of TDM of PV module under seven distinct operation conditions (different colors are experimental P-V data with different solar irradiance and cell temperature).

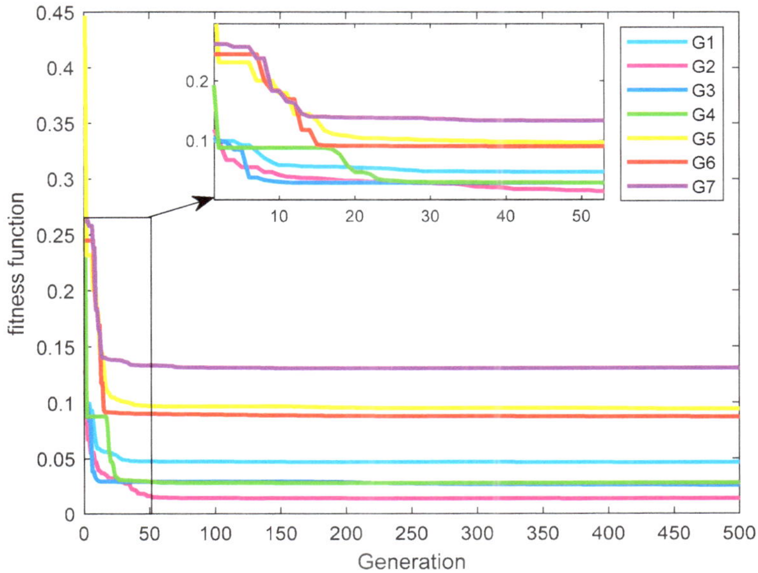

Figure 5. Fitness function evolution of TDM of PV module parameter estimation using DEIMA under seven operation conditions.

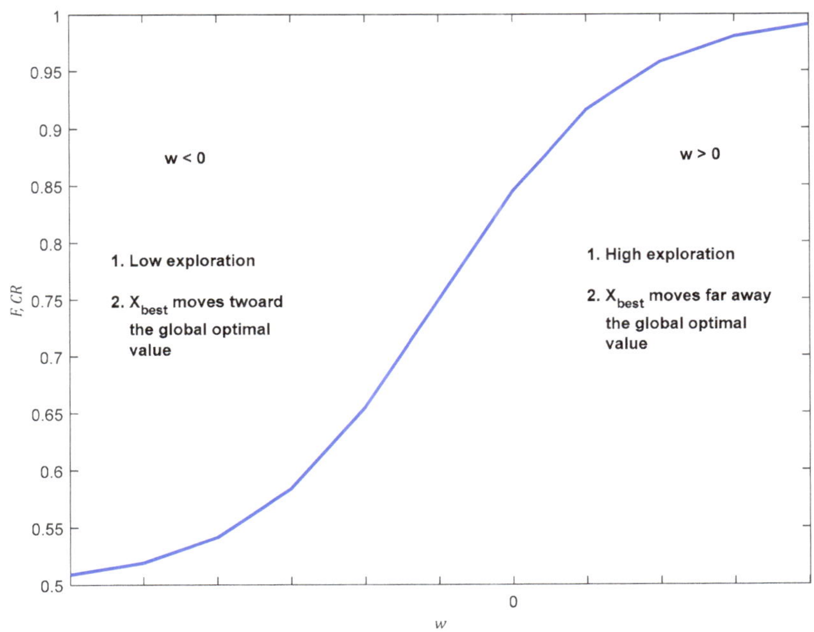

Figure 6. Possible F and CR values regarding w values.

The time computational complexity of the proposed DEIMA depends on the initialization, the evaluation of fitness functions (FE), the update of the solution (mutation and crossover stages), the population size (N_P), the number of decision variables (D_P), and the maximum number of iterations (G_{max}). Thus, the overall computational complexity of the DEIMA can be presented by

$$\begin{aligned} O(DEIMA) &= O(initialization) + O(function\ evaluation) + O(mutation) \\ &\quad + O(crossover) + O(selection) \\ &= N_P(1 + D_P + G_{max}(FE + 2D_P)) \end{aligned}$$

In order to highlight the superiority of the proposed PV modeling method based on the DEIMA, a fair comparison with other methods in the literature is presented. The penalty differential evolution algorithm (PDEA) [36], the improved adaptive differential evolution algorithm (IADEA) [31], the electromagnetism-like algorithm (EMA) [37], ImSMA_LW [29], and the ant lion optimizer with Lambert W-function (ALO_LW) [38] are all used as benchmarks for comparison purposes and thus for evaluating the effectiveness of the DEIMA. The PDEA and EMA are utilized for comparison purposes because the proposed DEIM algorithm is a hybridization between them. Thus, the comparison with the PDEA and EMA is important to show the superior performance of the DEIM algorithm, where the latter combines the useful aspects of various algorithms. Furthermore, the models based on the PDEA, EMA, IADE, ALO_LW, and ImSMA_LW algorithms use the same experimental I-V data that were used in the current research work. Therefore, for the sake of transparency and justice, the aforementioned models are used for a comparison with the performance of the DEIM algorithm. In addition, the PDEA, EMA, IADE, ALO_LW, and ImSMA_LW algorithms offer acceptable and accurate PV models according to the literature. The experimental conditions, including the size of the population, maximum number of generations, and search space of the decision variables, for all of the aforementioned methods consider the same metrics for all of the methods to ensure fair comparisons. It should be noted that the mutation factor and crossover rate of the DEIMA and IADEA are dynamically adaptive according to a formula. Meanwhile, the mutation factor and crossover rate of the PDEA are chosen to be 0.5 and 1, respectively [36].

Figure 7 shows the $RMSE$ of the proposed DEIMA and other compared algorithms under seven operation conditions. The proposed DEIMA offers the lowest $RMSE$ values under all of the operation conditions with an average value around 0.06024. The ALO_LW, ImSMA_LW, IADEA, and EMA provided the second-, third-, fourth-, and fifth-lowest $RMSE$ values, respectively. Meanwhile, the PDEA exhibited the worst $RMSE$ value as compared to the other methods. The DEIMA presents the lowest MBE as compared to other methods, with an average MBE of 0.00518 over the seven operation conditions, as shown in Figure 8. The PDEA still has the worst MBE values over all of the operation conditions.

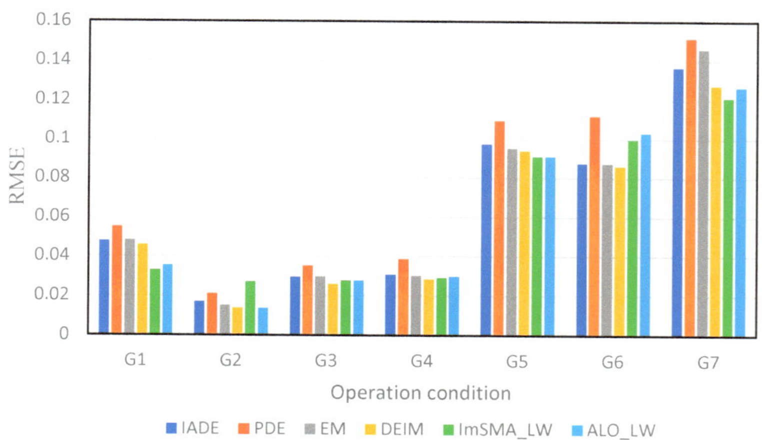

Figure 7. $RMSE$ of various optimization algorithms under different operation conditions.

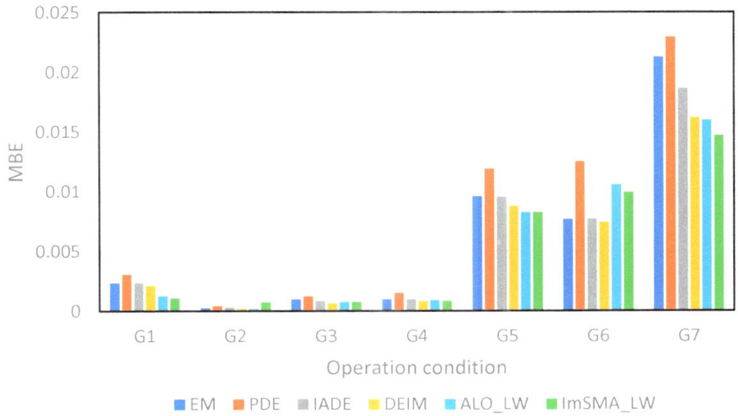

Figure 8. *MBE* of various optimization algorithms under different operation conditions.

Figure 9 shows the determination coefficient of various algorithms as a bar chart. The proposed DEIMA also offers compromised R^2 values, which are close to the value of one over the seven operation conditions. The average R^2 values of the DEIMA, ALO_LW, ImSMA_LW, IADEA, EMA, and PDEA are 0.9923, 0.994, 0.9935, 0.9915, 0.9914, and 0.9885, respectively.

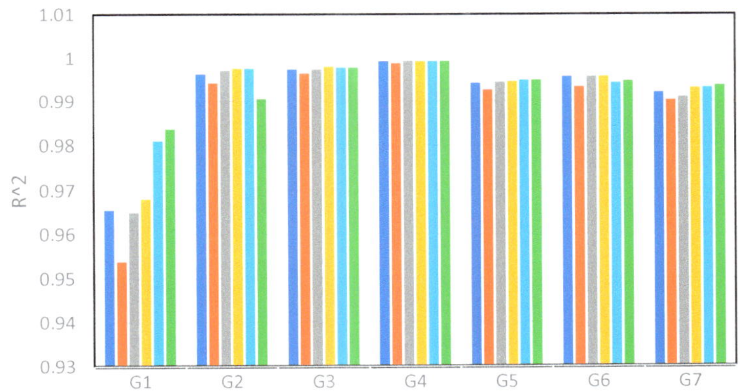

	G1	G2	G3	G4	G5	G6	G7
IADE	0.9655	0.9963	0.9974	0.9992	0.9942	0.9957	0.992
PDE	0.9538	0.9942	0.9964	0.9988	0.9926	0.9933	0.9902
EM	0.9649	0.9971	0.9974	0.9993	0.9944	0.9957	0.9909
DEIM	0.968	0.9976	0.998	0.9993	0.9946	0.9958	0.9931
ALO_LW	0.981	0.9976	0.9978	0.9993	0.9949	0.9943	0.9932
ImSMA_LW	0.9836	0.9905	0.9978	0.9993	0.9949	0.9947	0.9937

Figure 9. R^2 of various optimization algorithms under different operation conditions.

The CPU execution times of the various algorithms are tabulated in Table 4 over seven weather conditions. The average execution time of the DEIMA is around 27.69 s. The DEIMA needs less time to conduct execution as compared to the other methods. Although the algorithms ALO_LW and ImSMA_LW offered promising results in many evaluation criteria, they need a long execution time to compute the optimal TDM parameters of the PV module.

Table 4. Execution time of various algorithms under several operation conditions.

Algorithm	Operation Condition							Average
	G1	G2	G3	G4	G5	G6	G7	
IADE	30.16	28.25	27.5	30.063	30.031	30.75	31.25	29.71
PDE	32.8	29.78	27.13	29.58	31.33	28.02	28.11	29.54
EM	5152.51	5050.24	5623.41	6599.23	5811.21	5769.82	6487.71	5784.88
DEIM	24.94	26.50	26.56	28.75	28.69	29.19	29.20	27.69
ALO_LW	1827.64	553.27	1383.69	1882.3	5615.73	345.59	3644.34	2178.94
ImSMA_LW	1380.98	2828.44	5338.09	7827.3	256.86	3200.22	2090.84	3274.68

Figure 10 shows the *RMSE* deviation of each operation condition (d_m). The DEIMA offers the lowest d_m values in the G1, G2, G4, and G7 operation conditions. The EMA joins the proposed DEIMA in the G3, G5, and G6 weather conditions in offering the best d_m values. Small d_m values prove the capability of the proposed method regarding the effective estimation of PV module parameters. The *STD* value can be computed for various methods by using d_m over the seven operation conditions. The DEIMA presents 0.0426 as the lowest *STD* value, as compared to the IADEA, ALO_LW, EMA, and PDEA with 0.04416, 0.0446, 0.04684, and 0.04937, respectively.

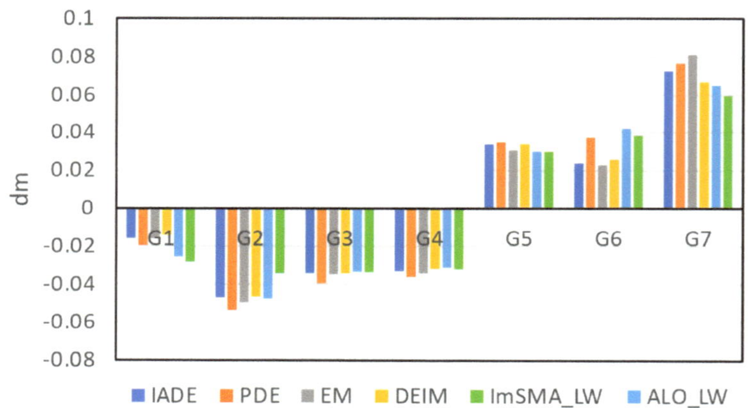

Figure 10. The d_m of various optimization algorithms under different operation conditions.

The last criterion used to compare the results of DEIMA with other algorithms is the average absolute error (*AAE*). Figure 11 shows the *AAE* of the proposed DEIMA and other methods. Based on Figure 10, the DEIMA presents the lowest *AAE* over the seven operation conditions. It is worth mentioning that the *AAE* is significantly increased when the operation conditions are changed from G1 to G7. This scenario occurred as the number of I-V points in the data set increased, as was stated in Table 2.

Based on the aforementioned discussions, the radar diagram in Figure 12 can be drawn by scoring each algorithm according to its performance regarding each criterion. It is noticed that the proposed DEIMA is claiming the first score among the other methods for most of the criteria. Meanwhile, the PDEA obtained the last score (6) for most of criteria, except for the CPU execution time.

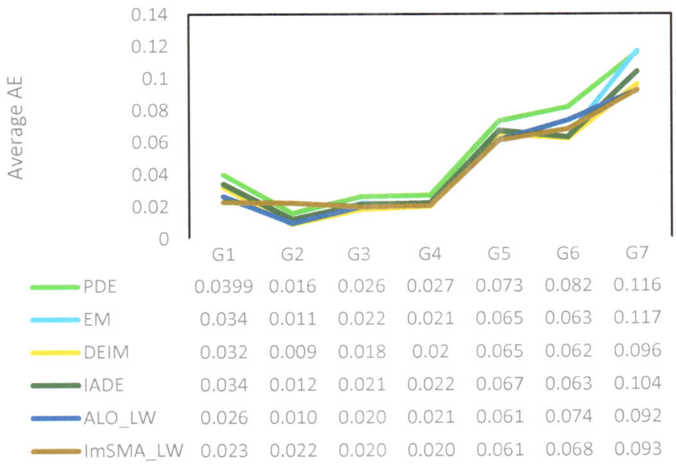

Figure 11. *AAE* of various optimization algorithms under different operation conditions.

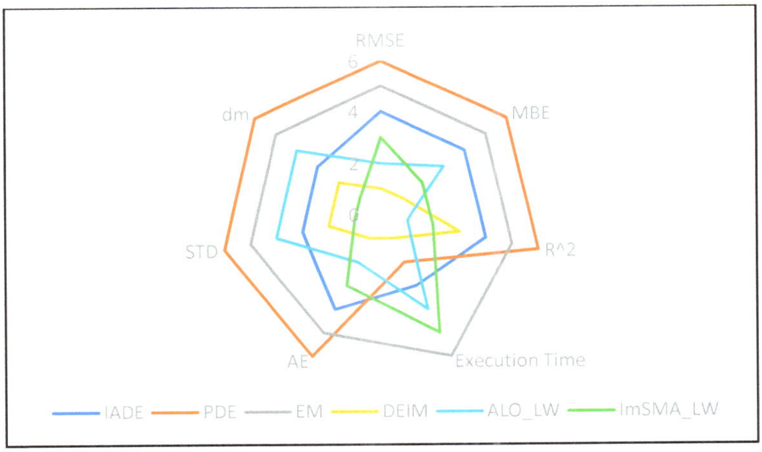

Figure 12. The radar diagram of various algorithms for different criteria.

6. Conclusions

A PV modeling technique based on a hybrid combined optimization algorithm is presented in this paper. The modeling method is utilized to find the unknown parameters of a triple-diode PV module. The proposed algorithm hybridized both the conventional DEA and EMA to obtain the DEIMA. The mutation stage of the DEA is boosted by the attraction–repulsion concept of the EMA. A novel formula is proposed for adapting the mutation factor and crossover rate control parameters for each iteration. The formula uses the sigmoid function and the evolution of fitness function. The proposed mutation factor and crossover rate formula improved the reliability and efficiency of the optimization algorithm. The fitness function is formulated by using the root mean square error between the experimental and computed PV output currents over n-points of the I-V characteristic with regards to seven operation conditions. The operation conditions are chosen with different solar irradiance and ambient temperatures in order to indicate the superiority of the proposed PV modeling method. The outcomes of the DEIMA are validated against other optimization algorithms by evaluating various statistical criteria. According to the results, the DEIMA offered a significant superiority as compared to the other compared methods in terms of accuracy and convergence. The proposed DEIM algorithm is characterized by simple execution, low-complexity computation, and guaranteed convergence. In future work, the authors will focus on using DEIMA-based multi-objective functions to estimate nine parameters of a triple-diode PV module, which will be integrated with a multicriteria decision-making method.

Author Contributions: Conceptualization, D.H.M., H.T.H. and Y.A.-N.; data curation, D.H.M. and H.T.H.; formal analysis, D.H.M., Y.A.-N. and H.T.H.; investigation, D.H.M. and H.T.H.; methodology, D.H.M. and H.T.H.; validation, H.T.H. and Y.A.-N.; writing—original draft preparation, D.H.M., H.T.H. and Y.A.-N.; writing—review and editing, D.H.M., H.T.H. and Y.A.-N. All authors have read and agreed to the published version of the manuscript.

Funding: This research received no external funding.

Data Availability Statement: The data presented in this study are available on request from the corresponding authors due to copyright issues.

Acknowledgments: The authors would like to thank Mustansiriyah University (www.uomustansiriyah.edu.iq) Baghdad-Iraq for its support in the present work.

Conflicts of Interest: The authors declare that they have no known competing financial interest or personal relationships that could have appeared to influence the work reported in this paper.

References

1. Mansoor, M.I.; Haider, H.T. A multi-objective residential load management based on self-adapting differential evolution. *Renew. Energy Focus* **2021**, *38*, 44–56. [CrossRef]
2. Kharchouf, Y.; Herbazi, R.; Chahboun, A. Parameter's extraction of solar photovoltaic models using an improved differential evolution algorithm. *Energy Convers. Manag.* **2022**, *251*, 114972. [CrossRef]
3. Fan, Y.; Wang, P.; Heidari, A.A.; Chen, H.; Mafarja, M. Random reselection particle swarm optimization for optimal design of solar photovoltaic modules. *Energy* **2022**, *239*, 121865. [CrossRef]
4. Rigo, P.D.; Siluk, J.C.M.; Lacerda, D.P.; Spellmeier, J.P. Competitive business model of photovoltaic solar energy installers in Brazil. *Renew. Energy* **2022**, *181*, 39–50. [CrossRef]
5. Ridha, H.M.; Hizam, H.; Gomes, C.; Heidari, A.A.; Chen, H.; Ahmadipour, M.; Muhsen, D.H.; Alghrairi, M. Parameters extraction of three diode photovoltaic models using boosted LSHADE algorithm and Newton Raphson method. *Energy* **2021**, *224*, 120136. [CrossRef]
6. Farah, A.; Belazi, A.; Benabdallah, F.; Almalaq, A.; Chtourou, M.; Abido, M.A. Abido. Parameter extraction of photovoltaic models using a comprehensive learning Rao-1 algorithm. *Energy Convers. Manag.* **2022**, *252*, 115057. [CrossRef]
7. Lappalainen, K.; Piliougine, M.; Spagnuolo, G. Experimental comparison between various fitting approaches based on RMSE minimization for photovoltaic module parametric identification. *Energy Convers. Manag.* **2022**, *258*, 115526. [CrossRef]
8. Kalliojärvi-Viljakainen, H.; Lappalainen, K.; Valkealahti, S. A novel procedure for identifying the parameters of the single-diode model and the operating conditions of a photovoltaic module from measured current–voltage curves. *Energy Rep.* **2022**, *8*, 4633–4640. [CrossRef]
9. Gude, S.; Jana, K.C.; Laudani, A.; Thanikanti, S.B. Parameter extraction of photovoltaic cell based on a multi-objective approach using nondominated sorting cuckoo search optimization. *Sol. Energy* **2022**, *239*, 359–374. [CrossRef]
10. Kalu, E.E. Numerical Methods with Applications: Abridged. Autar Kaw, 2008. Available online: https://nm.mathforcollege.com/NumericalMethodsTextbookUnabridged/ (accessed on 1 June 2024).
11. Navabi, R.; Abedi, S.; Hosseinian, S.H.; Pal, R. On the fast convergence modeling and accurate calculation of PV output energy for operation and planning studies. *Energy Convers. Manag.* **2015**, *89*, 497–506. [CrossRef]
12. Lineykin, S.; Averbukh, M.; Kuperman, A. An improved approach to extract the single-diode equivalent circuit parameters of a photovoltaic cell/panel. *Renew. Sustain. Energy Rev.* **2014**, *30*, 282–289. [CrossRef]
13. Hejri, M.; Mokhtari, H.; Azizian, M.R.; Ghandhari, M.; Söder, L. On the Parameter Extraction of a Five-Parameter Double-Diode Model of Photovoltaic Cells and Modules. *IEEE J. Photovolt.* **2014**, *4*, 915–923. [CrossRef]
14. Karamirad, M.; Omid, M.; Alimardani, R.; Mousazadeh, H.; Heidari, S.N. ANN based simulation and experimental verification of analytical four- and five-parameters models of PV modules. *Simul. Model. Pract. Theory* **2013**, *34*, 86–98. [CrossRef]
15. Ram, J.P.; Pillai, D.S.; Mathew, D.; Ha, J.; Kim, Y.J. A simple, reliable and adaptive approach to estimate photovoltaic parameters using spotted hyena optimization: A framework intelligent to predict photovoltaic parameters for any meteorological change. *Sol. Energy* **2022**, *236*, 480–498.
16. Khanna, V.; Das, B.K.; Bisht, D.; Singh, P.K. A three diode model for industrial solar cells and estimation of solar cell parameters using PSO algorithm. *Renew. Energy* **2015**, *78*, 105–113. [CrossRef]

17. Qais, M.H.; Hasanien, H.M.; Alghuwainem, S.; Nouh, A.S. Nouh. Coyote optimization algorithm for parameters extraction of three-diode photovoltaic models of photovoltaic modules. *Energy* **2019**, *187*, 116001. [CrossRef]
18. Shankar, N.; Saravanakumar, N. Solar photovoltaic module parameter estimation with an enhanced differential evolutionary algorithm using the manufacturer's datasheet information. *Optik* **2020**, *224*, 165700.
19. Kumar, C.; Raj, T.D.; Premkumar, M.; Raj, T.D. A new stochastic slime mould optimization algorithm for the estimation of solar photovoltaic cell parameters. *Optik* **2020**, *223*, 165277. [CrossRef]
20. Qais, M.H.; Hasanien, H.M.; Alghuwainem, S. Transient search optimization for electrical parameters estimation of photovoltaic module based on datasheet values. *Energy Convers. Manag.* **2020**, *214*, 112904. [CrossRef]
21. Yousri, D.; Abd Elaziz, M.; Oliva, D.; Abualigah, L.; Al-Qaness, M.A.; Ewees, A.A. Reliable applied objective for identifying simple and detailed photovoltaic models using modern metaheuristics: Comparative study. *Energy Convers. Manag.* **2020**, *223*, 113279. [CrossRef]
22. Qais, M.H.; Hasanien, H.M.; Alghuwainem, S. Parameters extraction of three-diode photovoltaic model using computation and Harris Hawks optimization. *Energy* **2020**, *195*, 117040. [CrossRef]
23. Chenouard, R.; El-Sehiemy, R.A. An interval branch and bound global optimization algorithm for parameter estimation of three photovoltaic models. *Energy Convers. Manag.* **2020**, *205*, 112400. [CrossRef]
24. Ibrahim, I.A.; Hossain, M.J.; Duck, B.C.; Nadarajah, M. An improved wind driven optimization algorithm for parameters identification of a triple-diode photovoltaic cell model. *Energy Convers. Manag.* **2020**, *213*, 112872. [CrossRef]
25. Houssein, E.H.; Zaki, G.N.; Diab, A.A.Z.; Younis, E.M. An efficient Manta Ray Foraging Optimization algorithm for parameter extraction of three-diode photovoltaic model. *Comput. Electr. Eng.* **2021**, *94*, 107304. [CrossRef]
26. Ćalasan, M.; Aleem, S.H.A.; Zobaa, A.F. A new approach for parameters estimation of double and triple diode models of photovoltaic cells based on iterative Lambert W function. *Sol. Energy* **2021**, *218*, 392–412. [CrossRef]
27. Zhou, W.; Wang, P.; Heidari, A.A.; Zhao, X.; Turabieh, H.; Mafarja, M.; Chen, H. Metaphor-free dynamic spherical evolution for parameter estimation of photovoltaic modules. *Energy Rep.* **2021**, *7*, 5175–5202. [CrossRef]
28. Wang, D.; Sun, X.; Kang, H.; Shen, Y.; Chen, Q. Heterogeneous differential evolution algorithm for parameter estimation of solar photovoltaic models. *Energy Rep.* **2022**, *8*, 4724–4746. [CrossRef]
29. El-Fergany, A.A. Parameters identification of PV model using improved slime mould optimizer and Lambert W-function. *Energy Rep.* **2021**, *7*, 875–887. [CrossRef]
30. Muhsen, D.H.; Ghazali, A.B.; Khatib, T.; Abed, I.A. Parameters extraction of double diode photovoltaic module's model based on hybrid evolutionary algorithm. *Energy Convers. Manag.* **2015**, *105*, 552–561. [CrossRef]
31. Jiang, L.L.; Maskell, D.L.; Patra, J.C. Parameter estimation of solar cells and modules using an improved adaptive differential evolution algorithm. *Appl. Energy* **2013**, *112*, 185–193. [CrossRef]
32. Muhsen, D.H.; Ghazali, A.B.; Khatib, T.; Abed, I.A. Extraction of photovoltaic module model's parameters using an improved hybrid differential evolution/electromagnetism-like algorithm. *Sol. Energy* **2015**, *119*, 286–297. [CrossRef]
33. Ishaque, K.; Salam, Z.; Taheri, H.; Shamsudin, A. A critical evaluation of EA computational methods for Photovoltaic cell parameter extraction based on two diode model. *Sol. Energy* **2011**, *85*, 1768–1779. [CrossRef]
34. Vitaliy, F. *Differential Evolution—In Search of Solutions*, 1st ed.; Springer: New York, NY, USA, 2006.
35. Benkercha, R.; Moulahoum, S.; Taghezouit, B. Extraction of the PV modules parameters with MPP estimation using the modified flower algorithm. *Renew. Energy* **2019**, *143*, 1698–1709. [CrossRef]
36. Ishaque, K.; Salam, Z.; Mekhilef, S.; Shamsudin, A. Parameter extraction of solar photovoltaic modules using penalty-based differential evolution. *Appl. Energy* **2012**, *99*, 297–308. [CrossRef]
37. Birbil, Ş.İ.; Fang, S.C. An electromagnetism-like mechanism for global optimization. *J. Glob. Optim.* **2003**, *25*, 263–282. [CrossRef]
38. Kanimozhi, G.; Kumar, H. Modeling of solar cell under different conditions by Ant Lion Optimizer with Lambert W function. *Appl. Soft Comput.* **2018**, *71*, 141–151.

MDPI AG
Grosspeteranlage 5
4052 Basel
Switzerland
Tel.: +41 61 683 77 34

Designs Editorial Office
E-mail: designs@mdpi.com
www.mdpi.com/journal/designs

www.ingramcontent.com/pod-product-compliance
Lightning Source LLC
Chambersburg PA
CBHW041425210926
51850CB00013B/3356
* 9 7 8 3 7 2 5 8 2 7 0 5 3 *